AF880580

Zoological Monographs

Volume 8

Series Editors

Heike Feldhaar
Tierökologie I, University of Bayreuth, Bayreuth, Germany

Andreas Schmidt-Rhaesa
Centrum für Naturkunde, Wirbellose Tiere 1, Zoologisches Museum, Hamburg, Germany

The book series *Zoological Monographs* is devoted to the study of the animal kingdom and covers all aspects of animals, both living and extinct, including taxonomy, systematics and evolution, biodiversity and biogeography, ecology, morphology and anatomy, physiology and behavior. The volumes in the series either discuss the biology of a single animal group, approach a topic from an interspecies level, or present recent methodologies. The series is intended for graduate students, university teachers, scientists and professionals.

John N. Maina

Current Perspectives on the Functional Design of the Avian Respiratory System

Springer

John N. Maina
Department of Zoology
University of Johannesburg
Johannesburg, South Africa

ISSN 2523-3904 ISSN 2523-3912 (electronic)
Zoological Monographs
ISBN 978-3-031-35179-2 ISBN 978-3-031-35180-8 (eBook)
https://doi.org/10.1007/978-3-031-35180-8

© The Editor(s) (if applicable) and The Author(s), under exclusive license to Springer Nature Switzerland AG 2023
This work is subject to copyright. All rights are solely and exclusively licensed by the Publisher, whether the whole or part of the material is concerned, specifically the rights of translation, reprinting, reuse of illustrations, recitation, broadcasting, reproduction on microfilms or in any other physical way, and transmission or information storage and retrieval, electronic adaptation, computer software, or by similar or dissimilar methodology now known or hereafter developed.
The use of general descriptive names, registered names, trademarks, service marks, etc. in this publication does not imply, even in the absence of a specific statement, that such names are exempt from the relevant protective laws and regulations and therefore free for general use.
The publisher, the authors, and the editors are safe to assume that the advice and information in this book are believed to be true and accurate at the date of publication. Neither the publisher nor the authors or the editors give a warranty, expressed or implied, with respect to the material contained herein or for any errors or omissions that may have been made. The publisher remains neutral with regard to jurisdictional claims in published maps and institutional affiliations.

This Springer imprint is published by the registered company Springer Nature Switzerland AG
The registered company address is: Gewerbestrasse 11, 6330 Cham, Switzerland

Preface

While integrating published works, this book captures four decades of own investigations on the functional design of the avian respiratory system. It has been written for a broad readership. Physiologists, morphologists, zoologists, evolutionary biologists, and ecologists will find it of interest. Those with particular interest in or are actively pursuing areas of comparative respiratory biology will find it useful. Researchers in fields of earth sciences who are interested in how the physical and the biological (natural) domains have interacted and 'coevolved' over the millennia and the transactions that lead to the eventuation of the broad range of extant life forms will find the book of interest and hopefully stimulating. To underscore the unique developmental, structural, and functional attributes of the avian respiratory system, comparative details are given to better explain why, how, and when amongst the broad spectrum of animal species, the lung-air sac system of birds evolved. The treatise details the evolutionary and existing uniqueness of the avian respiratory system. The illustrations were carefully chosen to enable even those readers without considerable understanding of the subject matter to follow the explanations and controversies.

In recent times, research activity in the area of avian respiratory biology has advanced rapidly. Molecular biological studies of the developmental aspects of the avian respiratory system stand out in having particularly generated extremely instructive information. Morphological, functional, and developmental details are, however, scattered across many publications, some of which especially for the older literature are regrettably difficult to access. By reducing isolation among researchers/investigators and in that way obviating costly duplication of effort, regular reviews of the area of avian respiratory biology should promote progress on the subject matter. Such appraisals should critically highlight the gaps in knowledge and point out the unresolved issues. Integrating past and existing findings, deductions and even speculations, outlining the development, the structure, and the function of the avian respiratory system, this book presents the most current account of what birds are and how and why they perform the various interesting things that they do. The peculiarity of the avian respiratory system and the processes and mechanisms that shaped it are delineated. It is hoped that the discourse will promote further discussion and investigation on the many still unresolved aspects.

Chapter 1 outlines among others the domestication and the sociocultural influences of birds on human societies, evolution of birds from reptilian stock, long distance migration, flight under the hypoxic conditions of the high altitude, and evolution of the avian respiratory system: in general, the section describes what birds are, how they came to be what they are, and why and how they do the things they do. Chapter 2 sketches the determinants and the principles that govern developments of biological structures especially the lung, the intricate process of branching morphogenesis by which the pulmonary airway and vascular systems development in vertebrates and the most important molecular factors (cues) that are involved in the elaboration of the lung. Chapter 3 runs through the structure and the complexity of the morphology of the lung-air sac system of birds which is compared with among others the mammalian lung. Documenting aspects such as the air flow dynamics across the airways and the air sacs, adaptations for coping with hypoxia and hypocapnia, strength of the lung and the air- and blood capillaries and morphometric specializations of the lung are delineated in Chap. 4: in general, the chapter explains the functional efficiency of the avian respiratory system.

In the main, this account draws attention to the instructive sentiments expressed by Jürgens and Gros (2002) that *There is no such thing as an ideal gas exchange system. The system that has evolved in each species depends to an impressive extent on environmental conditions, on body build and size, on the animal's patterns of movement, and on its energy consumption.*

Reference

Jürgens KD, Gros G (2002) Phylogeny of gas exchange systems. Anasthesiol Intensivmed Notfallmed Schmerzther 37:185–198

Acknowledgments

My interest in the comparative functional morphology of the gas exchangers and that of the avian respiratory system in particular began under the mentorship of the late Professor Anthony Stuart King at the University of Liverpool. I owe him an immense debt of gratitude for his guidance during my vacillating years. At various points of my journey, I have had the good fortune of interacting with colleagues who have unreservedly given me their time and shared ideas with me enthusiastically. The interactions have helpfully attuned my thinking and determined my research trajectory. I should particularly mention Drs. Mohamed A. Abdalla (Khartoum University), Geoffrey M.O. Maloiy (University of Nairobi), Stephen P. Thomas (Duquesne University), Dallas M. Hyde (University of California—Davis), Andrew N. Makanya and Stephen K Kiama (University of Nairobi), Chris Wood (University of British Columbia), and Jeremy D Woodward (University of Cape Town). I hasten to add that I bear full responsibility for this work and any oversights or lapse of judgment are utterly mine. I would be most grateful if such cases were pointed out to me.

I heartily thank my family for their patience during the long period of isolation when this book was under preparation.

Contents

1 Fascinating Natural and Biological Traits of Birds 1
1.1 Birds: A Profoundly Interesting Group of Animals 1
1.2 Domestication of Birds 8
1.3 Megapodes: "The Incubator Birds" 10
1.4 Hoatzin: "The Ruminant Bird" 10
1.5 Cognitive Capacities, Nest Making, Tool Use, and Dropping Objects on Hard Surface for Food 11
1.6 Vocalization in Birds: Sound Production 13
1.7 Avian Body Bauplan: Structure and Function 17
1.8 Avian Flight: A Novel Mode of Locomotion 19
1.9 Bird Migration 28
1.10 Impacts of Anthropogenic Activities on Birdlife 31
1.11 Evolution of Birds 34
1.12 Evolution of the Avian Respiratory System 38
References 45

2 Development of the Avian Respiratory System 99
2.1 Developmental Biology: The less Well-Known and Explored Aspects 99
2.2 Developments of Gas Exchangers 106
2.3 Branching Morphogenesis 112
2.4 Development of the Avian Lung 116
2.4.1 Development of the Airway (Bronchial) System 119
2.4.2 Development of the Pulmonary Blood Vessels 127
2.4.3 Development of the Air and the Blood Capillaries 131
2.5 Development of the Air Sacs 133
2.6 Molecular and Genetic Aspects of the Development of the Avian Respiratory System 136
2.6.1 Fibroblast Growth Factor (FGF) Signaling 141

2.6.2 WNT Genes Signaling . . . 144
2.6.3 Sonic Hedgehog SHH) Signaling . . . 148
2.6.4 Vascular Endothelial Growth Factor (VEGF) . . . 148
References . . . 152

3 Structure of the Avian Respiratory System . . . 191
3.1 Fundamental Aspects of the Structural Designs of Gas Exchangers . . . 191
3.2 Structure of the Avian Respiratory System: The Lung-Air Sac System . . . 198
3.2.1 Larynx . . . 198
3.2.2 Trachea . . . 200
3.2.3 Syrinx . . . 203
3.2.4 Lung . . . 207
3.2.5 Pulmonary Circulation . . . 236
3.2.6 Air Sacs . . . 241
References . . . 246

4 Function of the Avian Respiratory System . . . 269
4.1 Fundamental Aspects of the Functional Designs of Gas Exchangers . . . 269
4.2 Structural and Functional Correlations in Properties of Gas Exchangers: General Considerations . . . 278
4.3 Air Flow Dynamics in the Avian Respiratory Systems . . . 281
4.4 Gas Exchange: Coping with Hypoxia and Hypocapnia in Birds . . . 288
4.5 Structural Complexity and Functional Efficiency of the Avian Respiratory System . . . 290
4.6 Gas Exchange: The Inventory—Respiratory Mechanisms and Processes . . . 296
4.7 Strength of the Avian Lung and the Blood-Gas Barrier . . . 303
4.8 Morphometric Properties of the Avian Lung: Comparative Perspectives . . . 314
4.8.1 Lung Volume . . . 315
4.8.2 Respiratory Surface Area . . . 323
4.8.3 Blood-Gas Barrier . . . 332
4.8.4 Pulmonary Capillary Blood Volume . . . 334
4.8.5 Mathematical Modeling of the Diffusing Capacity of the Avian Lung for Oxygen . . . 340
References . . . 344

Index . . . 375

Chapter 1
Fascinating Natural and Biological Traits of Birds

Abstract The natural history of birds is summarized. Account of what contemporary birds are, when and how they came to be what they are, and why and how they evolved exceptional physiognomies are given. The evolution of birds from reptilian stock, their domestication that resulted in some of the species becoming leading food animals and the sociocultural impacts of birds on organizations of many human societies are outlined. The evolution of the lung-air sac system of birds, which among the air-breathing vertebrates is the most structurally complex and efficient gas exchanger, is described. Unique properties, capacities, and activities such as long distant migration, flight under the extremely hypoxic conditions of high altitude, anthropogenic impacts of climate change (global warming) on the ecology and biology of birds, sound production (vocalization), birds as bioindicator animals of environmental health, and the cognitive prowess of birds in exploits such as dropping hard food objects on firm surfaces to break them and that way access otherwise unobtainable food and caching of food in various ways and places and shrewdly accessing it for use during adverse conditions are presented. The biology of birds can only be well understood by considering them from various perspectives that include the habitats they occupy and the lifestyles that they lead.

1.1 Birds: A Profoundly Interesting Group of Animals

In abundant, extensively distributed and even cosmopolitan, birds are an instantly familiar and very attractive group of animals. In serving as predators, pollinators, seed dispersers, scavengers, and even ecosystem bioengineers, birds perform roles that are critical to the food chain, a modus operandi that is elemental to the existence of life. From their capacities of migrating across vast distances, Peter Hore and Henrick Mouritsen (Hore and Mouritsen 2022) designated birds as "*extraordinary creatures*" and in the opening statement of his essay, Gareth Dyke (Dyke 2009) ardently declared that "*everybody loves birds*." In folk stories, it is often told that when the first cave person walked out of their modest abode and cast eyes up to the sky, the first thing they saw were birds gracefully soaring high up above and that from that moment a bond was hermetically sealed between human beings and birds.

© The Author(s), under exclusive license to Springer Nature Switzerland AG 2023

J. N. Maina, *Current Perspectives on the Functional Design of the Avian Respiratory System*, Zoological Monographs 8, https://doi.org/10.1007/978-3-031-35180-8_1

Coveted by humans for a long time before machine-powered flight was achieved, volancy (powered flight) is the most captivating behavior of birds. With many well-documented disastrous outcomes, for a long time, humans had strived to achieve flight by firstly directly emulating the flapping actions of bird wings and later after attaching various types of contrivances to the body. Franz Reichelt's disastrous attempt in 1912 to fly off the Eiffel Tower (Paris) using a "parachute suit" is one of such unfortunate historic events that was captured on film (https://www.youtube.com/watch?v=v4Qp56roKsw—accessed on 23-09-2022).

The most notable morphological, physiological, and behavioral features of modern bird species include feathers that provide insulation and aerodynamically shaped winged body form that to a large extent permit powered flight: others include toothlessness (edentulism) and endothermy (Martin 1983; Brusatte et al. 2015; Ruaux et al. 2020). A diverse animal group (Hunter 2006, 2007; Gould and Gould 2007; Clucas and Marzluff 2012; Robbins 2017; Sault 2018; BirdLife International 2018), using conventional taxonomic methods, it is commonly reported that between ~10,000 and 11,000 species of birds exist on Earth (Sætre 2007; Brusatte et al. 2015; Saupe et al. 2019; Callaghan et al. 2021). Interestingly, by applying a diagnostic evolutionary species concept on a sample of 200 species of birds, Barrowclough et al. (2016) showed that the generally reported number of extant avian species may be an underestimate by at least a factor of two and that the probable worldwide average number of species may be as high as 20,470 species. Comprising ~6500 species and 140 families (i.e., three-quarters of the number of extant avian species), passerines (songbirds or perching birds) (Order: Passeriformes) are the most speciose group (Collias 1997; Sibley and Monroe 1990; Barker et al. 2002; Payevsky 2014; Ericson et al. 2003, 2014; Wu et al. 2015; Selvatti et al. 2015; Tietze 2018; Gill et al. 2020): they are the "cichlid equivalents" of fish (Salzburger 2009; Fan et al. 2012).

Populating practically every ecosystem on Earth, birds have culturally, philosophically, artistically, economically, and technologically influenced the "evolution" of human societies (Meighan 1966; Jackson 1997; Burt 2007; Cushman 2013; Cummings et al. 2014; Morelli et al. 2015): they feature prominently in the history of painting, poetry, commerce, and music (Shepherd 2005; Bourn and Bonine 2015) and even geopolitics! Across written human history, music composers and songwriters have been inspired by birdsongs (Doupe and Kuhl 1999; Düring et al. 2013). For example, it is told that the classical music virtuoso Wolfgang Amadeo Mozart (1756–1791) treasured the songs of his pet Starling (*Sturnus vulgaris*) so much that on its death, he gave the bird a ceremonial funeral (https://www.gresham.ac.uk/lectures-and-events/birdsong-and-music—accessed on 12-04-2022). Acknowledging how much observing birds influenced their perspicacity of the physics of flight, Orville Wright (of the Wright Brothers recognition) wrote that "*learning the secret of flight from a bird was a good deal like learning the secret of magic from a magician. After you once know the trick and know what to look for you see things that you did not notice when you did not know exactly what to look for*" and after studying them (birds), Wilbur Wright noted that "*we soon passed from the reading to the thinking stage, and finally to the working stage*" (Jakab and Young

2000). Regarding their form, lifestyle, and habitat occupied, every species of bird is unique in its own right. The extant avian species ranges in body mass from the tiny Cuban Bumblebee Hummingbird (*Mellisuga helenae*) that weighs as little as 1.6 g (Peters 2000; Dalsgaard et al. 2009; Smith and Lyons 2013) to the 160 kg body mass Ostrich (*Struthio camelus*) (Davies 2002; Simon 2015; Birdlife International 2022), a factorial difference of ~7.10^4. The size of the Bumblebee Hummingbird (~5 cm length) is about that of the eye of the Ostrich (Monfared and Bakhteyari 2013; Cerio and Witmer 2020)! From the fossil record, the largest bird that ever lived is the so-called "big-bird" (*Vorombe titan*), a member of the ratite family (Aepyornithidae) and one of the three genera of Elephant Birds (*Mullerornis*, *Vorombe*, and *Aepyornis*) that inhabited the Island of Madagascar during the Late Quaternary Period (~12,000 years ago): it stood at a height of 3 m and weighed ~800 kg (Hansford and Turvey 2018). Apparently, the elephant birds became extinct from human activities within recorded human history (Flacourt 1658). Birds such as penguins (Family: Sphenisdae) and the ratite group that comprise species such as the Ostrich (*Struthio camelus*), the Emu (*Dromaius novaehollandiae*), the Cassowary (Family: Casuariidae), the Kiwi (*Actinidia chinensis*), and the Rhea (*Rhea americana*) are flightless while others such as the Rufous Griffon Vulture (*Gyps rueppellii*), the Bar-Headed Goose (*Anser indicus*) (Hawkes et al. 2013), and the Andean Goose (*Chloephaga melanoptera*) (Maina et al. 2017; Laguë 2017) are extreme altitude residents and flyers (Milsom et al. 2021). Regarding geographical distribution, compelled largely by factors such as food availability and local environmental conditions, some species of birds exist in large numbers, others consist of small populations and certain species are predominantly sedentary (i.e., they inhabit small spaces for the best part of their life) while others migrate over staggering distances (Berthold 2001; Newton 2008; Alvarez-Alvarez et al. 2018).

Birds have considerably culturally, philosophically, artistically, economically, and scientifically shaped the organization and the intellectual progress of human societies (Mayr 2016; Wang and Zhou 2017): they feature conspicuously in the history of painting, poetry, commerce, and music. Studies of the so-called "Darwin's finches" of the Galápagos Islands greatly influenced Charles Darwin's (Darwin 1859) thinking in the formulation of the theory of evolution by natural selection: as well, his (Darwin's) breeding of pigeons as a pastime fostered his understanding on heritability and species variation. In particularly less accessible (remote) parts of the World, "new" species of birds continue to be 'discovered' (Barnett and Buzzetti 2014). Between the years 2000 and 2009, 73 additional species of birds were "discovered" (Pennak 2012): the Spectacled Flowerpecker (*Dicaeum dayakorum*) was found in the lowland forests of Borneo and scientifically named in 2009 (Edwards et al. 2009; Saucier et al. 2019). Recently, five bird species and five subspecies were located in three small island groups of Sulawesi (Indonesia) (Rheindt et al. 2020). From ecological, behavioral, morphological, physiological, and evolutionary studies (Hunt 2006; Ericson et al. 2014; Jaspers 2015; Seki et al. 2017; Oliveros et al. 2019), it has been shown that among the air-breathing vertebrates, birds are a unique group of animals. Species such as zebra finch (*Taeniopygia guttata*), budgerigar (*Melopsittacus undulatus*), canary (*Serinus canaria*), house sparrow (*Passer domesticus*), starling (*Sturnus vulgaris*), mallard duck (*Anas platyrhynchos*), and common quail (*Coturnix coturnix*) are commonly used in biological research. The domestic fowl (*Gallus gallus* variant *domesticus*) is more often used because of its availability and ease of keeping in laboratories rather than for its suitability in some of the experiments that are performed and the research question(s) propounded. To reduce the number of birds that are utilized in research and to ensure their appropriateness in the investigations performed, physiologically based species-specific models have been and continue to be explored and developed (Cortright et al. 2009; Nichols et al. 2010; Jaspers 2015; Wang et al. 2020; Lautz et al. 2020; Scanes et al. 2022). Birds are instructive animal models in studies of human biological aspects such as neural plasticity in early learning (McCabe and Horn 1994; Rose 2000), adult neurogenesis (Goldman and Nottebohm 1983; Nottebohm 2005), and autism spectrum disorder (ASD) that is one of the most occurring neurodevelopmental disorders that is associated with altered social behavior (Ergaz et al. 2016; Csillag et al. 2022).

From as far back as ~3 million years ago (mya) (Spottiswoode et al. 2016), birds have had long and close historical associations with human beings (Watson et al. 2015; George et al. 2018; Hill 2019; Kost and Hussain 2019; Kost 2019): socio-culturally, they have directly and indirectly influenced cultural organization and advancement of human civilizations. In the Old Testament part of the Holy Bible, birds are mentioned many times by name and where they are not have been termed "animals of the air" (Deysel 2017). For example, in Genesis, after the great flood, to examine whether the waters had receded, Noah freed a "raven" (from the Ark) and when it did not return, he sent a dove that returned with a leaf, showing that flood had abated, plant life had flourished on dry Earth and that land was once again hospitable (Gen. 8:6-7); God brought "birds" and beasts of the Earth before Adam so that he might give them names (Gen. 2:20); the "Law" of Moses prohibits "ravens" from being eaten (Leviticus 11:15; Deuteronomy 14:14); in the Book of Judges, one of the Midianite Kings killed by Gideon is named "Orev"—word of "raven"; in the Book of Kings (17:4-6), God commanded the "ravens" to feed prophet Elijah; King Solomon's hair was as black as a "raven" (Song of Songs 5:11); "ravens" show God's gracious grace to all creations (Psalm 147:9; Job 38:41); and in the New Testament, Jesus Christ depicts "ravens" as examples of God's protection (Luke 12: 24). In the Talmud, the "raven" is one of three creatures on the Noah's Ark that copulated during the flood and was reprimanded by God, and in the Holy Qur'an's account of the story of Cain and Abel, a "raven" is said to have showed Cain how to inter his slain brother (Surah 5: al-Mã'ida, Ayah 31). To the Germanic people, Odin was often related to the "ravens." Legend has it that the Kingdom of England will fall if and when the ravens of the Tower of London are taken away (Boria 2007). The mutually beneficial ("symbiotic") relationship between certain indigenous African peoples and the Honeyguide (*Indicator indicator*) has been well-scientifically studied (Isack and Reyer 1989; Spottiswoode et al. 2016; Wood et al. 2014; Bhadra 2019). Cormorant (*Phalacrocorax auritus*) fishing is a historic activity which was at various times in the past and for different reasons practiced in Japan, Greece, North Macedonia, and for a brief period in England and France (Jackson 1997; Manzi and

Coomes 2010): to a lesser extent, the tradition continues in some remote parts of Japan and China (Jackson 1997).

Compared with ~8 billion human beings that now inhabit Earth (Nygaard 1994; Roser et al. 2013), at a ratio of ~6 to ~40 birds to 1 person, it is estimated that ~300 to ~500 billion individual birds (American billion, i.e., 10^9 years) exist on the planet (Gaston and Blackburn 1997; Callaghan et al. 2021). About 20% of them live in urban areas (Isaksson 2018). Except for the Arctic Region, where there are few or no endemic species (Newton 2003; Clements 2007), birds have dispersed widely across the globe (Stiels and Schidelko 2018; Schweizer and Liu 2018). The largest number of birds inhabits the tropical rainforests (Moritz et al. 2000; Newton 2003). Leading to massive displacement of animal life from their natural habitats, by altering land use, human urbanization has adversely affected the spatial distribution of birds (Reynolds et al. 2019). From ~1961, the surface area of Earth that has been committed to agriculture has increased from 35.6 to 37.4% (World Bank 2020). About 17% of the migratory bird species which comprise ~2000 species (Blount et al. 2021) are threatened or nearly threatened with extinction (Horns and Şekercioğlu 2018) and according to the Birdlife International's State of the World's Birds Report for 2022, about half of all bird species are decreasing and more than one in eight of them are at risk of extinction (https://www.birdlife.org/wp-content/uploads/2022/09/SOWB2022_EN_compressed.pdf—accessed on 11-12-2022). Anthropogenic activities have been and will continue to be a major cause of habitat loss and fragmentation of animal communities, a process that has resulted in massive and extensive extinction of animal species, including birds (Brown and Brown 1992; Brooks et al. 2002; Uezu and Metzger 2011; Vale et al. 2018; Develey and Phalan 2021; Xu and Wang 2022; Lee et al. 2022a, b; Chapman et al. 2022; Angeler et al. 2022). Rosenberg et al. (2019) reported that over a period of 48 years, across much of North America, bird population losses, that constitute a net loss of ~3 billion or 29% of their 1970 abundance, have occurred, and d'Amour et al. (2017) estimated that with the greatest loss in Asia, ~80% of global agricultural land loss may occur from urban expansion in Asia and Africa.

It is expected that by 2030, China and India will be the foremost places of global urban expansion (Seto et al. 2012). According to Archer et al. (2018), by the year 2100, more than half of Africa's bird and mammal species could have vanished. Driven from their natural habitats, birds, in particular, have been forced to adapt to alien environment conditions and habitats such as the urban ecosystems that present foreign selective pressures such as high noise pollution, high light and electromagnetic radiation, chemical pollution, and unfamiliar foods (Shanahan et al. 2014; Sumasgutner et al. 2014; Biard et al. 2017; Isaksson 2020; Becker et al. 2020; Korner et al. 2022). Halfwerk et al. (2011) and Bernat-Ponce et al. (2021) showed that traffic noise, recreational noise pollution, and overlap of acoustic frequencies reduced the reproductive success of the Great Tit (*Parus major*). For the zebra finch (*Taeniopygia guttata*), Zollinger et al. (2019) noted that nestlings growing under noise exposure were lighter than those of the same parents exposed to control, i.e., noise-free conditions. After analyzing global community datasets, Cooper et al. (2022) noted that most common raptors (Family: Accipitridae) that occupy urban centers are smaller, habitat-generalist species and that the species without these traits (i.e., large, specialist taxa) may be at greater conservation peril as the range of global urbanization extends. For the Brazilian Atlantic Forest Blue Manakin (*Chiroxiphia caudate*), habitat loss and fragmentation have upset the movement of Manakins (Hansbauer et al. 2008; Barbosa et al. 2017) and may adversely affect gene flow between populations (Francisco et al. 2007; Assis et al. 2022). With no community and country spared, human beings have themselves not escaped from the injurious effects of pollution and environmental perturbations which they have caused (Lelieveld et al. 2015; Cohen et al. 2017a; Perera 2018; Heft-Neal et al. 2018; Chen et al. 2022). It has been determined that surpassing other major causes of loss of life that include smoking, malaria, and general violence, air pollution is now a leading cause of global death: Fuller et al. (2022) reported that pollution accounts for ~9 million deaths per year, a number that globally translates into one in six deaths.

In the most common and populous livestock species, the chickens are an important food animal (FAO 2008; Padhi 2016). Worldwide, as many as 65 billion chickens are raised for meat and 80 million metric tons of eggs are consumed annually (Lawal et al. 2018, 2020). In the United States of America, ~8.6 billion chickens are consumed every year (Rude 2016). By genetic and molecular comparison of the domesticated White Leghorn Chicken (*Gallus gallus* variant *domesticus*) with the Wild Red Jungle Fowl (*Gallus gallus*), Campler et al. (2009) noted that the former had relatively superior temperate human fear behavior and displayed less physiological stresses compared to the later. Generally, hummingbirds (Trochlids) are distinctive for their physical beauty. It is for that reason, some of the species bear highly desirable names such as Blue-Throated Sapphire Hummingbird (*Chrysuronia coeruleogularis*), Blue-Throated Goldentail Hummingbird (*Chlorestes eliciae*), Amethyst-Throated Hummingbird (*Lampornis amethystinus*), Short-Tailed Emerald Hummingbird (*Chlorostilbon poortmani*), Ruby-Throated Hummingbird (*Archilochus colubris*), Ruby-Topaz Hummingbird (*Chrysolampis mosquitus*), and Marvellous Spatuletail Hummingbird (*Loddigesia mirabilis*). Compared with other animal taxa, more images of bird species appear on national flags, coat of arms, and national emblems of many countries on Earth. For example, the East African Crested Crane (*Balearica regulorum gibbericeps*) appears on the flag of Uganda; the Bald Eagle (*Haliaeetus leucocephalus*) is displayed on the emblems of the United States of America; the Blue Crane (*Anthropoides paradisea*) appears on the coat of arms of the Republic of South Africa; the Red-Crowned Crane (*Grus japonensis*) is the national bird of China; and the White-Cheeked Bulbul (*Pycnonotus leucogenys*), the Hoopoe (*Upupa epops*), and the Kiwi (*Apteryx mantelli*) are, respectively, the national birds of Bahrain, Israel, and New Zealand. The Andean Condor (*Vultur gryphus*) appears on the flag of Equador, the Hoatzin (*Opisthocomus hoazin*) on the coat of arms of Guyana, and the male Bird-of-Paradise (*Paradisaea raggiana*) is displayed on the coat of arms of Papua New Guinea. The peoples of Papua New Guinea and beyond use the plumages of the Birds of Paradise (Family: Paradisaeidae) as ceremonial adornments during social and cultural activities (Andala 2017). For the indigenous American cultures, hummingbirds (Family: Trochlidae) have for a long time featured in traditional stories, symbols, music, art, and religion (Lake-Thom 1997; Tate 2008; Larson and Yorinks 2011). The Aztecs believed that hummingbirds represented the great sun god Huitzilopochtli and were emissaries between the people and their ancestors as well as healers and providers of love, good luck, and joy (Boone 1989; León-Portilla 2017). Compelled by the metabolic demands of powered flight, compared with most other animal taxa, birds are relatively morphologically more homogenous (Ricklefs 2012; Salewski et al. 2014). With colorfulness of plumage being the most perceptible and impressionable feature of many species, birds display wide variety of colors (Andersson 1986; Butcher and Rohwer 1989; Savalli 1995; Bennett and Théry 2007) that even rival those of butterflies and tropical reef fishes. The so-called "birds of paradise" (Family: Paradisaeidae), that are found in Papua New Guinea and surrounding islands as well as Eastern Australia, are particularly colorful and their courtship displays are spectacular (Diamond 1986; McNab 2005; Miles and Fuxjager 2018). The Malaysian catch-word for the birds of paradise, "manuq dewata" translates into "birds of the gods" (Scott 1897).

In the ancient religion practiced by the Romans known as "augury" an "augur" interpreted omens from observing behavior of birds (Troxler 2008) and "ornithomancy" by other pre-industrial societies living around the Mediterranean Sea Region, the omens or wills of gods were also interpreted by the "augur" after observing the behaviors of birds (Ingersoll 1923; Yasuhiko 2013; Mouton et al. 2013). Between the years 1500 and 800 BC, in the Vedas, a large body of religious texts that originated from Ancient India and were written in Vedic Sanskrit (an ancient language of the Indo-Aryan subgroup of the Indo-European language family), the habit of brood parasitism in the Asian Koel (*Eudynamys scolopacea*) is mentioned (Ali 1979). In his book entitled *Historia Animalium* (History of Animals) that is based on studies of ~170 species of birds, the great philosopher-polymath-scientist Aristotle (384–322 BC) (Pitskhelauri 1980; Dunn 2006) presents insightful observations on their behaviors, lifestyles, and physiology but made a wrong deduction that some species "disappeared" and others "changed into other species" during certain times of the year: the actual reason was that the species that were missing had migrated to other areas. Humankind has always considered birds to be special and have sort them for emotional support and companionship (Brooks et al. 2018; Van der Swan et al. 2019). Dubbed "lovebirds," the genus *Agapornis* (Family: Psittaculidae) comprises nine small colourful Old World (African) parrot species that are widely kept as pets (Dilger 1960; Dorrestein and van der Hage 1987). The name stems from the social and affectionate nature of the parrots that display strong monogamous pair-bonding and the long periods that paired birds spend sitting side-by-side (Jobling 2010). Bird watching is a historic human pursuit that continues to gain popularity in the present times (Cordell et al. 1999; Cordell and Herbert 2002). Collaboration between amateurs and scientists has popularized ornithology and advanced conservation science in general and that of birds in particular (Greenwood 2007). Most of the people who are enthusiastic and active in the field of ornithology, many who lack formal conventional scientific (biological) training, are purely driven by love of birds for what they are. The number of ornithologists surpasses that of people who are interested in other vertebrate groups, except perhaps for those attracted to domestic animals such as dogs, cats, and horses (Stevens 1996). Understanding the bird biology of birds and their place in ecosystems importantly contributes to formulating of meaningful conservation efforts: biology explains why animals behave the way they do and what threatens their survival. Long before formal animal welfare societies existed, it is told that the artist-engineer-scientist Leonardo da Vinci (1452–1519), who devoted much time studying bird flight and wrote a textbook entitled "Codex on the flight of birds" on the subject, bought caged birds from markets and later set them free. The area of scientific publishing, where few academic journals are named after individual animal species, birds are exceptional: many reputable, long-established journals and periodicals bear names of bird species or aspects related to birds. Among others, they include "Ostrich" which is managed by the Birdlife South Africa, "Emu" which is managed by the Royal Australasian Ornithologists Union, "Hirundo" which is managed by the Estonian Ornithological Society, "Loon" which is managed by the Minnesota Ornithologists' Union, "Raven" which is managed by the Virginia Society of Ornithology, "Sandgrouse" which is managed by the Ornithological Society of the Middle East, "Auk" which is managed by the American Ornithological Society, "Ibis" which is managed by the British Ornithological Union, "Honeyguide" which is managed by the Zimbabwean and Regional Ornithology, "Condor" which is managed by the American Ornithological Society, "Wingspan" which is managed by the Royal Australasian Ornithologists, and "Forktail" which is managed by the Oriental Bird Club.

To show how and why birds are a biologically distinctive and exceptionally fascinating animals, some sociobiological, evolutionary, morphological, physiological, behavioral, and genetic aspects are succinctly outlined below.

1.2 Domestication of Birds

Domestication is an evolutionary process in which organisms/animals adapt to existence in close proximity and association with human beings (Sætre 2007; Kaiser et al. 2015): the relationship results in harmless, submissive, and docile creatures/animals (Wilkins et al. 2014). Domestication that is multigenerational results in phenotypic shift in the physiology of an animal, and it differs from tameness in wild animals. It (domestication) provides a good model of understanding the genetic basis of stress sensitivity (Jensen 2014; Douxfils et al. 2015; Fallahsharoudi 2017). Ericsson et al. (2014) and Fallahsharoudi et al. (2015) showed that domesticated chickens have less marked acute corticosterone response than their ancestors but take longer to recover from the effect while Fallahsharoudi et al. (2017) reported that domestication has moderated the stress response in chickens by changes in parts of their genome that are involved in the control of the hypothalamic-pituitary-adrenal (HPA) axis, with the genes *SERPINA10* and *PDE1C* being strong candidates for the after-effect.

With the most recent, common ancestor of modern birds inhabiting South America ~95 million years ago, regarding their species diversity and global distribution, the modern birds (Neornithes) are the most diverse group of terrestrial vertebrates (Claramunt and Cracraft 2015). Definite historical and archaeological data are, however, lacking on when, where, how, places and the situations, and settings under which birds were first domesticated (Gifford-Gonzalez and Hanotte 2011; Peters et al. 2015, 2016; Honka et al. 2018). For chickens, which are the most important poultry species that now comprise ~93% of the world's poultry and which among the commonly raised livestock species constitute at least five times more than any mammalian stock, the history of their domestication has been highly controversial (Peters et al. 2015; Eda et al. 2016, 2022; Dodson and Guanghui 2016; Hirst 2019) and is largely unresolved. It is rather from their global abundance and not from hard scientific evidence that it is assumed that chickens were the first avian species to be domesticated. With a long history, the domestic fowl (*G. g.* variant *domesticus*) is nevertheless believed to be one of the earliest birds to be domesticated, ~8000 to 10,000 years ago (West and Zhou, 1988; Stevens 1991; Akishinonomiya et al. 1994, 1996; Liu et al. 2006; Sawai et al. 2010; Kyselý 2010; Tixier-Boichard et al. 2011; Storey et al. 2012; Peters et al. 2016; Lawal et al. 2018), presumably from the Wild Red Jungle Fowl (*G. gallus*), a species that still inhabits in some remote parts of the forests of South and South-East Asia (Collias et al. 1966; Collias 1987; Collias and Collias 1996; Liu et al. 2006; Eriksson et al. 2008; Lawal et al. 2020). In a rather controversial study, Xiang et al. (2014) reported that chickens were domesticated on the North China plain 10,000 years ago. A little earlier, Tixier-Boichard et al. (2011) had recorded that that chickens were domesticated from the Red Junglefowl (RJF) ~8000 years ago. Most recently, claiming it to be the oldest domesticated poultry species in history, Eda et al. (2022) dated the domestication of the goose to be 7000 years ago at a "cultivation" village in the lower Yangtze River (China). Sawai et al. (2010) contended that chickens may have diverged from the Red Jungle Fowl (*Gallus gallus*) from as far back as ~58,000 mya and their common ancestor may have separated from the Green Jungle Fowl (*Gallus gallus*) 3.6 million years ago. In their domestication, chickens underwent various important morphological, physiological and behavioral changes that included aspects such as faster growth rate, greater level and production of larger eggs, less frightfulness (especially of humans), and greater tolerance of stress (Price 1999; Sætre 2007; Kerje et al. 2003; Schütz et al. 2004; Campler et al. 2009; Wright et al. 2012). Other species of birds such pigeons (Family: Columbidae) could have been domesticated about the same time or even earlier than the chickens (Shapiro and Domyan 2013; Gilbert and Shapiro 2014; Blasco et al. 2014). From phylogenetic analyses of DNA sequences, the process of domestication of the chicken is reported to have occurred many times, with multiple genetic contributions from other birds (Nishibori et al. 2005; Zhang et al. 2017; Lawal 2017; Kanginakudru et al. 2008; Eriksson et al. 2008; Pitt et al. 2016; Tiley et al. 2018, 2020; Hirst 2019).

1.3 Megapodes: "The Incubator Birds"

Indigenous to the Oceania, the Megapodes (Family: Megapodiidae) are a group of birds that has been dubbed "incubator birds" or "mound-builders." A distinctive group of birds in the Order Galliformes (Starck and Sutter 2000; Birks and Edwards 2002; Göth and Booth 2005), they have evolved novel strategy of incubating eggs using externally generated heat and not their body heat, like other birds do. In some of the species, heat is generated from microbial decomposition of heaps of organic plant materials while for others, the burrow-nesters, eggs are laid in geothermal or solar-heated soils (Dekker 2007; Harris et al. 2014; Shute et al. 2017). Interestingly, some of the magapode species display temperature-dependent sex ratio (Göth and Booth 2005; Eiby et al. 2008; Eiby and Booth 2008, 2009), a feature that is rare in birds but common in reptiles (Rhen and Lang 1994; Doddamani et al. 2012; Merchant-Larios and Diaz-Hernández 2013). During hatching, the highly precocial, well-feathered chicks of the megapodes break out of the eggs and leave the mounds or burrows, instantly walking, running, and foraging (Harris et al. 2014): no kind of husbandry is provided by to the chicks by the adult birds (Jones et al. 1995). Mound building and egg incubation of the megapodes is not a random process but is a carefully and strategically executed process: the incubation mound is well-sized and before the eggs are laid, temperature is thoroughly checked to be ~36 °C (96 °F) and the depth at which the eggs are laid (in the mound) is painstakingly fixed and ventilatory shafts are built to regulate temperature (Frith 1957, 1962; Eiby and Booth 2008).

1.4 Hoatzin: "The Ruminant Bird"

The Hoatzin (*Opisthocomus hoazin*) is a species of tropical bird that inhabits the swamps, riparian forests and mangroves of the Amazon, and the Orinoco River Basins of South America (Sibley and Ahlquist 1973). From its external morphological, physiological, behavioral, and physical attributes, it has been dubbed "reptile bird," "skunk bird," "stinkbird," or "Canje pheasant." The phylogenetic position of the Hoatzin is enigmatic (Dos Santos et al. 2018): the species has been classified among the fowl-like birds (Order: Galliformes), placed among the cuckoos (Order: Cuculiformes) (Sibley and Ahlquist 1973, 1990; Hedges et al. 1995), categorized among the Turacos (Family: Musophagidae), a small family of arboreal, frugivorous birds that live in sub-Saharan Africa (Hughes and Baker 1999) and ordered on its own in the Order Opisthocomiformes (Thomas 1996). By exhibiting functional wing claws that the chicks use to hold onto and climb trees (Abourachid et al. 2019) and its poor flight capability at maturity, the Hoatzin was once deemed to represent the "missing link," i.e., the evolutionary connection between fossil birds such as *Archaeopteryx*, which had craws on its wings (Griffiths 1994), and the modern birds.

The Hoatzin is the only bird known to display fully functional foregut fermentation (Grajal et al. 1989; Grajal 1995; Wright et al. 2009) similar to that of the group of mammals designated "ruminants" (Godfrey 1961; Short 1964; Ruckebusch et al. 1983; Karasov and Douglas 2013; Munn et al. 2021; Dawson et al. 2021). Ruminants (Order: Artiodactyla) are hoofed (ungulate) herbivorous grazing or browsing mammals that obtain nutrients from plant-based food by fermenting and breaking it mainly by microbial activities in specialized foregut chambers before digestion occurs in the intestines. For the Hoatzin, an enlarged crop ferments plant matter (leaves and fruits) (Thomas 1996; Godoy-Vitorino et al. 2012). The mode of feeding of the Hoatzin, i.e., eating bulky plant materials that are of low nutritive value, differs from the high energy demands for powered flight that compels volant birds to consume foods of high nutritive value such as flower nectar for the case of hummingbirds (Miller 1985; Wester 2014; Ingersoll et al. 2018). The diverse animal taxa such as the so-called flying frogs, snakes, lizards, squirrels, and fish are strictly acrobatic gliders (Emerson and Koehl 1990; Vernes 2001; McCay 2001; Mendelson et al. 2008; Park and Choi 2010; Bahlman et al. 2013; Suzuki and Yanagawa 2019; Yeaton et al. 2020; Lu 2020; Tamisiea 2022). Evolutionary, they have not had to grapple with the extreme morphological and energetic demands that powered flight imposed directly on birds. For the Gliding Frog (*Rhacophorus kio*), Wu et al. (2022) recently determined that genes related to limb development showed evidence of accelerated evolution that led to formation of the webbed-feet that are used for parachuting. The question whether the next vertebrate animal taxon to evolve powered flight, after bats (Order: Chiroptera) that developed flight in the early Eocene, i.e., ~50 to 60 million years ago (Amador et al. 2019; Anderson and Ruxton 2020) is likely to come from these or such gliders is an interesting one (Dudley 2002; Dudley et al. 2012).

1.5 Cognitive Capacities, Nest Making, Tool Use, and Dropping Objects on Hard Surface for Food

Among the land-dwelling vertebrates, birds, especially the generally small passerine (perching or songbird) species, construct exquisitely designed nests (Collias 1997; Walsh et al. 2013; Mainwaring et al. 2014; Heenan 2013; Breen et al. 2016; Jessel et al. 2019) that are strategically placed in areas that are difficult for predators to access, ensuring safety of adults and that of their eggs and chicks (Quader 2005; Lima 2009; Heenan 2013; Jessel et al. 2019). Although avian brains lack a neo (cerebral) cortex, interestingly, their cognitive abilities appear to rival those of mammals (Reiner et al. 2005; Karten 1991, 2015; Stacho et al. 2020). For corvids (Family: Corvidae) and parrots (Family: Psittacidae), the forebrains compare in size with those of the primates (Farries 2004). Contrasted with those of other species of birds, the brains of these avian taxa are relatively cognitively superior and their level of intelligence rivals that of apes, if they are not better (Farabaugh et al. 1994; Emery and Clayton 2004a, b; Emery 2006). Corvids and parrots exhibit complex well-organized social grouping and display long developmental periods before the chicks become independent of their parents (Emery and Clayton 2004a, b; Emery et al. 2007; Clayton et al. 2007; Hobson et al. 2014).

Birdsong has fascinated scientists since Aristotle's (384–322 BC) time (Fishbein 2022). The mode of vocalization of birds has been investigated both in behavioral science (ethology) and neuroscience (Catchpole and Slater 2008). For the Canary (*Serinus canaria*), an overlying left hemispheric dominance of the brain suggests that it is the site of vocal control in birds (Nottebohm et al. 1976). The song system of birds is similar to the other neural systems in mammals (Jarvis 2004a, b; Farries 2004). The Eastern Scrub Jay (*Aphelocoma californica*) uses adaptable cognitive caching and recovery strategies to reduce kleptoparasitism from fellow birds (Emery et al. 2004; Balda and Kamil 2006).

The strategy of dropping hard food objects such as nuts and mollusc shells from the sky onto hard ground so as to break them and access food by birds was first documented over a century ago (Saunders 1891) and has since been investigated (Zach 1978, 1979; Whiteley et al. 1990; Switzer and Cristol 1999; Cristol and Switzer 1999). Object dropping is a smart method by which birds can access otherwise unavailable resources such as food and water. Together with tool making, it (object dropping) is performed by various species of birds in different places and diverse habitats (Bluff et al. 2007). The early investigators of object dropping by birds largely focused on trying to answer the following questions: a) how did some species of birds (e.g., crows) determine which surfaces were appropriately hard to break objects; b) how did birds determine the optimal heights from which to drop the objects and successfully break them; and, c) were birds such as gulls that failed to discriminate between hard and soft substrates onto which to drop objects less intelligent (Oldham 1930; Tinbergen 1960; Barash et al. 1975). More recently, interest in bird object dropping has been directed at understanding the activity from the perspective of optimal foraging theory (Zach 1978, 1979; Maron 1982; Whiteley et al. 1990; Davenport et al. 2014). By studying object dropping of the urban-dwelling American Crow (*Corvus brachyrhynchos*), Cristol and Switzer (1999) noted the following aspects: (a) when choosing heights from which to drop walnuts, the crows were mindful of the risk of kleptoparasitism; they reduced the dropping height when other birds were in close proximity; (b) the crows appeared to make a deliberate decision that repeated dropping of a walnut increased the possibility of breaking it on subsequent drops and then reduced the dropping heights during successive drops; (c) the crows did not alter height from which walnuts were dropped based on the weights of nuts; and (d) when faced with different sizes of nuts, surfaces onto which to drop them and high level of kleptoparasitism, the crows changed the drop heights in a manner that reduced possibility of kleptoparasitism. After investigating object dropping behavior of the New Caledonian Crow (*Corvus moneduloides*), a species that is deemed exceptional among birds from displaying superior capacity of tool manufacture and application both in the wild and in captivity, Chappell and Kacelnik (2002), Roelofs (2010), St Clair and Rutz (2013), Lambert et al. (2017), Jelbert et al. (2019) and Kacelnik et al. (2006) noted that in determining the appropriate heights from which to drop an object, the birds employed pre-existing experience gained on outcomes of objects of different weights previously released from various heights under conditions such as speed and wind direction: the birds selected the right object size for the prevailing environmental conditions.

While use of tools has for a long time been regarded a human attribute (Ambrose 2001; Lawrence 2002; Wadley et al. 2009; Susan et al. 2014), at various levels of sophistication, such activities have also been reported in animals ranging from insects to birds (Shumaker et al. 2011). Tool use has been reported in more than 105 species of birds that include gulls, crows, and eagles (Beck 1980; Iwaniuk et al. 2009). In birds, the wide range of tool making and application abilities are largely for food procurement, self-protection, and courtship (Roelofs 2010). Some of the exceptional cases of tool use in birds are the following: drumming by the Palm Cockatoos (*Probosciger aterrimus*) where nuts and sticks are employed to generate loud banging sounds that serve to assert territory and for courtship (Wood 1984, 1988); the Egyptian Vulture (*Neophron percnopterus*) chooses stones of appropriate size to smash Ostrich (*S. camelus*) egg shells (Thouless et al. 1987); and the Woodpecker Finch (*Camarhynchus pallidus*) can arrange items in a manner that fashions practical tools (Millikan and Bowman 1967; Tebbich and Bshary 2004). Interestingly, while for a long time, the skulls of woodpeckers were posited to serve as a shock absorbers that reduced the injurious deceleration of their brain upon impact into trees (Oda et al. 2006; Wang et al. 2011; Zhu et al. 2012, Zhu et al. 2014; Wang et al. 2013; Liu et al. 2015); recently, Wassenbergh et al. (2022) reported that their (woodpecker's) cranial skeleton functions as stiff hammers to boost the hitting action and not as a shock-absorbing system of protecting the brain.

Although successful tool use does not necessarily constitute intellectual understanding (Hunt et al. 2006; Teschke et al. 2011; Seed et al. 2015; Cochet and Byrne 2015), a strong correlation between tool-use and brain size and greater cognition power has been reported in birds (Lefebvre et al. 2002, 2004; Sol et al. 2005; Cnotka et al. 2008; Roelofs 2010). Suggesting that tool users have superior cognitive abilities (Roelofs 2010), the structure of the avian brain (Lefebvre et al. 2002; Iwaniuk et al. 2009; Auersperg et al. 2017) compares with the primate ones (Reader and Laland 2002). Taylor et al. (2012) observed that from their figuring out things both causally and analogically, the New Caledonian Crows (*Corvus moneduloides*) can solve complex physical and mental problems.

1.6 Vocalization in Birds: Sound Production

The capacity of mimicking complex sounds is unique and among birds has ostensibly only been realized in parrots, songbirds, and hummingbirds (Jarvis and Mello 2000; Gahr 2000; Jarvis 2004a; Chakraborty et al. 2015): parrots evolved vocal learning systems at least 29 million years ago (Chakraborty et al. 2015). In biology, vocalization is a process by which aerodynamic energy is converted into acoustic energy (Bouhuys et al. 1968; Hertegard et al. 1995; Titze et al. 2016; Titze 2018; Vamsi et al. 2019). Phonation is a complex process that evolved through changes in various central and peripheral neural control systems and development of suitable structures that serve as primary sites of sound production (Suthers and Zollinger 2004; Goller and Cooper 2004; Düring et al. 2013; Habib 2019). Pfenning et al. (2014) observed that human beings and song-learning birds share independently evolved parallels in brain pathways for vocal learning and speech, aspects, and features that are lacking in most other species of animals. Like in human beings, vocalization in birds involves three main motor systems, i.e., respiratory motor system, the vocal organ(s) [where primary (raw) sound is produced], and the vocal tract where sound is moderated (Riede and Goller 2010a). Except for human beings and a few other primates, birds are a highly vocal animal group (Schmidt and Martin 2014). Avian vocalization, which comprises production of different kinds of sounds that are categorized as "calls" or "songs" (MacDougall-Shackleton 1997; Zeigler and Marler 2008; Catchpole and Slater 2008), evolved under sexual and ecological selection pressures (Catchpole and Slater 2008; Derryberry et al. 2018; Wada 2010; Mainwaring et al. 2014; Fishbein 2022): songs generally last longer while calls are simple and short intonations. Bird vocalizations are associated with territorial assertion, mate attraction, and courtship (Read and Weary 2004; Catchpole and Slater 2008; Nowicki and Searcy 2014; Hill et al. 2018) while calls are largely used to set off alarms or to keep chicks and their parents and flocks together (Tefera 2012). The chicken (*G. g.* variant *domesticus*) has characteristic alarm calls that discriminate aerial from ground predators (Collias 1987; Evans et al. 1993; Tefera 2012). Detecting birds by sound and not visually (i.e., from plumage, colors and field markings), a method dubbed "birding by the ear," is a valuable skill for bird watchers/ornithologists. Most birds, however, produce infrasonic sounds that cannot be detected by the human ear (Brandes 2008).

In birds, sounds are largely generated from a morphologically highly specialized organ called syrinx (Warner 1972a, b; King 1989; Brackenbury 1989; Suthers and Zollinger 2004; Goller and Cooper 2004; Kabak et al. 2007; Miller et al. 2008; Riede and Goller 2010a, b; Prince et al. 2011; Düring et al. 2013): it serves as the "voice box" equivalent of the larynx of mammals. The most ancient fossil of an avian syrinx is that of a specimen of *Vegavis* that was dated back to the late Cretaceous, i.e., 66–69 million years ago, by Clarke et al. (2016). In most birds, the syrinx is located at the distal (terminal) part of the trachea where the airway branches into the extrapulmonary primary bronchi (Plate 1.1a–d). Depending on location and the cartilages that underwrite its body, three forms of syrinxes, namely, the tracheal, the tracheobronchial, and the bronchial syrinxes exist in birds (King 1989; Brackenbury 1989; Riede et al. 2010; Erdoğan et al. 2015). Sound is produced by air flowing through the syrinx, a process that triggers a self-oscillating mechanism by which airflow is modulated to create sound by vibration of some or all of the tympanic membranes (King 1989; Brackenbury 1989; Larsen and Goller 2002; Riede and Goller 2010b). The syrinx is located close to and in some cases directly contacts air sacs, especially the interclavicular air sacs (Plate 1.1d), and to a lesser extent, the cranial thoracic air sacs: pressure changes in the compliant structures may

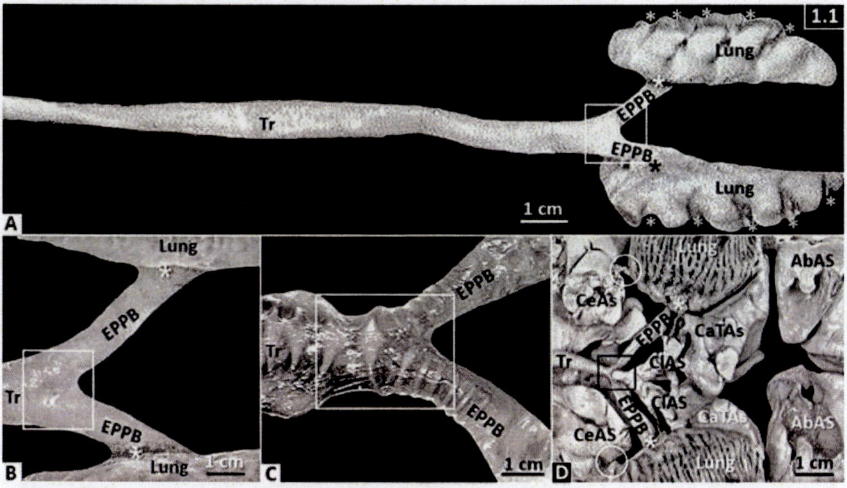

Plate 1.1 The preparations are from the respiratory system of the domestic fowl (*Gallus gallus* variant *domesticus*). (**a–c**) Dorsal view showing the trachea (Tr) and lungs (**a**) and close-up of the location of the syrinx (**b**, **c**). EP, extrapulmonary primary bronchi; stars (⋆), hila; asterisks (*), costovertebral impressions; boxed area (□), location of the syrinx. (**d**) Latex rubber cast of the lungs and air sacs showing trachea (Tr), extrapulmonary primary bronchi (EP), cervical air sacs (CS), hila (star, ⋆), clavicular air sacs (ClS), cardiothoracic air sacs (CaS), abdominal air sacs (AS), ostia (circle), syrinx (boxed area)

directly and/or indirectly influence sound production (Lockner and Murrish 1975; Mackelprang and Goller 2013). Showing that pressure balance within and without the syrinx is vital to sound production, a bird cannot vocalize if the interclavicular air sacs are punctured (Schmidt and Martin 2014). Kingsley et al. (2018) pointed out that the syrinx is a novel organ of which there is significant functional association between the different structures that comprise it. The tension and the diameter of the syringeal airway are regulated by a group of syringeal muscles (Warner 1972a, b; King 1989; Goller and Suthers 1996; Goller and Larsen 1997; Larsen and Goller 2002; Elemans et al. 2006). It has been suggested that the location of the syrinx at the base of the trachea, as opposed to the mammalian larynx that is located in the neck, may permit birds to vocalize more efficiently (Habib 2019): the anatomical arrangement makes the avian trachea serve as an additional sound modulating part. Bird songs are best developed in the so-called "songbirds" (Order: Passeriformes) while some species of birds such as stocks are nearly voiceless from lack of well-developed syrinx (Williams 2006, 2008; Domet 2015).

With the greatest sensitivity occurring between 1 Hz and 5 kHz, the hearing range of birds is generally ~12 kHz to just under 50 Hz (Dooling 1982). In a study performed on chickens, the birds produced sound frequencies ranging from 10 kHz to 50 kHz (Tefera 2012). It has, however, been reported that the domestic fowl (*G. g.* variant *domesticus*) can hear sounds ranging from 60 to 11,950 Hz (Broitman 2007): the hearing range is within that of humans of 20 Hz to 20,000 Hz, with peak sensitivity ranging between 1000 Hz and 4000 Hz (Purves et al. 2001). The Black Jacobin Hummingbird (*Florisuga fusca*) vocalizes at ~11.8 kHz, and other hummingbird species such as the Andean Ecuadorian Hillstar Hummingbird (*Oreotrochilus chimborazo*), the Buff-Tailed Coronet Hummingbird (*Boissonneaua flavescens*), the Speckled Hummingbird Hummingbird (*Adelomyia melanogenys*), and the Violet-Tailed Sylph Hummingbird (*Aglaiocercus coelestis*) hear at an average frequency of ~13.4 kHz (Duque et al. 2018). While Olson et al. (2018) suggested that birds may produce sounds of such a high frequency that they may not hear them, a more recent study by Duque et al. (2020) reported that hummingbirds can indeed hear such sounds, and they may even communicate at those levels of vocalization. The Blue-Throated Hummingbird (*Lampornis clemenciaes*) generates complex songs that extend into the ultrasonic frequency range of up to 30 kHz, but the high-frequency song domains are not used for communication between species (Pytte et al. 2004). According to Podos and Cohn-Haft (2019), at 125 dB, the White Bellbird (*Procnias albus*) makes the loudest call ever recorded for birds while among passerine species, the Screaming Piha (*Lipaugus vociferans*) makes loud calls of 116 dB (Erwin 2004). In urban areas, interestingly, because of background low-frequency noise, birds sing louder and at higher pitches (Slabbekoorn and Peet 2003; Brumm 2004). For urban dwelling birds, matching intensified levels of background noise, song volume, and frequency increase in minimum frequency (Brumm and Todt 2002; Brumm 2004; Luther and Derryberry 2012). Although specific details are at present lacking (Najar and Benedict 2019), it has been suggested that birds display increase of song complexity with latitudinal variation and migratory behavior (Catchpole 1982; Read and Weary 1992). Superiority of bird song appears to mirror fitness and health (Read and Weary 2004), and changes of song repertoire show a relationship with conditions and states such as parasitic infestation as well as existing pathologies and diseases (Redpath et al. 2000; Garamszegi et al. 2004). For penguins (Family: Spheniscidae), birds that commonly breed in very large compact colonies (Stokes and Boersma 2000; Pozzi et al. 2015), where there is great likelihood of confusion between adults themselves and between adults and chicks, identification and recognition occurs by acoustic signals (Proffitt and McLean 1991; Seddon and Van Heezik 1992): the birds appear to analyze beat amplitude induced by a two-voice system to identify individuals of interest (Lengagne et al. 2001; Aubin and Jouventin 2002; Aubin 2004). Because of their small body size, hummingbirds (Family: Trochilidae) have undergone extreme selection for some biological traits that set them well-apart from many other avian species, even from swifts (Family: Apodidae) to which they phylogenetically closely relate (McGuire et al. 2014). One of such traits is the remarkable diversity of their mode of vocalization and acoustic communication, adaptations that largely relate with their foraging strategies (Baptista and Schuchmann 1990; Ferreira et al. 2006; Araya-Salas and Wright 2013; Clark et al. 2018).

1.7 Avian Body Bauplan: Structure and Function

The classification of living organisms/animals specifies how all of them relate in nature (ICZN 1999). It is based on various factors among which are morphologies (forms, shapes, and sizes), physiologies, and behaviors (e.g., the lifestyles practiced), habitat occupied, type of food eaten, and, in the recent times, their genetic constitution. On cursory glance, although birds may appear relatively structurally feeble compared with mammals, contrary to the long held believe that their bones are hollow, thin-walled, light, and, therefore, weak, Dumont (2010) determined that for their sizes, bird bones are in particular denser, stiffer, and stronger. Internal struts possibly allow the skeletal framework of birds tolerate large physical and mechanical forces such as those generated during powered flight and when the animal is running fast on hard ground (Earls 2000; Askew and Marsh 2001; Askew et al. 2001; Biewener 2011; Chin and Lentink 2019). Interestingly, penguin-like birds have solid bones (https://penguinsblog.com/do-penguins-have-hollow-bones/—accessed on 14-08-2022), a feature that would be advantageous in their staying and swimming ("flying") underwater for long periods of time. Even the most aerial species of birds such as Swifts (Order: Apodiformes), Albatrosses (Order: Procellariiformes), Frigatebirds (Order: Pelecaniformes), Martins (Order: Passeriformes), Nighthawks (Order: Caprimulgiformes), Nightjars (Order: Caprimulgiformes), Petrels (Order: Procellariiformes), Shearwaters (Order: Procellariiformes), and Swallows (Order: Passeriformes) (Martins and Mead 2003; Heers et al. 2011; Weimerskirch et al. 2000, 2015; Heers and Dial 2015; Richardson et al. 2018) utilize bipedal terrestrial locomotion for times some during their life. Regarding muscle function, leg movement, stability, and agility during terrestrial locomotion, in birds, the relationship between force, work, and power has been well-studied (Fedak and Seeherman 1979; Rubenson et al. 2004; Heers et al. 2011; Daley and Birn-Jeffrey 2018). The Ostrich (*S. camelus*) can attain and sustain a speed of 75 km h^{-1} (Davies 2003; Sellers and Manning 2007; Abourachid and Renous 2008): it is one of the few animals that can outrun fast predators such as lions and cheetahs. For birds, the main flight muscles are the pectoralis and supracoracoideus (Plate 1.2a, b). During wing flapping, the flight muscles exert inordinately high pressures and stresses on the skeletal construction of birds, especially the rib cage (Bishop and Butler 2015; Usherwood 2016; Butler 2016) that has been adaptively strengthened by presence of structures such as the uncinate processes which interconnect the ribs along the mid-shaft parts (Welty and Baptista 1988; Coles 1997; Codd 2010; Kardong 2014) (Plate 1.2c): also, the structures (uncinate processes) allow dissipation of the forces that originate from wingbeats. Additionally, they serve as levers that create and augment mechanical advantage during the forward movement of the dorsal ribs, depressing the sternum (Codd et al. 2005; Tickle et al. 2007). The shapes of the ribs and designs of the rib cages of birds vary between species of birds, with the uncinate processes being small in walking birds, elongated in diving species such as guillemots (Family: Alcidae) and loons (Family: Gaviidae) and of medium size in ordinary ones (Tickle et al. 2007): uncinate processes are lacking in the screamers (Order: Anseriformes), that

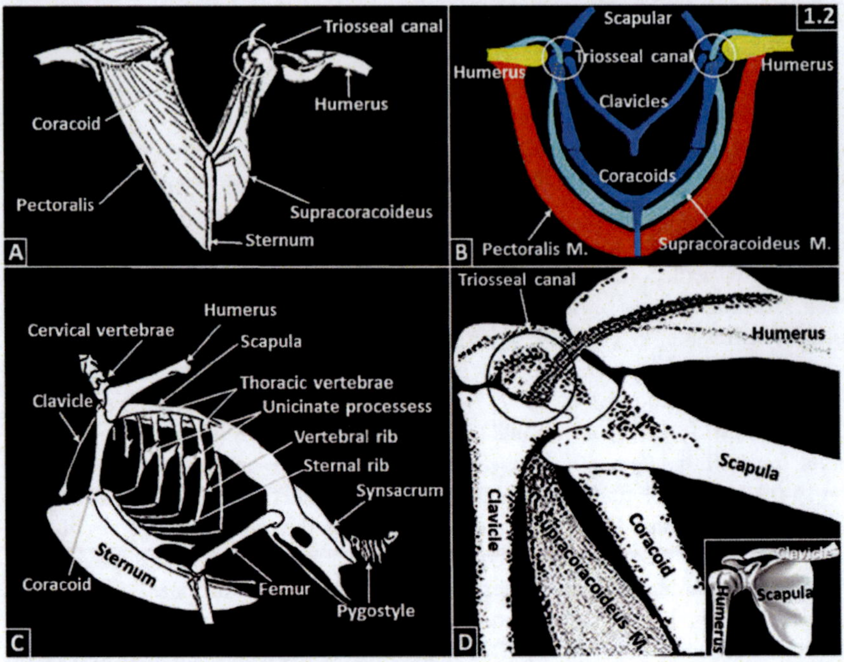

Plate 1.2 (**a, b**) Major wing flight muscles of birds, namely, the pectoralis (the depressor) and the supracoracoideus (elevator) muscles and associated bones. (**c**) Skeletal system of the trunk of a bird. The uncinate processes interconnect the ribs and strengthen the rib cage. (**d**) The complex pectoral girdle of birds which compared with the mammalian one (inset) comprises four bones, namely, humerus, clavicle, coracoid, and scapula

are South American water-fowl species (Fowler and Cubas 2001). The structural differences indicate variations in the respiratory biomechanics of birds and especially show a relationship with differences in locomotory characteristics (Tickle et al. 2007, 2009; Codd 2010).

Birds have a unique pectoral girdle that comprises four bones, namely, a scapula, a clavicle, a coracoid, and a humerus (Plate 1.2d): they strengthen and maintain the integrity of the skeletal system from certain bones acting as transverse props (e.g., the clavicle) and as braces (e.g., the coracoid) that direct forces of wing beats away from the rib cage (King and McLelland 1984) (Plate 1.2d). In divers, the uncinate processes ensure tolerance of the pressure which is applied on the thoracic cavity by water (Coles 1997): with water pressure increasing at a rate of 1 atmosphere per meter (which is ~1 kg.cm^2), Guillemots and Razorbills (Family: Alcidae) have been recorded to dive to a depth of 180 m (Piatt and Nettleship 1985) while Emperor Penguins (*Aptenodytes forsteri*) (Family: Spheniscidae) have been found to dive as deep as 534 m (Kooyman and Kooyman 1995). In mammals, the pectoral girdle comprises only three bones, namely, a humerus, a clavicle, and a scapula (Plate 1.2d—inset). The avian one that is relatively more complex (Plate 1.2d) allows the rib cage and the proximate bones to absorb and efficiently dissipate the enormous forces and stresses created by the beating (flapping) wings. In order to generate the power required to support the body in air and to overcome drag for propulsion during forward flight, birds have to flap their wings, an activity that entails contraction and relaxation of mainly the two primary flight muscles, where the pectoralis acts as the depressor and the supracoracoideus as the elevator of the wing (Plate 1.2a, b). Compared with minimum mechanical power demands, during avian flight, across a range of flight speeds, mechanical power increases by factors of 2.1–2.97 (Hedrick et al. 2003; Tobalske et al. 2003; Ellerby and Askew 2007; Morris and Askew 2010a, b; Morris et al. 2010). Extremely high wing beat frequencies of 90 times per second have been reported in hummingbirds (Rauzon 1997; Clark 2011; Steen et al. 2020; Schaefer and Huegel 2020). At powerful frequencies, the muscles contract over a large part of their resting lengths by as much as 33–42% (Biewener 2011), performing considerable amount of work (Suarez 1992; Butler 1991; Jackson 2009; Reiser et al. 2013).

Comprising as much as 17–25% of their total body mass, the pectoralis muscle is one of the largest anatomical structures in bodies of birds (Greenewalt 1962; Dietz et al. 2007). Responsible for the downstroke phase of wing beat (Torrella et al. 1998), the muscle provides lift and thrust during forward powered flight (Driedzic et al. 1993). Like for the mammalian respiratory muscles (Rowley et al. 2005; Gransee et al. 2012), across the species of birds, the pectoralis muscle displays remarkable structural and functional plasticity (Jimenez et al. 2019; Lindstrom et al. 2000; Jimenez 2020), especially with regard to the mass of the muscle between seasons and before birds embark on migratory excursions (Marsh 1984; Gaunt et al. 1990; O'Connor 1995; Swanson and Merkord 2013; Swanson et al. 2013; Vágási et al. 2016). For the small passerine birds, the pectoral muscle serves as an important thermogenic organ, where in promigratory species, its size about doubles to increase heat production (Evans et al. 1992; Jimenez et al. 2013; Jimenez 2020). Various energy saving changes that reduce energy consumption during terrestrial locomotion have evolved in birds (Roberts et al. 1997, 1998; Alexander 2002, 2005; Wilson and Lichtwark 2011; Holt et al. 2014; Peyré-Tartaruga and Coertjens 2018; Halsey and White 2019).

1.8 Avian Flight: A Novel Mode of Locomotion

For over the ~3.5 to 4 billion years that life has existed on Earth (Abramov and Mojzsis 2009; Nutman et al. 2016; Baumgartner et al. 2019), powered (active = flapping) flight has evolved in only four animal taxa. Chronologically, these are insects that evolved at the mid-Devonian Period (~400 mya) (Dudley and Yanoviak 2011; Grimaldi and Engel 2004, 2005); the now extinct pterosaurs that lived from the Late-Triassic to the End-Cretaceous (~220 mya) (Padian 1985; Hazlehurst and Rayner 1992; Barrett et al. 2008; Claessens et al. 2009; Witton and Habib 2010; Benson et al. 2012); birds that appeared around the Upper Jurassic Period (~160 mya)

(Ruben et al. 2003; Chin et al. 2017; Voeten et al. 2018) and bats (Order: Chiroptera) that appeared at the Early Eocene (~50 mya) (Yalden and Morris 1975; Simmons et al. 2008; Amador et al. 2019; Anderson and Ruxton 2020; Damaraju et al. 2021). The acquisition of powered flight in remarkably phylogenetically different animal taxa is an excellent example of convergent evolution of a common locomotory lifestyle that was critical to the survival and the evolutionary advancement of a group of animals (birds) (McGhee 2011; Losos 2011; Gleiss et al. 2011; Gabora 2013; Bale et al. 2015; Sackton et al. 2019). While all birds evolved from volant progenitors (Welty 1982; Welty and Baptista 1988; Coles 1997), for different reasons, many species have lost flight or currently have poor flight capability (Harshman et al. 2008; Sackton et al. 2019; Pan et al. 2019) (Plate 1.3). Nonvolant birds include a broad range of species that include groups such as penguins, megapodes, and ratites (i.e., the rhea, the kiwi, the ostrich, the cassowaries, and the emu) (Faux and Field 2017) (Plate 1.3b, i–v). Flightlessness appears to have been an "adaptation" for energy conservation, in response to low level of predation and response to changes in food availability (McNab 1994; Wright et al. 2016; Van de Crommenacker et al. 2019; Stervander et al. 2019). Dearth of predators, especially mammalian ones (predominantly in island settings) may have led to loss of flight (Quammen 1996; Worthy 2001; Steadman 2006). Flight (volancy) permitted birds to undergo first rate adaptive radiation, which among the tetrapods made them the most species-rich (speciose) taxon with global ecological distribution (Zhang et al. 2014; Ruaux et al. 2020). By adopting various strategies, the early birds and their closest relatives (Paraves) experimented with powered flight for a long time (Dyke et al. 2013; Palmer 2014; Dececchi et al. 2016; Pei et al. 2020; Motta et al. 2020).

At the extreme altitudes that some species of birds fly, the environmental conditions greatly threaten survival (Ward et al. 2002; Körner 2007; Storz et al. 2015; Yang et al. 2011; Hawkes et al. 2011; Meir and Milsom 2013; Butler 2016; Meir et al. 2019; Harrison 2019; Yu et al. 2022): the surroundings include low partial pressure of O_2 (hypoxia), low temperature (hypothermia), less dense (rarefied) and dry air, and high intensity of ultraviolet radiation. The respiratory physiologies of high elevation dwelling and flying birds markedly differ from those of low-altitude dwelling or operating ones (Weber 2007; McCracken et al. 2009, 2010; Scott 2011; Scott et al. 2015; Jendroszek et al. 2018). In birds, to lessen the energetic demands for powered flight, among other features, body mass has been lowered by anatomical reorganization of the skeletal system: certain bones have been eliminated or reduced in size while others have fused (King and McLelland 1984; Britt et al. 1998; Smith et al. 2004; Dumont 2010; Puttick et al. 2014; Sullivan et al. 2017).

Powered flight entails production of lift, acceleration and maneuver in air at various speeds in different directions (Jenni-Eiermann and Srygley 2017; Chilson et al. 2017): it involves delicate balance between thrust, drag, lift and weight (mass = gravity) (Norberg 1990; Alexander 2013; Dvořák 2016; Chin and Lentink 2019) (Plate 1.4a). Wings produce thrust and lift, aspects that overcome drag, generating forward travel through air (Dvořák 2016). Features such as wing morphology, wingbeat kinematics, generation of aerodynamic forces, and flight muscle

Plate 1.3 Variety of species of birds that display various capacities of volancy (flight capacity) and nonvolancy (flightlessness)

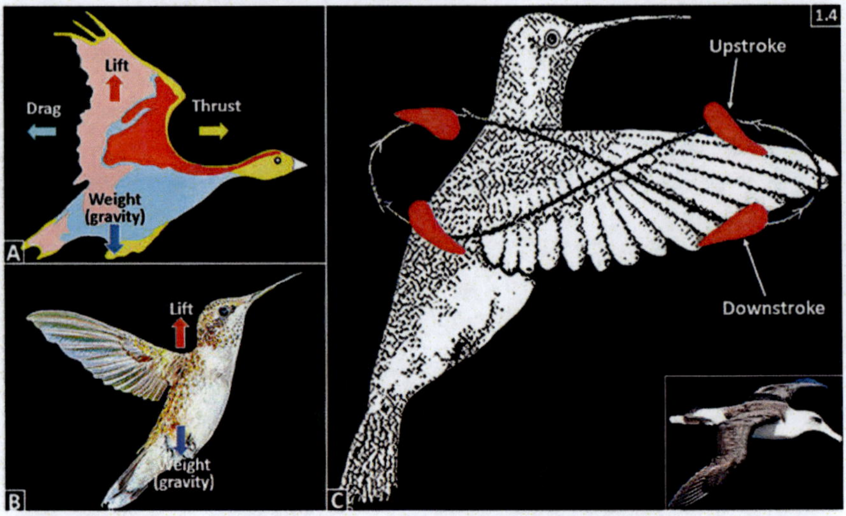

Plate 1.4 Flight. (**a**) Flying is a balance between forces, namely lift (red arrow), weight (gravity) (blue arrow), thrust (yellow arrow), and drag (cyan arrow): during forward flight, the lift is generated by the flow of air over the wings while drag is the resistance the air presents to forward movement. The most efficient wings are those that provide maximum lift with minimum drag. (**b**) In a hovering bird, lift is totally generated by flapping of wings, i.e., the body weight is supported by power generated by flight muscles. (**c**) For a hummingbird, during hovering, the wing traces a figure of eight: lift is generated on both upwing and downwing strokes. (**c**)—**Inset:** An albatross utilizing dynamic soaring as it flies over ocean

mechanics work together to produce the power needed to overcome gravity and provide forward thrust (Lasiewski et al. 1965; Greenewalt 1962; Chai and Millard 1997; Tobalske et al. 2007; Tobalske 2010). The elegance of powered avian flight is greatly deceptive of the demanding structural and energetic requisites for the mode of locomotion that has compelled complex structural and functional adaptations/specializations and structural and functional integration of various parts of the body, especially the cardiovascular, the respiratory, the nervous, the gastrointestinal, and the muscular systems (Swartz et al. 2003; Videler 2006; Norberg 2006; Brusatte et al. 2015; Vágási et al. 2016; Cao and Jin 2020). During flight, the heart and the respiratory system consume ~10% of the amount of O_2 that is obtained from outside: it (O_2) is utilized to meet the metabolic cost of flight (Pennycuick 1975, 1989; Tucker, 1973; Ward et al. 2001).

Birds display various forms of flight modalities that correspond with factors such as body mass, lifestyle and environment occupied. Compared with flapping flight (Plate 1.4a), soaring, and gliding are relatively energetically less costly (Williams et al. 2020) while during hovering, which is energetically the most costly form of avian flight (Powers and Naggy 1988; Bartholomew and Lighton 1986; Suarez 1998; Wells 1993; Altshuler et al. 2004; Tobalske 2010; Chen and Welch 2014; Groom et al. 2018), body mass is supported against gravity entirely by power generated by flight muscles (Chai and Millard 1997) (Plate 1.4b). In the hummingbirds (Family: Trochlidae), during hovering, the flight muscles contract at very high frequencies (Weis-Fogh 1972; Warrick et al. 2005, 2012; Tobalske et al. 2010) and the wingbeat rather traces a figure of eight shape (Plate 1.4c): lift is generated during both the downstroke and the upstroke beats (Norberg 1990; Askew et al. 2001; Askew and Marsh 2001; Tobalske 2010; Biewener 2011; Pennycuick 2015). Hummingbirds attain wingbeat frequencies of ~80 to 200 Hz and high aerobic power and muscle power output of ~100 to 120 W kg^{-1} at 9–11% mechanochemical efficiency (Wells 1993; Rauzon 1997; Schaefer and Huegel 2020; Warrick et al. 2012).

Among the endothermic-homeotherms, hummingbirds have the highest mass-specific metabolic rates (Suarez 1992, 1995, 1998), and the flight muscles comprise 21–30% of the body mass (Hartman 1954, 1955; Greenewalt 1962, 1990). In relative terms, birds have notably large hearts (Hartman 1955). While the human heart constitutes 0.4% of body mass (Antoł and Kozłowski 2020), the heart of a hummingbird comprises as much a 2.4% of the body mass (Hartman 1955). On analyzing the sizes of hearts of 915 species of birds, Nespolo et al. (2018) found that the residuals correlated with flight characteristics and capacity. Compared with those species of birds that utilize flapping and gliding as their primary modes of flight, the species of birds that hover (e.g., hummingbirds) have relatively larger hearts in relation to those of the poor flyers and flightless ones.

For birds, adaptive strategies such as coordination of locomotory activities with ventilatory ones and adjustments for environmental and ecological conditions lower overall energetic demands for powered flight (Schmidt-Nielsen 1972; Reilly et al. 2007; Maina 2000a; Gleiss et al. 2011; Halsey 2016; Halsey and White 2019). In birds, during powered flapping flight, O_2 consumption ($\dot{V}O_2$) exceeds that at the basal metabolic rate (BMR) by a factor of 3–15 (Lasiewski 1963; Tucker 1968; Bishop 1999; Speakman and Thomas 2003; Butler 2016) while during gliding flight O_2consumption is three times that of BMR (Ward et al. 2002; Butler 2010; Duriez et al. 2014; Cao and Jin 2020). Birds that soar, e.g., raptors, vultures, and storks utilize rising hot air masses (thermals) to gain and maintain altitude (Cone 1962; Pennycuick 2008; Reddy et al. 2016). A passive process by which birds figuratively "ride on the wind," soaring flight requires little, if any, energy outlay. With wing flapping occurring only 1% of the flight time, an Andean Condors (*Vultur gryphus*) weighing between 9.5 and 13.9 kg flew for more than five hours over a distance of ~172 km (Williams et al. 2020). By employing a process termed "dynamic soaring," propulsive energy is obtained totally from horizontal wind shears, allowing a bird to fly over long distances without having to beat its wings, i.e., without expending energy (Cone 1964; Shenstone 1968; Sachs et al. 2012; Pfeifhofer and Tributsch 2014) (Plate 1.4c—inset): dynamic soaring works best in settings where physical obstacles that generate wind turbulence are absent, e.g., flat ground and surface of the ocean. With a wing span of ~3.5 m, the Wandering Albatross (*Diomedea exulans*) (Plate 1.3a, vi) of the Southern Oceans, where permanently strong winds (the so-called "roaring forties") exist (Suryan et al. 2008), the bird employs dynamic soaring during its long journeys: the bird takes 46 days to fly around the world (Jouventin and Weimerskirch 1990; Catry et al. 2004; Croxall et al. 2005;

Bonadonna et al. 2005; Sachs et al. 2013). Expending minimal energy (Weimerskirch et al. 2000; Sachs et al. 2012; Richardson et al. 2018) and depending only on wind energy, the Wandering Albatross has been logged to cover distances of 5000 km/week (Weimerskirch et al. 2000; Catry et al. 2004; Sachs et al. 2012). Albatrosses possess wings with a joint locking mechanism that reduces tension on muscles of flight during their lengthy soaring flight (Videler 2006). Other species of birds utilize "intermittent flight" to lessen the mechanical power needed for horizontal movement (Tucker 1968, 1971, 1975; Rayner et al. 2001; Tobalske 2001). Compared with continuous level flight, occasioning significant energy saving, during undulating flight a bird glides with flexed wings that are interspersed by active wing flapping (Rayner et al. 2001).

While birds inhabit altitudinal gradients that range from sea level to montane regions up to higher elevations of more than 4000 m above sea level, some of them fly to extreme altitudes, particularly as they migrate across lofty landforms such as mountains (McCracken et al. 2009, 2010; Laguë et al. 2016, 2017; Ivy et al. 2019). Their capacity of flight at high altitude, where extreme environmental conditions of hypoxia and very low environmental temperatures exist, is beyond tolerance of most other animals. Dubbed "super birds" (Whiteman 2000), only a few species of birds normally fly under such conditions (Laguë 1985; Scott and Milsom 2006; Scott 2011; Scott et al. 2015; Scott and Dawson 2017). While at high-altitude drag is low from the thinness (rarefication) of air, a bird must expend more power to create thrust and maintain forward level flight (Tucker 1974; Harrison 2019) under hypoxic conditions (Chai and Dudley 1995; Altshuler and Dudley 2006). Some of the champion high-altitude flying/inhabiting birds are the Ruppell's Griffon Vulture (*Gyps rueppellii*) (Plate 1.3c, i), the Bar-Headed Goose (*Anser indicus*) (Plate 1.3c, ii), the Andean Goose (*Chloephaga melanoptera*) (Plate 1.3c, iii), and the Tibetan chicken (*G. gallus*) that lives in the high elevation Qinghai-Tibet Plateau of an average elevation above 4000 m (Qiao et al. 2016; Tang et al. 2021; Zhao et al. 2022; Niu et al. 2022). Because the actual time and elevation at which the collision occurred was precisely recorded, one of the most reliably documented high altitude bird flight is that of a Ruppell's Griffon Vulture that was sucked into the engine of a jet craft at an altitude of 11.3 km (37,000 feet) above sea level over Abidjan (Côte d'Ivoire, West Africa) (Laybourne 1974): at that elevation, the barometric pressure was ~24 kPa (180 mmHg) (i.e., ~24% of that at sea level) and the ambient temperature was ~−60 °C. Annually, the Bar-Headed Geese (*Anser indicus*) migrate across the Himalayas between their breeding grounds in the region of the Mongolian and the Tibetan Plateaus of South-Central Asia (that lie at an elevation of between 2000 and 4000 m above sea level) to overwinter for the most part at sea level (less than 1000 m altitude) in India and Bangladesh. Occasionally reaching an altitude of 9000 m, the birds crossover the mountains at altitudes between 4500 and 5500 m, elevations at which the PO_2 ranges between ~50% and 30% of the sea level values (Takekawa et al. 2009; Köppen et al. 2010; Butler 2010). Without displaying signs of respiratory distress, experimentally, the Bar-Headed Geese (*Anser indicus*) were found to tolerate hypoxia up to an altitude of 10.7 km (35,105 ft) (Black and Tenney 1980). Theoretically, during migration, if the Bar-Headed Geese maintained a constant body core temperature of ~41 °C and the temperature of the inhaled air is heated to that of the body (40–42 °C) and 100% saturated with water vapor, the PO_2 in the air reaching the respiratory surface (air capillaries) should not be more than 4.9 kPa (36.6 mmHg).

In accordance with the "Two-State Monod-Wyman-Changeaux Model," thermodynamic analyses of O_2 equilibrium curves showed that compared with the low elevation dwelling Greylag Goose (*Anser anser*), the greater R-state O_2affinities of the hemoglobin of the Bar-Headed Goose (*Anser indicus*) was characterized by decreased O_2 dissociation rate (Jendroszek et al. 2018): the very high affinity (R-state) increases O_2 saturation in the lungs during hypoxia without hampering O_2 delivery to the tissues. Comparing the Graylag Geese and the Pekin Ducks (*Anas domesticus*) that are lowland dwelling species of birds with the Bar-Headed Geese, during experimental exposure to extreme hypoxia (5% inspired O_2), the Bar-Headed Geese significantly increased both O_2-loading into the blood and arterial PO_2 by considerably increasing the tidal volume above those of other birds (Scott and Milsom 2007). For the Bar-Headed Geese, the effects of temperature and pH on blood O_2affinity in cold alkalotic lungs and warm acidotic muscle may increase O_2-transport two-fold during continuous wing flapping at high altitude (Meir and Milsom 2013). Compared with those raised at low elevation, Laguë et al. (2016) observed that the Bar-Headed Geese reared at high elevation showed lower O_2-consumption at rest and little but significant increase of O_2-acquisition and transfer with increase in hypoxia. Comparison of the respiratory physiologies of the lowland dwelling Greylag Geese with the Bar-Headed Geese (Petschow et al. 1977; Black and Tenney 1980; Fedde et al. 1989; Jessen et al. 1991; Weber et al. 1993; Meir and Milsom 2013; Natarajan et al. 2018) showed the following important effects: (a) the Bar-Headed Geese have higher blood-O_2 affinity (P_{50}) and (b) under hypoxic conditions, for the Bar-Headed Geese, muscle blood supply and O_2-loading from the muscle blood capillaries and not ventilation and pulmonary gas transfer are the limiting factors of O_2 delivery to the flight muscles. Laguë (2017) observed that the Andean Goose (*Chloephaga melanoptera*) and the Crested Duck (*Lophonetta specularioides*), two unrelated avian species that live in the Andes at elevations between 4000 to 5500 m and occasionally fly to altitudes of ~6000 m where O_2 levels are limiting, have evolved totally different respiratory adaptive stratagems of conserving O_2 supply under hypoxic conditions: while the Bar-Headed Goose responds to hypoxic conditions by increasing ventilation of the lungs and heart rate, the Andean high altitude flying species increases lung O_2 extraction and cardiac stroke volume. It appears like for short-term hypoxia exposed high-altitude flying birds such as the Bar-Headed Geese (that are exposed to extreme hypoxia only during seasonal migration), strong convective O_2 transport mechanisms are utilized while birds that permanently live at high elevation, such as the Andean Goose, utilize specializations of pulmonary and cardiac structures to provide large pulmonary diffusion capacity of O_2 and stroke volume (Laguë 2017). The respiratory strategies utilized by the Andean Ducks, where higher hematocrit and hemoglobin concentration increase the O_2 content of arterial blood, confer greater reserve for increasing O_2 transfer during hypoxia (Laguë et al. 2020): they differ from those of both the Andean Geese and the Bar-Headed Geese that do not have significantly higher hematocrit or hemoglobin concentration compared with those of low elevation dwelling avian species.

Comparative study of lungs of five species of high elevation dwelling ducks around the Lake Titicaca region of the Peruvian Andes with those of the high elevation migratory Bar-Headed Geese and the low-altitude living migratory Barnacle Geese (*Branta leucopsis*) showed that the mass-specific lung volumes and the mass-specific volume of the exchange tissue of the lung were greater in lungs of high elevation inhabiting ducks (Bakkeren et al. 2020). The hemoglobin of the Andean Geese has greater O_2affinity compared with that of the Bar-Headed Geese (Black and Tenney 1980; McCracken et al. 2010; Laguë et al. 2016). Together with ensuing cellular and molecular mechanisms that maintain O_2availability under shifting environmental conditions and metabolic demands, largely regarding the cardiovascular and the pulmonary systems, gasotransmitter signaling molecules such as hydrogen sulfide (H_2S) and nitric oxide (NO) have been shown to play important roles in physiological responses to hypoxia (Van Faassen et al. 2009; Flögel et al. 2010; Imbrogno et al. 2011; Fago et al. 2012). Compared with other species of birds that have been investigated, while the lung of the Andean Goose generally structurally resembles those of the other species of birds (Dubach 1981; Maina 1989, 1993, 2005, 2017; Maina et al. 1989a; York et al. 2018), certain important specializations were observed by Maina et al. (2017). In contrast to those of lungs of other species of birds that have been investigated, the infundibula of the lung of the Andean Goose are highly vascularized and the lung is morphometrically greatly specialized with mass-specific lung volume of 42.8 $cm^3\ kg^{-1}$, mass-specific respiratory surface area of 96.5 $cm^2\ g^{-1}$, mass-specific volume of the pulmonary capillary blood of 7.44 $cm^3\ kg^{-1}$, and mass-specific total pulmonary morphometric diffusing capacity of the lung for O_2 of 0.119 $mlO_2\ s^{-1}\ mbar^{-1}\ kg^{-1}$: values are some of the highest in lungs of species of birds that have been studied (Dubach 1981; Duncker and Guntert 1985a, b; Maina 1989, 1993, 2005, 2006; Maina et al. 1989a) (Table 4.3: Chap. 4). Recently, Zhao et al. (2022) observed that lungs of the Tibetan sheep that live in three altitudes (2500, 3500, and 4500 m) display progression of adaptive physiological and molecular specializations for hypoxia that correlate with increasing elevation. Different respiratory, cardiovascular, and flight muscle adaptations permit birds to acquire, distribute, and utilize the large amounts of O_2 that is crucial to powered flight, especially at high altitude (Black and Tenney 1980; Fedde et al. 1989; Jessen et al. 1991; Scott et al. 2015; Maina et al. 2017; Laguë et al. 2017; Hawkes et al. 2017; Jendroszek et al. 2018; Parr et al. 2019; Meir et al. 2019). The foremost adaptive features and properties that characterize volant birds are high metabolic capacities (Hedenström 2010; Piersma 2011; Conklin et al. 2017), a structurally, and functionally highly specialized respiratory system (Maina et al. 2017), high diffusion capacity of O_2 of the lung, tissues, and cells, and large wings that reduce the metabolic costs of flying, especially through the rarefied air of high elevation (Butler 2010; Scott 2011; Scott and Dawson 2017).

While pulmonary vasoconstriction occurs under hypoxia in mammalian (Sylvester et al. 2012; Dunham-Snary et al. 2017; Swenson 2013; Khan et al.

2022) and reptilian lungs (Skovgaard et al. 2005), for the high-altitude flying birds and those experimentally exposed to hypoxia, the effect is inconsequential (Faraci et al. 1984; Laguë 1985; Faraci 1991; Scott et al. 2011a, b, 2015; Ramirez et al. 2007). Under the hypoxic conditions of the high altitude, to acquire a sufficient quantity of O_2, a bird responds by hyperventilating the respiratory system (Laguë 1985; Monge and León-Velarde 1991; Altshuler and Dudley 2006; Dodd et al. 2007), a process that increases CO_2 elimination. For unclear reason(s) that remain unclear, the cerebral vasculature of birds appears to be insensitive to hypocapnia, i.e., the blood vessels do not constrict to limit blood flow to the brain, when a bird hyperventilates to compensate for a lack of O_2 at high elevation (Scheid 1990). In ducks (Family: Anatidae), at an arterial PO_2 of 4 kPa (30 mmHg), Grubb et al. (1978) determined that cerebral blood flow actually increased by a factor of more than 600% above normoxic levels. Also, interestingly, birds appear to be immune to conditions such as high-altitude pulmonary edema that occurs in mammalian lungs (Bärtsch et al. 1985; Bärtsch et al. 20032005; Swenson and Bärtsch 2012; Pennardt 2013; Wilkins et al. 2015).

In birds, the pectoralis and the supracoracoideus muscles, which are the main flight muscles (Tobalske and Biewener 2008; Biewener 2011), are unique in that unlike for the general anatomical arrangement of antagonistic muscles in quadrupedal animals, where muscles that perform converse functions (antagonistic muscles) originate and insert on opposite aspects of bones (across a joint), for the wings of birds, the two flight muscles originate from the same site, i.e., the sternum (keel) and insert on opposite parts of the humerus (Plate 1.2b, d): the supracoracoideus muscle passes through the triosseal canal to insert on the proximodorsal aspect of the humerus while the pectoralis inserts on the ventral aspect (Plate 1.2b, d). Showing that the upstroke phase of the wing beat is the less energetically costly, the supracoracoideus muscle is about one-sixth to about one-third of the size of the pectoralis muscle (Biewener 2011; Nan et al. 2019). The muscle fibers of the pectoralis and the supracoracoideus muscles contract by 33–42% (Biewener 2011). For hummingbirds (Family: Trochlidae), the flight muscles comprise entirely of fast oxidative-glycolytic type-IIa fibers and the mitochondria constitute ~50% of the total muscle volume (Suarez et al. 1991).

In the past, exceptionally large terrestrial animals lived and flew (Lawson 1975; Wing and Sues 1992; Mazzetta et al. 2004; Smith et al. 2010). Among these are birds such as the Super-Condor (*Teratornis incredibilis*) of the North American Pleistocene which is estimated to have weighed more than 20 kg and the Late Miocene (~6 mya) Giant Teratorn (*Argentavis magnificens*) that may have weighed between ~70 and ~120 kg and had a wingspan of between 5.8 and 8 m, a wing area of 7 m^2, and wing loading of ~11.5 kg m^{-2} (Campbell and Tonni 1983; Chatterjee et al. 2007; Ksepka 2014). Because of its estimated high wing loading, the flight capacity of the Giant Teratorn may have been poor and might only have occurred under warm conditions by soaring on thermals, especially those rising against steep land shapes such as cliffs (Chatterjee et al. 2007; Williams et al. 2020). The extinct pterosaur, *Quetzalcoatlus northropii*, that is known to have lived during the Late Cretaceous of North America, had a wingspan of 10–11 m and might have weighed between

200 and 250 kg (Witton and Habib 2010; Habib and Witton 2010): it is apparently the largest known animal that ever flew. While a supposed lower Earth's gravity (Carey 1976; Wesson 1980; Kennedy 2018) may explain the flight capacity huge animals, theoretical estimations of the maximum power generation by flight muscles of modern volant animals (Marden 1994; Ellington 1991) suggests that even under the present-day greater gravitational force, take-off, and short burst flights may have been possible by some of the giant now extinct birds, including the 250 kg heavy pterosaur, *Quetzalcoatlus northropi* (Paul 1991). Regarding the flightlessness birds, the rare Island rail (*Atlantisia rogersi*) that inhabits the remote, rocky Island of the Tristan da Cunha Archipelago in the Central South Atlantic Ocean, and weighs between 34 to 49 g is the world's smallest flightless bird (Plate 1.3b, vii): it is believed that the species lost volancy because it could acquire all its food by walking and there were no predators to avoid (Stervander et al. 2019).

Among the modern species of birds, the largest volant birds weigh ~20 kg (Bacon and Coleman 1986; Bird 1999; Snyder and Schmitt 2002; Alonso et al. 2009, 2010; Alarcón et al. 2017; Williams et al. 2020). Some of the heavy flying birds are the Kori Bustard (*Ardeotis kori*) (~21 kg) (Plate 1.3d, i), the Great Bustard (*Otis tarda*) (~20 kg), the Mute Swan (*Cygnus olor*) (~14.3 kg) (Plate 1.3d, ii), the Whooper Swan (*Cygnus cygnus*) (~14 kg), the Trumpeter Swan (*Cygnus buccinator*) (~17.2 kg), the Andean Condor (*Vultur gryphus*) (~15 kg) (Plate 1.3d, iii), American Golden Eagle (*Aquila chrysaetos*) (~9 kg) (Plate 1.3d, iv), and the Dalmatian Pelican (*Pelecanus crispus*) (~15 kg). Such birds rarely take off directly but have to first run on the ground or jump off from a height (e.g., a cliff) into the wind to get airborne: their landings are largely clumsy. With a flight muscle ratio, i.e., flight muscle mass per unit body mass, of 0.164, the Kori Bustard sprints on the ground before taking-off to make short, labored flights (Pennycuick 1969). Comparative molecular genetics study of 89 bone-associated ossification genes from 47 avian, 39 mammalian, and 20 reptilian genomes showed that during the evolution of birds, there was a period of intense selective pressure when flight efficiency was increased while increase in body mass was limited (Machado et al. 2016). The fastest flying bird on record is the Peregrine Falcon (*Falco peregrinus*) (Plate 1.3e) which during free dive attains a speed of 548 km h^{-1} (Tucker and Parrott 1970; Orton 1975; Savage 1992; Clark 1995; Tucker 1998).

1.9 Bird Migration

After undergoing various physiological preparations/adjustments/adaptations every year, billions of animals that include insects, fish, birds, and mammals migrate over various distances from their resident (primary) locations to new places/habitats (Hansson and Åkesson 2014; Hegemann et al. 2022). Migration is one of the most challenging functional tasks in the animal's life: it requires certain physiological changes and critical resource-balancing between energy and nutrients and metabolic energy production (Butler 2010, 2016; Laguë 2017; McWilliams et al. 2004, 2020;

Sapir et al. 2011; Hegemann et al. 2019; Maggini et al. 2022). Migrating birds traverse colossal geographical barriers, face changing weather and environmental conditions, endure unreliable resource availability, and encounter erratic predation pressure (Shochat et al. 2022). Hegemann et al. (2022) found that for migratory birds, the adaptive immunological states depend on migratory strategies and that early and late migrating birds may strategically commit resources in different ways and means. The impact of avian gut microflora on activities such as migratory travels of various avian taxa remain undetermined (Wang et al. 2022).

Starting in the 1980s (Strikwerda et al. 1986), Earth-based satellite tracking has modernized studies of animal mass movements and migration especially that of birds: by allowing precise routes to be followed and any stop overs and durations of such events to be accurately determined, avian migration is now a sound science (Ueta et al. 2000; Hake et al. 2001; Ueta and Higuchi 2002; Haines et al. 2003; Meyburg and Fuller 2007). Presently, at low scale, space-based tracking technology using low-cost miniature tags is providing instructive data on fine-scale animal movement at near-global scale (Jetz et al. 2022). Migratory strategies, distances, and speeds vary greatly between species, populations, sexes, and ages of birds (Coppack and Both 2002; Alerstam et al. 2003; Pulido 2007; Pulido and Berthold 2010; Newton 2011; Schmaljohann 2019). About 1800 species, i.e., ~20%, of bird species are long-distant migrants (Voelkl and Fritz 2017). Among them are those that make direct flights, especially when they have to fly across vast water masses or over inimical land masses such as deserts (Hedenström 2010; Alves et al. 2016). Migration over land is generally marked by stopovers when food and water are replenished (McWilliams et al. 2004; Gill et al. 2009a, b; Loonstra et al. 2019; Zinßmeister et al. 2022). Accessing better foraging conditions, escaping adverse environmental conditions and avoiding seasonal predators, pressures from seasonal upsurges of numbers of parasites and pathogens, and acquiring better reproductive opportunities in new more hospitable ecological locations are the foremost factors that drive or compel the long and perilous migratory bird flights (Alerstam et al. 2003; McKinnon et al. 2010; Gilg and Yoccoz 2010; Hedenström 2010; Battley et al. 2012; Bauer and Klaassen 2013; Dingle 2014; Somveille et al. 2015; Alves et al. 2016). Traveling at an average speed of 146 km.h^{-1}, the Bar-Tailed Godwit (*Limosa lapponica*) that weighs ~285 g (Gill et al. 2005), flies nonstop over a distance of 28,000 km from Asia (Siberia) across North America and South America to Australia in only 8 days: flying at an average speed of 3500 km/day, it covers the distance in only 192 h (Gill et al. 2009a). The American Golden Plovers (*Pluvialis dominica*) covers a distance of 4000 km between Nova Scotia and South America (Stoddard et al. 1983). The Ruddy Turnstones (*Arenaria interpres*) fly over a distance of 4000 km between the Pribilof Islands and Hawaii (Thompson 1974). Every autumn, the Amur Falcon (*Falco amurensis*), that breeds in Siberia, Mongolia, and Northeast China, embarks on a 11,000 km journey to South Africa, crossing the Indian Ocean (Ferguson-Lees and Christie 2001). After making two non-stop sea crossings and flying at speeds of up to 24 m s^{-1} (84 km h^{-1}), the fastest speed recorded for shorebirds flying across oceans, the Icelandic Whimbrels (*Numenius phaeopus*) complete a round-trip of 11,000 km (Alves et al. 2016). The Great Snipe

(*Gallinago media*) maintains flight speeds of 80 km h^{-1} for several days (Klaassen et al. 2011) and the Red Knots (*Calidris canutus*) cover a distance of 4800 km between the Wadden Sea and their breeding area on Taymyr (Piersma and Jukema 1990). Annually, travelling between its Arctic breeding grounds and Antarctica, covering a distance of 80,000 km, the Arctic Tern (*Sterna paradisea*) (Plate 1.3a, ii), a bird that weighs less than 125 g, covers the longest distance currently well-documented for a species of bird (Egevang et al. 2010).

The cues and mechanisms that migratory passerine birds in particular apply to steer to their breeding and wintering grounds continue to be fervently argued (Alerstam 2006; Thorup and Holland 2009). It has been shown that birds react to variations in the magnetic field (Wiltschko and Wiltschko 1995; Phillips et al. 2006) and probably have two sensory pathways by which they sense the magnetic field (Holland and Helm 2013). One hypothesis that has advanced is that during migration, birds appear to read the Earth's magnetic field lines using transient molecular fragments (radical pairs) in their eyes to form a "compass" that they follow (Schulten and Windemuth 1986; Ritz et al. 2000; Liedvogel et al. 2007; Hore and Mouritsen 2022). The process by which the "radical pairs" are photochemically formed (in the eyes) generates the information used to steer the course (Schulten et al. 1978; Timmel et al. 1998; Ritz et al. 2000, 2004; Rogers and Hore 2009; Hore and Mouritsen 2016, 2022). The transient radical pairs are generated by photo-induced electron transfer reactions that occur in cryptochrome proteins, with their spin dynamics being influenced by prevailing geomagnetic field that causes changes in the quantum yield of the signaling state of the protein (Maeda et al. 2012; Dodson et al. 2013; Kattnig et al. 2016). Among others, another hypothesis that has been proposed is that of the mechanism of magnetoreception that is based on a receptor that uses ferrimagnetic particles (Kirschvink and Gould 1981). While in some studies it has been reported that birds exposed to strong magnetic pulse, designed to change the magnetic sense, travelled in a direction that varied significantly from that of controls when investigated in natural conditions (Holland 2010; Holland and Helm 2013), more recently, for the migratory European Robins (*Erithacus rubecula*), magnetic pulsing was reported to have no effect on possibility of departure, nocturnal departure timing, departure direction, or consistency of flight direction (Karwinkel et al. 2022): they deduced that either no usage of the geomagnetic map by their birds, or that magnetic pulses did not affect the sensory system driving geomagnetic map detection. By studying experimental and control groups of the Great Tits (*Parus major*), birds that were relocated to a new place within a familiar environment, Mahr et al. (2022) observed that olfactory (i.e., scent) signals provide information that helps during navigation in the home direction, even in surroundings where familiar visual landmarks such as hills and mountains exist. While the return rates between olfaction-deprived (impaired) and control birds did not vary, the experimental birds took longer to arrive home. The observation supported the earlier finding made by Fiaschi et al. (1974), Wallraff et al. (1995) and Holland et al. (2009) that homing capacities in Starlings (*Sturnus vulgaris*) and Swifts (*Apus apus*) were considerably affected by transection of the olfactory nerve. Certain species of the Family Paridae, e.g., Blue Tits (*Cyanistes caeruleus*) and Great Tits (*Parus major*), appear to utilize olfaction to find food in their surroundings (Amo et al. 2011, 2013; Mäntylä et al. 2017; Hiltpold and Shriver 2018).

During hovering flight (Plate 1.4b, c), hummingbirds (Order Apodiformes; Family Trochilidae) exhibit mass-specific metabolic rates that are close to the theoretical limits of vertebrates (~40 $mlO_2\ g^{-1}\ h^{-1}$) (Bartholomew and Lighton 1986). Adaptively, hummingbirds possess exceptionally large hearts, large and highly metabolically efficient flight muscles, and extreme pulmonary morphological specializations (Dubach 1981; Suarez et al. 1991; Suarez 1992). In hummingbirds, during migration, fuel is metabolized at a rate of 2% of body mass per hour (Hedenström and Alerstam 1997; Hedenström 2010). Because of the high caloric value of fat, compared to carbohydrates or proteins, long distance flights of birds are largely powered by metabolism of fatty acids (Klaassen and Biebach 1994; Jenni and Jenni-Eiermann 2000; McWilliams et al. 2004; Pierce et al. 2005; Guglielmo 2010, 2018; Dick and Guglielmo 2019; Jensen et al. 2020). For the duration of direct migration flight, birds depend on premigratory body fuel reserves in the form of fats (Klaassen 1996; Weidensaul 1999; McWilliams et al. 2004; Guglielmo 2010; Araújo et al. 2019). Among migratory species of birds with different fuelling strategies, Lee et al. (2022a, b) reported evidence of selective utilization of fatty acid of different make-ups: early fuellers exploited higher n-3 polyunsaturated fatty acid while gradual fuellers operated on higher mono-unsaturated fatty acid. Among the avian species, particularly with regard to wing morphology, long distance migration appears to have compelled evolutionary convergence, with long distal and short proximal primaries existing and the wing tips lying closer to the leading edge of the wing (Mönkkönen 1995; Marchetti et al. 1995; Peiró 2003; Arizaga et al. 2006).

1.10 Impacts of Anthropogenic Activities on Birdlife

The human impacts on the biosphere, that are termed anthropogenic activities, denote a range of adverse changes that human beings have caused directly or indirectly and intentionally or unintentionally to environments, ecosystems, biodiversity, and natural resources (Leakey and Lewin 1996; Sahney et al. 2010; Pimm et al. 2014; Ceballos et al. 2015a, b, 2017; Cook 2016; Wuebbles et al. 2017; Trenberth 2018; Syvitski et al. 2020; Elhacham et al. 2020). Environmental pollution is one of the undesirable effects of the so-called "human development" (WHO—https://www.who.int/health-topics/air-pollution#tab=tab_1—accessed 16-04-2022; Manisalidis et al. 2020): it leads to destruction of habitats, increase in chemical contamination and electromagnetic discharges and light and noise emissions (Ferguson and Williams, 1993; Hölker et al. 2010; Murphy and King 2014; Robert et al. 2015; Khan et al. 2018).

Integral part of the global biological continuum, birds have been and continue to be adversely affected by the unregulated and uncontrolled anthropogenic activities (Whelan et al. 2008; Tabur and Ayvaz 2010; Shanahan et al. 2014; Green and Elmberg 2014; Garfinkel et al. 2020). Species diversity and numerical abundance of birds in a particular geographical area is an outcome of a delicate balance between speciation, extinction, and colonization (Rundle and Nosil 2005; Bellemain and Ricklefs 2008; Schweizer and Liu 2018; Price 2018). Globally, species density and diversity of birds are greatest in the tropical ecosystems of the southern continents (Barnagaud et al. 2012; Schweizer and Liu 2018). It has been conjectured that the Earth's changing climate, i.e., well-underway, heralds a very hot future that may greatly affect the geographic distribution of the range-bound groups of birds (Saupe et al. 2019). Climate change and habitat loss have been listed as the two greatest threats to biodiversity (Sala et al. 2000; Jetz et al. 2007). For the Harpy Eagle (*Harpia harpyja*), that subsists in the Southern Amazonia, an area that is undergoing dramatic decrease in range, combination of climate change and ensuing loss of habitat may bring about substantial losses in the dispersal of the bird (Sutton et al. 2022). Regarding aspects such as speeds they travel at, the migratory distances they cover and the routes they follow, changes in temperatures (from global warming) between winter and summer months is measurably affecting the distribution and the migratory activities and patterns of birds (Fiedler 2003; Lemoine and Böhning-Gaese 2003; Pulido and Widmer 2005; Marra et al. 2005; Jonzén et al. 2006; Both and Marvelde 2007; Schaefer et al. 2008; Visser et al. 2009; Pulido and Berthold 2010; Clairbaux et al. 2019; Horton et al. 2020; Covino et al. 2020). Higher temperatures are compelling birds to expend more energy on thermoregulation, making their bodies smaller while their wingspans are increasing (Weeks et al. 2019). Global warming, i.e., high environmental temperatures from anthropogenic activities, is driving massive and extensive reorganization of biotic assemblages across the Planet (Mota et al. 2022; du Toit et al. 2022). For organisms/animals, elevated external temperature beyond the physiological range constraints adaptive capability and may indirectly threaten survival from unavailability of life-supporting factors such as food and water and increase risks from predators and pathogens (Pattinson et al. 2022; du Toit et al. 2022). For human beings, Rahman et al. (2022) observed that brief exposure to extreme heat and air pollution correlated with increased risk of mortality but exposure to both properties had greater impact that surpassed the sum of their separate effects. Regarding birds, global warming has become an important concern that has adversely affected poultry production in tropical and subtropical parts as well as temperate regions of the world (Renaudeau et al. 2012; Brugaletta et al. 2022). From their distinctive physiology, compared with other terrestrial animal taxa, from features such as lack of sweat glands and presence of feather covering, poultry is more sensitive to high ambient temperature that severely affects their health, physiology, and productivity (Emami et al. 2020). Elevated temperature is a very important stressor of poultry birds: it causes endocrine imbalances while provoking electrolyte disruptions that negatively upset production performance and lower antioxidant and anti-inflammatory competency (Thaxton and Siegel 1970; Yang et al. 2010; Slawinska et al. 2016; Akbarian et al. 2016; Del Vesco et al. 2017; Du et al. 2022); these properties cause weight loss, lower egg production and result in deaths. The impact of global warming is already having severe impact on poultry husbandry and production and is causing huge financial losses (Emami et al. 2020). The sudden death of more than 700,000 birds that was reported in California by Nienaber and Hahn (2007) was attributed to heat stress. In the United States of America, St-Pierre et al. (2003) showed that for the poultry industry, the financial encumbrance caused by heat stress is 128–165$ million per year. Birds lack sweat glands (King and McLelland 1984) and therefore to dissipate heat, they mainly rely on evaporative cooling from the respiratory system. Recently, Baldwin et al. (2022) observed that species of birds with larger brains and hence greater cognition ability may respond better to climate change, alleviating the adverse environmental effects caused by global temperatures. The fertility (i.e., laying and successful incubation and hatching of eggs) of Ostriches (*S. camelus*) is adversely affected by extreme environmental temperatures in their habitats (Schou et al. 2021).

Birds are highly sensitive to environmental perturbations (Temple and Wiens 1989; Feduccia 2003; Morelli et al. 2014; Alexandrino et al. 2016): this is particularly because of their capacity of flight that permits them to physically vacate polluted habitats and relocate to more hospitable ones. Birds are therefore a reliable bioindicator or sentinel animals of environmental health and level of contamination (Dyke 2009; Roché et al. 2010; Amat and Green 2010; Steyn and Maina 2015; Mekonen 2017; Egwumah et al. 2017; Dutta 2017; Hartfield et al. 2018; Tietze 2018). Specific diversity and numerical abundance of birds in a habitat is a good indicator of environmental quality (O'Connell et al. 2000; Stolen et al. 2005; Butler et al. 2012a; Wade et al. 2014). Showing the degree of pollution of planet Earth, albeit their capacity of easily and quickly reacting to adverse environmental conditions, as many as 1313 (~13%) of the extant species of birds are listed on the IUCN Red List of Threatened Species (IUCN 2020). Human activities have directly and indirectly negatively affected many species of animals, including birds (Kerr and Currie 2002; Steidl and Powell 2006; DU 2006; Otto 2018; Cooke et al. 2019; Çağan et al. 2019; Einfeldt et al. 2020). In particular, domestic cats (*Felis catus*) have caused extinction of many vertebrate animals, including species of birds (Loss and Marra 2017; Loss et al. 2013). Migratory birds that have to make stop-overs (to rest, feed, and drink water) during their strenuous journeys have been greatly affected by habitat loss from anthropogenic activities (Culp et al. 2017; Cohen et al. 2017b; Ward et al. 2018). Their journeys are also affected by construction of physical structures such as lofty buildings (skyscrapers), wind turbines and high-tension electric cables along their migratory routes (Johnston and Haines 1957; Osborn et al. 2000; Manville 2009; Machtans et al. 2013; Loss et al. 2014).

During the last ~500 years, anthropogenic activities have resulted in habitat degradation and industrial discharge of toxic chemicals and global warming gases, with the main ones being water vapor, CO_2, methane (CH_4), nitrous oxide (N_2O), fluorinated gases such as hydrofluorocarbons (HFCs), perfluorocarbons (PFCs), and sulfur hexafluoride (SF_6). Overexploitation of resources and increase in the coverage of invasive species have considerably changed the ecological balance and diversity of animal life (Hoover and Schelsky 2020). According to authorities such as Leakey and Lewin (1996), Wake and Vredenburg (2008), Barnosky et al. (2011) and Ceballos et al. (2015a, b, 2017), a "sixth mass extinction" is well-underway. Decline of species diversity is escalating at a rate more than 100 times that occurred during all the past extinction events (Ceballos et al. 2017), namely, the Cambrian Period's Ordovician-Silurian (444 mya), the Late Devonian (383–359 mya), the Permian-Triassic (252 mya), the Triassic-Jurassic (201 mya), and the most recent Cretaceous-Paleogene (66 mya) extinctions (Barnosky et al. 2011; Dishon et al. 2020; Randall et al. 2014; Dirzo et al. 2014; Ceballos et al. 2015a, b, 2017). According to Dirzo et al. (2014), among the terrestrial vertebrates, 322 species have become extinct during the last 1500 years, and the populations of the rest of the species have declined by 25%. Showing the tenacity/resiliency of life, by a process termed "filtering," animal and plant species have invaded, exploited and thrived in erstwhile inhospitable environments (Bonier et al. 2007; Lowry et al. 2013; Wingfield et al. 2015; Gallien et al. 2015; Cooper et al. 2020).

1.11 Evolution of Birds

The origin and evolution of birds is one of the longest studied and vehemently debated area in paleobiology (Darwin 1859; Chiappe 2004, 2009; Feduccia et al. 2005; Chiappe and Dyke 2006; Xu et al. 2011; Godefroit et al. 2014; Donoghue and Keating 2014; Brusatte et al. 2014, 2015; Wang and Zhou 2017; Foth and Rauhut 2020; Rauhut and Foth 2020). Initially, until the discovery of *Archeopteryx lithographica*, elucidation of the process was curtailed by lack of transitional fossil forms of the highly derived avian morphology (Owen 1863; Xu et al. 2011; Godefroit et al. 2013). It is now generally accepted that birds evolved from a group of theropod dinosaurs during the Jurassic Period, ~150–165 mya (Sereno 2004; O'Connor and Claessens 2005, Kurochkin 2006; Dyke and Kaiser 2011; Makovicky and Zanno 2011; Xu et al. 2014; Benson et al. 2014; Brusatte et al. 2015; Feduccia 2016; Rauhut et al. 2019). It is not explicitly known whether birds originated directly from highly derived theropods or from a common progenitor that lacked novel anatomical traits (Xu et al. 2014). Since the Late Mesozoic, from early stages, among the air-breathing vertebrates, birds progressed to dominate the aerial ecotope (Brocklehurst et al. 2012; Brusatte et al. 2015). Notwithstanding the fact that the Cretaceous-Paleogene (K-Pg) mass extinction event (Feduccia 2003; Longrich et al. 2011; Jetz et al. 2012; Jarvis et al. 2014; Schweizer and Liu 2018) annihilated ~80% of the extant species of animals, subsequently, within a short period of evolutionary time, birds went through striking accelerated morphological transformations that resulted in exceptional speciation and colossal ecological dispersal (Newton 2003; Jarvis et al. 2014; Brusatte et al. 2015; Ksepka et al. 2017; Schweizer and Liu 2018). The skeletal morphology of *Archeopteryx*, a species estimated to have been the size of the Common Raven (*Circus corax*) (Yalden 2008; Erickson et al. 2009), strongly suggests that it was possibly volant and certainly an arboreal bird (Mayr et al. 2007; Voeten et al. 2018). From an assortment of dinosaurian-avian traits that include feathers, a long tail, and a toothed skull, the iconic *Archeopteryx* is believed to occupy a pivotal transitional position in the evolution of birds (Ostrom 1974, 1976; Martin 1983; Cracraft 1986; Feduccia 2001; Zhou 2004; Witmer 2009;

Chiappe 2009; Xu et al. 2014). Some investigators, however, still consider *Archeopteryx* to have been, rather that occupying the branching point, more dinosaurian than avian (Zhou 2004; Erickson et al. 2009; Lee and Worthy 2011; Foth et al. 2014; Wang and Zhou 2017).

The evolution of birds has been studied directly from fossil remains and indirectly from extrapolation of morphological, functional, developmental, and genetic/molecular data acquired from related fossilized remains as well as closely related extant animal taxa (Holland and Chen 2001; Hall 2002; Livezey and Zusi 2007; Holmes et al. 2016). While the former approach produces what are termed "primary data" and the later one yields "secondary data" (Parry et al. 2018), integration of facts acquired by the two approaches provides robust results (Hall 2002; Hutchinson and Allen 2009). Recently, insights into the evolutionary origin of birds have progressed at an unprecedented rate, especially with discoveries of well-preserved and highly instructive fossils remains from critical evolutionary time-points (O'Connor and Zhou 2015; Bailleul and Horner 2016; Zheng et al. 2017; Bailleul et al. 2020; O'Connor et al. 2021). Application of molecular phylogenetic (Sanderson and Shaffer 2002; Cassini et al. 2020) and quantitative macroevolutionary analyses (Horiike 2016; Jablonski 2017a, b; Felice et al. 2018; Sarwar et al. 2019) of materials acquired from protobirds and from non-avian dinosaurs have greatly contributed toward understanding of the origin and evolution of birds (Zhou and Zhang 2006; Xu et al. 2014; Jarvis et al. 2014). The most instructive specimens that have contributed to understanding the dinosaur-bird transition have been "discovered" from the Cretaceous terrestrial sediments of North-Eastern China (Zhou et al. 2013; Mayr 2016; Yang et al. 2020). The fossil record of birds and their dinosaurian progenitors is, however, still too fragmentary to completely determine the dinosaurian-avian evolutionary conundrum (Fountaine et al. 2005; Dalsätt 2012; Brocklehurst et al. 2012). This is largely because birds and possibly their antecedents lacked especially hard structures such as the enamel of the teeth and most of their long bones were largely hollow and the cortices of the shafts very thin for good preservation (fossilization) (Gill 2007; Dumont 2010). The process of fossilization occurs by complex biomineralization of biological structures/tissues under special geological settings: it is a rare occurrence, especially for soft tissues (Iniesto et al. 2016; Muscente et al. 2017; Parry et al. 2018). The body parts of organisms/animals that are most resistant to degradation such as shells, bones, and teeth stand a better chance of at least some of their parts being preserved (Martín-González et al. 2009; Iniesto et al. 2016; Wiemann et al. 2018; Parry et al. 2018). Seldom, soft structures, i.e., the non-biomineralized parts of the animal's body, are preserved and may fossilize together with the hard skeleton (Tang et al. 2019). During the last three decades, spectacularly well-fossilized remains of birds with well-fossilized soft tissues/structures such as feathers (Zhang et al. 2006; Heers et al. 2014; Navalón et al. 2015; O'Connor and Zhou 2015; Xu et al. 2015) and reproductive (Zheng et al. 2013; Bailleul et al. 2020), digestive (Zheng et al. 2011, 2014; O'Connor et al. 2018), musculoskeletal (Jiang et al. 2017), and respiratory systems (Wang et al. 2018) (Plate 1.5) have been discovered. Although uncommon, fossil lungs have been reported in coelacanths (Brito et al. 2010; Cupello et al. 2017, 2019), a salamander

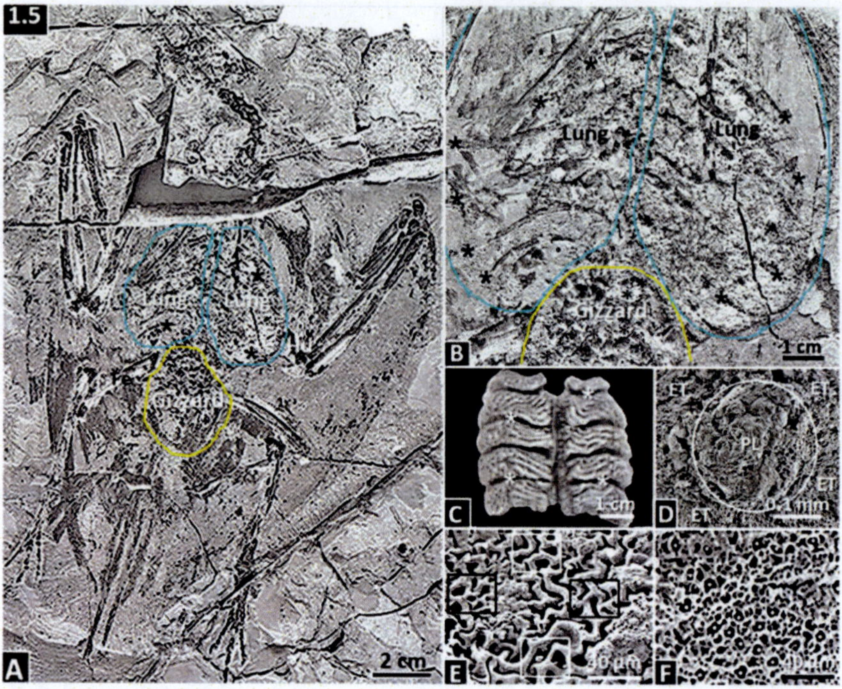

Plate 1.5 (**a, b**) A fossilized specimen of an extinct bird, *Archaeorhynchus spathula*, showing putative preserved lungs (enclosed by cyan lines) and the gizzard (enclosed by a yellow line). Stars (⋆), costovertebral impressions on the lung; asterisks (*) (**b**), gastroliths in the gizzard and Hu, humerus and Fe, femur (**a**). (**c**) Dorsal view of the lung of a modern bird, the Domestic Fowl (*Gallus gallus* variant *domesticus*), showing the costovertebral impressions (stars, ⋆). (**d**) Parabronchus of the fossilized lung showing a parabronchial lumen (PL) surrounded by gas exchange tissue (ET). (**e, f**) Gas exchange tissue of the lung that comprises blood capillaries (enclosed areas, □) (**e**) and air capillaries (white dots, •) (**f**)

(Tissier et al. 2017), and possibly a tadpole (Rossi et al. 2019). The interpretation of thoracic paired masses in the antiarch placoderm, *Bothriolepis canadensis*, as a lung by investigators such as Denison (1941) and Arsenault et al. (2004), an observation which if correct would suggest that lungs originated from among early jawed vertebrates dating back to at least the Devonian Period, is problematic and controversial (Goujet 2011; Janvier et al. 2007; Béchard et al. 2014; Cupello et al. 2022).

During the early stages of the evolution of birds, the Early Cretaceous was a critical time when most of the major clades arose (Tuinen and Hedges 2001; Brusatte et al. 2015; Mayr 2016; Field et al. 2018). By calibrating the avian molecular clock, Tuinen and Hedges (2001) and Weir and Schluter (2008) observed that modern birds may have coexisted with other avian lineages for a long period of time during the Cretaceous. While *Archeopteryx* is highly instructive in understanding the evolution of birds, questions still remain on how and for how long the dinosaurian-avian evolutionary transformation lasted (Chiappe 2009; Xu et al. 2011; Brusatte et al.

2015). Fossil remains of bird-like theropod dinosaurs and basal avialans (Xu 2006; Foth et al. 2014) have, however, allowed accurate positioning of *Archaeopteryx* (the "Urvogel" = the first bird) into its proper evolutionary status. Modern birds (Neornithes) are the extant direct descendants of dinosaurs (Chiappe 2004; Vargas and Fallon 2005; Brusatte et al. 2015). Therefore dubbed "living dinosaurs" (Dyke and Kaiser 2011), they arose from small maniraptoran theropods that appeared during the Mezozoic Era (Hou et al. 1996; Gauthier and Gall 2001; Turner et al. 2012; Kawakami and Eda 2018). Belonging to 40 orders and more than 140 families, birds are the most speciose extant tetrapods (Monroe and Sibley, 1993; Cooney et al. 2017; del Hoyo et al. 2020). Although once considered a "diagnostic" avian trait (Padian 1998; Clarke and Middleton 2006), feathers are now known to have existed in dinosaurs well-before birds appeared (Chen et al. 1998; Kundrát 2004; Hanson 2011; Clarke 2013; Godefroit et al. 2014; Prado et al. 2016; Yang et al. 2019; Benton et al. 2019; Unwin and Martill 2020; Foth and Rauhut 2020). Feathered and protofeathered non-avian theropods provide further evidence of the origin of birds from a dinosaurian stock (Feduccia et al. 2005; Brusatte et al. 2015). Showing that feathers and scales once coexisted during the evolutionary history of birds, Godefroit et al. (2014) described a neornithischian dinosaur, *Kulindadromeus zabaikalicus*, from the Jurassic Period of Siberia that had both structures (feathers and scales) around the humeral region. Feathers are envisaged to have originated ~50 to 100 million years or more before *Archaeopteryx* appeared (Godefroit et al. 2014; Benton et al. 2019). While they may have initially evolved for insulation and/or sexual display (Dimond et al. 2011; Zhang et al. 2010), later, they provided large surface areas of wings and streamlined the body, reducing the energy required for flight (Clarke 2013; Lee et al. 2014). While lack of teeth (edentulism) is a rare feature in vertebrates (Berkovitz and Shellis 2018), for birds, toothlessness is a distinctive morphological trait that occurred very early in the evolution of birds (Hou et al. 1995, 1996; Zimmer 1996; Sanz et al. 1997). Meredith et al. (2014) determined that the common progenitor of modern birds lacked mineralized teeth and that loss of teeth or at least that of enamel caps occurred ~116 million years ago. Loss of teeth in birds is supposed to have originated from selective pressures for optimization of the body form in order to promote activities such as powered flight (Louchart and Viriot 2011): it was purported that in birds, edentulism occurred after the development of a prominent horny beak and a muscular gizzard, features that traded off necessity of dentition for food acquisition and processing. According to Brooke and Birkhead (1991), Bühler (1981), Feduccia (1996), Zimmer (1996) and Ackermann (1998), by reducing the weight of the avian head (with loss of teeth), existence of a top-heavy body form was obviated. Comparative studies do not, however, show compelling correlation between bipedalism and volancy with toothlessness (Homberger 1999). In birds, toothlessness may simply have evolved from the fact that in contrast to mammals, birds do not need to break up their food before swallowing even when they feed on large food/prey objects: food is broken up in the muscular gizzard. Furthermore, the complex functional morphology of the avian jaw apparatus, that comprises among other structures the jugal arch, the quadrate, the pterygoid, and the palatine bones as well as a craniofacial hinge joint (in most birds) generate cranial kinetic and prokinetic movements that create large gape that permits large pieces of food to be swallowed intact (King and McLelland 1984).

As environmental changes occurred and morphological innovations developed, compared with the other groups of dinosaurs, the theropod lineage underwent a gradual decrease in body size over a period of ~50 million years, a time when evolutionary pace increased four times, compared with that during the earlier time (Brusatte et al. 2014). Showing notable evolutionary stasis at the level of nucleotide sequence and displaying gene loss and chromosomal structural change, while birds are the most speciose group of tetrapod vertebrates, the mean genome size, and the variance in genome sizes between species are generally relatively smaller in birds (Class: Aves) (Griffin et al. 2007; Hughes and Friedman 2008; Gregory 2009; Nam et al. 2010; Zhang et al. 2014; Lovell et al. 2014). Ranging from 0.9 Gbp (Giga base pairs) in the Black-Chinned Hummingbird (*Archilochus alexandri*) to 2.2 Gbp in the Ostrich (*Struthio camelus*) (Scanes 2015; Kapusta et al. 2017; Trense and Tietze 2018), among the amniotes, birds, especially hummingbirds (Family: Trochilidae), have the smallest genomes (Gregory et al. 2009): bony fishes (Osteichthyes), amphibian, reptilian, and mammalian genomes, respectively, range from 0.35 to 133, 0.95 to 121, 1.1 to 5.44, and 1.6 to 6.3 Gbps (Kapusta et al. 2017). From their extremely small genomic size, Lovell et al. (2014) birds have been termed "natural knockouts." Gregory (2002) and Hughes and Friedman 2008) considered the genomic size reduction in birds and bats to have been an adaptation for their high metabolic capacities under powered flight (Gregory 2005; Kapusta et al. 2017; Cao and Jin 2020): compared with their nonvolant relatives, massive DNA loss and not paucity of transposable element (TE), accumulation has been the main driver for the greater genomic density in the volant birds and bats. Many non-neutral evolutionary changes in protein-coding genes and non-coding regions explain how genomic diversity in birds changed with adaptations to different lifestyles in various ecological settings (Ellegren 2013; Zhang et al. 2014).

1.12 Evolution of the Avian Respiratory System

Pagano and Márquez (2022) stated that "*breathing is among the most important physiological functions*." For terrestrial air-breathing vertebrates, evolution of the lung was a seminal event (Jürgens and Gros 2002; Arratia et al. 2004; Willmer et al. 2005; Ashley-Ross et al. 2013; Lambertz 2017). Acquisition of O_2 from air opened new ecological opportunities that accelerated the adaptive radiation of animal taxa into different habitats and permitted accomplishment of superior activities and lifestyles such as flight (Abe 1991; Graham 1997; Maina 2002a, 2011; Graham and Lee 2004; Belão et al. 2015; Moreira et al. 2021; Milsom et al. 2022). Air-breathing allowed transition of animal life from water onto land, a process that mostly happened during the Late Devonian (Randall et al. 1981; Dejours 1981; Little 1990; Maina 1998; Long and Gordon 2004; Lee 2011; Ashley-Ross et al. 2013). Change from breathing water to air, fluid media that are physiochemically very different, obligated profound changes in respiratory morphology, physiology, and blood chemistry (Rahn 1966; Dejours 1981, 1989; Maina 1998). Questions on how, why and when lungs evolved have been asked, investigated, and intensely debated (Maina 1998, 2002b; Jürgens and Gros 2002; Farmer 2010, 2017; Ashley-Ross et al. 2013; Bayley et al. 2019; Perry et al. 2019). Lungs are envisaged to have evolved during the Devonian Period, i.e., ~385 to 360 mya, in fishes ancestral to ray-finned (Actinopterygii) and lobe-finned (Sarcopterygii) fishes (Todd 1980; Graham 1997; Hedrick and Katz 2016; Lambertz 2017; Zaccone et al. 2018) that occupy the basal most status in the vertebrate evolutionary continuum (Graham and Lee 2004; Johanson et al. 2006; Cavin and Guinot 2014; Lambertz 2017; Clement and Long 2010). Originally, the lungs were simple structures that were lined by a single low parenchymal division (Perry and Sander 2004; Graham and Lee 2004). Showing that pulmonary morphogenetic organization was well-underway in the common ancestors of the actinopterygian and the sarcopterygian fishes, Tatsumi et al. (2016) observed that the development of the lung in the Senegal Bichir (*Polypterus senegalus*) was similar to that of modern tetrapods. After transcriptomic analyses of the tissues that comprise the swim-bladder of the Zebrafish (*Danio rerio*) and those of the lung, Zheng et al. (2011) showed structural similarity between the two organs. Taking into account that cartilaginous fish (e.g., sharks and rays), that diverged from bone fishes ~420 million years ago (mya) lack lungs and swim-bladders, it seems like lungs evolved after the two fish taxa separated (Farmer 1997): lungs evolved as preadaptations for air-breathing in hypoxic waters (Thompson 1969; Packard 1974; Graham et al. 1978a, b). The development and the structure of the lungs of the extant lobe-finned fish, is highly instructive in cognizing the evolution of the vertebrate lung (Maina and Maloiy 1985; Maina 1987; Farmer 1999, 2017; Lambertz 2017). For the coelacanths, fish that for a long time were presumed extinct before the first discovery of a live specimen, *Latimeria chalumnae* (Smith 1939), lungs have been reported in an embryonic specimen while in an adult one they were vestigial and surrounded by hard, flexible plates (Brito et al. 2010; Cupello et al. 2015; Meunier et al. 2019).

The lungfishes (Sarcopterygii: Dipnoi) are an ancient osteichthyes fish lineage that first appeared in the Early Devonian (Chang and Yu 1984; Joss 2006): phylogenetically, lungfishes are close relatives of tetrapods (Takezaki et al. 2004; Takezaki and Nishihara 2016, 2017). The vertebrate lungs are considered to be the particular anatomical structures that largely allowed the tetrapods to occupy and thrive on the terra firma (Roux 2002; Perry and Sander 2004; Graham and Lee 2004; Duncker 2004; Hoffman et al. 2016; Brocklehurst et al. 2020). Interestingly, two-thirds of the ~700 species of extant salamanders (Family: Salamandridae) are lungless (Lewis and Hanken 2017). Given that their amphibian progenitors possessed lungs (Cabej 2011), lack of lungs in the plethodontid salamanders is an example of extirpation, i.e., loss of an organ when it becomes functionally redundant. Amphibian species that live in cold, well-oxygenated water use the skin, that is well-vascularized and perfused with blood, to meet O_2 requirements (de Saint-Aubain 1982; Czopek 1983; Burggren and Moalli 1984; Feder and Burggren 1985a, b; Tattersall 2010; Kardong 2014). Overexpression of the transforming growth factor beta (TGF-β), a multifunctional cytokine belonging to the TGF superfamily that drives cell differentiation, proliferation, apoptosis, and cell matrix production in the region of the primitive pharynx (Rohde and Friedland 2022; Pagano and Márquez 2022; Jankowski 2022) showed that plethodontidae salamanders have transitional lungs that recede very early during embryonic development (Lewis et al. 2011; Lewis 2016). It is envisaged that the modern structurally complex vertebrate lungs appear to have originated from an ancestral simple, saccular, gas-filled structure (an air-bladder) that might have served roles such as gas exchange (respiration), hydrostatic regulation, hearing, and sound production (Perry 1989a, b; Schulz-Mirbach et al. 2012; Ganias et al. 2015; Perry et al. 2019; Vetter and Sisneros 2020; de Souza et al. 2021).

While evolution of swim-bladders in fish inhabiting hypoxic waters points to evolution of respiratory role (Liem 1991), for those dwelling in the well-oxygenated waters of the open sea, it suggests a primary hydrostatic function (Liem 1987a, b). Questions on which structure, i.e., the "swim-bladder" or the "lung," derived first, which gave rise to which, whether the organs evolved independently or whether they are homologous poses a cause-and-effect conundrum of the so-called "chicken and egg developmental paradox" (Longo et al. 2013; Icardo 2018; Funk et al. 2020, 2021). While some investigators consider the lung to be ancestral to the swim-bladder (Perry 2001; Torday and Rehan 2012; Torday 2013; Icardo 2018), other investigators consider the swim-bladder to have derived from the lung (Liem 1988; Hoffman et al. 2016). The general morphological similarities between the swim-bladders and the lungs are that they are both (a) innervated by the vagus and the sympathetic nerves; (b) they originate from the primitive pharynx region of the foregut (Girosi et al. 2006; Rohde and Friedland 2022; Pagano and Márquez 2022; Jankowski 2022); and c) the epithelial surfaces are covered by a surface active agent (surfactant) (Neumayer 1930; Wassnetzov 1932; Robertson et al. 2014; Chen et al. 2018; Kunchala et al. 2022). The differences are (a) generally, the swim-bladder is a single structure while lungs are ordinarily paired; (b) developmentally, the swim-bladders originate from the distal, dorsal, or lateral aspects of the walls of the pharyngeal part of the foregut while the lungs originate from the distal-ventral aspect (Packard 1974; Perry et al. 2001; Perry 2001; Graham 2001; Torday et al. 2007; Hsia et al. 2013; Frisdal and Trainor 2014; Hoffman et al. 2016; Perry et al. 2019); (c) the lungs connect to the left and right pulmonary arteries that originate from the last pair of epibranchial arteries while swim-bladders are supplied with blood from the aorta; and d) lungs remain connected to the pharynx for life whilst some of the swim-bladders are separated from it (the physoclistous ones) while the phytostromatic ones are permanently connected to the pharyngeal region (Goodrich 1930; Nelsen 1953; Nilsson 1983; Fänge 1983; Maina et al. 1996; Maina 2000b; Perry et al. 2001; Daniels et al. 2004).

The "oldest" vertebrate lungs are ostensibly those of the primitive Silurian to Late Devonian Period placoderm armored fish, *Bothriolepis* (Janvier et al. 2007; Young 2010; Goujet 2011; Johanson and Trinajstic 2014). The fossil specimen examined by Denison (1941) shows "lungs" that are paired sacs which connect to the oropharynx. The lungs of the bichirs (a family of freshwater ray-finned fishes) also originate from the ventral aspect of the primitive pharynx (Lauder and Liem 1983). Genetic and molecular instructions appear to have been hardwired into the developmental system of the lung from as early as the time of existence of the common ancestor of actinopterygian and sarcopterygian fishes (Tatsumi et al. 2016). In the lungfish, during the early stages of lung development, mesenchymal expression of growth (molecular) factors such as Fibroblast Growth Factor-10 (FGF-10), T-box Transcription Factor-4 (Tbx-4), and T-box Transcription Factor-5 (Tbx-5) match those expressed in lungs of modern tetrapods (Sekine et al. 1999; Sakiyama et al. 2003; Arora et al. 2012; Tatsumi et al. 2016). A Tbx-4 core lung mesenchyme-specific enhancer (C-LME) was localized in the genomes of the basal fish, the Bichir (*Polypterus senegalus*), and the Coelacanth (*Latimeria chalumnae*) (Tatsumi et al. 2016): the molecular factor (Tbx-4) also drives the development of lungs of the extant tetrapods.

Lungs mainly appear to have developed to supplement the amount of O_2 obtained through other pathways such as gills, skin, and buccal cavity (Liem 1988; Duncker 2004; O'Rourke and Jenkins 2008; Mylniczenko 2009; Taylor et al. 2010; Graham 2011; Vitt and Caldwell 2013). Morphologically, the vertebrate lungs vary greatly (Plate 1.6). Simple (unicameral) lungs exist in the amphibian, the lungfish (Dipnoan) and less-derived reptilian taxa (Perry 1983, 1989a, b, 1998; Maina and Maloiy 1988; Maina 1998, 2011; Duncker 2004) (Plate 1.6a–c): they are saccular structures that are scantily internally subdivided into a single height of large air cells (spaces) or foveolar (Plate 1.6a–d) while the multicamera lungs such as those of more derived reptiles, birds, and mammals (Perry 1988; Maina et al. 1989a, b; Schachner et al. 2022) are structurally more complex (Plate 1.6d–j). For the African Caecilian (*Boulengerula taitanus*), while the right lung is well-developed the left one is vestigial (Maina and Maloiy 1988) (Plate 1.6a—inset); the African Lungfish (*Protopterus aethiopicus*) has paired lungs of similar size (Maina and Maloiy 1985) (Plate 1.6d—inset) and reptilian (Maina et al. 1989a, b) (Plate 1.6e–h), avian (Maina 1989, 2005) (Plate 1.6i) and mammalian (Maina 1990) (Plate 1.6j) lungs are relatively structurally complex. Although deeply marked by costovertebral impressions (Plate 1.6i), the avian lungs are not lobulated like the mammalian ones (Plate 1.6j). The snake lungs are rather tubular in shape (Plate 1.6g), and the parenchyma of multicamera lungs is intensely internally subdivided (compartmentalized) into respiratory units (air cells = air spaces) (Duncker 1978a, b; Perry 1988, 1989a, b, 1998; Maina et al. 1989a, b; Powell and Hopkins 2004; Duncker 2004) (Plate 1.6e–h: Plate 1.7). The process of internal subdivision of the lung increases the strength and the respiratory surface of the organ. Subdivision of the parenchyma of the volume constant parabronchial lung of birds (Plate 1.6i; Plate 1.7a) (Duncker 1971, 2004; Scheid 1979; Fedde 1980; Scheid and Piiper 1989; Maina 2005) and the mammalian (bronchioalveolar) one (Mead 1961; Crosfill and Widdicombe 1961; Bennett and Tenney 1982; Leith 1983; Fahlman et al. 2017) (Plate 1.6j; Plate 1.7b) are envisaged to have evolved from the reptilian multicamera lung (Perry 1983, 1988, 1989a, b) (Plate 1.6e–h: Plate 1.7c–e). The lung of the lungfish (*Protopterus aethiopicus*) is as well internally subdivided (Plate 1.7f) as those of the Pancake Tortoise (*Malacochersus tornieri*) (Plate 1.6e; Plate 1.7c), the Black Mamba (*Dendroaspis polylepis*) (Plate 1.7e), and the Monitor Lizard (*Varanus

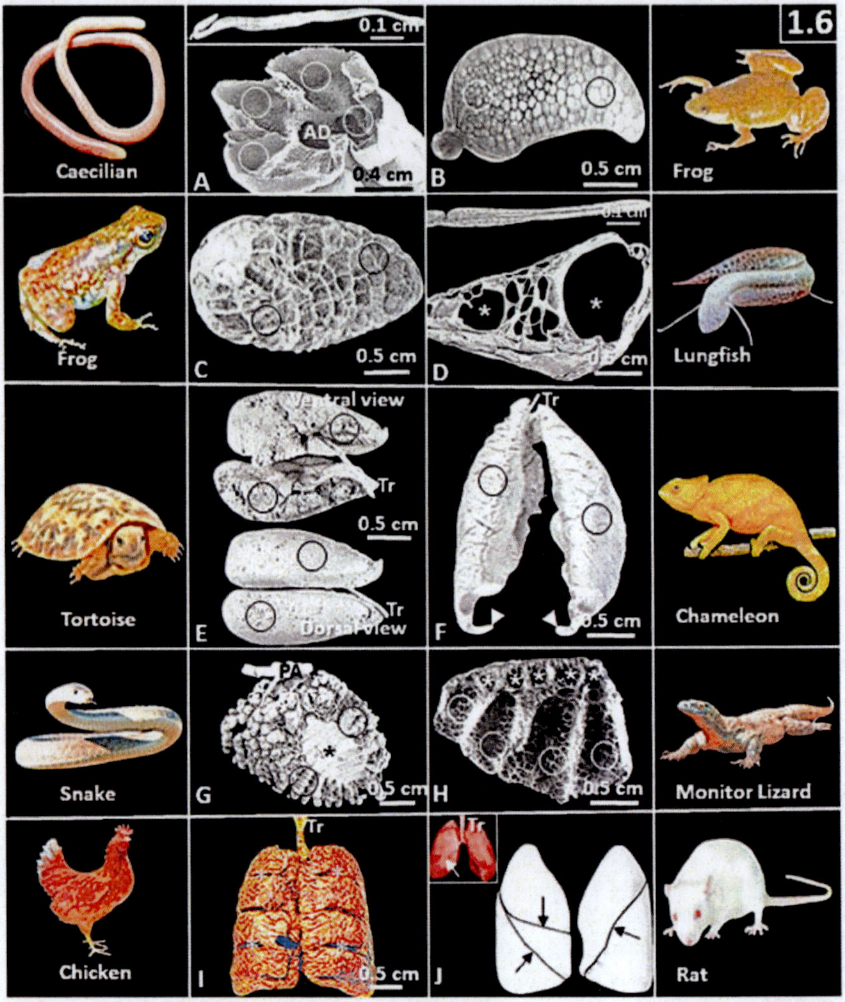

Plate 1.6 Various morphologies of lungs of different air-breathing vertebrates. Lungs of less-derived groups are structurally simpler and internally poorly subdivided with a single-level air cells (**a–d**) while those of more derived groups are relatively structurally more complicated (**e–h**). The lungs of serpentine animal forms such as caecilians display differences between sizes of left- and right lungs, with the left one generally being vestigial or lacking (**a—inset**) The lungs of lungfish (e.g., *Protoptrus aethiopicus*) are about equal in size (**d—inset**) Whilst the avian lung is marked by deep costovertebral impressions (asterisks, *) (**i**), the organ is not lobulated like the mammalian lungs (arrows, ↓) (**j**). Circles (O), air cells (**a–h**); stars (⋆), air ducts (**d, g, h**); PA, pulmonary artery (**g**); Tr, trachea (**e, f**); arrowheads, air sac-like extensions of a chameleon's lung (**f**)

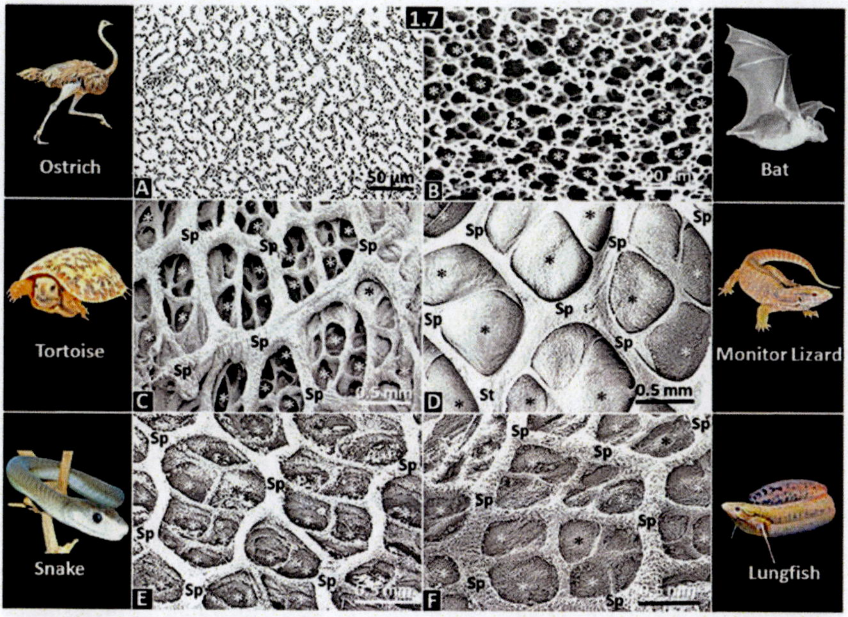

Plate 1.7 Internal compartmentalization (subdivision) of the lungs of some air-breathing vertebrates. (**a**) In the avian lung, the terminal respiratory units (air spaces) are the extremely small air capillaries (asterisks, *) while for the mammalian one (**b**) alveoli (asterisks, *) are the terminal respiratory units. (**c**, **d**, **e**) For lungs of the more derived reptilian species, the respiratory units that are much larger than those of the parabronchial lung of birds and the bronchioalveolar one of mammals have generally termed faveoli (*). Similar structures exist in the lung of the lungfish (**f**). Sp, septa which compartmentalize the lung into air spaces

exanthematicus) (Plate 1.6h: Plate 1.7d) (Maina and Maloiy 1985; Maina et al. 1989a, b).

Showing that birds committed considerable resources in the evolution of the exceptionally complex respiratory system (the lung-air sac system), for animals of equivalent body mass, except for bats that have remarkably larger lungs (Maina et al. 1982; Maina and King 1984; Maina et al. 1991; Maina 2006), the avian respiratory system comprises a volume of the body that is about three times larger compared with that of lungs of terrestrial (nonvolant) mammals (Jürgens and Gros 2002). In certain morphological aspects, the crocodilian lung resembles the avian one (Schachner et al. 2022): patterning and bifurcation of the main airways, especially the primary bronchus and the secondary bronchi, resembles those of the bird lung and saccular (air sac-like) blind-ended laterodorsally secondary exist. Furthermore, crocodilian species present a unidirectional airflow pattern in various parts of the bronchial system (Farmer 2015c; Farmer and Sanders 2010; Schachner et al. 2013b). From having relatively much smaller abdominal air sacs and a lung that may have lacked a dedicated ventilatory mechanism, it has been argued that it is unlikely that theropod dinosaurs had an avian-like respiratory system (Perry 2001; Quick and

Ruben 2009). Hicks and Farmer (1999) observed that for the early endotherms, non-avian lungs could not have been a limiting factor to gas exchange efficiency and that factors such as diffusive gas exchange were more important in driving the evolution of highly effective avian and mammalian lungs. The reported existence of unidirectional air-flow in lungs of archosaurian species and in some extant reptilian ones (Schachner et al. 2013a, b, 2014; Farmer and Sanders 2010; Cieri et al. 2014; Farmer 2015a, b, c, d, 2017; Cieri and Farmer 2016, 2019), a feature on which the exceptional gas exchange efficiency of the avian respiratory system is much attributed (Scheid 1979; Scheid and Piiper 1989; Powell and Scheid 1989; Powell 2000, 2015; Maina 2005), shows that the process is plesiomorphic (ancestral) to the Archosauria. The process of unidirectional and continuous air flow in the avian lung may not have evolved in birds as a direct adaptation for powered flight but only later became a critical feature to it when all the other structural features that were fundamental to active flight such as development of powerful flight muscles and stream-lined body shape were in place. Recently, a well-preserved fossil lung of an ancient bird, *Archaeorhynchus spathula*, that coexisted with the nonavian dinosaurs during the Cretaceous Period (~120 mya) was described by Wang et al. (2018): *A. spathula* is the basal most member of the Ornithuromorpha, the most inclusive clade containing all modern birds (neornithines) except for the Mesozoic enantiornithes (Zhou and Zhang 2006; Zhou et al. 2013; Zhou 2014; Chiappe et al. 2014; Wang et al. 2015; Wang and Zhou 2016). The specimen displayed well-preserved bones and feathers and internal soft tissues particularly those of the gastrointestinal system, with gizzard stones (gastroliths) and lungs that morphologically resembled those of the modern birds being conspicuous (Wang et al. 2018) (Plate 1.5a–c). Like for the lungs of the modern birds (Maina 2005) (Plate 1.5c), those of *A. spathula* were firmly attached to the ribs (Plate 1.5a, b), the airways (parabronchi) displayed a lumen that was surrounded by a gas exchange mantle (Plate 1.5d) that comprised minute respiratory units, that resembled the air capillaries of the lungs of the modern birds (Plate 1.5e, f). These observations showed that the structure of the avian lung has been conserved for at least the last ~120 million years. Zhou et al. (2013) and Wang et al. (2015) noted that the novel morphologies of the modern birds were already established in the basal ornithuromorphs from the Cretaceous Period while Böhmer et al. (2019) observed that the avian body plan has not changed much for the last ~160 million years of the existence of birds and the foundation of volancy. In a specimen of a non-avian theropod, *Majungasaurus crenatissimus* (Krause et al. 2007), a theropod dinosaur that lived between 70 to 66 million years ago, postcranial pneumaticity showed cervical and abdominal air-sacs, anatomical structures that are similar to those of the extant birds (O'Connor and Claessens 2005). According to Wedel (2001), postcranial pneumaticity in sauropods may have allowed evolution of exceptionally long necks in some of the sauropod lineages and the thoracoabdominal air sacs should have compensated for the respiratory dead space and the pneumatization of the axial skeleton would have reduced body mass. As more fossilized materials are discovered and extrapolation of data acquired from comparative studies made, it is becoming clearer that an avian-style respiratory system possibly existed in the basal dinosauriform archosaurs and that the lung-air sac system of the modern birds is a more recent evolutionary innovation on an ancient structure (O'Connor and Claessens 2005; Schachner et al. 2009, 2011; Perry et al. 2009a; Butler et al. 2009; Brocklehurst et al. 2020). Butler et al. (2012b) observed that pulmonary air sacs existed in the common ancestor of Ornithodira and might have been lost later or drastically reduced in some members of the clade, especially in ornithischian dinosaurs.

Existence of a heterogeneous avian-like respiratory system in the dinosaurian ancestors of birds could have greatly influenced evolution of gigantism in sauropod dinosaurs (O'Connor and Claessens 2005; Wedel 2003, 2006, 2009; Perry et al. 2009a, b, 2011; O'Connor 2006, 2009; Yates et al. 2012; Sander 2013; Lambertz 2016; Lambertz et al. 2017; Melstrom et al. 2016). From modeling the respiratory traits of a 11-m long and 30 ton weight sauropod dinosaur, *Mamenchsaurus hochuanensis*, and factoring into it the bellows-like ventilatory mechanism of the lung-air sac system of the birds, Benton (1984) and Daniels and Pratt (1992) deduced that the greater efficiency and the less work of breathing that characterizes the avian respiratory system might have existed in an animal with a long neck, large body mass, and reptilian level of metabolism. Pierson (2009) argued that solution of the breathing challenges that sauropod dinosaurs may have faced from their having very long necks could be explained by likelihood that they possessed an avian-type respiratory system, i.e., a lung-air sac respiratory system, that is more efficient than the bronchioalveolar one of mammals. According to Sander et al. (2011) and Sander (2013), important evolutionary innovations such as an avian-style respiratory system could have reduced respiratory cost, lowered the overall specific gravity of the body and might even have been important in elimination of excess body heat. Large avian-like heterogeneous lungs with a crosscurrent gas exchange system are supposed to have existed in saurischian dinosaurs (Perry et al. 2009a). Lambertz et al. (2015) argued that since the first amniote lungs were complex and multichambered, the single chambered lungs of the modern lizards and snakes (Duncker 1978a; Perry 1983, 1989a, b, 1998; Perry and Sander 2004; Kardong 2014) present a secondarily simplified rather than a plesiomorphic condition.

References

Abe AS (1991) Effect of the environment on ventilation in reptiles. In: Bicudo JEPW (ed) The vertebrate gas transport cascade. CRC Press, Boca Raton (FL), pp 87–93

Abourachid A, Herrel A, Decamps T, Pages F, Fabre AC, Van Hoorebeke L, Adriaens D, Garcia Amado MA (2019) Hoatzin nestling locomotion: acquisition of quadrupedal limb coordination in birds. Sci Adv 5:eaat0787. https://doi.org/10.1126/sciadv.aat0787

Abourachid A, Renous S (2008) Bipedal locomotion in ratites (Paleognatiform): examples of cursorial birds. Ibis 142:538–549

Abramov O, Mojzsis S (2009) Microbial habitability of the Hadean Earth during the late heavy bombardment. Nature 459:419–422

Ackermann J (1998) Dinosaurs take wing. Natl Geogr 194:74–99

Akbarian A, Michiels J, Degroote J, Majdeddin M, Golian A, De Smet S (2016) Association between heat stress and oxidative stress in poultry; mitochondrial dysfunction and dietary interventions with phytochemicals. J Anim Sci Biotechnol 7:1–14. https://doi.org/10.1186/s40104-016-0097-5

Akishinonomiya F, Mityake T, Sumi SI, Takada M, Ohno S, Kondo N (1994) One subspecies of the red jungle fowl (*Gallus gallus gallus*) suffice as the matriarchic ancestor of all domestic chickens. Proc Natl Acad Sci (USA) 91:12505–12509

Akishinonomiya F, Mityake T, Takada M, Shingu R, Endo T, Gojobori T, Kondo N, Ohno S (1996) Monophyletic origin and unique dispersal patterns of domestic fowls. Proc Natl Acad (USA) 93: 6792–6795

Alarcón PAE, Morales JM, Donázar JA, Sánchez-Zapata JA, Hiraldo F, Lambertuci SA (2017) Sexual-size dimorphism modulates the trade-off between exploiting food and wind resources in a large avian scavenger. Sci Rep 7:11461. https://doi.org/10.1038/s41598-017-11855-0

Alerstam T (2006) Conflicting evidence about long-distance animal navigation. Science 313:791–794

Alerstam T, Hedenström A, Åkesson S (2003) Long-distance migration: evolution and determinants. Oikos 103:247–260

Alexander RMCN (2002) Tendon elasticity and muscle function. Comp Biohem Physiol A Mol Integr Physiol 133:1001–1011

Alexander RMCN (2005) Models and the scaling of energy costs for locomotion. J Exp Biol 208: 1645–1652

Alexander RMCN (2013) Principles of animal locomotion. Princeton University Press, Princeton (NJ)

Alexandrino E, Buechley E, Piratelli A, Ferraz KMPMB, Moral RDA, Şekercioğlu ÇH, Silva W, do Couto HTZ (2016) Bird sensitivity to disturbance as an indicator of forest patch conditions: an issue in environmental assessments. Ecol Indic 66:369–381

Ali S (1979) Bird study in India: its history and its importance. Indian Council for Cultural Relations, New Delhi

Alonso JC, Magaña M, Alonso JA, Palacín C, Martín CA, Martín B (2009) The most extreme sexual size dimorphism among birds: allometry, selection, and early juvenile development in the great bustard (*Otis tarda*). Auk 126:657–665

Alonso JC, Magaña M, Palacín C, Martín C (2010) Correlates of male mating success in great bustard leks: the effects of age, weight, and display effort. Behav Ecol Sociobiol 64:1589–1600

Altshuler DL, Dudley R (2006) The physiology and biomechanics of avian flight at high altitude. Intergr Comp Biol 46:62–71

Altshuler DL, Dudley R, McGuire JA (2004) Resolution of a paradox: Hummingbird flight at high elevation does not come without a cost. Proc Natl Acad Sci (USA) 101:17731–17736

Alvarez-Alvarez EA, Pablo C, Almazán-Núñez RC (2018) Spatiotemporal variation in the structure and diet types of bird assemblage in tropical dry forest in southwestern Mexico. Wilson J Ornithol 130:457–469

Alves J, Dias M, Méndez V, Katrínardóttir B, Gunnarsson TG (2016) Very rapid long-distance sea crossing by a migratory bird. Sci Rep 6:38154. https://doi.org/10.1038/srep38154

Amador LI, Simmons NB, Giannini NP (2019) Aerodynamic reconstruction of the primitive fossil bat *Onychonycteris finneyi* (Mammalia: Chiroptera). Biol Lett 15:10–14

Amat JA, Green AJ (2010) Waterbirds as bioindicators of environmental conditions. In: Hurford C, Schneider M (eds) Conservation monitoring in freshwater habitats: a practical guide and case studies. Springer Science, Berlin, pp 45–52

Ambrose SH (2001) Paleolithic technology and human evolution. Science 291:1748–1753

Amo L, Jansen JJ, van Dam NM, Dicke M, Visser ME (2013) Birds exploit herbivore-induced plant volatiles to locate herbivorous prey. Ecol Lett 16:1348–1355

Amo L, Visser ME, Oers KV (2011) Smelling out predators is innate in birds. Ardea 99:177–184

Andala LY (2017) Flights of fancy: the bird of paradise and its cultural impact. J Southeast Asian Stud 48:372–389

Andersson M (1986) Evolution of condition-dependant sex ornaments and mating preference: sexual selection based on viability differences. Evolution 40:804–816

Anderson SC, Ruxton GD (2020) The evolution of flight in bats: a novel hypothesis. Mamm Rev 50:426–439

Angeler DG, Roberts CP, Twidwell D, Allen CR (2022) The role of rare avian species for spatial resilience of shifting biomes in the great plains of north america. Front Ecol Evol 10:849944. https://doi.org/10.3389/fevo.2022.849944

Antoł A, Kozłowski J (2020) Scaling of organ masses in mammals and birds: phylogenetic signal and implications for metabolic rate scaling. ZooKeys 982:149–159

Araya-Salas M, Wright T (2013) Open-ended song learning in a hummingbird. Biol Lett 9: 20130625. https://doi.org/10.1098/rsbl.2013.0625

Araújo PM, Viegas I, Rocha AD, Villegas A, Jones JG, Mendonça L, Ramos JA, Masero JA, Alves JA (2019) Understanding how birds rebuild fat stores during migration: insights from an experimental study. Sci Rep 9:10065. https://doi.org/10.1038/s41598-019-46487-z

Archer E, Dziba L, Mulongoy KJ, Maoela MA, Walters M, Biggs R, Cormier-Salem MC, DeClerck F, Diaw MC, Dunham AE et al (2018) The regional assessment report on biodiversity and ecosystem services for Africa: summary for policymakers. Intergovernmental Science-Policy Platform on Biodiversity and Ecosystem Services, Bonn. 52. https://www.ipbes.net/system/tdf/spm_africa_2018_digital.pdf?file=1&type=node&id=28397

Arizaga J, Campos F, Alonso D (2006) Variations in wing morphology among subspecies might reflect different migration distances in bluethroat. Ornis Fenn 83:162–169

Arora R, Metzger RJ, Papaioannou VE (2012) Multiple roles and interactions of Tbx4 and Tbx5 in development of the respiratory system. PLos Genet 8:e1002866. https://doi.org/10.1371/journal.pgen.1002866

Arsenault M, Desbiens S, Janvier P, Kerr J (2004) New data on the soft tissues and external morphology of the antiarch *Bothriolepis canadensis* (Whiteaves, 1880) from the Upper Devonian of Miguasha, Québec.

Arratia G, Wilson MVH, Cloutier R (eds) (2004) Recent advances in the origin and early radiation of vertebrates. Verlag Dr. Friedrich Pfeil, Munich (Germany), pp 439–454

Ashley-Ross MA, Hsieh ST, Gibb AC, Blob RW, Notes A (2013) Vertebrate invasions—past, present, and future: an introduction to the symposium. Integr Comp Biol 53:192–196

Askew GN, Marsh RL (2001) The mechanical power output of the pectoralis muscle of blue-breasted quail (*Coturnix chinensis*): the in vivo length cycle and its implications for muscle performance. J Exp Biol 204:3587–3600

Askew GN, Marsh RL, Ellington CP (2001) The mechanical power output of the flight muscles of blue-breasted quail (*Coturnix chinensis*) during take-off. J Exp Biol 204:3601–3619

Assis APA, Galetti M, Maia KP, Guimarães PR (2022) Reduced evolutionary potential of a frugivorous bird species in fragmented forests. Front Ecol Evol 10:804138. https://doi.org/10.3389/fevo.2022.804138

Aubin T (2004) Penguins and their noisy world. Ann Brazilian Acad Sci 76:279–283

Aubin T, Jouventin P (2002) Localisation of an acoustic signal in a noisy environment: the display call of the king penguin *Aptenodytes patagonicus*. J Exp Biol 205:3793–3798

Auersperg A, Teschke I, Tebbich S (2017) Physical cognition and tool use in birds. In: Ten Cate C, Healy S (eds) Avian cognition. Cambridge University Press, Cambridge, pp 163–183

Bacon PJ, Coleman AE (1986) An analysis of weight changes in the mute swan, *Cygnus olor*. Bird Study 33:145–158

Bahlman JW, Swartz SM, Riskin DK, Breuer KS (2013) Glide performance and aerodynamics of non-equilibrium glides in northern flying squirrels (*Glaoucomys sabrinus*). J R Soc Interface 10: 20120794. https://doi.org/10.1098/rsif.2012.0794

Bailleul AM, Horner JR (2016) Comparative histology of some craniofacial sutures and skull-base synchondroses in non-avian dinosaurs and their extant phylogenetic bracket. J Anat 229:252–285

Bailleul AM, O'Connor J, Li Z, Wu Q, Zhao T, Monleon MAM, Wang M, Zheng X (2020) Confirmation of ovarian follicles in an enantiornithine (Aves) from the Jehol biota using soft tissue analyses. Commun Biol 3:399. https://doi.org/10.1038/s42003-020-01131-9

Bakkeren C, Smith E, York JM, Chua B, McCracken KG, Milsom WK (2020) A morphometric analysis of the lungs of high-altitude ducks and geese. J Anat 237:188–196

Balda R, Kamil A (2006) The ecology and life history of seed caching corvids. In: Brown MF, Cook RG (eds) Animal spatial cognition: comparative, neural, and computational approaches. Comparative Cognition Press. available on pigeon.psy.tufts.edu/asc/balda/ - accessed on 14011-22

Baldwin JW, Garcia-Porta J, Botero CA (2022) Phenotypic responses to climate chage are significantly dampened in big-brained birds. Ecol Lett 25:939–947

Bale R, Neveln ID, Bhalla APS, MacIver MA, Patankar NA (2015) Convergent evolution of mechanically optimal locomotion in aquatic invertebrates and vertebrates. PLoS Biol 13: e1002123. https://doi.org/10.1371/journal.pbio.1002123

Baptista LF, Schuchmann KL (1990) Song learning in Anna hummingbird (*Calypte anna*). Ethology 84:15–26

Barash DP, Donovan P, Myrick R (1975) Clam dropping behavior of the glaucous-winged gull (*Larus glaucescens*). Wilson Bull 87:60–84

Barbosa KVC, Knogge C, Develey PF, Jenkins CN, Uezu A (2017) Use of small Atlantic Forest fragments by birds in Southeast Brazil. Perspect Ecol Conserv 15:42–46

Barker FK, Barrowclough GF, Groth JG (2002) A phylogenetic hypothesis for passerine birds: taxonomic and biogeographic implications of an analysis of nuclear DNA sequence data. Proc Biol Sci 269:295–308

Barnagaud J-Y, Devictor V, Jiguet F, Barbet-Massin M, Le Viol I, Archaux F (2012) Relating habitat and climatic niches in birds. PLoS ONE 7:e32819. https://doi.org/10.1371/journal.pone.0032819

Barnett JM, Buzzetti DRC (2014) A new species of *Cichlocolaptes* Reichenbach 1853 (Furnariidae), the 'gritador-do-nordeste', an undescribed trace of the fading bird life of northeastern Brazil. Rev Brasileira Ornitologia 22:75–94

Barnosky AD, Matzke N, Tomiya S, Wogan GOU, Schwartz B, Quental TB, Marshall C, McGuire JL, Lindsey EL, Maguire KC et al (2011) Has the Earth's sixth mass extinction arrived? Nature 471:51–57

Barrett PM, Butler RJ, Edwards NP, Milner AR (2008) Pterosaur distribution in time and space: an atlas. Zitteliana B 28:61–107

Barrowclough GF, Cracraft J, Klicka J, Zink RM (2016) How many kinds of birds are there and why does it matter? PLoS ONE 11:e0166307. https://doi.org/10.1371/journal.pone.0166307

Bartholomew GA, Lighton JRB (1986) Oxygen consumption during hover-feeding in free-ranging Anna hummingbirds. J Exp Biol 123:191–199

Bärtsch P, Mairbäurl H, Maggiorini M, Swenson ER (1985) Physiological aspects of high-altitude pulmonary edema. J Appl Physiol 98:1101–1110

Bärtsch P, Mairbäurl H, Swenson ER, Maggiorini M (2003) High altitude pulmonary oedema. Swiss Med Wkly 133:377–384

Bärtsch P, Mairbäurl H, Maggiorini M, Swenson ER (2005) Physiologial aspects of high-altitude pulmonary edema. J Appl Physiol 98:1101–1110

Battley PF, Warnock N, Tibbitts TL, Gill RE, Piersma T, Hassell CJ, Douglas DC, Mulcahy DM, Gartrell BD, Schuckard R et al (2012) Contrasting long-distance migration patterns in bar-tailed godwits *Limosa lapponica*. J Avian Biol 43:21–32

Bauer S, Klaassen M (2013) Mechanistic models of animal migration behaviour—their diversity, structure and use. J Anim Ecol 82:498–508

Baumgartner RJ, Van Kranendonk MJ, Wacey D, Fiorentini ML, Saunders M, Caruso S, Pages A, Homann M, Guagliardo P (2019) Nano-porous pyrite and organic matter in 3.5-billion-year-old stromatolites record primordial life. Geology 47:1039–1043

Bayley M, Damsgaard C, Thomsen M, Malte H, Wang T (2019) Learning to air-breathe: the first steps. Physiology 34:14–29

Beck BB (1980) Animal tool behavior: the use and manufacture of tools by animals. Garland STMP, New York

Béchard I, Arsenault F, Cloutier R, Kerr J (2014) The Devonian placoderm fish Bothriolepis canadensis revisited with three-dimensional digital imagery. Palaeontol Electronica 17:1–19

Becker DJ, Singh D, Pan Q, Montoure JD, Talbott KM, Wanamaker SM, Ketterson ED (2020) Artificial light at night amplifies seasonal relapse of haemosporidian parasites in a widespread songbird. Proc R Soc (Lond) B 287. https://doi.org/10.1098/rspb.2020.1831

Belão TC, Zeraik VM, Florindo LH, Kalinin AL, Leite CAC, Rantin FT (2015) Control of cardiorespiratory function in response to hypoxia in an air-breathing fish, the African sharptooth catfish, *Clarias gariepinus*. Comp Physiol Biochem A 187:130–140

Bellemain E, Ricklefs RE (2008) Are islands the end of the colonization road? Trends Ecol Evol 23: 461–468

Bennett ATD, Théry M (2007) Simultaneous crypsis and conspicuousness in color patterns: comparative analysis of a neotropical rainforest bird community. Amer Natur 169:S42–S61

Bennett M, Tenney SM (1982) Comparative mechanics of mammalian respiratory system. Respir Physiol 49:131–140

Benson RBJ, Butler RJ, Carrano MT, O'Connor PM (2012) Air-filled postcranial bones in theropod dinosaurs: physiological implications and the 'reptile'-bird transition. Biol Rev 87:168–193

Benson RBJ, Campione NE, Carrano MT, Mannion PD, Sullivan C, Upchurch P, Evans DC (2014) Rates of dinosaur body mass evolution indicate 170 million years of sustained ecological innovation on the avian stem lineage. PLoS Biol 12:e1001853. https://doi.org/10.1371/journal.pbio.1001853

Benton MJ (1984) The dinosaur encyclopedia. Simon and Schuster, New York

Benton MJ, Dhouailly D, Jiang B, McNamara M (2019) The early origin of feathers. Trends Ecol Evol 34:856–869

Berkovitz BKB, Shellis RP (2018) The teeth of mammalian vertebrates, 1st edn. Academic Press, New York

Bernat-Ponce E, Gil-Delgado JA, López-Iborra GM (2021) Recreational noise pollution of traditional festivals reduces the juvenile productivity of an avian urban bioindicator. Envronmental Pollution 286:117247. https://doi.org/10.1016/j.envpol.2021.117247

Berthold P (2001) Bird migration: a general survey, 2nd edn. Oxford University Press, USA

Bhadra A (2019) An african tale of two species. Resonance May 2019:547–559

Biard C, Brischoux F, Meillère A, Michaud B, Nivière M, Ruault S, Vaugoyeau M, Angelier F (2017) Growing in cities: an urban penalty for wild birds? A study of phenotypic differences between urban and rural great tit chicks (*Parus major*). Front Ecol Evol 5:79. https://doi.org/10.3389/fevo.2017.00079

Biewener AA (2011) Muscle function in avian flight: achieving power and control. Philos Trans R Soc (Lond) B Biol Sci 366:1496–1506

Bird DM (1999) The bird almanac: the ultimate guide to essential facts and figures of the world's birds. Firefly Books, Buffalo (NY)

BirdLife International (2022) Species factsheet: Mellisuga helenae. Downloaded from http://www.birdlife.org on 24/04/2022.

BirdLife International (2018) State of the world's birds: taking the pulse of the planet. BirdLife International, Cambridge (UK)

Birks SM, Edwards SV (2002) A phylogeny of the megapodes (Aves: Megapodiidae) based on nuclear and mitochondrial DNA sequences. Mol Phylogen Evol 23:408–421

Bishop CM (1999) The maximum oxygen consumption and aerobic scope of birds and mammals: getting to the heart of the matter. Proc R Soc (Lond) B 266:2275–2281

Bishop CM, Butler PJ (2015) Flight. In: Scanes CG (ed) Sturkie's avian physiology, 6th edn. Academic Press, New York (NY), pp 919–974

Black CP, Tenney SM (1980) Oxygen transport during progressive hypoxia in high-altitude and sea-level waterfowl. Respir Physiol 39:217–239

Blasco R, Finlayson C, Rosell J, Marco AS, Finlayson S, Finlayson G, Negro JJ, Pacheco FG, Vidal JR (2014) The earliest pigeon fanciers. Sci Rep 4:5971. https://doi.org/10.1038/srep05971

Blount JD, Horns J, Kittelberger KD, Neate-Clegg MHC, Şekercioğlu CH (2021) Avian use of agriculatural areas as migration stopovers sites: a review of crop management practices and ecological correlates. Front Ecol Evol 9:650641. https://doi.org/10.3389/fevo.2021.650641

Bluff LA, Weir AAS, Rutz C, Wimpenny JH, Kacelnik A (2007) Tool-related cognition in New Caledonian crows. Comp Cogn Behav Rev 2:1–25

Böhmer C, Plateau O, Cornette R, Abourachid A (2019) Correlated evolution of neck length and leg length in birds. R Soc Open Sci 6:181588. https://doi.org/10.1098/rsos.181588

Bonadonna F, Bajzak C, Benhamou S, Igloi K, Jouventin P, Lipp HP, Dell'Omo G (2005) Orientation in the wandering albatross: interfering with magnetic perception does not affect orientation performance. Proc Biol Sci 272:489–495

Bonier F, Martin PR, Wingfield JC (2007) Urban birds have broader environmental tolerance. Biol Lett 3:670–673

Boone EH (1989) Incarnations of the Aztec supernatural: the image of Huitzilopochtli in Mexico and Europe. Transac Amer Philos Soc 49, @inproceedings{Boone1989IncarnationsOT. https://doi.org/10.2307/1006524

Boria S (2007) Medievalism, paganism and the Tower Ravens. Int J Pagan Studies 9:272–274

Both C, Marvelde L (2007) Climate change and timing of avian breeding and migration throughout Europe. Clim Res 35:93–105

Bouhuys A, Mead J, Proctor DF, Stevens KN (1968) Pressure-flow events during singing. Ann NY Acad Sci 155:165–176

Bourn A, Bonine M (2015) The elegant plume: Ostrich feathers, African commercial networks, and European capitalism. J North Africa Stud 20:5–26

Brackenbury JH (1989) Functions of the syrinx and the control of sound production. In: King AS, McLelland J (eds) Form and function in birds, vol 1. Academic Press, London, pp 193–220

Brandes TS (2008) Automated sound recording and analysis techniques for bird surveys and conservation. Bird Conserv Int 18:163–173

Breen AJ, Guillette LM, Healy SD (2016) What can nest-building birds teach us? Comp Cognit Behav Rev 11:83–102

Brito PM, Meunier FJ, Clement G, Geffard-Kuriyama D (2010) The histological structure of the calcified lung of the fossil coelacanth *Axelrodichthys araripensis* (Actinistia: Mawsoniidae). Palaeontology 53:1281–1290

Britt BB, Makovicky PJ, Gauthier J, Bonde N (1998) Postcranial pneumatization in *Archaeopteryx*. Nature 395:374–376

Brocklehurst RJ, Schachner ER, Codd JR, Sellers WI (2020) Respiratory evolution in archosaurs. Philos Trans R Soc (Lond) B Biol Sci 375:20190140. https://doi.org/10.1098/rstb.2019.0140

Brocklehurst N, Upchurch P, Mannion PD, O'Connor J (2012) The completeness of the fossil record of mesozoic birds: implications for early avian evolution. PLoS ONE 7:e39056. https://doi.org/10.1371/journal.pone.0039056

Broitman IAE (2007) Bird song formal language modeling based on acoustic syllable detection. PhD Dissertation, Instituto Tecnológico Y De Estudios Superiores De Monterrey Campus Estado De México.

Brooke M, Birkhead T (1991) The Cambridge encyclopedia of ornithology. Cambridge University Press, Cambridge

Brooks HL, Rushton K, Lovell K, Bee P, Walker L, Grant L, Rogers A (2018) The power of support from companion animals for people living with mental health problems: a systematic review and narrative synthesis of the evidence. BMC Psychiatry 18:31

Brooks TM, Mittermeier RA, Mittermeier CG, da Fonseca GAB, Rylands AB, Konstant WR, Flick P, Pilgrim J, Oldfield S, Magin G et al (2002) Habitat loss and extinction in the hotspots of biodiversity. Conserv Biol 16:909–923

Brown KS, Brown GG (1992) Habitat alteration and species loss in Brazilian forests. In: Whitmore TC, Sayer J (eds) Tropical deforestation and species extinction. Chapman and Hall, London (UK), pp 129–142

Brugaletta G, Teyssier J-R, Rochell SJ, Dridi S, Sirri F (2022) A review of heat stress in chickens. Part I: insights into physiology and gut health. Front Physiol 13:934381. https://doi.org/10.3389/fphys.2022.934381

Brumm H (2004) The impact of environmental noise on song amplitude in a territorial bird. J Anim Ecol 73:434–440

Brumm H, Todt D (2002) Noise-dependent song amplitude regulation in a territorial songbird. Anim Behav 63:891–897

Brusatte SL, Lloyd GT, Wang SC, Norell MA (2014) Gradual assembly of avian body plan culminated in rapid rates of evolution across the dinosaur-bird transition. Curr Biol 24:2386–2392

Brusatte SL, O'Connor JK, Jarvis ED (2015) The origin and diversification of birds. Curr Biol 25: R888–R898

Bühler P (1981) Functional anatomy of the avian jaw apparatus. In: King AS, McLelland J (eds) Form and function in birds, vol 2. Academic Press, London, pp 439–468

Burggren WW, Moalli R (1984) Active regulation of cutaneous gas exchange by capillary recruitment in amphibians: experimental evidence and a revised model for skin respiration. Respir Physiol 55:379–392

Burt DW (2007) Emergence of the chicken as a model organism: implications for agriculture and biology. Poult Sci 86:1460–1471

Butcher GS, Rohwer S (1989) The evolution of conspicuous and distinctive coloration for communication in birds. In: Power DM (ed) Current Ornithoogy, vol 6. Springer, Boston (MA), pp 51–108

Butler PJ (1991) Exercise in birds. J Exp Biol 160:233–262

Butler PJ (2010) High fliers: the physiology of bar-headed geese. Comp Biochem Physiol A: Mol Integr Physiol 156:325–329

Butler PJ (2016) The physiological basis of bird flight. Philos Trans R Soc (Lond) B 371:20150384. https://doi.org/10.1098/rstb.2015.0384

Butler RJ, Barrett PM, Gower DJ (2009) Postcranial skeletal pneumaticity and air-sacs in the earliest pterosaurs. Biol Lett 5:557–560

Butler RJ, Barrett PM, Gower DJ (2012a) Reassessment of the evidence for postcranial skeletal pneumaticity in triassic archosaurs, and the early evolution of the avian respiratory system. PLoS ONE 7:e34094. https://doi.org/10.1371/journal.pone.0034094

Butler SJ, Renwick AR, Norris K (2012b) An objective, niche-based approach to indicator species selection. Methods Ecol Evol 3:317–326

Cabej NR (2011) Epigenetic principles of evolution. Elsevier Science Direct Inc., Amsterdam

Çağan HŞ, Mendenhall CD, Oviedo-Brenes F, Horns JJ, Ehelich PR, Daily GC (2019) Long-term declines in bird populations in tropical agricultural countryside. Proc Natl Acd Sci USA 116: 9903–9912

Callaghan CT, Nakagawa S, Cornwella WK (2021) Global abundance estimates for 9,700 bird species. Proc Natl Acad Sci USA 118(21):e2023170118. https://doi.org/10.1073/pnas.2023170118

Campbell KE, Tonni EP (1983) Size and locomontion in teratorns (Aves: Teratornithidae). Auk 100:390–403

Campler M, Jöngren M, Jensen P (2009) Fearfulness in red junglefowl and domesticated white leghorn chickens. Behav Processes 81:39–43

Cao T, Jin JP (2020) Evolution of flight muscle contractility and energetic efficiency. Front Physiol 11:1038. https://doi.org/10.3389/fphys.2020.01038

Carey SW (1976) The expanding earth. Elsevier, New York

Cassini CS, Taucce PPG, de Carvalho TR, Fouquet A, Solé M, Haddad CFB, Garcia PCA (2020) One step beyond a broad molecular phylogenetic analysis: species delimitation of *Adenomera marmorata* Steindachner, 1867 (Anura: Leptodactylidae). PLoS One 15(2):e0229324. https://doi.org/10.1371/journal.pone.0229324

Catchpole CK (1982) The evolution of bird sounds in relation to mating and spacing behavior. In: Kroodsma DE, Miller EH (eds) Acoustic communication in birds, vol 1. Academic Press, New York, pp 297–319

Catchpole C, Slater PJB (2008) Bird song: biological themes and variation. Cambridge University Press, Cambridge

Catry P, Phillips RA, Croxall JP (2004) Sustained fast travel by a grey-headed albatross (*Thalassarchie chrysostoma*) riding an Antarctic storm. Auk 121:1208–1213

Cavin L, Guinot G (2014) Coelocanths as almost living fossils. Front Ecol Evol 2:49. https://doi.org/10.3389/fevo.2014.00049

Ceballos G, Ehrlich PR, Dirzo R (2017) Biological annilation via ongoing sixth mass extinction signaled by vertebrate population losses and declines. Proc Natl Acad Sci (USA) 114:6089–6096

Ceballos G, Ehrlich AH, Ehrlich PR (2015a) The annihilation of nature: human extinction of birds and mammals. Johns Hopkins University Press, Baltimore (MD)

Ceballos G, Ehrlich PR, Paul R, Barnosky AD, Garcia A, Pringle RM, Palmer TM (2015b) Accelerated modern human-induced species losses: entering the sixth extinction. Sci Adv 1(5):e1400253. https://doi.org/10.1126/sciadv.1400253

Cerio DG, Witmer LM (2020) Modeling visual fields using virtual ophthalmoscopy: incorporating geometrical optics, morphometrics, and 3D visualization to validate an interdisciplinary technique. Vision Res 167:70–86

Chai P, Dudley R (1995) Limits to vertebrate locomotor energetics suggested by hummingbirds hovering in heliox. Nature 377:722–725

Chai P, Millard M (1997) Flight and size constraints: hovering performance of large hummingbirds under maximal loading. J Exp Biol 200:2757–2763

Chakraborty M, Walløe S, Nedergaard S, Fridel EE, Dabelsteen T, Pakkenberg B, Bertelsen MF, Dorrestein GM, Brauth SE, Durand SE et al (2015) Core and shell song systems unique to the parrot brain. PLoS ONE 10:e0118496. https://doi.org/10.1371/journal.pone.0118496

Chang MM, Yu X (1984) Structure and phylogenetic significance of *Diabolichthys speratus* gen. et sp. nov., a new dipnoan-like form from the Lower Devonian of eastern Yunnan, China. Proc Linn Soc NSW 107:171–184

Chapman CA, Abernathy K, Chapman LJ, Downs C, Effiom EO, Gogarten JF, Golooba M, Kalbitzer U, Lawes MJ, Mekonnen A et al (2022) The future of sub-Saharan Africa's biodiversity in the face of climate and societal change. Front Ecol Evol 10:790552. https://doi.org/10.3389/fevo.2022.790552

Chappell J, Kacelnik A (2002) Tool selectivity in a non-primate, the New Caledonian crow (*Corvus moneduloides*). Anim Cogn 5:71–78

Chatterjee S, Templin RJ, Campbell KEJ (2007) The aerodynamics of *Argentavis*, the world's largest flying bird from the Miocene of Argentina. Proc Natl Acad Sci USA 104:12398–12403

Chen CCW, Welch KC (2014) Hummingbirds can fuel expensive hovering flight completely with either exogenous glucose or fructose. Funct Ecol 28:589–600

Chen P, Dong Z, Zhen S (1998) An exceptionally well-preserved theropod dinosaur from the Yixian formation of China. Nature 391:147–152

Chen S, Duan P, Yu X (2022) Ecological aspiration and the income of farmers aroused by Grain for Green Project. Front Ecol Evol 10:961490. https://doi.org/10.3389/fevo.2022.961490

Chen T, Song G, Yang H, Mao L, Cui Z, Huang K (2018) Development of the swimbladder surfactant system and biogenesis of lysosome-related organelles is regulated by BLOS1 in zebrafish. Genetics 208:1131–1146

Chiappe LM (2004) Closest relatives of birds. Ornitol Neotrop 15:101–116

Chiappe LM (2009) Downsized dinosaurs: the evolutionary transition to modern birds. Evol Educ Outreach 2:248–256

Chiappe LM, Dyke G (2006) The early evolutionary history of birds. J Paleontol Soc Korea 22: 133–151

Chiappe LM, Zhao B, O'Connor JK, Chunling G, Wang X, Habib M, Marugan-Lobon J, Meng Q, Cheng X (2014) A new specimen of the Early Cretaceous bird *Hongshanornis longicresta*: insights into the aerodynamics and diet of a basal ornithuromorph. PeerJ 2:e234. https://doi.org/10.7717/peerj.234

Chilson PB, Frick WF, Kelly JF, Liechti F (eds) (2017) Aeroecology. Springer, Cham (Switzerland)

Chin DD, Lentink D (2019) Birds repurpose the role of drag and lift to take off and land. Nat Commun 10:5353. https://doi.org/10.1038/s41467-019-13347-3

Chin DD, Matloff LY, Stowers AK, Tucci ER, Lentink D (2017) Inspiration for wing design: how forelimb specialization enables active flight inn modern vertebrates. J R Soc Interface 14: 20170240. https://doi.org/10.1098/rsif.2017.0240

Cieri RL, Craven BA, Schachner ER, Farmer CG (2014) New insight into the evolution of the vertebrate respiratory system and the discovery of unidirectional airflow in iguana lungs. Proc Natl Acad Sci USA 111:17218–17223

Cieri RL, Farmer CG (2016) Unidirectional pulmonary airflow in vertebrates: a review of structure, function, and evolution. J Comp Physiol B 185:541–552

Cieri RL, Farmer CG (2019) Computational fluid dynamics reveals a unique net unidirectional pattern of pulmonary airflow in the savannah monitor lizard (*Varanus exanthematicus*). Anat Rec 303:1768–1791

Claessens LPAM, O'Connor PM, Unwin DM (2009) Respiratory evolution facilitated the origin of pterosaur flight and aerial gigantism. PLoS One 4:e4497. https://doi.org/10.1371/journal.pone.0004497

Clairbaux M, Fort J, Matherson P, Porter W, Strøm GD (2019) Climate change could overturn bird migration: transarctic flights and high-latitude residency in a sea ice free Arctic. Sci Rep 9: 17767. https://doi.org/10.1038/s41598-019-54228-5

Clark CJ (2011) Wing, tail, and vocal contributions to the complex acoustic signals of courting Calliope hummingbirds. Curr Zool 57:187–196

Clark CJ, McGuire JA, Bonaccorso E, Berv JS, Prum RO (2018) Complex coevolution of wing, tail, and vocal sounds of courting male bee hummingbirds. Evolution 72:630–646

Clark WS (1995) How fast is the fastest bird? WildBird 9:42–43

Clarke JA (2013) Paleontology. Feathers before flight. Science 340:690–692

Clarke JA, Chatterjee S, Zhiheng L, Riede T, Agnolin F, Goller F, Isasi MP, Martinioni DR, Mussel FJ, Novas FE (2016) Fossil evidence of the avian vocal organ from the Mesozoic. Nature 538: 502–505

Clarke JA, Middleton K (2006) Bird evolution. Curr Biol 16:R350–R354

Clayton NS, Dally JM, Emery NJ (2007) Social cognition by food-caching corvids: the western scrub-jay as a natural psychologist. Phil Trans R Soc (Lond) B 362:507–522

Clement AM, Long JA (2010) Air-breathing adaptation in a marine Devonian lungfish. Biol Lett 6: 509–512

Clements JF (2007) Birds of the world: a checklist. Cornell Laboratory of Ornithology, Cornell University Press

Clucas B, Marzluff JM (2012) Attitudes and actions toward birds in urban areas: human cultural differences influence bird behavior. Auk 129:8–16

Cnotka J, Güntürkün O, Rehkämper G, Gray RD, Hunt GR (2008) Extraordinary large brains in tool-using New Caledonian crows (*Corvus moneduloides*). Neurosci Lett 433:241–245

Cochet H, Byrne R (2015) Complexity in animal behaviour: towards common ground. Acta Ethol 18:237–241

Codd JR (2010) Uncinate processes in birds: morphology, physiology and function. Comp Biochem Physiol A 156:303–308

Codd JR, Boggs DF, Perry SF, Carrier DR (2005) Activity of three muscles associated with the uncinate processes of the giant Canada goose *Branta canadensis maximus*. J Exp Biol 208:849–857

Cohen AJ, Brauer M, Burnett R, Anderson HR, Frostad J, Estep K, Balakrishnan K, Brunekreef B, Dandona L, Dandona R et al (2017a) Estimates and 25-year trends of the global burden of disease attributable to ambient air pollution: an analysis of data from the Global Burden of Diseases Study 2015. Lancet 389:1907–1918

Cohen EB, Buler JJ, Deppe JL, Farnsworth A, Marra PP, McWilliams SR, Mehlman DW, Wilson RR, Woodrey MS, Moore FR (2017b) How do en route events around the Gulf of Mexico influence migratory landbird populations? Condor 119:327–343

Coles BH (1997) Essentials of avian medicine and surgery, 3rd edn. Blackwell Publishing, Oxford

Collias NE (1987) The vocal repertoire of the red junglefowl: a spectrographic classification and the code of communication. Condor 89:510–524

Collias NE (1997) On the origin and evolution of nest building passerine birds. Condor 99:253–270

Collias NE, Collias EC (1996) Social organization of a red junglefowl, *Gallus gallus*, population related to evolution theory. Anim Behav 51:1337–1354

Collias NE, Collias EC, Hunsaker D, Minning L (1966) Locality fixation, mobility and social organization within an unconfined population of red jungle fowl. Anim Behav 14:550–559

Cone CD (1962) Thermal soaring of birds. Am Sci 50:180–209

Cone CD (1964) A mathematical analysis of the dynamic soaring flight of the albatross with ecological interpretations. Virginia Institute of Marine Science, No. 50. Virginia Institute of Marine Science, College of William and Mary (VA). Open Library OL6010988M

Conklin JR, Senner NR, Battley PF, Piersma T (2017) Extreme migration and the individual quality spectrum. J Avian Biol 48:19–36

Cook J (2016) Consensus on consensus: a synthesis of consensus estimates on human-caused global warming. Environ Res Lett 11:048002. https://doi.org/10.1088/1748-9326/11/4/048002

Cooke RSC, Eigenbrod F, Bates AE (2019) Projected losses of global mammal and bird ecological strategies. Nat Commun 10:2279. https://doi.org/10.1038/s41467-019-10284-z

Cooney CR, Bright JA, Capp EJR, Chira AM, Hughes EC, Moody CJA, Nouri LO, Varley ZK, Thomas GH (2017) Mega-evolutionary dynamics of the adaptive radiation of birds. Nature 542: 344–346

Cooper DS, Shultz AJ, Blumstein DT (2020) Temporally separated data sets reveal similar traits of birds persisting in a United States megacity. Front Ecol Evol 8:251. https://doi.org/10.3389/fevo.2020.00251

Cooper DS, Shultz AJ, Şekercioğlu ÇH, Osborn FM, Blumstein DT (2022) Community science data suggest the most common raptors (Accipitridae) in urban centres are smaller, habitat-generalist species. Ibis. https://doi.org/10.1111/ibi.13047

Coppack T, Both C (2002) Predicting life-cycle adaptations of migratory birds to global climate change. Ardea 90:369–378

Cordell HK, Herbert NG (2002) The popularity of birding is still growing. Birding 2002:54–61

Cordell HK, Herbert NG, Pandolfi F (1999) The growing popularity of birding in the United States. Birding 31:168–176

Cortright KA, Wetzlich SE, Craigmill AL (2009) A PBPK model for midazolam in four avian species. J Vet Pharmacol Ther 32:552–565

Covino KM, Horton KG, Morris SR (2020) Seasonally specific changes in migration phenology across 50 years in the black-throated blue warbler. Auk 137:1–11

Cracraft J (1986) The origin and early diversification of birds. Paleobiology 12:383–399

Claramunt S, Cracraft J (2015) A new time tree reveals Earth history's imprint on the evolution of modern birds. Sci Adv 1:e1501005. https://doi.org/10.1126/sciadv.1501005

Cristol DA, Switzer PV (1999) Avian prey-dropping behavior. II. American crows and walnuts. Behav Ecol 10:220–226

Crosfill ML, Widdicombe JG (1961) Characteristics of the chest and lungs and the work of breathing in different mammalian species. J Physiol 158:1–14

Croxall JP, Silk JRD, Phillips RA, Afanasyev V, Briggs DR (2005) Global circumnavigations: tracking year-round ranges of nonbreeding albatrosses. Science 307:249–250

Csillag A, Ádám Á, Zachar G (2022) Avian models for brain mechanisms underlying altered social behavior in autism. Front Physiol 13:1032046. https://doi.org/10.3389/fphys.2022.1032046

Culp LA, Cohen EB, Scarpignato AL, Thogmartin WE, Marra PP (2017) Full annual cycle climate change vulnerability assessment for migratory birds. Ecosphere 8:1–23

Cummings V, Jordan P, Zvelebil M (eds) (2014) The Oxford handbook of archeology and anthropology of hunter-gatherers. Oxford University Press, Oxford. ISBN 978-0-19-102526-6

Cupello C, Brito P, Herbin M, Meunier FJ, Janvier P, Dutel H, Clément G (2015) Allometric growth in the extant coelacanth lung during ontogenetic development. Nat Commun 6:8222. https://doi.org/10.1038/ncomms9222

Cupello C, Clément G, Brito PM (2019) Evolution of air breathing and lung distribution among fossil fishes. In: Johanson Z (ed) Evolution and development of fishes. Cambridge University Press, Cambridge, pp 252–262

Cupello C, Hirasawa T, Tatsumi N, Yabumoto Y, Gueriau P, Isogai S, Matsumoto R, Saruwatari T, King A, Hoshino et al (2022) Lung evolution in vertebrates and the water-to-land transition. eLife 11:e77156. https://doi.org/10.7554/eLife.77156

Cupello C, Meunier FJ, Herbin M, Janvier P, Clément G, Brito PM (2017) The homology and function of the lung plates in extant and fossil coelacanths. Sci Rep 7:1–8. https://doi.org/10.1038/s41598-017-09327-6

Cushman GT (2013) Guano and the opening of the Pacific World: a global ecological history. Cambridge University Press, Cambridge

Czopek J (1983) Distribution of capillaries in the respiratory surfaces in two species of Batrachophrynus (Amphibian, Anura, Leptodactylidae). Zoologica Poloniae 30:211–226

Daley MA, Birn-Jeffrey A (2018) Scaling of avian bipedal locomotion reveals independent effects of body mass and leg posture on gait. J Exp Biol 221:jeb152538. https://doi.org/10.1242/jeb.152538

Dalsätt J (2012) Fossil birds: contributions to the understanding of avian evolution. PhD Thesis, Stockholm University.

Dalsgaard B, Martín González AM, Olesen JM, Ollerton J, Timmermann A, Andersen LH, Tossas AG (2009) Plant-hummingbird interactions in the West Indies: Floral specialization gradients associated with environment and hummingbird size. Oecologia 159:757–766

Damaraju M, Pita MRT, Frankini EL, Granatosky MC (2021) Megachiroptera locomotion. In: Vonk J, Shackelford T (eds) Encyclopedia of animal cognition and behavior. Springer, Cham (Switzerland), pp 1–8. https://doi.org/10.1007/978-3-319-47829-6_1159-1

Daniels CB, Orgeig S, Sullivan LC, Ling NL, Bennett MB, Schürch S, Luis-Val A, Brauner CJ (2004) The origin and evolution of the surfactant system in fish: insights into the evolution of lungs and swimbladder. Physiol Biochem Zool 77:732–749

Daniels CB, Pratt J (1992) Breathing in long necked dinosaurs; did the sauropods have bird lungs? Comp Biochem Physiol A 101:43–46

d'Amour BC, Reitsma F, Baiocchi G, Barthel S, Güneralp B, Erb KH, Haberl H, Creutzig F, Seto KC (2017) Future urban land expansion and implications for global croplands. Proc Natl Acad Sci USA 114:8939–8944

Darwin C (1859) On the origin of species hy means of natural selection, or the preservation of favoured races in the struggle for life, 2nd edn. D. Appleton and Company, New York

Davenport J, O'Callaghan MJA, Davenport JL, Kelly TC (2014) Mussel dropping by Carrion and Hooded crows: biomechanical and energetic considerations. J Field Ornithol 85:196–205

Davies S (2002) Ratites and tinamous. Oxford University Press, New York

Davies SJJF (2003) Birds I: Tinamous and Rratites to hoatzins. In: Hutchins M (ed) Grzimek's animal life encyclopedia, 2nd edn. Gale Group, Farmington Hills (MI), pp 99–101

Dawson TJ, Norton MA, Rodoreda S, Abbott SK, McLeod SR (2021) The burden of size and growth for the juveniles of large mammalian herbivores: structural and functional constraints in the feeding biology of juveniles relative to adults in red kangaroos, *Osphranter rufus*. Ecol Evol 11:9062–9078

Dececchi TA, Larsson HCE, Habib MB (2016) The wings before the bird: an evaluation of flapping-based locomotory hypotheses in bird antecedents. Peer J 4:e2159. https://doi.org/10.7717/peerj.2159

Dejours P (1981) Principles of comparative respiratory physiology. Elsevier, Amsterdam

Dejours P (1989) From comparative physiology of respiration to several problems of environmental adaptations and to evolution. J Physiol 410:1–19

Dekker RWRJ (2007) Distribution and speciation of megapodes (Megapodiidae) and subsequent development of their breeding. In: Renema W (ed) Biogeography, time, and place: distributions, barriers, and islands. Springer, Dordrecht, pp 93–102

del Hoyo J, Elliott A, Sargatal J, Christie DA, de Juana E (eds) (2020) Birds of the World. Cornell Lab of Ornithol, Ithaca (NY). https://doi.org/10.2173/bow.gillar1.01

Del Vesco A, Khatlab A, Goes E, Utsunomiya K, Vieira J, Neto AO, Gasparino E (2017) Age-related oxidative stress and antioxidant capacity in heatstressed broilers. Animal 11: 1783–1790

Denison RH (1941) The soft anatomy of *Bothriolepis*. J Paleontol 15:553–561

Derryberry EP, Seddon N, Derryberry GE, Claramunt S, Seeholzer GF, Brumfield RT, Tobias JA (2018) Ecological drivers of song evolution in birds: disentangling the effects of habitat and morphology. Ecol Evol 8:1890–1905

de Saint-Aubain ML (1982) The morphology of amphibian skin vascularization before and after metamorphosis. Zoomorphology 100:55–63

de Souza RBB, Bonfim VMG, Rios VP, Klein W (2021) Allometric relations of respiratory variables in Amniota: effects of phylogeny, form, and function. Comp Biochem Physiol A 252:110845. https://doi.org/10.1016/jcbpa.2020.110845

Develey PF, Phalan BT (2021) Bird extinction in Brazil's Atlantic forest and how they can be prevented. Front Ecol Evol 9:624587. https://doi.org/10.3389/fevo.2021.624587

Deysel LCF (2017) Animal names and categorization in the Hebrew Bible: a textual and cognitive approach. Doctor of Letters Thesis, University of Pretoria.

Diamond J (1986) Biology of birds of paradise and bowerbirds. Annu Rev Ecol Syst 17:17–37

Dick MF, Guglielmo GG (2019) Dietary polyunsaturated fatty acids influence flight muscle oxidative capacity but not endurance flight performance in a migratory songbird. Am J Physiol Regul Integr Comp Physiol 316:R362–R375

Dietz MW, Piersma T, Hedenström A, Brugge M (2007) Intraspecific variation in avian pectoral muscle mass: constraints on maintaining manoeuvrability with increasing body mass. Funct Ecol 21:317–326

Dilger WC (1960) The comparative ethology of the african parrot genus Agaporni. Zeitschr Tierpsychol 17:649–685

Dingle H (2014) Migration, 2nd edn. Oxford University Press, Oxford (UK)

Dimond CC, Cabin RJ, Brooks JS (2011) Feathers, dinosaurs, and behavioural cues: defining the visual display hypothesis for the adaptive function of feathers in non-avian theropods. Bios 82: 58–63

Dirzo R, Young HS, Galetti M, Ceballos G, Isaac NJB, Collen B (2014) Defaunation in the Anthropocene. Science 345:401–406

Dishon G, Grossowicz M, Krom M, Guy G, Gruber DF, Tchernov D (2020) Evolutionary traits that enable scleractinian corals to survive mass extinction events. Sci Rep 10:3903. https://doi.org/10.1038/s41598-020-60605-2

Dodd GAA, Scott GR, Milsom WK (2007) Ventilatory roll off during sustained hypercapnia is gender specific in pekin ducks. Respir Physiol Neurobiol 156:47–60

Doddamani LS, Vani V, Seshagiri PB (2012) Atropical oviparous lizard, *Calotes versicolor*, exhibiting a potentially novel FMFM pattern of temperature-dependent sex determination. J Exp Zool A Ecol Genet Physiol 317:32–46

Dodson CA, Hore PJ, Wallace MI (2013) A radical sense of direction: signalling and mechanism in cryptochrome magnetoreception. Trends Biochem Sci 38:435–446. https://doi.org/10.1016/j.tibs.2013.07.002

Dodson J, Guanghui D (2016) what do we know about domestication in Eastern Asia? Quaternary Internat 426:2–9

Domet MJ (2015) Voiceless: tuning into natures messages. Balboa Press, Bloomington (IN)

Donoghue PCJ, Keating JN (2014) Early vertebrate evolution. Palaeontology 57:879–893

Dooling RJ (1982) Auditory perception in birds. In: Kroodsma DE, Miller EH (eds) Acoustic communication in birds, Vol. 1, vol 1. Academic Press, New York, pp 95–130

Dorrestein GM, van der Hage MH (1987) Aviculture and veterinary problems in lovebirds. Proc 1st Int Conf Zoo and Avian Med, Oahu (HI), pp. 243–262.

Dos Santos MS, Furo IO, Tagliarini MM, Kretschmer R, O'Brien PCM, Ferguson-Smith MA, de Oliveira EHC (2018) The karyotype of the hoatzin (*Opisthocomus hoazin*)—A phylogenetic enigma of the neornithes. Cytogenet Genome Res 156:158–164

Doupe AJ, Kuhl PK (1999) Birdsong and human speech: common themes and mechanisms. Annu Rev Neurosci 22:567–631

Douxfils J, Mandiki SNM, Mathieu C, Milla S, Kestemont P (2015) Domestication and responses to stress. In: Kestemont P, Dabrowski K, Summerfelt R (eds) Biology and culture of percid fishes. Springer, Dordrecht, pp 743–760

Driedzic WR, Crowe HL, Hicklin PW, Sephton DH (1993) Adaptations in pectoralis muscle, heart mass, and energy metabolism during premigratory fattening in semipalmated sandpipers (*Calidris pusilla*). Can J Zool 71:1602–1608

DU (2006) Duke University: birds going extinct faster due to human activities: a new study looks back and forward in time over a span of centuries. Science Daily 5th July 2006. https://today.duke.edu/2006/07/nobirds.html

Du J, Shi Y, Zhou C, Guo L, Hu R, Huang C, Hu G, Gao X, Guo X (2022) Antioxidative and anti-inflammatory eects of vitamin C on the liver of laying hens under chronic heat stress. Front Vet Sci 9:1052553. https://doi.org/10.3389/fvets.2022.1052553

Dubach M (1981) Quantitative analysis of the respiratory system of the house sparrow, budgerigar, and violet-eared hummingbird. Respir Physiol 46:43–60

Dudley R (2002) The biomechanics of insect flight. Form, function, evolution. Princeton University Press, Princeton (NJ)

Dudley R, Byrnes G, Yanoviak SP, Borrell B, Brown RM, McGuire JA (2012) Gliding and the functional origins of flight: biomechanical novelty or necessity? Ann Rev Ecol Evol Syst 38: 179–201

Dudley R, Yanoviak SP (2011) Animal aloft: the origins of aerial behavior and flight. Integr Comp Biol 51:926–936

Dumont ER (2010) Bone density and the lightweight skeletons of birds. Proc Biol Sci 277:2193–2198

Duncker HR (1971) The lung-air sac system of birds. A contribution to the functional anatomy of the respiratory apparatus. Ergeb Anat Entwicklung 45:1–171

Duncker HR (1978a) General morphological principles of amniotic lungs. In: Piiper J (ed) Respiratory function in birds, adult and embryonic. Springer, Berlin, pp 1–15

Duncker HR (1978b) Functional morphology of the respiratory system and coelomic subdivisions in reptiles, birds and mammals. Vehr Dtsch Zool Ges 1978:99–132

Duncker HR (2004) Vertebrate lungs: structure, topography and mechanics. A comparative perspective of the progressive integration of respiratory system, locomotor apparatus and ontogenetic development. Respir Physiol Neurobiol 144:111–124

Duncker HR, Guntert M (1985a) The quantitative design of the atian respiratory system-from hummingbird to swan. In: Nachtigall W (ed) Bird Flight - BIONA Report No. 3. Fischer Verlag, Stuttgart, pp 361–378

Duncker HR, Guntert M (1985b) Morphometric analysis of the avian respiratory system. In: Duncker HR, Fleischer G (eds) Vertebrate morphology. Gustav Fischer Verlag, Stuttgart, pp 383–387

Dunham-Snary KJ, Wu D, Sykes EA, Thakrar A, Parlow L, Mewburn JD, Parlow JL, Archer SL (2017) Hypoxic pulmonary vasoconstriction: from molecular mechanisms to medicine. Chest 151:181–192

Dunn PM (2006) Aristotle (384-322 BC): philosopher and scientist of ancient Greece. Arch Dis Child Fetal Neonatal Ed 91:F75–F77

Duque FG, Rodriguez-Saltos CA, Wilczynski W (2018) High frequency vocalizatiobs in Andean hummingbirds. Curr Biol 28:R927–R928

Duque FG, Rodriguez-Saltos CA, Umar S, Nasir I, Monteros MF, Wilcyynski W, Carruth LL (2020) High frequency hearing in a hummingbird. Sci Adv 6:eabb9393. https://doi.org/10.1126/sciadv.abb9393

Duriez O, Kato A, Tromp C, Dell'Omo G, Vyssotski AL, Sarrazin F, Ropert-Coudert Y (2014) How cheap is soaring flight in raptors? A preliminary investigation in freely-flying vultures. PLoS One 9:e84887. https://doi.org/10.1371/journal.pone.0084887

Düring DN, Ziegler A, Thompson CK, Ziegler A, Faber C, Müller J, Scharff C, Elemans CP (2013) The songbird syrinx morphome: a three-dimensional, high-resolution, interactive morphological map of the zebra finch vocal organ. BMC Biol 11. https://doi.org/10.1186/1741-7007-11-1

du Toit JT, Hetem RS, Dearing MD (2022) Editorial: mammalian responses to climate change: from organisms to communities. Front Ecol Evol 10:1023587. https://doi.org/10.3389/fevo.2022.1023587

Dutta H (2017) Insights into the impacts of four current environmental problems on flying birds. Energ Ecol Environ 2:329–349

Dvořák R (2016) Aerodynamics of bird flight. EPJ Web of Conferences 114:01001. https://doi.org/10.1051/epconf/20161140100

Dyke G (2009) Birds are sensitive indicators of climate change: they have been for 140 million years. https://www.ucd.ie/t4cms/cc8_13feb09_gd_paper.pdf. Accessed on 15th February 2022.

Dyke G, de Kat R, Palmer C, van der Kindere J, Naish D, Ganapathisubramani B (2013) Aerodynamic performance of the feathered dinosaur Microraptor and the evolution of feathered flight. Nat Commun 4:2489. https://doi.org/10.1038/ncomms3489

Dyke GJ, Kaiser GW (2011) Living dinosaurs: the evolutionary history of modern birds. Wiley-Blackwell, Hoboken (NJ)

Earls KD (2000) Kinematics and mechanics of ground take-off in the starling, *Sturnus vulgaris*, and the quail, *Coturnix coturnix*. J Exp Biol 203:725–739

Eda M, Itahashi Y, Kikuchi H, Sun G, Hsu KH, Gakuhari T, Yoneda M, Jiang L, Yang G, Nakamura S (2022) Multiple lines of evidence of early goose domestication in a 7,000-y-old rice cultivation village in the lower Yangtze River, China. Proc Natl Acad Sci USA 119: e2117064119. https://doi.org/10.1073/pnas.2117064119

Eda M, Lu P, Kikuchi H, Li Z, Li F, Yuan J (2016) Reevaluation of early Holocene chicken domestication in northern China. J Archaeol Sci 67:25–31

Edwards DP, Webster RE, Rowlett RA (2009) Spectacled flowerpecker: a new species new to science discovered in Borneo. BirdingAsia 12:38–41

Egevang C, Stenhuse IJ, Phillips RA, Petersn A, Fox JW, Silk JRD (2010) Tracking of Arctic terns, *Sterna paradise* reveals longest animal migration. Proc Natl Acad Sci USA 107:2078–2081

Egwumah FA, Egwumah PO, Edet D (2017) Paramount roles of wild birds as bioindicators of contamination. Int J Avian Wildlife Biol 2:194–199

Eiby YA, Booth DT (2008) Embryonic thermal tolerance and temperature variation in mounds of the Australian Brush-turkey (*Alectura lathami*). Auk 125:594–599

Eiby YA, Booth DT (2009) The effects of incubation temperature on the morphology and composition of Australian Brush-turkey (*Alectura lathami*) chicks. J Comp Physiol B Biochem, Syst Environ Physiol 179:875–882

Eiby YA, Worthington WJ, Booth DT (2008) Temperature dependent sex-biased embryo mortality in a megapode bird. Proc R Soc (Lond) B 275:2703–2706

Einfeldt AL, Jesson LK, Addison JA (2020) Historical human activities reshape evolutionary trajectories across both native and introduced ranges. Ecol Evol 10:6579–6592

Elemans CP, Spierts IL, Hendriks M, Schipper H, Müller UK, van Leeuwen JL (2006) Syringeal muscles fit the trill in ring doves (*Streptopelia risoria* L.). J Exp Biol 209:965–977

Elhacham E, Ben-Uri L, Grozovski J, Bar-On YM, Milo R (2020) Global human-made mass exceeds all living biomass. Nature 588:442–444

Ellegren H (2013) The evolutionary genomics of birds. Annu Rev Ecol Evol Syst 44:239–259

Ellerby DJ, Askew GN (2007) Modulation of flight muscle power output in budgerigars *Melopsittacus undulatus* and zebra finches *Taeniopygia guttata*: *in vitro* muscle performance. J Exp Biol 210:3780–3788

Ellington CP (1991) Limitations on animal flight performance. J Exp Biol 160:71–91

Emami NK, Jung U, Voy B, Dridi S (2020) Radical response: effects of heat stressinduced oxidative stress on lipid metabolism in the avian liver. Antioxidants 10:35. https://doi.org/10.3390/antiox10010035

Emerson SB, Koehl MAR (1990) The interaction of behavioural and morphological change in the evolution of a novel locomotor type: flying frogs. Evolution 44:1931–1946

Emery NJ (2006) Cognitive ornithology: the evolution of avian intelligence. Philos Trans R Soc B (Lond) 361:23–43

Emery NJ, Clayton NS (2004a) Comparing the complex cognition of birds and primates. In: Rogers LJ, Kaplan G (eds) Comparative vertebrate cognition: are primates superior to nonprimates? Kluwer Academic Press, New York, pp 3–5

Emery NJ, Clayton NS (2004b) The mentality of crows: convergent evolution of intelligence in corvids and apes. Science 306:1903–1907

Emery NJ, Dally JM, Clayton NS (2004) Western scrub-jays (*Aphelocoma californica*) use cognitive strategies to protect their caches from thieving conspecifics. Anim Cogn 7:37–43

Emery NJ, Seed AM, von Bayern AMP, Clayton NS (2007) Cognitive adaptations of social bonding in birds. Philos Trans R Soc (Lond) B 362:489–505

Erdoğan S, Sağsöz H, Paulsen F (2015) Functional anatomy of the syrinx of the chukar partridge (Galliformes: *Alectoris chukar*) as a model for phonation research. Anat Rec 298:602–617

Ergaz Z, Weinstein-Fudim L, Ornoy A (2016) Genetic and non-genetic animal models for autism spectrum disorders (ASD). Reprod Toxicol 64:116–140. https://doi.org/10.1016/j.reprotox.2016.04.024

Erickson GM, Rauhut OWM, Zhou Z, Turner AH, Inouye BD, Hu D, Norell MA (2009) Was dinosaurian physiology inherited by birds? Reconciling slow growth in *Archaeopteryx*. PLoS ONE 4:e7390. https://doi.org/10.1371/journal.pone.0007390

Ericson PGP, Klopfstein S, Irestedt M, Nguyen JMT, Nylander JAA (2014) Dating the diversification of the major lineages of Passeriformes (Aves). BMC Evol Biol 14:8. https://doi.org/10.1186/1471-2148-14-8

Ericson PGP, Irestedt M, Johansson US (2003) Evolution, biogeography, and patterns of diversification in passerine birds. J Avian Biol 34:3–15

Ericsson M, Fallahsharoudi A, Bergquist J, Kushnir MM, Jensen P (2014) Domestication effects on behavioural and hormonal responses to acute stress in chickens. Physiol Behav 133:161–169

Eriksson J, Larson G, Gunnarsson U, Bed'hom B, Tixier-Boichard M, Strömstedt L, Wright D, Jungerius A, Vereijken A, Randi E et al (2008) Identification of the yellow skin gene reveals a hybrid origin of the domestic chicken. PLoS Genet 4:e1000010. https://doi.org/10.1371/journal.pgen.1000010

Erwin N (2004) Measuring the sound pressure level of the song of the screaming piha, *Lipaugus vociferans*: one of the loudest birds in the World? Bioacoustics 14:225–228

Evans PR, Davidson NC, Uttley JD, Evans RD (1992) Premigratory hypertrophy of flight muscles: an ultrastructural study. Ornis Scand 23:238–243

Evans CS, Macedonia JM, Marler P (1993) Effects of apparent size and speed on the response of chickens, *Gallus gallus*, to computer-generated simulations of aerial predators. Anim Behav 46:1–11

Fago A, Jensen FB, Tota B, Feelisch M, Olson KR, Helbo S, Lefevre S, Mancardi D, Palumbo A, Sandvik GK et al (2012) Integrating nitric oxide, nitrite and hydrogen sulfide signaling in the physiological adaptations to hypoxia: a comparative approach. Comp Biochem Physiol A 162:1–6

Fahlman A, Moore MJ, Garcia-Parraga D (2017) Respiratory function and mechanics in pinnipeds and cetaceans. J Exp Biol 220:1761–1773

Fallahsharoudi A (2017) Domestication effects on the stress response in chickens: genetics, physiology, and behavior. PhD Thesis, Linköping University (Sweden).
Fallahsharoudi A, de Kock N, Johnsson M, Ubhayasekera SJ, Bergquist J, Wright D, Jensen P (2015) Domestication effects on stress induced steroid secretion and adrenal gene expression in chickens. Sci Rep 5:15345. https://doi.org/10.1038/srep15345
Fallahsharoudi A, de Kock N, Johnsson M, Bektic J, Ubhayasekera SJKA, Bergquist J, Wright D, Jensen P (2017) Genetic and targeted eqtl mapping reveals strong candidate genes modulating the stress response during chicken domestication. G3: Genes, Genomes, Genetics 7:497–504
Fan S, Elmer KR, Meyer A (2012) Genomics of adaptation and speciation in cichlid fishes: recent advances and analyses in African and Neotropical lineages. Philos Trans R Soc (Lond) B 367: 385–394
Fänge R (1983) Gas exchange in fish swimbladder. Rev Physiol Biochem Pharmacol 97:111–158
FAO (2008) Poultry in the 21st century: Avian influenza and beyond. In: Thieme O and Pilling D (Eds.), Proceedings of the International Poultry Conference (5-7 November 2007). FAO Animal Production and Health Proceedings, No. 9 Rome, Bangkok (Thailand).
Farabaugh SM, Linzenbold A, Dooling RJ (1994) Vocal plasticity in budgerigars (*Melopsittacus undulatus*): evidence for social factors in the learning of contact calls. J Comp Psychol 108:81–92
Faraci FM (1991) Adaptations to hypoxia in birds: how to fly high. Annu Rev Physiol 53:59–70
Faraci FM, Kilgore DL, Fedde MR (1984) Attenuated pulmonary pressor response to hypoxia in bar-headed geese. Am J Physiol 247:R402–R403
Farmer CG (1997) Did lungs and intracardiac shunt evolve to oxygenate the heart in vertebrates? Paleobiology 23:358–372
Farmer CG (1999) Evolution of the vertebrate cardio-pulmonary system: new insights. Annu Rev Physiol 61:573–592
Farmer CG (2010) The provenance of alveolar and parabronchial lungs: insights from paleoecology and the discovery of cardiogenic, unidirectional airflow in the American alligator (*Alligator mississippiensis*). Physiol Biochem Zool 83:561–575
Farmer CG (2015a) Evolution of vertebrate lung morphology and function. FASEB J 29
Farmer CG (2015b) The evolution of unidirectional pulmonary air flow. Physiology 30:260–272
Farmer CG (2015c) Similarity of crocodilian and avian lungs indicates unidirectional flow is ancestral for archosaurs. Integr Comp Biol 55:962–971
Farmer CG (2015d) Unidirectional flow in lizard lungs: a paradigm shift in our understanding of lung evolution in Diapsida. Zoology 118:299–301
Farmer CG (2017) Pulmonary transformations of vertebrates. In: Maina JN (ed) The biology of the avian respiratory system. Springer-Verlag, Berlin, pp 99–112
Farmer CG, Sanders K (2010) Unidirectional airflow in the lungs of alligators. Science 327:338–340
Farries MA (2004) The avian song system in comparative perspective. Ann NY Acad Sci 1016:61–76
Faux C, Field DJ (2017) Distinct developmental pathways underlie independent losses of flight in ratites. Biol Lett 13:20170234. https://doi.org/10.1098/rsbl.2017.0234
Fedak M, Seeherman H (1979) Reappraisal of energetics of locomotion shows identical cost in bipeds and quadrupeds including ostrich and horse. Nature 282:713–716
Fedde MR (1980) The structure and gas flow pattern in the avian lung. Poult Sci 59:2642–2653
Fedde MR, Orr JA, Shams H, Scheid P (1989) Cardiopulmonary function in exercising bar-headed geese during normoxia and hypoxia. Respir Physiol 77:239–252
Feder ME, Burggren WW (1985a) Skin breathing in vertebrates. Sci Am 253:126–142
Feder ME, Burggren WW (1985b) Cutaneous gas exchange in vertebrates: design, patterns, control and implications. Biol Rev 60(1):45
Feduccia A (1996) The origin and evolution of birds. Yale University Press, New Haven
Feduccia A (2001) The problem of bird origins and early avian evolution. J Ornithol 42:139–147
Feduccia A (2003) Big bang for tertiary birds? Trends Ecol Evol 18:172–176

Feduccia A (2016) Fantasy vs reality: a critique of Smith et al.'s bird origins. Open Ornithol J 9:14–38

Feduccia A, Lingham-Solia T, Hinchliffe JR (2005) Do feathered dinosaurs exist? Testing the hypothesis on neotological and paleontological evidence. J Morphol 266:125–166

Felice RN, Randau M, Goswami A (2018) A fly in a tube: macroevolutionary expectations for integrated phenotypes. Evolution 72:2580–2594

Ferguson-Lees J, Christie DA (2001) Raptors of the world. Houghton Mifflin Harcourt, Boston (MA)

Ferguson RS, Williams T (1993) Safety and EMC. In: Fraidoon M (ed) Telecommunications engineer's reference book. Butterworth-Heinemann, New York, pp 1–27

Ferreira AR, Smulders TV, Sameshima K, Mello CV, Jarvis ED (2006) Vocalizations and associated behaviors of the sombre hummingbird (*Aphantochroa cirrhochloris*) and the rufous-breasted hermit (*Glaucis Hirsutus*). Auk 123:1129–1148

Fiaschi V, Farina A, Ioalé P (1974) Homing experiments on swifts *Apus apus* (L.) deprived of olfactory perception. Monit Zool Ital 8:235–244

Fiedler W (2003) Recent changes in migratory behaviour of birds: a compilation of field observations. In: Berthold P, Gwinner E, Sonnenschein E (eds) Avian migration. Springer-Verlag, Berlin, pp 21–38

Field DJ, Bercovici A, Berv JS, Dunn R, Fastovsky DE, Lyson TR, Vajda V, Gauthier JA (2018) Early evolution of modern birds structured by global forest collapse at the End-Cretaceous mass extinction. Curr Biol 28:1825–1831

Fishbein A (2022) What birds really listen for in birdsong—it is not what you think. Sci Am 326: 38–43

Flacourt E (1658) Historie de la grande isle Madagascar. P. L'Amy, Paris

Flögel U, Fago A, Rassaf T (2010) Keeping the heart in balance: the functional interactions of myoglobin with nitrogen oxides. J Exp Biol 213:2726–2733

Foth C, Rauhut OWM (2020) The evolution of feathers: from their origin to the present. Springer, Fascinating Life Sciences Series, Cham (Switzerland), pp. 1–243.

Foth C, Tischlinger H, Rauhut OWM (2014) New specimen of *Archeopteryx* provides insights into the evolution of pennaceous feathers. Nature 511:79–82

Fountaine TMR, Benton MJ, Dyke GJ, Nudds RL (2005) The quality of the fossil record of Mesozoic birds. Proc Biol Sci 272:289–294

Fowler ME, Cubas ZS (2001) Biology, medicine, and surgery of South American wild animals. Iowa State U

Francisco MR, Gibbs HL, Galetti M, Lunardi VO, Junior PMG (2007) Genetic structure in a tropical lek-breeding bird, the blue manakin (*Chiroxiphia caudata*) in the Brazilian Atlantic Forest. Mol Ecol 16:4908–4918

Frisdal A, Trainor PA (2014) Development and evolution of the pharyngeal apparatus. Rev Dev Biol 3:403–416

Frith HJ (1957) Experiments on the control of temperature in the mound of the malee fowl. CISRO Wildlije Res 2:101–110

Frith HJ (1962) The mallee fowl: the bird that builds an incubator. Angus and Robertson, Sydney

Fuller R, Landrigan PJ, Balakrishnan K, Batham G, O'Reilley SB, Brauer M, Caravanos JC, Chiles T, Cohen A, Corra L et al. (2022) Pollution and health: a progress update. www.thelancet.com/planetary-health, published online May 17, 2022. https://doi.org/10.1016/S2542-5196(22)00090-0

Funk EC, Birol EB, McCune AR (2021) Does the bowfin gas bladder represent an intermediate stage during the lung-to-gas bladder evolutionary transition? J Morphol 282:600–611

Funk E, Lencer E, McCune A (2020) Dorsoventral inversion of the air-filled organ (lungs, gas bladder) in vertebrates: RNAsequencing of laser capture microdissected embryonic tissue. J Exp Zool B Mol Dev Evol 334:325–338

Gabora L (2013) Convergent evolution. In: Maloy S, Hughes K (eds) Brenner's encyclopedia of genetics, 2nd edn. Elsevier Inc., New York, pp 178–180

Gahr M (2000) Neural song control system of hummingbirds: comparison to swifts, vocal learning (songbirds) and nonlearning (suboscines) passerines, and vocal learning (budgerigars) and nonlearning (dove, owl, gull, quail, chicken) nonpasserines. J Comp Neurol 426:182–196

Gallien L, Mazel F, Lavergne S, Renaud J, Douzet R, Thuiller W (2015) Contrasting the effects of environment, dispersal and biotic interactions to explain the distribution of invasive plants in alpine communities. Biol Invasions 17:1407–1423

Ganias K, Michou S, Nunes C (2015) A field based study of swimbladder adjustment in a physostomous teleost fish. Peer J 3:e892. https://doi.org/10.7717/peerj.892. eCollection 2015

Garamszegi LZ, Møller AP, Török J, Michl G, Péczely P, Richard M (2004) Immune challenges mediates vocal communication in a passerine bird: an experiment. Behav Ecol 15:148–157

Garfinkel MB, Minor ES, Whelan CJ (2020) Birds suppress pests in corn but release them in soybean crops within a mixed prairie/agriculture system. Condor 122:1–12. https://doi.org/10.1093/condor/duaa009

Gaston KJ, Blackburn TM (1997) How many birds are there? Biodivers Conserv 6:615–625

Gaunt AS, Hikida RS, Jehl JR, Fenbert L (1990) Rapid atrophy and hypertrophy of an avian flight muscle. Auk 107:649–659

Gauthier J, Gall LF (2001) Phylogenetic relationships among coelurosaurian theropods. In: Gauthier J, Gall LF (eds) New perspectives on the origin and early evolution of birds. Yale University Press, New Haven, pp 49–67

George RJ, Plog S, Watson AS, Schmidt KL, Culleton BJ, Harper TK, Gilman RA, LeBlanc SA, Amato G, Whiteley P et al (2018) Archaeogenomic evidence from the southwestern US points to a preHispanic scarlet macaw breeding colony. Proc Natl Acad Sci USA 115:8740–8745

Gifford-Gonzalez D, Hanotte O (2011) Domesticating animals in Africa: implications of genetic and archaeological findings. J World Prehist 24:1–23

Gilbert MTP, Shapiro MD (2014) Pigeons: domestication. In: Smith C (ed) Encyclopedia of global archaeology. Springer, New York (NY), pp 212–276

Gilg O, Yoccoz NC (2010) Explaining bird migration. Science 327:276–277

Gill FB (2007) Ornithology, 3rd edn. WH Freeman, New York

Gill FB, Donsker D, Rasmussen P (Eds.) (2020) Family index: IOC World Bird List (v10.1). https://doi.org/10.14344/IOC.ML.10.1

Gill RE, Piersma T, Hufford G, Servranckx R, Riegen A (2005) Crossing the ultimate ecological barrier: evidence for an 11 000-km-long nonstop flight from Alaska to New Zealand and eastern Australia by bar-tailed godwits. Condor 107:1–20

Gill RE, Piersma T, Hufford G, Servranckx R, Riegen A (2009a) Crossing the ultimate ecological barrier: evidence for a 11,000 km long nonstop flight from Alaska to New Zealand and eastern Australia by bar-tailed godwits. Condor 107:1–20

Gill RE, Tibbitts TL, Douglas DC, Handel CM, Mulcahy DM, Gottschalck JC, Warnock N, McCaffery BJ, Battley PF, Piersma T (2009b) Extreme endurance flights by landbirds crossing the Pacific Ocean: ecological corridor rather than barrier? Proc Biol Sci 276:447–457

Girosi D, Bellodi S, Sabatini F, Rossi GA (2006) The lung and the gut: common origins, close links. Paediatr Resp Rev 7S:S235–S239

Gleiss A, Jorgensen S, Liebsch N, Sala JE, Norman B, Hays GC, Quintana F, Grundy E, Campagna C, Trites AW et al (2011) Convergent evolution in locomotory patterns of flying and swimming animals. Nat Commun 2:352. https://doi.org/10.1038/ncomms1350

Godefroit P, Cau A, Hu DY, Escuillié F, Wu W, Dyke G (2013) A Jurassic avialan dinosaur from China resolves the early phylogenetic history of birds. Nature 498:359–362

Godefroit P, Sinitsa SM, Dhouailly D, Bolotsky YL, Sizov AV, McNamara ME, Benton MJ, Spagna P (2014) A Jurassic ornithischian dinosaur from Siberia with both feathers and scales. Science 345:451–455

Godfrey NW (1961) The functional development of the calf II. Development of rumen function in the calf. J Agric Sci 57:177–183

Godoy-Vitorino F, Golgfarb KC, Karaoz U, Leal S, Garcia-Amado MA, Hugenholtz P, Tringe SH, Brodie EL, Dominguez-Bello MG (2012) Comparative analyses of foregut and hindgut bacterial communities in hoatzins and cows. ISME J 6:531–541

Goldman SA, Nottebohm F (1983) Neuronal production, migration, and differentiation in a vocal control nucleus of the adult female canary brain. Proc Natl Acad Sci USA 80:2390–2394

Goller F, Cooper BG (2004) Peripheral motor dynamics of song production in the zebra finch. Ann NY Acad Sci 1016:130–152

Goller F, Larsen ON (1997) *In situ* biomechanics of syrinx and sound generation in pigeons. J Exp Biol 200:2165–2176

Goller F, Suthers RA (1996) Role of syringeal muscles in gating airflow and sound production in singing brown thrashers. J Neurophysiol 75:867–876

Goodrich ES (1930) Studies on the structure and development of vertebrates. Macmillan, London

Göth A, Booth DT (2005) Temperature-dependent sex ratio in a bird. Biol Lett 1:31–33

Goujet D (2011) Lungs in placoderms, a persistent palaeobiological myth related to environmental preconceived interpretations. Comptes Rendus Palevol 10:323–329

Gould JL, Gould CG (2007) Animal architects: building and the evolution of intelligence. Basic Books, New York

Graham JB (1997) Air-breathing fishes: evolution, diversity and adaptation. Academic Press, San Diego (CA)

Graham A (2001) The development and evolution of the pharyngeal arches. J Anat 199:133–141

Graham JB (2011) Air-breathing fishes: respiratory adaptations for air-breathing fishes. In: Farrell AP (ed) Encyclopedia of fish physiology. Academic Press, San Diego, pp 1861–1874

Graham JB, Kramer DL, Pineda E (1978a) Comparative respiration of an air-breathing and nonair-breathing characoid fish and the evolution of aerial respiration in characins. Physiol Zool 51: 279–288

Graham JB, Lee HJ (2004) Breathing air in air: in what ways might extant amphibious fish biology relate to prevailing concepts about early tetrapods, the evolution of vertebrate air breathing, and the vertebrate land transition? Physiol Biochem Zool 77:720–731

Graham JB, Rosenblatt RH, Gans C (1978b) Vertebrate air breathing arose in fresh waters and not in the ocean. Evolution 32:459–463

Grajal A (1995) Structure and function of the digestive tract of the hoatzin (*Opisthocomus hoazin*): a folivorous bird with foregut fermentation. Auk 112:20–28

Grajal A, Strahl SD, Parra R, Domijnguf MG, Neher A (1989) Foregut fermentation in the hoatzin, a Neotropical avian folivore. Science 245:1131–1134

Gransee HM, Mantilla CB, Sieck GC (2012) Respiratory muscle plasticity. Compr Physiol 2:1441–1462

Green AJ, Elmberg J (2014) Ecosystem services provided by waterbirds. Biol Rev 89:105–122

Greenewalt CH (1962) Dimensional relationships for flying animals. Smithson misc Collns 144:1–46

Greenewalt CH (1990) Hummingbirds. Dover, New York (NY)

Greenwood JJD (2007) Citizens, science and bird conservation. J Ornithol 148:S77–S124

Gregory TR (2002) A bird's-eye view of the C-value enigma: genome size, cell size, and metabolic rate in the Class Aves. Evolution 56:121–130

Gregory TR (2005) Genome size evolution in animals. In: Gregory TR (ed) The evolution of the genome. Elsevier, San Diego (CA), pp 3–87

Gregory TR (2009) Animal genome size database. http://www.genomesize/com. Accessed on 25-04-22

Gregory TR, Andrews CB, McGuire JA, Witt CC (2009) The smallest avian genomes are found in hummingbirds. Proc Biol Sci 276:3753–3757

Griffin DK, Robertson LBW, Tempest HG, Skinner BM (2007) The evolution of the avian genome as revealed by comparative molecular cytogenetics. Cytogenet Genome Res 117:64–77

Griffiths PJ (1994) The claws and digits ofArchaeopteryx lithographica. Geobios 27:101–106

Grimaldi DA, Engel MS (2004) New light shed on the oldest insect. Nature 427:627–630

Grimaldi D, Engel MS (2005) Evolution of the insects. Cambridge University Press, New York (NY)

Groom DJE, Toledo MCB, Powers DR, Tobalske BW, Welch KC (2018) Integrating morphology and kinematics in the scaling of hummingbird hovering metabolic rate and efficiency. Proc R Soc (Lond) B 285:20172011. https://doi.org/10.1098/rspb.2017.2011

Grubb B, Colacino M, Schmidt-Nielsen K (1978) Cerebral blood flow in birds: effect of hypoxia. Am J Physiol 234:H230–H243

Guglielmo CG (2010) Move that fatty acid: fuel selection and transport in migratory birds and bats. Integr Comp Biol 50:336–345

Guglielmo CG (2018) Obese super athletes: fat-fueled migration in birds and bats. J Exp Biol 221: jeb165753. https://doi.org/10.1242/jeb.165753

Habib MB (2019) New perspectives on the origins of the unique vocal tract of birds. PLos Biol 17: e3000184. https://doi.org/10.1371/journal.pbio.3000184

Habib MB, Witton MP (2010) Soaring efficiency and long distance travel in giant pterosaurs. Acta Geosci Sin 31:27–28

Haines AM, Mcgrady MJ, Martell MS, Dayton BJ, Henke MB, Seear WS (2003) Migration routes and wintering locations of Broad-winged Hawks tracked by satellite telemetry. Wilson Bull 115: 166–169

Hake M, Kjellén N, Alerstam T (2001) Satellite tracking of Swedish Ospreys Pandion haliaetus: autumn migration routes and orientation. J Avian Biol 32:47–56

Halfwerk W, Holleman LJM, Lessells CM, Slabbekoorn H (2011) Negative impact of traffic noise on avian reproductive success. J Appl Ecol 48:210–219

Hall BK (2002) Paleobiology and evolutionary developmental biology: a science of the nineteenth and twenty-first centuries. Paleobiology 45:647–669

Halsey LG (2016) Terrestrial movement energetics: current knowledge and its application to the optimising animal. J Exp Biol 219:1424–1431

Halsey LG, White CR (2019) Terrestrial locomotion energy costs vary considerably between species: no evidence that this is explained by rate of leg force production or ecology. Sci Rep 9:656. https://doi.org/10.1038/s41598-018-36565-z

Hansbauer MM, Storch I, Leu S, Nieto-Holguin JP, Pimentel RG, Knauer F, Metzger JP (2008) Movements of neotropical understory passerines affected by anthropogenic forest edges in the Brazilian Atlantic rainforest. Biol Conserv 141:782–791

Hansford JP, Turvey ST (2018) Unexpected diversity within the extinct elephant birds (Aves: Aepyornithidae) and a new identity for the world's largest bird. R Soc Open Sci 5:181295. https://doi.org/10.1098/rsos.181295

Hanson T (2011) Feathers: the evolution of a natural miracle. Basic Books, New York

Hansson LA, Åkesson S (eds) (2014) Animal movement across scales. Oxford University Press, Oxford (UK)

Harris RB, Birks SM, Leache AD (2014) Incubator birds: biogeographical origins and evolution of underground nesting in megapodes (Galliformes: Megapodiidae). J Biogeogr 41:2045–2056

Harrison J (2019) The highs and lows of bird flight: bar-headed geese lower their flight metabolic rates to fly in low-oxygen conditions. eLife 8:e50626. https://doi.org/10.7554/eLife.50626

Harshman J, Braun EL, Braun MJ, Huddleston CJ, Bowie RC, Chojnowski JL, Hackett SJ, Han KL, Kimball RT, Marks BD et al (2008) Phylogenomic evidence for multiple losses of flight in ratite birds. Proc Natl Acad Sci USA 105:13462–13467

Hartfield JH, Orme CDL, Tobias JA, Banks-Leite C (2018) Trait-based indicators of bird species sensitivity to habitat loss are effective within but not across data sets. Ecol Appl 28:28–34

Hartman FA (1954) Cardiac and pectoral muscles of trochlids. Auk 71:467–469

Hartman FA (1955) Heart weight in birds. Condor 57:221–238

Hawkes LA, Balachandaran S, Batgayar N, Butler PJ, Chua B, Douglas DC, Frappell PB, Hou Y, Milsom WK, Newman SH et al (2013) The paradox of extreme high-altitude migration in bar-headed geese, *Anser indicus*. Proc Biol Sci 280:20122114. https://doi.org/10.1098/rspb.2012.2114

Hawkes LA, Balachandran S, Batbayar N, Butler PJ, Frappell PB, Milsom WK, Tseveenmyadag N, Newman SH, Scott GR, Sathiyaselvam P et al (2011) The trans-Himalayan flights of bar-headed geese (*Anser indicus*). Proc Natl Acad Sci USA 108:9516–9519

Hawkes LA, Batbayar N, Butler PJ, Chua B, Frappell PB, Meir JU, Milsom WK, Natsagdorj T, Parr N, Scott GR et al (2017) Do bar-headed geese train for high altitude flights? Integr Comp Biol 57:240–251

Hazlehurst GA, Rayner JMV (1992) Flight characteristics of Triassic and Jurassic Pterosauria: an appraisal based on wing shape. Paleobiology 18:447–463

Hedenström A (2010) Extreme endurance migration: what is the limit to non-stop flight? PLoS Biol 8:e1000362. https://doi.org/10.1371/journal.pbio.1000362

Hedenström A, Alerstam T (1997) Optimum fuel loads in migratory birds: distinguishing between time and energy minimization. J Theor Biol 189:227–234

Hedges SB, Simmons MD, van Dijk MA, Caspers GJ, de Jong WW, Sibley CG (1995) Phylogenetic relationships of the hoatzin, an enigmatic South American bird. Proc Natl Acad Sci USA 92:11662–11665

Hedrick MS, Katz SL (2016) Control of breathing in primitive fishes. In: Zaccone G, Dabrowski K, Hedrick MS, Fernandes JMO, Icardo JM (eds) Phylogeny, anatomy and physiology of ancient fishes. CRC Press (Taylor and Francis Group), Boca Raton, pp 179–200

Hedrick TL, Tobalske BW, Blewener AA (2003) How cockatiels (*Nymphicus hollandicus*) modulate pectoralis power output across flight speeds. J Exp Biol 206:1363–1378

Heenan CB (2013) An overview of the factors influencing the morphology and thermal properties of avian nests. Avian Biol Res 6:104–118

Heers AM, Dial KP (2015) Wings versus legs in the avian bauplan: development and evolution of alternative locomotor strategies. Evolution 69:305–320

Heers AM, Dial KP, Tobalske BW (2014) From baby birds to feathered dinosaurs: incipient wings and the evolution of flight. Paleobiol 40:459–476

Heers AM, Tobalske BW, Dial KP (2011) Ontogeny of lift and drag production in ground birds. J Exp Biol 214:717–725

Heft-Neal S, Burney J, Bendavid E, Burke M (2018) Robust relationship between air quality and infant mortality in Africa. Nature 559:254–258

Hegemann A, Birberg C, Hasselquist D, Nilsson J-Å (2022) Early and late migrating avian individuals differ in constitutive immune function and blood parasite infections—but patterns depend on the migratory strategy. Front Ecol Evol 10:880426. https://doi.org/10.3389/fevo.2022.880426

Hegemann A, Fudickar AM, Nilsson JÅ (2019) A physiological perspective on the ecology and evolution of partial migration. J Ornithol 160:893–905

Hertegard S, Gauffin J, Lindestad PA (1995) A comparison of subglottal and intraoral pressure measurements during phonation. J Voice 9:149–155

Hicks JW, Farmer CG (1999) Gas exchange potential in reptilian lungs: implications for the dinosaur—avian connection. Respir Physiol 117:73–83

Hill E (2019) Humans, birds and burial practices at Ipiutak, Alaska: perspectivism in the Western Arctic. Environ Archaeol 24:434–448

Hill SD, Brunton DH, Anderson MA, Weihong J (2018) Fighting talk: complex song elicits more aggressive responses in a vocally complex songbird. Ibis 160:257–268

Hiltpold I, Shriver WG (2018) Birds bug on indirect plant defenses to locate insect prey. J Chem Ecol 44:576–579

Hirst KK (2019) The domestication history of chickens (*Gallus domesticus*): who gets the credit for taming the wild jungle fowl? https://www.thoughtco.com/the-domestication-history-of-chickens-170653. Accessed on 04-07-2022

Hobson EH, Avery ML, Wright TF (2014) The socioecology of monk parakeets: insights into parrot social complexity. Auk 131:756–775

Hölker F, Wolter C, Perkin EK, Tockner K (2010) Light pollution as a biodiversity threat. Trends Ecol Evol 25:681–682

Hoffman M, Taylor BE, Harris MB (2016) Evolution of lung breathing from a lungless primitive vertebrate. Respir Physiol Neurobiol 224:11–16

Holland N, Chen J (2001) Origin and early evolution of vertebrates: new insights from advances in molecular biology, anatomy, and paleontology. BioEssays 23:142–151

Holland RA (2010) Differential effects of magnetic pulses on the orientation of naturally migrating birds. J R Soc Interface 7:1617–1625

Holland RA, Helm B (2013) A strong magnetic pulse affects the precision of departure direction of naturally migrating adult but not juvenile birds. J R Soc Interface 10:2012104720121047. https://doi.org/10.1098/rsif.2012.1047

Holland RA, Thorup K, Gagliardo A, Bisson IA, Knecht E, Mizrahi D, Wikelski M (2009) Testing the role of sensory systems in the migratory heading of a songbird. J Exp Biol 212:4065–4071

Holmes MW, Hammond TT, Wogan GOU, Walsh RE, LaBarbera K, Wommack EA, Martins FM, Crawford JC, Mark KL, Bloch LM, Nachman MW (2016) Natural history collections as windows on evolutionary processes. Mol Ecol 25:864–881

Holt N, Roberts TJ, Askew G (2014) The energetic benefits of tendon springs in running: is the reduction of muscle work important? J Exp Biol 217:4365–4371

Homberger DG (1999) The avian tongue and larynx: multiple functions in nutrition and vocalisation. In: Adams NJ, Slotow RH (eds) Proc 22nd Int Ornithol Congr. Durban, South Africa, pp 94–113

Honka J, Heino MT, Kvist L, Askeyev IV, Shaymuratova DN, Askeyev OV, Askeyev AO, Heikkinen ME, Searle JB, Aspi J (2018) Over a thousand years of evolutionary history of domestic geese from Russian archaeological sites, analysed using ancient DNA. Genes 9:367. https://doi.org/10.3390/genes9070367

Hoover JP, Schelsky WM (2020) Warmer April temperatures on breeding grounds promote earlier nesting in a long-distance migratory bird, the prothonotary warbler. Front Ecol Evol 8:580725. https://doi.org/10.3389/fevo.2020.580725

Hore PJ, Mouritsen H (2016) The radical pair mechanism of magnetoreception. Annu Rev Biophys 45:299–344

Hore PJ, Mouritsen H (2022) The quantum nature of bird migration. Sci Am 4(April):27–31

Horns JJ, Şekercioğlu ÇH (2018) Conservation of migratory species. Curr Biol 28:R980–R983

Horiike T (2016) An introduction to molecular phylogenetic analysis. Rev Agric Sci 4:36–45

Horton KG, La Sorte FA, Sheldon D, Lin TY, Winner K, Bernstein G, Maji S, Hochachka WM, Farnsworth A (2020) Phenology of nocturnal avian migration has shifted at the continental scale. Nat Clim Change 10:63–68

Hou L, Martin M, Zhou Z, Feduccia A (1996) Early adaptive radiation of birds: evidence from fossils from Northeastern China. Science 274:1164–1167

Hou LH, Zhou Z, Martin LD, Feduccia A (1995) A beaked bird from the Jurassic of China. Nature 877:616–618

Hsia CC, Schmitz A, Lambertz M, Perry SF, Maina JN (2013) Evolution of air breathing: oxygen homeostasis and the transitions from water to land and sky. Compr Physiol 3:849–915

Hughes AL, Friedman R (2008) Genome size reduction in the chicken has involved massive loss of ancestral protein-coding genes. Mol Biol Evol 25:2681–2688

Hughes JM, Baker AJ (1999) Phylogenetic relationships of the enigmatic hoatzin (*Opisthocomus hoazin*) resolved using mitochondrial and nuclear gene sequences. Mol Biol Evol 16:1300–1307

Hunt RG, Rutledge RB, Gray RD (2006) The right tool for the job: what strategies do wild New Caledonian crows use? Anim Cogn 9:307–316

Hunt V (2006) Art informing science education: the potential contributions of ornithological illustration to ecology education. Doctoral Dissertation, Louisiana State University and Agricultural and Mechanical College.

Hunter P (2006) Defining species: the indirect impact of human on biodiversity. EMBO Rep 7:763–766

Hunter P (2007) The nature of flight: the molecules and mechanics of flight in animals. EMBO Rep 8:811–813

Hutchinson JR, Allen V (2009) The evolutionary continuum of limb function from early theropods to birds. Naturwissenschaften 96:423–448

Icardo JM (2018) Lungs and gas bladders: morphological insights. Acta Histochem 120:605–612

ICZN (1999) International code of zoological nomenclature, 4th edn. International Trust for Zoological Nomenclature, The Natural History Museum, Cromwell Road (London). ISBN 0 85301 006 4

Imbrogno S, Tota B, Gattuso A (2011) The evolutionary functions of cardiac NOS/NO in vertebrates tracked by fish and amphibian paradigms. Nitric Oxide 25:1–10

Ingersoll E (1923) Birds in legend fable and folklore. Green and Co, Longmans

Ingersoll R, Haizman L, Lentink D (2018) Biomechanics of hover performance in Neotropical hummingbirds versus bats. Sci Adv 4(9):eaat2980. https://doi.org/10.1126/sciadv.aat2980

Iniesto M, Buscalioni AD, Guerrero MC, Benzerara K, Moreira D, López-Achilla AI (2016) Involvement of microbial mats in early fossilization by decay delay and formation of impressions and replicas of vertebrates and invertebrates. Sci Rep 6:25716. https://doi.org/10.1038/srep25716

Isack HA, Reyer HU (1989) Honeyguides and honey gatherers: interspecific communication in a symbiotic relationship. Science 243:1343–1346

Isaksson C (2018) Impact of urbanization on birds. In: Tietze D (ed) Bird species. Fascinating Life Sciences Series. Springer, Cham (Switzerland), pp 235–257

Isaksson C (2020) Urban ecophysiology: beyond costs, stress and biomarkers. J Exp Biol 223: jeb203794. https://doi.org/10.1242/jeb.203794

IUCN 2020 Red list of threatened species. Version 2020-2. https://iucnredlist.org. ISSN 2307-8235 - Accessed on 24-04-22

Ivy CM, Lague SL, York JM, Chua BA, Alza L, Cheek R, Dawson NJ, Frappell PB, McCracken KG, Milsom WK, Scott GR (2019) Control of breathing and respiratory gas exchange in high-altitude ducks native to the Andes. J Exp Biol 222:jeb198622. https://doi.org/10.1242/jeb.198622

Iwaniuk AN, Lefebvre L, Wylie DR (2009) The comparative approach and brain-behaviour relationships: a tool for understanding tool use. Can J Exp Psychol 63:150–159

Jablonski D (2017a) Approaches to macroevolution: 1. General concepts and origin of variation. Evol Biol 44:427–450

Jablonski D (2017b) Approaches to macroevolution: 2. Sorting of variation, some overarching issues, and general conclusions. Evol Biol 44:451–475

Jakab PI, Young R (eds) (2000) Published writings of Wilbur- and Orville Wright. Smithonian Institution, Philadelphia

Jackson CE (1997) Fishing with cormorants. Arch Nat Hist 24:189–211

Jackson BE (2009) The allometry of bird flight performance. PhD Thesis, University of Montana, Missoula.

Jankowski R (2022) The evo-devo origins of the nasopharynx. Anat Rec 305:1857–1870

Janvier P, Desbiens S, Willett JA (2007) New evidence for the controversial "lungs" of the late Devonian Antiarch *Bothriolepis canadensis* (Whiteaves, 1880) (Placodermi: Antiarcha). J Vertebrate Paleontol 27:709–710

Jarvis ED (2004a) Brains and birdsong. In: Marler P, Slabbekoorn H (eds) Nature's music: the science of bird song. Elsevier/Academic Press, San Diego, pp 226–271

Jarvis ED (2004b) Learned birdsong and the neurobiology of human language. Ann NY Acad Sci 1016:749–777

Jarvis ED, Mello CV (2000) Molecular mapping of brain areas involved in parrot vocal communication. J Comp Neurol 419:1–31

Jarvis ED, Mirarab S, Aberer AJ, Li B, Houde P, Li C, Ho SYW, Faircloth BC, Nabholtz B, Howard JT, Suh A et al (2014) Whole genome analyses resolve early branches in the tree of life of modern birds. Science 346:1320–1331

Jaspers VLB (2015) Selecting the right bird model in experimental studies on endocrine disrupting chemicals. Front Environ Sci 3:35. https://doi.org/10.3389/fenvs.2015.00035

Jelbert SA, Miller R, Schiest M, Boeckle M, Cheke LG, Gray RD, Taylor AH, Clayton NS (2019) New Caledonian crows infer the weight of objects from observing their movements in a breeze. Proc Biol Sci 286:20182332. https://doi.org/10.1098/rspb.2018.2332

Jendroszek A, Malte H, Overgaard CB, Beedholm K, Natarajan C, Weber ER, Storz JF, FagoJo A (2018) Allosteric mechanisms underlying the adaptive increase in hemoglobin-oxygen affinity of the bar-headed goose. J Exp Biol 221:jeb185470. https://doi.org/10.1242/jeb.185470

Jenni L, Jenni-Eiermann (2000) Fuel supply and metabolic constraints in migrating birds. Comp Biochem Physiol B 126. https://doi.org/10.1016/S0305-0491(00)80104-2

Jenni-Eiermann S, Srygley RB (2017) Physiological Aeroecology: anatomical and physiological adaptations for flight. In: Chilson P, Frick W, Kelly J, Liechti F (eds) Aeroecology. Springer, Cham, pp 87–118

Jensen JK, Andersson N, Isaksson C, Eikenaar C, Andersson MI (2020) Migrant blackbirds, *Turdus merula*, have higher plasma levels of polyunsaturated fatty acids compared to residents, but not enhanced fatty acid unsaturation index. Ecol Evol 2020:10196–10206

Jensen P (2014) Behavior genetics and the domestication of animals. Annu Rev Anim Biosci 2:85–104

Jessel HR, Aharoni L, Efroni S, Bachelet I (2019) A modeling algorithm for exploring the architecture and construction of bird nests. Sci Rep 9:14772. https://doi.org/10.1101/600718

Jessen TH, Weber RE, Fermi G, Tame J, Braunitzer G (1991) Adaptation of bird hemoglobins to high altitudes: demonstration of molecular mechanism by protein engineering. Proc Natl Acad Sci USA 88:6519–6522

Jetz W, Tertitski G, Kays R, Mueller U, Wikelski M (2022) Biological Earth observation with animal sensors. Trends Ecol Evol 37:293–298

Jetz W, Thomas GH, Hartmann JJB, Mooers AA (2012) The global diversity of birds in space and time. Nature 491:444–448

Jetz W, Wilcove D, Dobson A (2007) Projected impacts of climate and land-use change on the global diversity of birds. PLoS Biol 5(6):e157. https://doi.org/10.1371/journal.pbio.0050157

Jiang B, Zhao T, Regnault S, Edward NP, Kohn SC, Li Z, Wogelius RA, Benton MJ, Hutchison JR (2017) Cellular preservation of musculoskeletal specializations in the Cretaceous bird *Confuciusornis*. Nat Commun 8:14779. https://doi.org/10.1038/ncomms14779

Jimenez AG (2020) Structural plasticity of the avian pectoralis: a case for geometry and the forgotten organelle. J Exp Biol 223:jeb234120. https://doi.org/10.1242/jeb.234120

Jimenez AG, Dillaman RM, Kinsey ST (2013) Large fibre size in skeletal muscle is metabolically advantageous. Nat Commun 4:2150. https://doi.org/10.1038/ncomms3150

Jimenez AG, O'Connor ES, Brown KJ, Briggs CW (2019) Seasonal muscle ultrastructure plasticity and resistance of muscle structural changes during temperature increases in resident black-capped chickadees and rock pigeons. J Exp Biol 222:jeb201855. https://doi.org/10.1242/jeb.201855

Jobling JA (2010) The helm dictionary of scientific bird names. Christopher Helm, London

Johanson Z, Long JA, Talent JA, Janvier P, Warren JW (2006) Oldest coelocanths, from the early Devonian of Australia. Biol Lett 2:443–446

Johanson Z, Trinajstic K (2014) Fossilized ontogenies: the contribution of placoderm ontogeny to our understanding of the evolution of early gnathostomes. Palaeontology 57:505–516

Johnston DW, Haines TP (1957) Analysis of mass bird mortality in October, 1954. Auk 74:447–458

Jones D, Dekker R, Roselaar C (1995) The megapodes. Oxford University Press Inc, New York

Jonzén N, Lindén A, Ergon T, Knudsen E, Vik JO, Rubolini D, Piacentinic D, Brinch C, Spina F, Karlsson L et al (2006) Rapid advance of spring arrival dates in long distance migratory birds. Science 312:1959–1961

Joss JM (2006) Lungfish evolution and development. Gen Comp Endocrinol 148:285–289

Jouventin P, Weimerskirch H (1990) Satellite tracking of wandering albatrosses. Nature 343:746–748

Jürgens KD, Gros G (2002) Phylogeny of gas exchange systems. Anasthesiol Intensivmed Notfallmed Schmerzther 37:185–198

Kabak M, Orhan O, Haziroglu RM (2007) The gross anatomy of larynx, trachea and syrinx in the long-legged buzzard (*Buteo rufinus*). Anat Histol Embryol 36:27–32

Kacelnik A, Chappell J, Weir AAS, Kenward B (2006) Cognitive adaptations for tool-related behaviour in New Caledonian crows. In: Wasserman EA, Zentall TR (eds) Comparative cognition: experimental explorations of animal intelligence. Oxford University Press, Oxford, pp 515–528

Kaiser S, Hennessy MB, Sachser N (2015) Domestication affects the structure, development and stability of biobehavioural profiles. Front Zool 12:S19. https://doi.org/10.1186/1742-9994-12-S1-S19

Kanginakudru S, Metta M, Jakati RD, Nagaraju J (2008) Genetic evidence from Indian red jungle fowl corroborates multiple domestication of modern day chicken. BMC Evol Biol 8:174. https://doi.org/10.1186/1471-2148-8-174

Kapusta A, Suh A, Feschotte C (2017) Dynamics of genome size evolution in birds and mammals. Proc Natl Acad Sci USA 114:E1460–E1469

Karasov WH, Douglas AE (2013) Comparative digestive physiology. Compr Physiol 3:741–783

Kardong KV (2014) Vertebrates: comparative anatomy, function, evolution, 7th edn. McGraw Hill Inc, New York

Karten HJ (1991) Homology and evolutionary origins of the 'neocortex'. Brain Behav Evol 38:264–272

Karten HJ (2015) Vertebrate brains and evolutionary connectomics: on the origins of the mammalian 'neocortex'. Philos Trans R Soc (Lond) B Biol Sci 370:20150060. https://doi.org/10.1098/rstb.2015.0060

Karwinkel T, Winklhofer M, Janner LE, Brust V, Hüppop O, Bairlein F, Schmaljohann H (2022) A magnetic pulse does not affect free-flight navigation behaviour of a medium-distance songbird migrant in spring. J Exp Biol 225:jeb244473. https://doi.org/10.1242/jeb.244473

Kattnig DR, Solov'yov IA, Hore PJ (2016) Electron spin relaxation in cytochrome-based magnetoreception. Phys Chem Chem Phys 18:12443–12456

Kawakami K, Eda M (2018) Dinosaurs as ancestors of birds, and birds as descendants of dinosaurs. Jap J Ornithol 67:7–23

Kennedy JR (2018) Changes in Earth's gravity reveal changes in groundwater storage: U.-S. Geological Survey Fact Sheet 2018-3032, 4 p. Doi: https://doi.org/10.3133/fs20183032

Kerje S, Carlborg O, Jacobsson L, Schütz K, Hartmann C, Jensen P, Andersson L (2003) The twofold difference in adult size between the red junglefowl and White Leghorn chickens is largely explained by a limited number of QTLs. Anim Genet 34:264–274

Kerr J, Currie DJ (2002) Effects of human activity on global extinction risk. Conserv Biol 9:1528–1538

Khan J, Ketzel M, Kakosimos K, Sørensen M, Jensen SS (2018) Road traffic air and noise pollution exposure assessment—A review of tools and techniques. Sci Total Environ 634:661–676

Khan M, Bordes S, Murray I, Sharma S (2022) Physiology, Pulmonary vasoconstriction. [Updated 2021 Apr 25]. In: StatPearls [Internet]. Treasure Island (FL): StatPearls Publishing; 2022 Jan-. Available from: https://www.ncbi.nlm.nih.gov/books/NBK499962/. Accessed on 24-04-22

King AS (1989) Functional anatomy of the syrinx. In: King AS, McLelland J (eds) Form and function in birds, vol 1. Academic Press, London, pp 105–192

King AS, McLelland J (1984) Birds: their structure and function. Baillière Tindall, Eat Sussex (England)

Kingsley EP, Eliason CM, Riede T, Li Z, Hiscock TW, Farnsworth M, Thompson SL, Goller F, Tabin CJ, Clark JA (2018) Identity and novelty in the avian syrinx. Proc Natl Acad Sci USA 115:10209–10217

Kirschvink JL, Gould JL (1981) Biogenetic magnetite as a basis for magnetic field detection in animals. BioSystems 13:181–201

Klaassen M (1996) Metabolic constraints on long-distance migration in birds. J Exp Biol 199:57–64

Klaassen M, Biebach H (1994) Energetics of fattening and starvation in the long-distance migratory garden warbler, *Sylvia borin*, during the migratory phase. J Comp Physiol B 164:362–371

Klaassen RHG, Alerstam T, Carlsson P, Fox JW, Lindström Å (2011) Great flights by great snipes: long and fast non-stop migration over benign habitats. Biol Lett 7:833–835

Kooyman GL, Kooyman TG (1995) Diving behavior of emperor penguins nurturingchicks at Coulman Island, Antarctica. Condor 97:536–549

Köppen U, Yakovlev A, Barth R, Kaatz M, Berthold P (2010) Seasonal migrations of four individual bar-headed geese *Anser indicus* from Kyrgyzstan followed by satellite telemetry. J Ornithol 151:703–712

Körner C (2007) The use of 'altitude' in ecological research. Trends Ecol Evol 22:569–574

Korner P, von Maravic I, Haupt H (2022) Birds and the 'Post Tower' in Bonn: a case study of light pollution. J Ornithol 163:827–841

Kost C (2019) Chasing the Halcyon Light—human-kingfisher relations in Eastern HanDynasty China (CE 25-220) and their material, sociocultural and ecological articulations. Environ Archaeol 4:411–433

Kost C, Hussain ST (2019) Archeo-ornithology: towards archaeology of human-bird interfaces. Environ Archeol 24:337–358

Krause DW, Sampson SD, Carrano MT, O'Connor PM (2007) Overview of the history of discovery, taxonomy, phylogeny, and biogeography of *Majungasaurus crenatissimus* (Theropoda: Abelisauridae) from the Late Cretaceous of Madagascar. In: Sampson SD, Krause DW (eds) *Majungasaurus crenatissimus* from the Late Cretaceous of Madagascar. Society of Vertebrate Paleontology Memoir. 8, pp 1–20

Ksepka DT (2014) Flight performance of the largest volant bird. Proc. Natl Acad Sci USA 111: 10624–10629

Ksepka DT, Stidham TA, Williamson TE (2017) Early Paleocene landbird supports rapid phylogenetic and morphological diversification of crown birds after the K-Pg mass extinction. Proc Natl Acad Sci USA 114:8047–8052

Kunchala SR, van Dijk A, Veldhuizen EJA, Donnellan SC, Haagsman HP, Orgeig S (2022) Avian surfactant protein (SP)-A2 first arose in an early tetrapod before the divergence of amphibians and gradually lost the collagen domain. Dev Comp Immunol 104582. https://doi.org/10.1016/j.dci.2011.104582

Kundrát M (2004) When did theropods become feathered?—evidence for pre-archaeopteryx feathery appendages. J Exp Zool: Mol Dev Evol 302B:355–364

Kurochkin EN (2006) Parallel evolution of theropod dinosaurs and birds. Entomol Rev 86:S45–S58

Kyselý R (2010) Review of the oldest evidence of domestic fowl *Gallus gallus* f. *domestica* from the Czech Republic in its European context. Acta Zool Cracov 53A:9–34

Laguë SL (1985) High-altitude champions: birds that live and migrate at altitude. J Appl Physiol 123:942–950

Laguë SL (2017) High-altitude champions: birds that live and migrate at altitude. J Appl Physiol 123:942–950. https://doi.org/10.1152/japplphysiol.00110.2017

Laguë SL, Chua B, Farrell AP, Wang Y, Milsom WK (2016) Altitude matters: differences in cardiovascular and respiratory responses to hypoxia in bar-headed geese reared at high and low altitudes. J Exp Biol 219:1974–1984

Laguë SL, Chua B, Alza L, Scott GR, Frappell PB, Zhong Y, Farrell AP, McCracken KG, Wang Y, Milsom WK (2017) Divergent respiratory and cardiocascular responses to hypoxia in bar-headed geese and Andean birds. J Exp Biol 220:4186–4194

Laguë SL, Ivy CM, York JM, Chua BA, Alza L, Cheek R, Dawson NJ, Frappell PB, Farrell AP, McCracken KG, Scott GR, Milsom WK (2020) Cardiovascular responses to progressive hypoxia in ducks native to high altitude in the Andes. J Exp Biol 223:jeb211250. https://doi.org/10.1242/jeb.211250

Lake-Thom B (1997) Spirits of the Earth: a guide to Native American nature symbols, stories, and ceremonies. Penguin Books (USA), Plume

Lambert ML, Schiestl M, Schwing R, Taylor AH, Gajdon GK, Slocombe KE, Seed AM (2017) Function and flexibility of object exploration in kea and New Caledonian crows. R Soc Open Sci 4:170652

Lambertz M (2016) Recent advances on the functional and evolutionary morphology of the amniote respiratory apparatus. Ann NY Acad Sci 1365:100–113

Lambertz M (2017) The vestigial lung of the coelacanth and its implications for understanding pulmonary diversity among vertebrates: new perspectives and open questions. R Soc Open Sci 4:171518. https://doi.org/10.1098/rsos.171518. eCollection

Lambertz M, Bertozzo F, Sander PM (2017) Data from: bone histological correlates for air sacs and their implications for understanding the origin of the dinosaurian respiratory system. Dryad Digital Repository. https://doi.org/10.5061/dryad.h103j

Lambertz M, Grommes K, Kohlsdorf T, Perry SF (2015) Lungs of amniotes: why simple if they can be complex? Biol Lett 11:20140848. https://doi.org/10.1098/rsbl.2014.0848

Larsen ON, Goller F (2002) Direct observation of syringeal muscle function in songbirds and a parrot. J Exp Biol 205:25–35

Larson J, Yorinks A (2011) Hummingbirds: facts and folklore from the Americas. Charlesbridge, Texas (USA)

Lasiewski RC (1963) Oxygen consumption of torpid, resting, active and flying hummingbirds. Physiol Zool 36:122–140

Lasiewski R, Galey F, Vasquez C (1965) Morphology and physiology of the pectoral muscles of humming-birds. Nature 206:404–405

Lauder GV, Liem KF (1983) The evolution and interrelationships of actinopterygian fishes. Bull Mus Comp Zool 150:95–197

Lautz LS, Nebbia C, Hoeks S, Oldenkamp R, Hendriks AJ, Ragas AMJ, Dome JLCM (2020) An open source physiologically based kinetic model for the chicken (*Gallus gallus domesticus*): calibration and validation for the prediction residues in tissues and eggs. Environ Int 136: 105488. https://doi.org/10.1016/j.envint.2020.105488

Lawal RA (2017) Signatures of selection and introgression in the genus *Gallus*. PhD Thesis, University of Nottingham, Nottingham.

Lawal RA, Al-Atiyat RM, Aljumaah RS, Silva P, Mwacharo JM, Hanotte O (2018) Whole-genome resequencing of red junglefowl and indigenous village chicken reveal new insights on the genome dynamics of the species. Front Genet 9:264. https://doi.org/10.3389/fgene.2018.00264

Lawal RA, Martin SH, Vanmechelen K, Vereijken A, Silva P, Al-Atiyat RM, Aljumaah RS, Mwachanro JM, Wu DD, Zhang YP et al (2020) The wild species genome ancestry of domestic chickens. BMC Biol 18:13. https://doi.org/10.1186/s12915-020-0738-1

Lawrence B (2002) Backed tools in Middle Pleistocene central Africa and their evolutionary significance. J Hum Evol 43:585–603

Lawson DA (1975) Pterosaur from the Latest Cretaceous of West Texas: discovery of the largest flying creature. Science 187:947–948

Laybourne RC (1974) Collision between a vulture and an aircraft at an altitude of 37,000 ft. Wilson Bull 86:461–462

Leakey R, Lewin R (1996) The sixth extinction: patterns of life and the future of humankind. Phoenix, London

Lee CJ (2011) An investigation of dynamic soaring and its applications to the albatross and RC sailplanes. Aerospace Engineering, California Polytechnic State University, California (CA).

Lee M, Viegas I, Norte AC, Ramos JA, Araúujo PM (2022a) Assessing the fatty acid profile of migratory birds with different fuelling strategies. Ibis 2022. https://doi.org/10.1111/ibi.13114

Lee MB, Chen D, Zou F (2022b) Winter bird diversity and abundance in small farmlands in a megacity of Southern China. Front Ecol Evol 10:859199. https://doi.org/10.3389/fevo.2022.859199

Lee MSY, Cau A, Naish D, Dyke GJ (2014) Sustained miniaturization and anatomical innovation in dinosaurian ancestors of birds. Science 345:562–566

Lee MSY, Worthy TH (2011) Likelihood reinstates *Archaeopteryx* as a primitive bird. Biol Lett 12: 1–6

Lefebvre L, Nicolakakis N, Boire D (2002) Tools and brains in birds. Behaviour 139:939–973

Lefebvre L, Reader SM, Sol D (2004) Brains, innovations and evolution in birds and primates. Brain Behav Evol 63:233–246

Leith DE (1983) Comparative mammalian respiratory mechanics. Am Rev Respir Dis 128:S77–S82

Lelieveld J, Evans JS, Fnais M, Giannadaki D, Pozzer A (2015) The contribution of outdoor air pollution sources to premature mortality on a global scale. Nature 525:367–371

Lemoine N, Böhning-Gaese K (2003) Potential impact of global change on species richness of long-distance migrants. Conserv Biol 17:577–586

Lengagne T, Lauga J, Aubin T (2001) Intra-syllabic acoustic signatures used by the king penguin in parent -chick recognition: an experimental approach. J Exp Biol 204:663–672

León-Portilla M (2017) La filosofía náhuatl estudiada en sus fuenteso. Instituto de Investigaciones Históricas Disponible, Universidad Nacional Autónoma de México, Serie Cultura Náhuatl: Monografías / 10.

Lewis ZR (2016) Causes and consequences of lung loss in salamanders. Doctoral dissertation, Harvard University, Graduate School of Arts and Sciences. http://nrs.harvard.edu/urn-3:HUL.InstRepos:26718748.

Lewis ZR, Hanken J (2017) Convergent evolutionary reduction of atrial septation in lungless salamanders. J Anat 230:16–29

Lewis ZR, Kerney R, Hanken J (2011) Lung development in lungless salamanders. Dev Biol 356: 250–251

Liedvogel M, Maeda K, Henbest K, Schleicher E, Simon T, Timmel CR, Hore PJ, Mouritsen H (2007) Chemical magnetoreception: bird cryptochrome 1a is excited by blue light and forms long-lived radical-pairs. PLoS ONE 2:e1106. https://doi.org/10.1371/journal.pone.0001106

Liem KF (1987a) Functional design of the air ventilation apparatus and overland excursions by teleosts. Fieldiana (NS) 37:1–29

Liem KF (1987b) Form and function of lungs: the evolution of air-breathing mechanisms. Am ZooI 28:739–759

Liem KF (1988) Form and function of lungs: the evolution of air breathing mechanisms. Am Zool 28:739–759

Liem KF (1991) Towards a new morphology: pluralism in research and education. Am Zool 31: 759–767

Lima SL (2009) Predators and the breeding bird: behavioral and reproductive flexibility under the risk of predation. Biol Rev 84:485–513

Lindstrom A, Kvist A, Piersma T, Dekinga A, Dietz MW (2000) Avian pectoral muscle size rapidly tracks body mass changes during flight, fasting and fueling. J Exp Biol 203:913–919

Little C (1990) The terrestrial invasion: an ecophysiological approach to the origins of land animals. Cambridge University Press, Cambridge

Liu Y, Qiu X, Zhang X, Yu TX (2015) Response of woodpecker's head during pecking process simulated by material point method. PLoS ONE 10:e0122677. https://doi.org/10.1371/journal.pone.0122677

Liu YP, Wu GS, Yao YG, Miao YW, Luikart G, Baig M, Beja-Pereira A, Ding ZL, Palanichamy MG, Zhang YP (2006) Multiple maternal origins of chickens: out of the Asian jungles. Mol Phylogenet Evol 38:12–19

Livezey BC, Zusi RL (2007) Higher-order phylogeny of modern birds (Theropoda, Aves: Neornithes) based on comparative anatomy. Zool J Linn Soc 149:1–95

Lockner FR, Murrish DE (1975) Interclavicular air sac pressures and vocalization in mallard ducks *Anas platyrhynchos*. Comp Biochem Physiol A: Physiol 52:183–187

Long JA, Gordon MS (2004) The greatest step in vertebrate history: a paleobiological review of the fish-tetrapod transition. Physiol Biocheml Zool 77:700–719

Longo S, Riccio M, McCune AR (2013) Homology of lungs and gas bladders: insights from arterial vasculature. J Morphol 274:687–703

Longrich NR, Tokaryk T, Field DJ (2011) Mass extinction of birds at the Cretaceous-Paleogene (K-Pg) boundary. Proc Natl Acad Sci (USA) 108:15253–15257

Loonstra AHJ, Verhoeven MA, Senner NR, Both C, Piersma (2019) Adverse wind conditions during northward Sahara crossings increase the in-flight mortality of black-tailed godwits. Ecol Lett 22:2060–2066

Losos JB (2011) Convergence, adaptation, and constraint. Evolution 65:1827–1840

Loss SR, Marra PP (2017) Population impacts of free-ranging domestic cats on mainland vertebrates. Front Ecol 15:502–509

Loss SR, Will T, Loss SS, Marra PP (2014) Bird-building collisions in the United States: estimates of annual mortality and species vulnerability. Condor 116:8–23

Loss SR, Will T, Marra P (2013) The impact of free-ranging domestic cats on wildlife of the United States. Nat Commun 4:1396. https://doi.org/10.1038/ncomms2380

Louchart A, Viriot L (2011) From snout to beak: the loss of teeth in birds. Trends Ecol Evol 26: 663–673

Lovell PV, Wirthlin M, Wilhelm L, Minx P, Lazar NH, Carbone L, Warren WC, Mello CV (2014) Conserved syntenic clusters of protein coding genes are missing in birds. Genome Biol 15(12): 565. https://digitalcommons.wustl.edu/open_access_pubs/3702

Lowry H, Lill A, Wong BBM (2013) Behavioural responses of wildlife to urban environments. Biol Rev 88:537–549

Lu D (2020) Flying snakes wiggle their bodies to glide down smoothly from trees. New Scientist accesses on 01-09-2022. https://www.newscientist.com/article/2247227-flying-snakes-wiggle-their-bodies-to-glide-down-smoothly-from-trees/

Luther DA, Derryberry EP (2012) Birdsongs keep pace with city life: changes in song over time in an urban songbird affects communication. Anim Behav 83:1059–1066

MacDougall-Shackleton SA (1997) Sexual selection and the evolution of song repertoires. In: Nolan VJ, Ketterson ED, Thompson CF (eds) Current Ornithology, vol 14. Plenum Press, New York, pp 81–124

Machado JP, Johnson WE, Gilbert MT, Zhang G, Jarvis ED, O'Brien SJ, Antunes A (2016) Bone-associated gene evolution and the origin of flight in birds. BMC Genomics 17:371. https://doi.org/10.1186/s12864-016-2681-7

Machtans CS, Wedeles CHR, Bayne EM (2013) A frst estimate for Canada of the number of birds killed by colliding with building windows. Avian Conserv Ecol 8:6. https://doi.org/10.5751/ACE-00568-080206

Mackelprang R, Goller F (2013) Ventilation patterns of the songbird lung/air sac system during different behaviors. J Exp Biol 216:3611–3619

Maeda K, Robinson AJ, Henbest KB, Hogben HJ, Biskup T, Ahmad M, Schleicher E, Weber S, Timmel CR, Hore PJ (2012) Magnetically sensitive light-induced reactions in cryptochrome are consistent with its proposed role as a magnetoreceptor. Proc Natl Acad Sci USA 109:4774–4779

Maggini I, Noakes MJ, Hawkes LA, Hegemann A (2022) Editorial: ecophysiological adaptations associated with animal migration. Front Ecol Evol 10:1022173. https://doi.org/10.3389/fevo.2022.1022173

Mahr K, Nowack L, Knauer F, Hoi H (2022) Songbirds use scent cues to relocate to feeding sites after displacement: an experiment in great tits (*Parus major*). Front Ecol Evol 10:858981. https://doi.org/10.3389/fevo.2022.858981

Maina JN (1987) The morphology of the lung of the African lungfish, *Protopterus aethiopicus*. Cell Tissue Res 250:191–196

Maina JN (1989) The morphometry of the avian lung. In: King AS, McLelland J (eds) Form and function in birds, Vol. 4. Academic Press, London, pp 307–368

Maina JN (1990) A morphological and morphometric study of the prosimian lung: the lesser bushbaby, *Galago senegalensis*. J Anat 172:129–148

Maina JN (1993) Morphometries of the avian lung: the structural-functional correlations in the design of the lungs of birds. Comp Biochem Physiol A 105:397–410

Maina JN (1998) The gas exchangers: structure, function, and evolution of the respiratory processes. Springer-Verlag, Berlin

Maina JN (2000a) What it takes to fly: the structural and functional respiratory refinements in birds and bats. J Exp Biol 203:3045–3064

Maina JN (2000b) Functional morphologogy of the gas-gland cells of the air-bladder of *Oreochromis alcalicus grahami* (Teleosti: Cichlidae): an ultrastructural study on a fish adapted to a severe, highly alkaline environment. Tissue Cell 32:117–132

Maina JN (2002a) Fundamental structural aspects and features in the bioengineering of the gas exchangers: comparative perspectives. Adv Anat Embryol Cell Biol 163:1–108

Maina JN (2002b) Some recent advances on the study and understanding of the functional design of the avian lung: morphological and morphometric perspectives. Biol Rev 77:97–152

Maina JN (2005) The design of the lung-air sac system of birds: development, structure, and function. Springer, Heidelberg

Maina JN (2006) Development, structure, and function of a novel respiratory organ, the lung-air sac system of birds: to go where no other vertebrate has gone. Biol Rev 81:545–579

Maina JN (2011) Bioengineering aspects in the design of gas exchangers: comparative evolutionary, morphological, functional, and molecular perspectives. Springer-Verlag, Berlin

Maina JN (2017) Functional design of the mature avian respiratory system. In: Maina JN (ed) The biology of the avian respiratory system: evolution, development, structure and function. Springer, Heidelberg, pp 191–218

Maina JN, King AS (1984) The structural functional correlation in the design of the bat lung. A morphometric study. J Exp Biol 111:43–63

Maina JN, King AS, King DZ (1982) A morphometric analysis of the lung of a species of bat. Respir Physiol 50:1–11

Maina JN, King AS, Settle JG (1989a) An allometric study of the pulmonary morphometric parameters in birds, with mammalian comparison. Philos Trans R Soc (Lond) B 326:1–57

Maina JN, Maloiy GMO (1985) The morphometry of the lung of the African lungfish (*Protopterus aethiopicus*): its structural-functional correlations. Proc R Soc (Lond) B 224:399–420

Maina JN, Maloiy GMO (1988) A scanning and transmission electron microscopic study of the lung of a caecilian, *Boulengerula taitanus*. J Zool (Lond) 215:739–751

Maina JN, Maloiy GMO, Warui CN, Njogu EK, Kokwaro ED (1989b) A scanning electron microscopic study of the morphology of the reptilian lung: the savanna monitor lizard *(Varanus exanthematicus)* and the pancake tortoise *(Malacochersus tornieri)*. Anat Rec 224:514–522

Maina JN, McCracken KG, Chua B, York JM, Milsom WK (2017) Morphological and morphometric specializations of the lung of the Andean goose, *Chloephaga melanoptera*: a lifelong high-altitude resident. PLoS ONE 12:e0174395. https://doi.org/10.1371/journal.pone.0174395

Maina JN, Thomas SP, Hyde DM (1991) A morphometric study of te lungs of different sized bats: correlations between structure and function of the chiropteran lung. Philos Trans R Soc B (Lond) 333:31–50

Maina JN, Wood CM, Narahara AB, Bergman HL, Laurent P, Walsh PJ (1996) Morphology of the swim (air) bladder of a cichlid teleost: *Oreochromis alcalicus grahami* (Trewavas, 1983), a fish adapted to a hyperosmotic, alkaline and hypoxic environment: a brief outline of the structure and function of the swim-bladder. In: Dutta HH, Munshi JSD (eds) Recent advances in fish biology. Oxford University Press and BHD Press, New Delhi, pp 27–45

Mainwaring MC, Hartley IR, Lambrechts MM, Deeming DC (2014) The design and function of birds' nests. Ecol Evol 4:3909–3928

Makovicky PJ, Zanno LE (2011) Theropod diversity and the refinement of avian characteristics. In: Dyke G, Kaiser G (eds) Living dinosaurs: the evolutionary history of modern birds, 1st edn. John Wiley and Sons Ltd, Chichester (England), pp 9–29

Manisalidis I, Stavropoulou E, Stavropoulos A, Bezirtzoglou E (2020) environmental and health impacts of air pollution: a Review. Front Public Health 8:14. https://doi.org/10.3389/fpubh.2020.00014

Mäntylä E, Kleier S, Kipper S, Hilker M (2017) The attraction of insectivorous tit species to herbivore-damaged Scots pines. J Ornithol 158:479–491

Manville AM (2009) Towers, turbines, power lines, and buildings-steps being taken by the U.S. Fish and Wildlife Service to avoid or minimize take of migratory birds at these structures. In: Ralph CJ, Rich TD (eds) Proc Fourth Intern Partners Flight Conference: Tundra to Tropics. Texas, McAllen, pp 262–272

Manzi M, Coomes OT (2010) Cormorant fishing in Southwestern China: a traditional fishery under siege. Geograph Rev 92:597–603

Marchetti K, Price T, Richman A (1995) Correlates of wing morphology with foraging behavior distance in the genus Phylloscopus. J Avian Biol 26:177–181

Marden JH (1994) From damselflies to pterosaurs: how burst and sustainable flight performance scale with size. Am J Physiol 266:R1077–R1984

Maron JL (1982) Shell-dropping behavior of Western Gulls (*Larus occidentalis*). Auk 99:565–569

Marra PP, Francis CM, Mulvihill RS, Moore FR (2005) The influence of climate on the timing and rate of spring bird migration. Oecologia 142:307–315

Marsh RL (1984) Adaptations of the gray catbird *Dumetella carolinensis* to long-distance migration: flight muscle hypertrophy associated with elevated body mass. Physiol Zool 57:105–117

Martin L (1983) The origin of birds and of avian flight. In: Johnston R (ed) Current ornithology, vol 1. Plenum, New York, pp 105–129

Martins T, Mead CJ (2003) Swifts. In: Perrins C (ed) The firefly encyclopedia of birds. Firefly Books, Buffalo (NY), pp 346–350

Martín-González A, Wierzchos J, Gutiérrez JC, Alonso J, Ascaso C (2009) Double fossilization in eukaryotic microorganisms from Lower Cretaceous amber. BMC Biology 7:9. https://doi.org/10.1186/1741-7007-7-9

Mayr G (2016) The origin of birds: the fossil records of birds and its paleobiological significance. John Wiley and Sons Ltd, New York

Mayr G, Pohl B, Hartman S, Peters DS (2007) The tenth skeletal specimen of *Archeopteryx*. Zool J 49:97–116

Mazzetta GV, Christiansen P, Farin RA (2004) Giants and bizarres: body size of some Southern South American cretaceous dinosaurs. Historical Biol 2004:1–13

McCabe BJ, Horn G (1994) Learning-related changes in Fos-like immunoreactivity in the chick forebrain after imprinting. Proc Natl Acad Sci USA 91:11417–11421

McCay MG (2001) Aerodynamic stability and maneuverability of the gliding from, *Polypedates dennysi*. J Exp Biol 204:2817–2826

McCracken KG, Barger CP, Bulgarella M, Johnson KP, Sonsthagen SA, Trucco J, Valqui TH, Wilson RE, Winker K, Sorenson MD (2009) Parallel evolution in the major haemoglobin genes of eight species of Andean waterfowl. Mol Ecol 18:3992–4005

McCracken KG, Barger CP, Sorenson MD (2010) Phylogenetic and structural analysis of the HbA (αA /βA) and HbD (αD/βA) hemoglobin genes in two high-altitude waterfowl from the Himalayas and the Andes: Barheaded goose (*Anser indicus*) and Andean goose (*Chloephaga melanoptera*). Mol Phylogenet Evol 56:649–658

McGhee GR (2011) Convergent evolution: limited forms most beautiful. MIT Press, Cambridge (MA)

McGuire JA, Witt CC, Remsen JV, Corl A, Rabosky DL, Atshuler DL, Dudley R (2014) Molecular phylogenetics and the diversification of hummingbirds. Curr Biol 24:906–916

McKinnon L, Smith PA, Nol E, Martin JL, Doyle FI, Abraham KF, Gilchrist HG, Morrison RIG, Bêty J (2010) Lower predation risk for migratory birds at high latitudes. Science 327:326–327

McNab BK (1994) Energy conservation and the evolution of flightlessness in birds. Am Nat 144: 628–642

McNab BK (2005) Food habits and the evolution of energetics in birds of paradise (Paradisaeidae). J Comp Physiol B 175:117–132

McWilliams SR, Guglielmo C, Pierce B, Klaassen M (2004) Flying, fasting, and feeding in birds during migration: a nutritional and physiological ecology perspective. J Avian Biol 35:377–393

McWilliams S, Pierce B, Wittenzellner A, Langlois L, Engel S, Speakman JR, Fatica O, DeMoranville K, Goymann W, Trost L et al (2020) The energy savings-oxidative cost tradeoff for migratory birds during endurance flight. eLife 9:e60626. https://doi.org/10.7554/eLife.60626

Mead J (1961) Mechanical properties of lungs. Physiol Rev 41:281–330

Meighan CW (1966) Prehistoric rock paintings in Baja California. Am Antiq 31:372–392

Meir JU, Milsom WK (2013) High thermal sensitivity of blood enhances oxygen delivery in the high-flying bar-headed goose. J Exp Biol 216:2172–2175

Meir JU, York JM, Chua BA, Jardine W, Hawkes LA, Milsom WK (2019) Reduced metabolism supports hypoxic flight in the high-flying bar-headed goose (*Anser indicus*). Elife 8:e44986. https://doi.org/10.7554/elife.44986

Mekonen S (2017) Birds as biodiversity and environmental indicator. J Nat Sci Res 7:28–34

Melstrom KM, D'emic MD, Chure D, Wilson JA (2016) A juvenile sauropod dinosaur from the Late Jurassic of Utah, U.S.A., presents further evidence of an avian style air-sac system. J Vertebr Paleontol 36:e1111898. https://doi.org/10.1080/02724634.2016.1111898

Mendelson JR, Savage JM, Griffith E, Ross H, Kubicki B, Gagliardo R (2008) Spectacular new gliding species of *Ecnomiojyla* (Anura: Hylidae) from Central Panama. J Herpetol 42:750–759

Merchant-Larios H, Diaz-Hernández V (2013) Environmental sex determination mechanisms in reptiles. Sex Dev 7:95–103

Meredith RW, Zhang G, Gilbert MT, Jarvis ED, Springer MS (2014) Evidence for a single loss of mineralized teeth in the common avian ancestor. Science 12:1254390. https://doi.org/10.1126/science.1254390

Meunier F, Cupello C, Clément G (2019) The skeleton and the mineralized tissues of the living coelacanths. Bull Kitakyushu Mus Nat Hist Hum Hist, Ser A 17:37–48

Meyburg BU, Fuller MR (2007) Satellite traking. In: Bird DM, Bildstein KL (eds) Raptor research and management techniques. Hancock House Publishers, Surrey (Canada)

Miles MC, Fuxjager MJ (2018) Synergistic selection regimens drive the evolution of display complexity in birds of paradise. J Anim Ecol 87:1149–1159

Miller EH, Seneviratne SS, Jones IL, Robertson GJ, Wilhelm SI (2008) Syringeal anatomy and allometry in murres (Alcidae: Uria). J Ornithol 149:545–554

Miller RS (1985) Why hummingbirds hover. Auk 102:722–726

Millikan GC, Bowman RI (1967) Observations on Galapagos tool-using finches in captivity. Living Bird 6:23–41

Milsom WK, Gilmour KM, Perry S, Gargaglioni LH, Hedrick MS, Kinkead R, Wang T (2022) Control of breathing in ectothermic vertebrates. Compr Physiol 12:1–120

Milsom WK, Scott GR, Frappell PB, McCracken KG (2021) Different strategies for convective O_2 transport in high altitude birds: a graphical analysis. Comp Biochem Physiol A 253:110871

Monfared AL, Bakhteyari Z (2013) Some gross anatomical features on ostriches'eye and its applicationfor clinical treatment of ocular abnormalities. Global Veterinaria 11:76–79

Monge CM, León-Velarde F (1991) Physiological adaptation to high altitude: oxygen transport in mammals and birds. Physiol Rev 71:1135–1172

Mönkkönen M (1995) Do migrant birds have more pointed wings? A comparative study. Evol Ecol 9:520–528

Monroe BL, Sibley CG (1993) A world checklist of birds. Yale University Press, New Haven

Moreira MO, Qu YF, Wiens JJ (2021) Large-scale evolution of body temperatures in land vertebrates. Evol Lett 5:484–494

Morelli F, Jerzak L, Tryjanowski P (2014) Birds as useful indicators of high nature value (HNV) farmland in Central Italy. Ecol Indicators 38:236–242

Morelli F, Kubicka A, Tryjanowski P, Nelson N (2015) The vulture in the sky and the hominin on the land: three million years of human-vulture interaction. Anthrozoös 28:449–468

Moritz C, Patton JL, Schneider CJ, Smith TB (2000) Diversification of rainforest faunas: an integrated molecular approach. Annu Rev Ecol Syst 31:533. https://www.jstor.org/stable/221742

Morris CR, Askew GN (2010a) The mechanical power output of the pectoralis muscle of the cockatiel (*Nymphicus hollandicus*): the *in vivo* muscle length trajectory and activity patterns and their implications for power modulation. J Exp Biol 213:2770–2780

Morris CR, Askew GN (2010b) Comparison between mechanical power requirements of flight estimated using an aerodynamic model and in vitro muscle performance in the cockatiel (*Nymphicus hollandicus*). J Exp Biol 213:2781–2787

Morris CR, Nelson FE, Askew GN (2010) The metabolic power requirements of flight and estimations of flight muscle efficiency in the cockatiel (*Nymphicus hollandicus*). J Exp Biol 213:2788–2796

Mota FMM, Heming NM, Morante-Filho JC, Talora DC (2022) Amount of bird suitable areas under climate change is modulated by morphological, ecological and geographical traits. Front Ecol Evol 10:987204. https://doi.org/10.3389/fevo.2022.987204

Motta MJ, Agolin FL, Egli FB, Novas FE (2020) New theropod dinosaur from the Upper Cretaceous of Patagonia sheds light on the paravian radiation in Gondwana. Naturwissenschafen 107:24. https://doi.org/10.1007/s00114-020-01682-1

Mouton A, Rutherford I, Yakubovich I (eds) (2013) Luwian identities culture, language and religion between Anatolia and Aegean. Brill, Leiden

Munn AJ, Snelling EP, Taggart DA, Seymour RS (2021) Ontogenetic scaling of the gastrointestinal tract of a marsupial foregut fermenter, the western grey kangaroo Macropus fuliginosus melanops. J Comp Physiol B 191:371–383

Murphy E, King EA (2014) Environmental noise pollution: noise mapping, public health and policy. Elsevier, Boston (MA). https://doi.org/10.1016/C2012-0-13587-0

Muscente AD, Schiffbauer JD, Broce J, LaFlamme M (2017) Exceptionally preserved fossil assemblages through geologic time and space. Gondwana Res 48:164–188

Mylniczenko N (2009) Amphibians. In: Mitchell MA, Tully TN (eds) Manual of exotic pet practice. WB Saunders, St Louis (MO), pp 73–111

Najar N, Benedict L (2019) The relationship between latitude, migration and the evolution of bird song complexity. Ibis 161:1–12

Nam K, Mugal C, Nabholz B, Schielzeth H, Wolf JBW, Backström N, Künster A, Balakrishnan C, Heger A, Ponting CP et al (2010) Molecular evolution of genes in avian genomes. Genome Biol 11:R68. https://doi.org/10.1186/gb-2010-11-6-r68

Nan Y, Peng B, Chen Y, McGlinchey D (2019) From studying real hummingbirds to designing hummingbird-like robots—a literature review. IEEE Access 7:131785

Natarajan C, Jendroszek A, Kumar A, Weber RE, Tame JRH, Fago A, Storz JF (2018) Molecular basis of hemoglobin adaptation in the high-flying bar-headed goose. PLoS Genet 14:e1007331. https://doi.org/10.1371/journal.pgen.1007331. eCollection

Navalón G, Marugán-Lobón J, Chiappe LM, Luis Sanz J, Buscalioni AD (2015) Soft tissue and dermal arrangement in the wing of an Early Cretaceous bird: implications for the evolution of avian flight. Sci Rep 5:14864. https://doi.org/10.1038/srep14864

Nespolo RF, González-Lagos C, Solano-Iguaran JJ, Elfwing M, Garitano-Zavala A, Mañosa S, Alonso JC, Altimiras J (2018) Aerobic power and flight capacity in birds: a phylogenetic test of the heart-size hypothesis. J Exp Biol 221:jeb175208. https://doi.org/10.1242/jeb.162693

Neumayer L (1930) Die Entwicklung des Darms von Acipenser. Acta Zool 39:1–151

Nelsen OE (1953) Comparative embryology of vertebrates. McGraw-Hill Book Company, New York

Newton I (2003) The speciation and biogeography of birds. Academic Press, London

Newton I (2008) The migration ecology of birds. Elsevier, NewYork

Newton I (2011) Migration within the annual cycle: species, sex and age differences. J Ornithol 152:S169–S185

Nichols JW, Bennett RS, Rossmann R, French JB (2010) A physiologically based toxicokinetic model for methylmercury in female American kestrels. Environ Toxicol Chem 29:1854–1867

Nienaber J, Hahn G (2007) Livestock production system management responses to thermal challenges. Int J Biometeorol 52:149–157

Nilsson S (1983) Autonomic nerve function in the vertebrates. Springer-Verlag, Berlin

Nishibori M, Shimogiri T, Hayashi T, Yasue H (2005) Molecular evidence for hybridization of species in the genus *Gallus* except for *Gallus varius*. Anim Genet 36:367–375

Niu Y, Zhang X, Xu T, Li X, Zhang H, Wu A, Storey KB, Chen Q (2022) Physiological and biochemical adaptations to high altitude in Tibetan Frogs, Nanorana parkeri. Front Physiol 13: 942037. https://doi.org/10.3389/fphys.2022.942037

Norberg UM (1990) Vertebrate flight: mechanics, physiology, morphology, ecology and evolution. Springer-Verlag, New York (NY)

Norberg UM (2006) Flight and scaling of flyers in nature. In: WIT Transactions on State of the Art in Science and Engineering, Vol 3. WIT Press, pp. 120–154. https://doi.org/10.2495/1-84564-001-2/2d

Nottebohm F (2005) The neural basis of birdsong. PLoS Biol 3(5):e164. https://doi.org/10.1371/journal.pbio.0030164

Nottebohm F, Stokes TM, Leonard CM (1976) Central control of song in the canary, *Serinus canaries*. J Comp Neurol 165:457–486

Nowicki S, Searcy WA (2014) The evolution of vocal learning. Curr Opin Neurobiol 28:48–53

Nutman AP, Bennett VC, Friend CRL, Van Kranendonk MJ, Allan R (2016) Rapid emergence of life shown by discovery of 3,700-million-year-old microbial structures. Nature 537:535–538

Nygaard DF (1994) World population projections, 2020. 2020 vision briefs 5. International Food Policy Research Institute (IFPRI).

O'Connell TJ, Jackson LE, Brooks RP (2000) Bird guilds as indicators of ecological condition in the central Appalachians. Ecol Appl 10:1706–1721

O'Connor PM (2006) Postcranial pneumaticity: an evaluation of soft-tissue influences on the postcranial skeleton and the reconstruction of pulmonary anatomy in archosaurs. J Morphol 267:1199–1226

O'Connor PM (2009) Evolution of archosaurian body plans: skeletal adaptations of an air-sac-based breathing apparatus in birds and other archosaurs. J Exp Zool A 311:629–646

O'Connor JK, Field DJ, Sullivan C (2021) Editorial: early avian evolution. Front Earth Sci 9: 730214. https://doi.org/10.3389/feart.2021.730214

O'Connor JK, Wang X, Sulivan C, Wang Y (2018) First report of gastroliths in the Early Cretaceous basal bird Jeholornis. Cretac Res 84:200–208

O'Connor JK, Zhou Z (2015) Early evolution of the biological bird: perspectives from new fossil discoveries in China. J Ornithol 156:333–342

O'Connor PM, Claessens LP (2005) Basic avian pulmonary design and flow-through ventilation in non-avian theropod dinosaurs. Nature 436:253–256

O'Connor TP (1995) Metabolic characteristics and body composition in house finches: effects of seasonal acclimatization. J Comp Physiol B 165:298–305

Oda J, Sakamoto J, Sakano K (2006) Mechanical evaluation of the skeletal structure and tissue of the woodpecker and its shock absorbing system. JSME Int J Ser A 49:390–396

Oldham C (1930) The shell-smashing habit of gulls. Ibis 6:239–243

Oliveros CH, Field DJ, Ksepka DJ, Barkere FK, Aleixo A, Andersen MJ, Alström P, Benz BW, Braun EL, Braun MJ et al (2019) Earth history and the passerine superradiation. Proc Natl Acad Sci USA 116:7916–7925

Olson CR, Fernández VM, Portfors CV, Mello CV (2018) Black Jacobin hummingbirds vocalize above the known hearing range of birds. Curr Biol 28:R204–R205

O'Rourke DP, Jenkins AL (2008) Anaesthesia and analgesia in amphibians. In: Fish RE, Danneman PJ, Karas AZ (eds) Anaesthesia and analgesia in laboratory animals, 2nd edn. Academic Press, San Diego, pp 511–518

Orton DA (1975) The speed of a peregrine's dive. The Field 9:588–590

Osborn RG, Higgins KF, Usgaard RE, Dieter CD, Neiger RD (2000) Bird mortality associated with wind turbines at the Buffalo Ridge Wind Resource, Minnesota. Am Midl Nat 143:41–52

Ostrom JH (1974) A*rchaeopteryx* and the origin of flight. Q Rev Biol 49:27–47

Ostrom JH (1976) Archeopteryx and the origin of birds. Biol J Linn Soc 8:91–182

Otto SP (2018) Adaptation, speciation and extinction in the Anthropocene. Proc Biol Sci 285: 20182047. https://doi.org/10.1098/rspb.2018.2047

Owen R (1863) On the Archaeopteryx of von Meyer, with a description of the fossil remains of a long-tailed species from the lithographic stone of Solenhofen. Philos Trans R Soc (Lond) 153: 33–47

Packard GC (1974) The evolution of air-breathing in Paleozoic gnathostome fishes. Evolution 28: 320–325

Padhi MK (2016) Importance of indigenous breeds of chicken for rural economy and their improvements for higher production performance. Scientifica 2016:2604685. https://doi.org/10.1155/2016/2604685

Padian K (1985) The origins and aerodynamics of flight in extinct vertebrates. Palaeontology 28: 413–433

Padian K (1998) When is a bird not a bird? *Nature*393:729–730

Pagano AS, Márquez S (2022) The nasopharynx as a window to half a billion years of evolutionary change to the upper respiratory tract. Anat Rec 305:1829–1841

Park H, Choi H (2010) Aerodynamic characteristics of flying fish in gliding flight. J Exp Biol 213: 3269–3279

Palmer C (2014) The aerodynamics of gliding flight and its application to the arboreal flight of the Chinese feathered dinosaur Microraptor. Biol J Linn Soc 113:828–835

Pan S, Lin Y, Liu Q, Duan J, Lin Z, Wang Y, Wang X, Lam SM, Zou Z, Shui G et al (2019) Convergent genomic signatures of flight loss in birds suggest a switch of main fuel. Nat Commun 10:2756. https://doi.org/10.1038/s41467-019-10682-3

Parr N, Wilkes M, Hawkes LA (2019) Natural climbers: insights from avian physiology at high altitude. High Alt Med Biol 20:427–437

Parry LA, Smithwick F, Nordén KK, Saitta KT, Lozano-Fernandez J, Tanner AR, Caron JB, Edgecombe GD, Briggs DGE, Vinther J (2018) Soft-bodied fossils are not simply rotten carcasses—toward a holistic understanding of exceptional fossil preservation. BioEssays 40: 1700167. https://doi.org/10.1002/bies.201700167

Pattinson NB, van den Ven TMFN, Finnie MJ, Nupen LJ, McKechnie AE, Cunningham SJ (2022) Collapse of breeding success in desertdwelling hornbills evident within a single decade. Front Ecol Evol 10:842264. https://doi.org/10.3389/fevo.2022.842264

Paul GS (1991) The many myths, some old, some new, of dinosaulogy. Mod Geol 16:69–99

Payevsky VA (2014) Phylogeny and classification of passerine birds, Passeriformes. Biol Bull Rev 4:143–156

Pei R, Pittman M, Goloboff PA, Drecchi TA, Habib MB, Kaye TG, Larsson HCE, Norell MA, Brusatte SL, Xu X (2020) Potential for powered flight neared by most close avialan relatives, but few crossed its thresholds. Curr Biol 30:1–14

Peiró IG (2003) Intraspecific variation in the wing shape of the long-distance migrant reed warbler, *Acrocephalus scirpaceus*: effects of age and distance of migration. Ardeola 50:31–37

Pennak S (2012) State of observed species: a decade of species discovery in review. International Institute for Species Exploration, Arizona State University Press, Arizona (USA).

Pennardt A (2013) High-altitude pulmonary edema: diagnosis, prevention, and treatment. Curr Sports Med Rep 12:115–119

Pennycuick CJ (1969) The mechanics of bird migration. Ibis 111:525–556

Pennycuick CJ (1975) Mechanics of flight. In: Farmer DS, King JR (eds) Avian biology, Vol. 5. Academic Press, London, pp 1–75

Pennycuick CJ (1989) Bird flight performance: a practical calculation manual. Oxford University Press, Oxford

Pennycuick CJ (2008) Soaring behaviour and performance of some East African birds, observed from a motor-glider. Ibis 114:178–218

Pennycuick CJ (2015) The flight of birds and other animals. Aerospace 2:5050–5523

Perera F (2018) Pollution from fossil-fuel combustion is the leading environmental threat to global pediatric health and equity: solutions exist. Int J Environ Res Public Health 15(1):16. https://doi.org/10.3390/ijerph15010016

Perry SF (1983) Reptilian lungs: functional anatomy and evolution. Adv Anat Embryol Cell Biol 79:1–81

Perry SF (1988) Functional morphology of the lungs of the Nile crocodile, *Crocodylus niloticus*: non-respiratory parameters. J Exp Biol 34:61–81

Perry SF (1989a) Mainstreams in the evolution of vertebrate respiratory structures. In: King AS, McLelland J (eds) Form and function in birds, Vol. 4. Academic Press, London, pp 1–67

Perry SF (1989b) Structure and function of the reptilian respiratory system. In: Wood SC (ed) Comparative pulmonary physiology current concepts. Marcel Dekker, New York, pp 369–416

Perry SF (1998) Lungs: comparative anatomy, functional morphology, and evolution. In: Gans C, Gaunt AS (eds) Biology of the reptilia, Vol. 19. Society for the Study of Amphibians and Reptiles, Ithaca (NY), pp 1–92

Perry SF (2001) Functional morphology of the reptilian and avian respiratory systems and its implications for theropod dinosaurs. In: Gauthier J, Gall LF (eds) Proceedings of the International Symposium in Honor of John H. Ostrom (February 13-14, 1999): new perspectives on the origins and evolution of birds. Yale Peabody Museum of Natural History. Yale University (NH), pp 429–441

Perry SF, Breuer T, Pajor N (2011) Structure and function of the sauropod respiratory system. In: Klein N, Remes K, Gee CT, Sander PM (eds) Biology of the sauropod dinosaurs: understanding the life of giants. University of Indiana Press, Bloomington (IN), pp 83–93

Perry SF, Christian A, Breuer T, Pajor N, Codd JR (2009a) Implications of an avian-style respiratory system for gigantism in sauropod dinosaurs. J Exp Zool 311A:600–610

Perry SF, Lambertz M, Schmitz A (2019) Respiratory biology of animals: evolutionary and functional morphology. Oxford University Press, Oxford

Perry SF, Pajor N, Breuer T (2009b) Did lungs constrain size or size constrain lungs in sauropod dinosaurs? In: Perry SF, Morris S, Breuer T, Pajor N, Lambertz M (eds) 2nd International Congress of Respiratory Science 2009 -Abstracts and Scientific Program. Tharax Verlag, Hildesheim/Berlin, pp 69–70

Perry SF, Sander PM (2004) Reconstructing the evolution of respiratory apparatus in tetrapods. Respir Physiol Neurobiol 144:125–139

Perry SF, Wilson RJA, Straus C, Harris MB, Remmers JE (2001) Which came first, the lung or the breath? Comp Biochem Physiol A, Mol Integr Physiol 129:37–47

Peters S (2000) Bumblebee hummingbird of Cuba. Welschner Books Inc., New York (NY)

Peters J, Lebrasseur O, Best J, Miller H, Fothergill T, Dobney K, Thomas RM, Maltby M, Sykes N, Hanotte O et al (2015) Questioning new answers regarding Holocene chicken domestication in China. Proc Natl Acad Sci USA 112:E2415. https://doi.org/10.1073/pnas.1503579112

Peters J, Lebrasseur O, Deng H, Larson G (2016) Holocene cultural history of red jungle fowl (*Gallus gallus*) and its domestic descendent in East Asia. Quart Sci Rev 142:102–119

Petschow D, Wurdinger I, Baumann R, Duhm J, Braunitzer G, Bauer C (1977) Causes of high blood O_2 affinity of animals living at high altitude. J Appl Physiol 42:139–143

Peyré-Tartaruga LA, Coertjens M (2018) Locomotion as a powerful model to study integrative physiology: efficiency, economy, and power relationship. Front Physiol 9:1789. https://doi.org/10.3389/fphys.2018.01789

Pfeifhofer G, Tributsch H (2014) The flight of the albatross—how to transfrom it into aerodynamic engineering? Engineeering 6:427–438

Pfenning AR, Hara E, Whitney O, Rivas MV, Wang R, Roulhac PL, Howard JT, Wirthlin M, Lovell PV, Ganapathy G et al (2014) Convergent transcriptional specializations in the brains of humans and song-learning birds. Science 346:1256846. https://doi.org/10.1126/science.1256846

Phillips JB, Schmidt-Koenig K, Muheim R (2006) True navigation: sensory bases of gradient maps. In: Brown MF, Cook RG (eds) Animal spatial cognition: comparative, neural, and computational approaches. Comparative Cognition Press. Available in https://pigeon.psy.tufts.edu/asc/Phillips/Default_Old_08_21.htm. Accessed on 12 July 2023

Piatt JF, Nettleship DN (1985) Diving depths of four alcids. Auk 102:293–297

Pierce BJ, McWilliams SR, O'Connor TP, Place AR, Guglielmo CG (2005) Effect of dietary fatty acid composition on depot fat and exercise performance in a migrating songbird, the red-eyed vireo. J Exp Biol 208:1277–1285

Piersma T (2011) Why marathon migrants get away with high metabolic ceilings: towards an ecology of physiological restraint. J Exp Biol 214:295–302

Piersma T, Jukema J (1990) Budgeting the flight of a long-distance migrant: changes in nutrient reserve levels of bar-tailed godwits at successive spring staging sites. Ardea 78:315–337

Pierson DJ (2009) The physiology of dinosaurs: circulatory and respiratory function in the largest animals ever to walk the earth: the Donald F Egan Scientific Memorial Lecture. Respir Care 54: 887–911

Pimm SL, Jenkins CN, Abell R, Brooks TM, Gittleman JL, Joppa LN, Raven PH, Roberts CM, Sexton JO (2014) The bioversity of species and their rates of extinction, distribution, and protection. Science 344:1246752. https://doi.org/10.1126/science.1246752

Pitskhelauri GZ (1980) Aristotle's knowledge in the field of natural and medical sciences (on celebrating 2300 years since his death). Sante Publique (Bucur) 23:375–379

Pitt J, Gillingham PK, Maltby M, Stewart JR (2016) New perspectives on the ecology of early domestic fowl: an interdisciplinary approach. J Archaeol Sci 74:1–10

Podos J, Cohn-Haft M (2019) Extremely loud mating songs at close range in white bellbirds. Curr Biol 29:R1068–R1069

Powell FL (2000) Respiration. In: Whittow GC (ed) Sturkie's avian physiology, 5th edn. Academic Press, San Diego, pp 233–264

Powell FL (2015) Respiration. In: Scanes CG (ed) Sturkie's avian physiology, 6th edn. Academic Press, San Diego, pp 301–336

Powell FL, Hopkins SR (2004) Comparative physiology of lung complexity: implications for gas exchange. News Physiol Sci 19:55–60

Powell FL, Scheid P (1989) Physiology of gas exchange in the avian respiratory system. In: King AS, McLelland J (eds) Form and function in birds, Vol 4. Academic Press, London, pp 393–437

Powers DR, Naggy KA (1988) Field metabolic rate and food consumption by free-living Anna's hummingbirds *Calypte anna*. Physiol Zool 61:500–506

Pozzi LM, Borboroglu PG, Boersma PD, Pascual MA (2015) Population regulation in Magellanic penguins: what determines changes in colony size? PLoS ONE 10:e0119002. https://doi.org/10.1371/journal.pone.0119002

Prado GMEM, Anelli LE, Petri S, Romero GR (2016) New occurrences of fossilized feathers: systematics and taphonomy of the Santana Formation of the Araripe Basin (Cretaceous), NE, Brazil. PeerJ 4:e1916. https://doi.org/10.7717/peerj.1916. eCollection 2016

Price EO (1999) Behavioral development in animals undergoing domestication. Appl Anim Behav Sci 65:245–271

Price T (2018) Foreword. In: Tietze DT (ed) Bird species, how they arise, modify and vanish. Springer Nature, Switzerland, pp vii–x

Prince B, Riede T, Goller F (2011) Sexual dimorphism and bilateral asymmetry of syrinx and vocal tract in the European starling (*Sturnus vulgaris*). J Morphol 272:1527–1536

Proffitt FM, McLean IG (1991) Recognition of parents calls by chicks of the snares crested penguin. Bird Behav 9:103–113

Pulido F (2007) The genetics and evolution of avian migration. Bioscience 57:165–174

Pulido F, Berthold P (2010) Current selection for lower migratory activity will drive the evolution of residency in a migratory bird population. Proc Natl Acad Sci USA 107:7341–7346

Pulido F, Widmer M (2005) Are long-distance migrants constrained in their evolutionary response to environmental change? Causes of variation in the timing of autumn migration in a blackcap (*Sylvia atricapilla*) and two garden warbler (*Sylvia borin*) populations. Ann NY Acad Sci 1046: 228–241

Purves D, Augustine GJ, Fitzpatrick D, Katz LC, LaMantia AS, McNamara JO, Williams SM (eds) (2001) Neuroscience, 2nd edn. Sinauer Associates, Sunderland (MA)

Puttick MN, Thomas GH, Benton MJ (2014) High rates of evolution preceded the origin of birds. Evolution 68:1497–1510

Pytte CL, Ficken MS, Moiseff A (2004) Ultrasonic singing by the blue-throated hummingbird: a comparison between production and perception. J Comp Physiol A Neuroethol Sens Neural Behav Physiol 190:665–673

Qiao Q, Huang Y, Qi J, Qu M, Jiang C, Lin P, Li R, Song L, Yonezawa T, Hasegawa M et al (2016) The genome and transcriptome of trichormus sp. nmc-1: insights into adaptation to extreme environments on the Qinghai-Tibet Plateau. Sci Rep 6:29404. https://doi.org/10.1038/srep294041-10

Quader S (2005) Elaborate nests in a weaverbird: a role for female choice? Ethology 111:1073–1088

Quammen D (1996) The song of the Dodo: Island biogeography in an age of extinction. Hutchinson, London

Quick DE, Ruben JA (2009) Cardio-pulmonary anatomy in theropod dinosaurs: implications from extant archosaurs. J Morphol 270:1232–1246

Rahman MM, McConnell R, Schlaerth H, Ko J, Silva S, Lurmann FW, Palinkas L, Johnston J, Hurburt M, Yin H, Ban-Weiss G, Garcia E (2022) The effects of co-exposure to extremes of heat and particulate air pollution on mortality in California: implications for climate change. Am J Reaspir Crit Care Med. https://doi.org/10.1164/rccm.202204-0657OC

Rahn H (1966) Aquatic gas exchange: theory. Respir Physiol 1:1–12

Ramirez JM, Folkow LP, Blix AS (2007) Hypoxia tolerance in mammals and birds: from the wilderness to the clinic. Annu Rev Physiol 69:113–143

Randall DJ, Burggren WW, Farell AP, Haswell MS (1981) The evolution of air breathing vertebrates. Cambridge University Press, Cambridge

Randall DJ, Rummer JL, Wilson JM, Wang S, Brauner CJ (2014) A unique mode of tissue oxygenation and the adaptive radiation of teleost fishes. J Exp Biol 217:1205–1214

Rauhut OWM, Foth C (2020) The origin of birds: current consensus, controversy, and the occurrence of feathers. In: Foth C, Rauhut OWM (eds) The evolution of feathers from their origin to the present. Springer Nature, Cham (Switzerland)

Rauhut OWM, Tischlinger H, Foth C (2019) A non-archaeopterygid avialan theropod from the late Jurassic of southern Germany. eLife 8:e43789. https://doi.org/10.7554/eLife.43789

Rauzon MJ (1997) Hummingbirds. F Watts, New York

Rayner JMV, Viscardi PW, Ward S, Speakman JR (2001) Aerodynamics and energetics of intermittent flight in birds. Am Zool 41:188–204

Read AF, Weary DM (1992) The evolution of bird song: comparative analyses. Philos Trans R Soc (Lond) B 338:165–187

Read AW, Weary DW (2004) Sexual selection and the evolution of bird song: a test of the Hamilton-Zuk hypothesis. Behav Ecol Sociobiol 26:47–56

Reader SM, Laland KN (2002) Social intelligence, innovation, and enhanced brain size in primates. Proc Natl Acad Sci USA 99:4436–4441

Reddy G, Celani A, Sejnowski TJ, Vergassola M (2016) Learning to soar in turbulent environments. Proc Natl Acad Sci USA 113:E4877–E4884

Redpath SM, Appleby BM, Petty SJ (2000) Do male hoots betray parasite loads in twany owls? J Avian Biol 31:457–462

Reilly SM, Mcelroy E, Biknevicius AR (2007) Posture, gait and the ecological relevance of locomotor costs and energy-saving mechanisms in tetrapods. Zoology 110:271–289

Reiner A, Yamamoto K, Karten HJ (2005) Organization and evolution of the avian forebrain. Anat Rec A Discov Mol Cell Evol Biol 287:1080–1102

Reiser PJ, Welch KC, Suarez RK, Altshuler DL (2013) Very low force-generating ability and unusually high temperature dependency in hummingbird flight muscle fibers. J Exp Biol 216: 2247–2256

Renaudeau D, Collin A, Yahav S, de Basilio V, Gourdine JL, Collier RJ (2012) Adaptation to hot climate and strategies to alleviate heat stress in livestock production. Animal 6:707–728. https://doi.org/10.1017/S1751731111002448

Reynolds SJ, Ibáñez-Álamo JD, Sumasgutner P, Mainwaring MC (2019) Urbanisation and nest building in birds: a review of threats and opportunities. J Ornithol 160:841–860

Rheindt RE, Prawiradilaga DM, Ashari H, Suparno, Gwee CY, Lee GWX, Wu MY, Ng NSR (2020) A lost world in Wallacea: description of a montane archipelagic avifauna. Science 367: 167–170

Rhen T, Lang JW (1994) Temperature-dependent sex determination in the snapping turtle: manipulation of the embryonic sex steroid environment. Gen Comp Endocrinol 96:243–254

Richardson PL, Wakefield ED, Phillips RA (2018) Flight speed and performance of the wandering albatross with respect to wind. Mov Ecol 6:3. https://doi.org/10.1186/s40462-018-0121-9

Ricklefs RE (2012) Species richness and morphological diversity of passerine birds. Proc Natl Acad Sci USA 109:14482–14487

Riede T, Fisher JH, Goller F (2010) Sexual dimorphism of the zebra finch syrinx indicates adaptation for high fundamental frequencies in males. PloS One 5:e11368. https://doi.org/10.1371/journal.pone.0011368

Riede T, Goller F (2010a) Peripheral mechanisms for vocal production in birds—differences and similarities to human speech and singing. Brain Lang 115:69–80

Riede T, Goller F (2010b) Functional morphology of the sound-generating labia in the syrinx of two songbird species. J Anat 216:23–36

Ritz T, Adem S, Schulten K (2000) A model for photoreceptor-based magnetoreception in birds. Biophys J 78:707–718

Ritz T, Thalau P, Wiltschko W (2004) Resonance effects indicate a radical-pair mechanism for avian magnetic compass. Nature 429:177–180

Robbins J (2017) The wonder of birds: what they tell us about ourselves, the world, and a better future. Spiegel and Grau, Penguin Random House, New York

Robert KA, Lesku JA, Partecke J, Chambers B (2015) Artificial light at night desynchronizes strictly seasonal reproduction in a wild mammal. Proc R Soc (Lond) B 282:20151745. https://doi.org/10.1098/rspb.2015.1745

Roberts TJ, Marsh RL, Weyand PG, Taylor CR (1997) Muscular force in running turkeys: the economy of minimizing work. Science 275:1113–1115

Roberts TJ, Chen MS, Taylor CR (1998) Energetics of bipedal running. II. Limb design and running mechanics. J Exp Biol 201:2753–2762

Robertson GN, Croll RP, Smith FM (2014) The structure of the caudal wall of the zebrafish (*Danio rerio*) swim bladder: evidence of localized lamellar body secretion and a proximate neural plexus. J Morphol 275:933–948

Roché J, Godinho C, Rabaça JE, Frochot B, Faivre B, Mendes A, Dias PC (2010) Birds as bio-indicators and as tools to evaluate restoration measures. Proc 7th European Conference on Ecological Restoration, Avignon, France, 23-27/08/2010, p. 6.

Roelofs YC (2010) Tool use in birds: an overview of reported cases, ontogeny and underlying cognitive abilities. MSc Thesis, University of Groningen, The Netherlands.

Rogers CT, Hore PJ (2009) Chemical magnetoreception in birds: the radical pair mechanism. Proc Natl Acad Sci USA 106:353–360

Rohde R, Friedland DR (2022) Clinical perspectives on nasopharyngeal morphology in humans. Anat Rec 305:2065–2074

Rose SP (2000) God's organism? The chick as a model system for memory studies. Learn Mem 7: 1–17

Rosenberg KV, Dokter AM, Blancher PJ, Sauer JR, Smith AC, Smith PA, Stanton JC, Panjabi A, Helft L, Parr M et al (2019) Decline of the North American avifauna. Science 366:120–124

Roser M, Ritchie H, Ortiz-Ospina E (2013) World population growth. OurWorldinData.org. https://ourworldindat.org/world.population.growth [Online resource].

Rossi V, McNamara ME, Webb SM, Ito S, Wakamatsu K (2019) Tissue-specific geometry and chemistry of modern and fossilized melanosomes reveal internal anatomy of extinct vertebrates. Proc Natl Acad Sci (USA) 116:17880–17889

Roux E (2002) Origin and evolution of the respiratory tract in vertebrates. Rev Mal Respir 19:601–615

Rowley KL, Mantilla CB, Sieck GC (2005) Respiratory muscle plasticity. Respir Physiol Neurobiol 147:235–251

Ruaux G, Lumineau S, de Margerie E (2020) The development of flight behaviours in birds. Proc R Soc (Lond) B 287:20200668. https://doi.org/10.1098/rspb.2020.0668

Ruben JA, Jones TD, Geist NR (2003) Respiratory and reproductive paleophysiology of dinosaurs and early birds. Physiol Biochem Zool 76:141–164

Rubenson J, Heliams DB, Lloyd DG, Fournier PA (2004) Gait selection in the ostrich: mechanical and metabolic characteristics of walking and running with and without an aerial phase. Proc Biol Sci 271:1091–1099

Rude E (2016) Tastes like chicken: a history of America's favourite bird. Pegasus Books Ltd, New York

Ruckebusch Y, Dardillat C, Guilloteau P (1983) Development of digestive functions in the newborn ruminant. Ann Rech Vét 14:360–374

Rundle HD, Nosil P (2005) Ecological speciation. Ecol Lett 8:336–352

Sackton TB, Grayson P, Cloutier A, Hu Z, Liu JS, Wheeler NE, Gardner PP, Clarke JA, Baker AJ, Clamp M (2019) Convergent regulatory evolution and loss of flight in paleognathous birds. Science 364:74–78

Sachs G, Traugott J, Nesterova AP, Bonadonna F (2013) Experimental verification of dynamic soaring in albatrosses. J Exp Biol 216:4222–4232

Sachs G, Traugott J, Nesterova AP, Dell'Omo G, Kümmeth F, Heidrich W, Vyssotski AL, Bonadonna F (2012) Flying at no mechanical energy cost: disclosing the secret of wandering albatrosses. PLoS ONE 7:e41449. https://doi.org/10.1371/journal.pone.0041449

Sætre GP (2007) Speciation in birds—Trevor Price 2007. Roberts and Company Publishers, Greenwood Village CO. https://doi.org/10.1080/10635150802164611

Sahney S, Benton MJ, Ferry PA (2010) Links between global taxonomic diversity, ecological diversity and the expansion of vertebrates on land. Biol Let 6:544–547

Sakiyama J, Yamagishi A, Kuroiwa A (2003) Tbx4-Fgf-10 system controls lung bud formation during chicken embryonic development. Development 130:1225–1234

Sala OE, Chapin FS, Armesto JJ, Berlow E, Bloomfield J, Dirzo R, Huber-Sanwald E, Huenneke LF, Jackson RB, Kinzig A et al (2000) Global biodiversity scenarios for the year 2100. Science 287:1770–1774

Salewski V, Siebenrock K-H, Hochachka WM, Woog F, Fiedler W (2014) Morphological change to birds over 120 years is not explained by thermal adaptation to climate change. PLoS ONE 9: e101927. https://doi.org/10.1371/journal.pone.0101927

Salzburger W (2009) The interaction of sexually and naturally selected traits in the adaptive radiations of cichlid fishes. Mol Ecol 18:169–185

Sander PM (2013) An evolutionary cascade model for sauropod dinosaur gigantism—overview, update and tests. PLoS One 8:e78573. https://doi.org/10.1371/journal.pone.0078573

Sander PM, Christian A, Clauss M, Fechner R, Gee CT, Griebeler EM, Gunga HC, Hummel J, Mallison H, Perry SF et al (2011) Biology of the sauropod dinosaurs: the evolution of gigantism. Biol Rev 86:117–155

Sanderson MJ, Shaffer HB (2002) Trouble shooting molecular phylogenetic analyses. Annu Rev Ecol System 1829:49–72

Sanz JL, Chiappe LM, Pérez-Moreno BP, Moratalla JJ, Hernándex-Carrasquilla F, Buscalioni AD, Ortega F, Poyato-Ariza FJ, Rasskin-Gutmann D, Martínez-Delclõs X (1997) A nestling bird from the lower Cretaceous of Spain: implications for avian skull and neck evolution. Science 276:15421546. https://doi.org/10.1126/SCIENCE.276.5318.1543

Sapir N, Butler PJ, Hedenström A, Wikelski M (2011) Energy gain and use during animal migration. In: MilnerGulland J, Fryxell JM, Sinclair ARE (eds) Animal migration: a synthesis. Oxford University Press, Oxford (UK), pp 52–67

Sarwar S, Firdous Q, Khalid AN (2019) Importance of molecular and phylogenetic analyses for identification of basidiomycetes. In: Yousaf Z (ed) Recent advances in phylogenetics. IntechOpen. https://doi.org/10.5772/intechopen.80671

Saucier JR, Milensky CM, Caraballo-Ortiz MA, Ragai R, Dahlan NF, Edwards DP (2019) A distinctive new species of flowerpecker (Passeriformes: Dicaeidae) from Borneo. Zootaxa 4: 451–464

Sault N (2018) The wonder of birds: what they tell us about ourselves, the World and a better future. Ethnobiol Lett 9:319–320. https://doi.org/10.14237/ebl.9.2.2018.1404

Saunders H (1891) Notes on birds observed in Switzerland; chiefly in the cantons of Vaud and Neuchâtel. Ibis 6:157–188

Saupe EE, Farnsworthb A, Lunt DJ, Sagooc N, Phamd KV, Field DJ (2019) Climatic shifts drove major contractions in avian latitudinal distributions throughout the Cenozoic. Proc Natl Acad Sci USA 116:12895–12900

Savage C (1992) Peregrine falcons. Sierra Club, San Francisco

Savalli UM (1995) The evolution of bird coloration and plumage elaboration. In: Power DM (ed) Current Ornithology, vol 12. Springer, Boston (MA), pp 141–190

Sawai H, Kim HL, Kuno K, Suzuki S, Gotoh H, Takada M, Takahata N, Satta N, Akishinonomiya F (2010) The origin and genetic variation of domestic chickens with special reference to jungle fowls *Gallus g. gallus* and *G. varius*. PLoS ONE 5:e10639. https://doi.org/10.1371/journal.pone.0010639

Scanes CG (ed) (2015) Sturkie's avian physiology, 6th edn. Academic Press/Elsevier, Amsterdam

Scanes CG, Witt J, Ebeling M, Schaller S, Baier V, Bone AJ, Preuss TG, Heckmann D (2022) Quantitative morphometric, physiological, and metabolic characteristics of chickens and mallards for physiologically based kinetic model development. Front Physiol 13:858283. https://doi.org/10.3389/fphys.2022.858283

Schachner ER, Cieri RL, Butler JP, Farmer CG (2013a) Unidirectional pulmonary airflow patterns in the savannah monitor lizard. Nature 506:367–370

Schachner ER, Cieri RL, Butler JP, Farmer CG (2014) Unidirectional pulmonary airflow patterns in the savannah monitor lizard. Nature 506:367–370

Schachner ER, Diaz RE, Coke R, Echols S, Osborn ML, Hedrick BP (2022) Architecture of the bronchial tree in Cuvier's caiman (*Paleosuchus palpebrosus*). Anat Rec 305:3037–3054

Schachner ER, Farmer CG, McDonald AT, Dodson P (2011) Evolution of the dinosauriform respiratory apparatus: new evidence from the postcranial axial skeleton. Anat Rec 294:1532–1547

Schachner ER, Hutchinson JR, Farmer CG (2013b) Pulmonary anatomy in the Nile crocodile and the evolution of unidirectional airflow in Archosauria. PeerJ 1:e60. https://doi.org/10.7717/peerj.60

Schachner ER, Lyson TR, Dodson P (2009) Evolution of the respiratory system in nonavian theropods: evidence from rib and vertebral morphology. Anat Rec 292:1501–1513

Schaefer HC, Jetz W, Böhning-Gaese K (2008) Impact of climate change on migratory birds: community reassembly versus adaptation. Global Ecol Biogeogr 17:38–49

Schaefer J, Huegel CN (2020) Hummingbirds of Florida. https://edis.ifas.ufl.edu/publication/UW059

Scheid P (1979) Mechanisms of gas exchange in bird lungs. Rev Physiol Biochem Pharmacol 86

Scheid P (1990) Avian respiratory system and gas exchange. In: Sutton JR, Coates G, Remmers JE (eds) Hypoxia: the adaptations. BC Decker, Toronto, pp 4–7

Scheid P, Piiper J (1989) Respiratory mechanics and air flow in birds. In: King AS, McLelland J (eds) Form and function in birds, vol 4. Academic Press, London, pp 369–391

Schmaljohann H (2019) The start of migration correlates with arrival timing, and the total speed of migration increases with migration distance in migratory songbirds: a cross-continental analysis. Mov Ecol 7:25. https://doi.org/10.1186/s40462-019-0169-1

Schmidt MF, Martin WJ (2014) The respiratory-vocal system of songbirds: anatomy, physiology, and neural control. Prog Brain Res 212:297–335

Schmidt-Nielsen K (1972) Locomotion: energy cost of swimming, flying, and running. Science 177:222–228

Schou MF, Bonato M, Engelbrecht A, Brand Z, Svensson EI, Melgar J, Muvhali PT, Cloete SWP, Cornwallis CK (2021) Extreme temperatures compromise male and female fertility in a large desert bird. Nat Commun 12:666. https://doi.org/10.1038/s41467-021-20937-7

Schulten K, Swenberg CE, Weller A (1978) A biomagnetic sensory mechanism based on magnetic field modulated coherent electron spin motion. Z Phys Chem NF111:1–5

Schulten K, Windemuth A (1986) Model for a physiological magnetic compass. In: Maret G, Boccara N, Kiepenheuer J (eds) Biophysical effects of steady magnetic fields. Springer, Berlin, pp 99–106

Schütz KE, Kerje S, Jacobsson L, Forkman B, Carlborg O, Andersson L, Jensen P (2004) Major growth QTLs in fowl are related to fearful behavior: possible genetic links between fear responses and production traits in a red junglefowl x white leghorn intercross. Behav Genet 34:121–130

Schulz-Mirbach T, Metscher B, Ladich F (2012) Relationship between swim bladder morphology and hearing abilities—a case study on Asian and African cichlids. PLoS ONE 7:e42292. https://doi.org/10.1371/journal.pone.0042292

Schweizer M, Liu Y (2018) Avian diversity and distributions and their evolution through space and time. In: Tietze D (ed) Bird species. Springer, Cham (Switzerland), pp 129–146

Scott CPG (1897) The Malayan words in English. J Am Oriental Soc 18:49–124

Scott GR (2011) Elevated performance: the unique physiology of birds that fly at high altitudes. J Exp Biol 214:2455–2462

Scott GR, Dawson NJ (2017) Flying high: the unique physiology of birds that fly at high altitude. In: Maina JN (ed) The biology of the avian respiratory system. Springer, Berlin, pp 113–126

Scott GR, Hawkes LA, Frappell PB, Butler PJ, Bishop CM, Milsom WK (2015) How bar-headed geese fly over the Himalayas. Physiology 30:107–115

Scott GR, Meir JU, Hawkes LA, Frappell PB, Milsom WK (2011a) Point: counterpoint: high altitude is/is not for the birds. J Appl Physiol 111:1514–1519

Scott GR, Milsom WK (2006) Flying high: a theoretical analysis of the factors limiting exercise performance in birds at altitude. Respir Physiol Neurobiol 154:284–301

Scott GR, Milsom WK (2007) Control of breathing and adaptation to high altitude in the bar-headed goose. Am J Regul Integr Comp Physiol 293:R379–R391

Scott GR, Schulte PM, Egginton S, Scott AL, Richards JG, Milsom WK (2011b) Molecular evolution of cytochrome C oxidase underlies high-altitude adaptation in the bar-headed goose. Mol Biol Evol 28:351–363

Seddon PJ, Van Heezik Y (1992) Parent–offspring recognition in the jackass penguin. J Field Orn 64:27–31

Seed A, Hanus D, Call J (2015) Causal knowledge in corvids, primates, and children more than meets the eye? In: McCormack T, Hoerl C, Butterfill S (eds) Tool use and causal cognition. Oxford Scholarship Online, Oxford, pp 1–35

Seki R, Li C, Fang Q, Hayashi S, Egawa S, Hu J, Xu L, Pan H, Kondo M, Sat T et al (2017) Functional roles of Aves class-specific cis-regulatory elements on macroevolution of bird-specific. Nat Commun 8:14229. https://doi.org/10.1038/ncomms14229

Sekine K, Ohuchi H, Fujiwara M, Yamasaki M, Yoshizawa T, Sato N, Nagishita N, Matsui D, Koga Y, Itoh N, Kato S (1999) Fgf10 is essential for limb and lung formation. Nat Genet 21: 138–141

Sellers WI, Manning PL (2007) Estimating dinosaur maximum running speeds using evolutionary robotics. Proc R Soc (Lond) B 274:2711–2716

Selvatti AP, Gonzaga LP, Russo CA, de Moraes CA (2015) A Paleogene origin for crown passerines and the diversification of the Oscines in the New World. Mol Phylogenet Evol 88: 1–15

Sereno PC (2004) Birds as dinosaurs. Acta Zool (Sinica) 50:991–1001

Seto KC, Güneralp B, Hutyra LR (2012) Global forecasts of urban expansion to 2030 and direct impacts on biodiversity and carbon pools. Proc Natl Acad Sci USA 109:16083–16088

Shanahan DF, Strohbach MW, Warren PS, Fuller RA (2014) The challenges of urban living. In: Gil D, Brumm H (eds) Avian urban ecology: behavioural and physiological adaptations. Oxford University Press, Oxford, pp 3–20

Shapiro MD, Domyan ET (2013) Domestic pigeon. Curr Biol 23:R302–R303

Shenstone B (1968) Unconventional flight. Aeronaut J 72:655–666

Shepherd K (2005) Birds and Eskimos. In: King JCH, Pauksztat B, Storrie R (eds) Arctic clothing. McGill-Queen's Press, Montreal, pp 64–65

Shochat E, Nilsson C, Lisovski S, Chernetsov N (2022) Editorial: optimal bird migration: implications for navigation, physiology, and stopover ecology. Front Ecol Evol 10:1029958. https://doi.org/10.3389/fevo.2022.1029958

Short HL (1964) Postnatal stomach development of white-tailed deer. J Wildl Manag 28:445–458

Shumaker RW, Walkup KR, Beck BB (2011) Animal tool behavior: the use and manufacture of tools by animals. Johns Hopkins University Press, Baltimore (MD)

Shute S, Prideaux GJ, Worthy TH (2017) Taxonomic review of the late Cenozoic megapodes (Galliformes: Megapodiidae) of Australia. R Soc Open Sci 4:170233. https://doi.org/10.1098/rsos.170233

Sibley CG, Ahlquist JE (1973) The relationships of the hoatzin. Auk 90:1–13

Sibley CG, Ahlquist JE (1990) Phylogeny and classification of birds: a study in molecular evolution. Yale University Press, New Haven (CT)

Sibley CG, Monroe BL (1990) Distribution and taxonomy of birds of the world. Yale University Press, New Haven (CT)

Simon M (2015) Absurd creature of the week: the world's tiniest bird weighs less than a dime. https://www.wired.com/2015/07/absurd-creature-of-the-week-bee-hummingbird/ - accesed on 24-04-22

Simmons NB, Seymour KL, Habersetzer J, Gunnell GF (2008) Primitive Early Eocene bat from Wyoming and the evolution of flight and echolocation. Nature 451:818–821

Skovgaard N, Abe AS, Andrade DV, Wang T (2005) Hypoxic pulmonary vasoconstriction in reptiles: a comparative study of fourspecies with different lung structures and pulmonary blood pressures. Am J Physiol Regul Integr Comp Physiol 289:R1280–R1288

Slabbekoorn H, Peet M (2003) Birds sing at a higher pitch in urban noise. Nature 424:267

Slawinska A, Hsieh JC, Schmidt CJ, Lamont SJ (2016) Heat stress and lipopolysaccharide stimulation of chicken macrophage-like cell line activates expression of distinct sets of genes. PLoS ONE 11:e0164575. https://doi.org/10.1371/journal.pone.0164575

Smith FA, Boyer AG, Brown JH, Costa DP, Dayan T, Ernest SKM, Evans AR, Fortelius M, Gittleman JL, Hamilton MJ et al (2010) The evolution of maximum body size of terrestrial mammals. Science 330:1216

Smith FA, Brown JH, Haskell JP, Lyons SK, Alroy J, Charnov EL, Dayan T, Enquist BJ, Ernest SK, Hadly EA (2004) Similarity of mammalian body size across the taxonomic hierarchy and across space and time. Am Nat 163:672–691

Smith FA, Lyons KS (eds) (2013) Animal body size: linking pattern and process across space, time, and taxonomic group. University of Chicago Press, Chicago (IL)

Smith JA (1939) Living fish of mesozoic type. Nature 143:455–456

Snyder NFR, Schmitt NJ (2002) In: Poole A (ed) California Condor (*Gymnogyps californianus*). Cornell Lab of Ornithology, Ithaca, New York. [Online.] Available at na.birds.cornell.edu/bna/species/610. Retrieved on 28-09-2020

Sol D, Duncan RP, Blackburn TM, Cassey P, Lefebvre L (2005) Big brains, enhanced cognition, and response of birds to novel environments. Proc Natl Acad Sci USA 102:5460–5465

Somveille M, Rodrigues ASL, Manica A (2015) Why do birds migrate? A macroecological perspective. Global Ecol Biogeogr 24:664–674

Speakman JR, Thomas DW (2003) Physiological ecology and energetics of bats. In: Kunz TH, Fenton MB (eds) Bat ecology. University of Chicago Press, Chicago, pp 430–490

Spottiswoode CN, Begg KS, Begg CM (2016) Reciprocal signaling in honeyguide-human mutualism. Science 353:387–389

Stacho M, Herold C, Rook N, Wagner H, Axer M, Amunts K, Güntürkün O (2020) A cortex-like canonical circuit in the avian forebrain. Science 369:eabc5534. https://doi.org/10.1126/science.abc5534

Starck JM, Sutter E (2000) Patterns of growth and heterochrony in moundbuilders (Megapodiidae) and fowl (Phasianidae). J Avian Biol 31:527–547

St Clair JJ, Rutz C (2013) New Caledonian crows attend to multiple functional properties of complex tools. Philos Trans R Soc (Lond) B Biol Sci 368:20120415. https://doi.org/10.1098/rstb.2012.0415

St-Pierre NR, Cobanov B, Schnitkey G (2003) Economic losses from heat stress by US livestock industries. J Dairy Sci 86:E52–E77

Steadman DW (2006) Extinction and biogeography of tropical pacific birds. University of Chicago Press, Chicago (IL)

Steen R, Kagge EO, Lilleengen P, Lindemann JP, Midtgaard F (2020) Wingbeat frequencies in free-ranging hummingbirds in Costa Rica and Ecuador. Cotinga 42:3–8

Steidl RJ, Powell BF (2006) Assessing the effects of human activities on wildlife. The George Wright forum, Vol. 23, no. 2, 2006. pp. 50-58. Published by the George Wright Society. http://www.jstor.org/stable/43598938.

Stervander M, Ryan PG, Melo M, Hansson B (2019) The origin of the world's smallest flightless bird, the Inaccessible Island Rail *Atlantisia rogersi* (Aves: Rallidae). Mol Phylogen Evol 130: 92–98

Stevens L (1991) Genetic and evolution of the domestic fowl. Cambridge University Press, New York

Stevens L (1996) Avian biochemistry and molecular biology. Cambridge University Press, Cambridge

Steyn L, Maina JN (2015) Comparison of the numbers of free (surface) macrophages in the respiratory systems of three species of birds in an urban and a rural area of South Africa. J Ornithol 156:1085–1093

Stiels D, Schidelko (2018) Modeling avian disttributions and niches: insights into invasions and speciation in birds. In: Tietze D (ed) Bird species. Fascinating life sciences, Springer, Cham (Switzerland), pp 147–162

Stoddard PK, Marsden JE, Williams TC (1983) Computer simulation of autumnal bird migration over the western north Atlantic. Anim Behav 31:173–180

Stokes DL, Boersma PD (2000) Nesting density and reproductive success in a colonial seabird, the Magellanic penguin. Ecology 81:2878–2891

Stolen ED, Breininger DR, Frederick PC (2005) Using waterbirds as indicators in estuarine systems: successes and perils. In: Bortone SA (ed) Estuarine indicators. CRC Press, Boca Raton, pp 409–422

Storey AA, Athens JS, Bryant D, Carson M, Emery K, deFrance S, Higham C, Huynen L, Intoh M, Jones S et al (2012) Investigating the global dispersal of chickens in prehistory using ancient

mitochondrial DNA signatures. PLoS ONE 7(7):e39171. https://doi.org/10.1371/journal.pone.0039171
Storz JF, Bridgham JT, Kelly SA, Garland T (2015) Genetic approaches in comparative and evolutionary physiology. Am J Physiol Reg Integr Comp Physiol 306(3):R197–R214
Strikwerda TE, Fuller MR, Seegar WS, Howey PW, Black HD (1986) Bird born satellite transmitter and location program. Johns Hopkins APL Tech Digest 7:203–208
Suarez RK (1992) Hummingbird flight: sustaining the highest mass-specific metabolic rates among vertebrates. Experentia 48:565–570
Suarez RK (1995) Sustaining the highest mass-specific metabolic rates in the animal kingdom. Physiol Zool 68:23–25
Suarez RK (1998) Oxygen and the upper limits to animal design and performance. J Exp Biol 201: 1065–1072
Suarez RK, Lighton JR, Brown GS, Mathieu-Costello O (1991) Mitochondrial respiration in hummingbird flight muscles. Proc Natl Acad Sci USA 88:4870–4873
Sullivan TN, Wang B, Espinosa HD, Meyers MA (2017) Extreme lightweight structures: Avian feathers and bones. Materials Today 20:377–391
Sumasgutner P, Nemeth E, Tebb G, Krenn HW, Gamauf A (2014) Hard times in the city—attractive nest sites but insufficient food supply lead to low reproduction rates in a bird of prey. Front Zool. 11:48. https://doi.org/10.1186/1742-9994-11-48
Suryan RM, Anderson DJ, Shaffer SA, Roby DD, Tremblay Y, Costa DP, Sievert PR, Sato F, Ozaki K, Balogh GR et al (2008) Wind, waves, and wing loading: morphological specialization may limit range expansion of endangered albatrosses. PLoS ONE 3:e4016. https://doi.org/10.1371/journal.pone.0004016
Susan AC, Potts R, Aeillo LC (2014) Evolution of early Homo: an integrated biological perspective. Science 345:45–59
Suthers RA, Zollinger SA (2004) Producing song: the vocal apparatus. Ann N Y Acad Sci 1016: 109–129
Sutton LJ, Anderson DL, Franco M, McClure CJW, Miranda EBP, Vargas FH, Gonzázalez DJV, Puschendorf R (2022) Reduced range size and important bird and biodiversity area coverage for the Harpy eagle predicted from future climate charge scenarios. Ibis. https://doi.org/10.1111/ibi.13046
Suzuki KK, Yanagawa H (2019) Gliding pattersns of Siberian flying squirrels in relation to forest structure. IForest—Biogeosci Forestry 12:114–117
Syvitski JW, Day CN, Colin ND, Milliman JD, Summerhayes C, Steffen W, Zalasiewicz J, Cearreta A, Gałuszka A, Hajdas I et al (2020) Extraordinary human energy consumption and resultant geological impacts beginning around 1950 CE initiated the proposed Anthropocene Epoch. Commun Earth Environ 1:32. https://doi.org/10.1038/s43247-020-00029-y
Swanson DL, Merkord C (2013) Seasonal phenotypic flexibility of flight muscle size in small birds: a comparison of ultrasonography and tissue mass measurements. J Ornithol 154:119–127
Swanson DL, Zhang Y, King MO (2013) Individual variation in thermogenic capacity is correlated with flight muscle size but not cellular metabolic capacity in American goldfinches (*Spinus tristis*). Physiol Biochem Zool 86:421–431
Swartz SM, Freeman PW, Stockwell EF (2003) Ecomorphology of bats: comparative and experimental approaches relating structural design to ecology Papers in Natural Resources. 10. https://digitalcommons.unl.edu/natrespapers/10
Swenson ER (2013) Hypoxic pulmonary vasoconstriction. High Alt Med Biol 14:101–110
Swenson ER, Bärtsch P (2012) High altitude pulmonary oedema. Comp Physiol 2:2753–2773
Switzer PV, Cristol DA (1999) Avian prey-dropping behavior. I. The effects of prey characteristics and prey loss. Behav Ecol 10:213–219
Sylvester JT, Shimoda LA, Aaronson PI, Ward JPT (2012) Hypoxic pulmonary vasoconstriction. Physiol Rev 92:367–520

Tabur MA, Ayvaz Y (2010) Ecological importance of birds. Second International symphosium on sustainable development, June 8-9 2010. Sarajevo, Bosnia-Herzegovina. pp. 560–565. https://omeka.ibu.edu.ba/items/show/3136/1/issd2010_science_book

Takekawa JY, Heath SR, Douglas DC, Perry WM, Javed S, Newman SH, Suwal RN, Rahmani AR, Choudhury BC, Prosser DJ et al (2009) Geographic variation in bar-headed geese *Anser indicus*: connectivity of wintering areas and breeding grounds across a broad front. Wildfowl 59:100–123

Takezaki N, Figueroa F, Zaleska-Rutczynska Z, Takahata N, Klein J (2004) The phylogenetic relationship of tetrapod, coelacanth, and lungfish revealed by the sequence of forty-four nuclear genes. Mol Biol Evol 21:1512–1524

Takezaki N, Nishihara H (2016) Resolving the phylogenetic position of coelacanth: the closest relative is not always the most appropriate outcrop. Genome Biol Evol 8:1208–1221

Takezaki N, Nishihara H (2017) Support for lungfish as the closest relative of tetrapods by using slowly evolving ray-finned fish as the outgroup. Genome Biol Evol 9:93–101

Tamisiea J (2022) Flying frogs. Sci Am 326:6. https://www.scientificamerican.com/article/how-parachute-frogs-took-to-the-sky/. Accessed on 12 July 2023

Tang Q, Wan B, Yuan X, Muscente AD, Xiao S (2019) Spiculogenesis and biomineralization in early sponge animals. Nat Commun 10:3348. https://doi.org/10.1038/s41467-019-11297-4

Tang Q, Xu Q, Ding C, Zhang H, Ling Y, Wu C, Fang M (2021) HIF-1 regulates energy metabolism of the Tibetan chicken brain during embryo development under hypoxia. Am J Physiol Regul Integr Comp Physiol 320:R704–R713

Tate P (2008) Flights of fancy: birds in myth, legend, and superstition. Arrow Books, UK

Tatsumi N, Kobayashi R, Yano T, Noda M, Fujimura K, Okada N, Okabea M (2016) Molecular developmental mechanism in polypterid fish provides insight into the origin of vertebrate lungs. Sci Rep 6:30580. https://doi.org/10.1038/srep30580

Tattersall GJ (2010) Skin breathing in amphibians. In: Aird WC (ed) Endothelial biomedicine. Cambridge University Press, Cambridge, pp 85–91. https://doi.org/10.1017/CBO9780511546198.010

Taylor EW, Leite CAC, McKenzie DJ, Wang T (2010) Control of respiration in fish, amphibians and reptiles. Braz J Med Biol Res 43:409–424

Taylor AH, Miller R, Gray RD (2012) New Caledonian crows reason about hidden causal agents. Proc Natl Acad Sci USA 109:16389–16391

Tebbich S, Bshary R (2004) Cognitive abilities related to tool use in the woodpecker finch, *Cactospiza pallida*. Anim Behav 67:689–697

Tefera M (2012) Acoustic signals in domestic chicken (*Gallus gallus*): a tool for teaching veterinary ethology and implication for language learning. Ethiop Vet J 16:77–84

Temple SA, Wiens JA (1989) Bird populations and environmental changes: can birds be bio-indicators. Am Birds 45:260–270

Teschke I, Cartmill EA, Stankewitz S, Tebbich S (2011) Sometimes tool-use is not the key: no evidence for cognitive adaptive specializations in tool-using woodpecker finches. Anim Behav 82:945–956

Thaxton P, Siegel H (1970) Immunodepression in young chickens by high environmental temperature. Poul Sci 49:202–205

Thomas BT (1996) Family Opisthocomidae (Hoatzins). In: del Hoyo J, Jordi A, Christie JA (eds) Handbook of the birds of the world, Vol. 3, Hoatzins to Auks. Lynx Edicions, Barcelona, pp 24–32

Thompson KS (1969) The environment and distribution of Paleozoic sarcopterygian fishes. Am J Sci 267:457–464

Thompson MG (1974) Migratory patterns of ruddy turnstones in the central Pacific region. Living Bird 12:5–23

Thorup K, Holland R (2009) The bird GPS. J Exp Biol 212:3597–3604

Thouless CR, Fanshawe JH, Bertram CR (1987) Egyptian vultures *Neophron percnopterus* and ostrich *Struthio camelus* eggs: the origins of stone-throwing behaviour. Ibis 131:9–15

Tickle PG, Ennos AR, Lennox LE, Perry SF, Codd JR (2007) Functional significance of the uncinate processes in birds. J Exp Biol 210:3955–3961
Tickle P, Nudds R, Codd J (2009) Uncinate process length in birds scales with resting metabolic rate. PLoS One 4:e5667. https://doi.org/10.1371/journal.pone.0005667
Tietze DT (2018) Introduction: studying birds in time and space. In: Tietze DT (ed) Bird species how they arise, modify and vanish. Fascinating Life Sciences Series. Springer, Cham (Switzerland), pp 1–7
Tiley GP, Kimball RT, Braun EL, Burleigh JG (2018) Comparison of the Chinese bamboo partridge and red junglefowl genome sequences highlights the importance of demography in genome evolution. BMC Genomics 19:336. https://doi.org/10.1186/s12864-018-4711-0
Tiley GP, Pandey A, Kimball RT, Braun EL, Burleigh JG (2020) Whole genome phylogeny of *Gallus*: introgression and data-type effects. Avian Res 11:7. https://doi.org/10.1186/s40657-020-00194-w
Timmel CR, Till U, Hore PJ (1998) Effects of weak magnetic fields on free radical recombination reactions. Mol Phys 95:71–89
Tinbergen N (1960) The herring gull's world. Basic Books, New York
Tissier J, Rage JC, Laurin M (2017) Exceptional soft tissues preservation in a mummified frog-eating Eocene salamander. PeerJ 5:e3861. https://doi.org/10.7717/peerj.3861
Titze IR (2018) Where has all the power gone? Energy production and loss in vocalization. Speech Commun 101:26–33
Titze IR, Maxfield L, Palaparthi A (2016) An oral pressure conversion ratio as a predictor of vocal efficiency. J Voice 30:398–406
Tixier-Boichard M, Bed'horn B, Rognon X (2011) Chicken domestication: from archeology to genomics. CR Biol 334:197–204
Tobalske BW (2001) Morphology, velocity, and intermittent flight in birds. Am Zool 41:177–187
Tobalske BW (2010) Hovering and intermittent flight in birds. Bioinspir Biomim 5:045004. https://doi.org/10.1088/1748-3182/5/4/045004
Tobalske BW, Biewener AA (2008) Contractile properties of the pigeon supracoracoideus during different modes of flight. J Exp Biol 211:170–179
Tobalske BW, Biewener AA, Warrick DR, Hedrick TL, Powers DL (2010) Effects of flight speed upon muscle activity in hummingbirds. J Exp Biol 213:2515–2523
Tobalske BW, Hedrick TL, Dial KP, Biewener AA (2003) Comparative power curves in bird flight. Nature 421:363–366
Tobalske BW, Warrick DR, Clark CJ, Powers DR, Hedrick TL, Hyder GA, Biewener AA (2007) Three-dimensional kinematics of hummingbird flight. J Exp Biol 210:2368–2382
Todd GT (1980) Evolution of the lung and the origin of bony fishes—a casual relationship? Am Zool 20:757A
Torday JS (2013) Evolutionary biology redux. Perspect Biol Med 56:455–484
Torday JS, Rehan VK (2012) Evolutionary biology, cell-cell communication and complex disease. Wiley, New Jersey
Torday JS, Rehan VK, Hicks JW, Wang T, Maina J, Weibel ER, Hsia CW, Sommer RJ, Perry SF (2007) Deconvoluting lung evolution: from phenotypes to gene regulatory networks. Integr Compar Biol 47:601–609
Torrella JR, Fouces V, Palomegue J, Viscor G (1998) Comparative skeletal muscle morphometry among wild birds with different locomotor behavior. J. Anat 192:211–222
Trenberth KE (2018) Climate change caused by human activities is happening and it already has major consequences. J Energy Natur Res Law 36:463–481
Trense D, Tietze DT (2018) Studying speciation: genomic essentials and approaches. In: Tietze D (ed) Bird species. Fascinating Life Sciences. Springer, Cham (Switzerland), pp 39–61
Troxler H (2008) The electoral abuse in the late Roman Republic. Masters of Science Thesis, University of South Florida (FL).
Tucker VA (1968) Bird metabolism during flight: evaluation and theory. J Exp Biol 58:689–709
Tucker VA (1971) Flight energetics in birds. Amer Zool 11:115–124

Tucker VA (1973) Bird metabolism during flight—evaluation of a theory. J Exp Biol 58:689–709

Tucker VA (1974) Energetics of natural avian flight. In: Paynter RA (ed) Avian energetics. Nuttal Ornithological Club, Cambridge (MA), pp 298–333

Tucker VA (1975) The energetic cost of moving about: walking and running are extremely inefficient forms of locomotion. Much greater efficiency is achieved by birds, fish—and bicyclists. Amer Sci 63:413–419

Tucker V (1998) Gliding flight: speed and acceleration of ideal falcons during diving and pull out. J Exp Biol 414:403–414

Tucker VA, Parrott GC (1970) Aerodynamics of gliding flight in a falcon and other birds. J Exp Biol 52:345–367

Tuinen M, Hedges SM (2001) Calibration of avian molecular clocks. Mol Biol Evol 18:206–213

Turner AH, Makovicky PJ, Norell MA (2012) A review of dromaeosaurid systematics and paravian phylogeny. Bull Am Mus Nat Hist 371:1–206

Ueta M, Higuchi H (2002) Difference in migration pattern between adult and immature birds using satellites. Auk 119:832–835

Ueta M, Sato F, Nakagawa J, Mita N (2000) Migration routes and differences of migration schedule between adult and young Steller's Sea Eagles *Haliaeetus pelagicus*. Ibis 142:35–39

Uezu A, Metzger JP (2011) Vanishing bird species in the Atlantic Forest: relative importance of landscape configuration, forest structure and species characteristics. Biodivers Conserv 20: 3627–3643

Unwin DM, Martill DM (2020) No protofeathers on pterosaurs. Nat Ecol Evol 4:1590–1591

Usherwood JR (2016) Physiological, aerodynamic and geometric constraints of flapping account for bird gaits, and bounding and flapping flight strategies. J Theor Biol 408:42–52

Vágási CI, Pap PL, Vincze O, Osváth G, Erritzøe J, Møller AP (2016) Morphological adaptations to migration in birds. Evol Biol 43:48–59

Vale MM, Tourinho L, Lorini ML, Rajão H, Figueiredo MSL (2018) Endemic birds of the Atlantic Forest: traits, conservation status, and patterns of biodiversity. J Field Ornithol 89:193–206

Vamsi D, Ryosuke OT, Cooper BG, Hahnloser RHR, Kojima S, Sober SJ, Todd RF (2019) Transitioning between preparatory and precisely sequenced neuronal activity in production of a skilled behavior. eLife 8:e43732. https://doi.org/10.7554/eLife.43732.001

Van De Crommenacker J, Bunbury N, Jackson HA, Nupen LJ, Wanless R, Fleischer-Dogley F, Groombridge JJ, Warren BH (2019) Rapid loss of flight in the Aldabra white-throated rail. PLoS ONE 14(12):e0226064. https://doi.org/10.1371/journal.pone.0226064

Van der Swan H, Visser C, van der Sluis R (2019) Plumage colour variations in the *Agapornis* genus: a review. J Afric Ornithol 90:1–10

Van Faassen EE, Bahrami S, Feelisch M, Hogg N, Kelm M, Kim-Shapiro DB, Kozlov AV, Li H, Lundberg JO, Mason R, Nohl H et al (2009) Nitrite as regulator of hypoxic signaling in mammalian physiology. Med Res Rev 59:683–741

Vargas AO, Fallon JF (2005) Birds have dinosaur wings: the molecular evidence. J Exp Zool B Mol Dev Evol 304:86–90

Vernes K (2001) Gliding performance of the northern flying squirrel (*Glaucomys sabrinus*) in mature mixed forest of Earstern Canada. J Mammal 82:1026–1033

Vetter BJ, Sisneros JA (2020) Swim bladder enhances lagenar sensitivity to sound pressure and higher frequencies in female plainfin midshipman (*Porichthys notatus*). J Exp Biol 223: jeb225177. https://doi.org/10.1242/jeb.225177

Videler JJ (2006) Avian flight. Oxford University Press, Oxford (UK)

Visser ME, Perdeck AC, van Balen JH, Both C (2009) Climate change leads to decreasing bird migration distances. Glob Chang Biol 15:1859–1865

Vitt LJ, Caldwell JP (2013) Water balance and gas exchange. In: Vitt LJ, Caldwell JP (eds) Herpetology, 4th edn: an introductory biology of amphibians and reptiles. Academic Press, San Diego, pp 181–202

Voelkl B, Fritz J (2017) Relation between travel strategy and social organization of migrating birds with special consideration of formation flight in the northern bald ibis. Philos Trans R Soc (Lond) B 372:20160235. https://doi.org/10.1098/rstb.2016.0235

Voeten DFAE, Cubo J, de Margerie E, Röper M, Beyrand V, Stanislav B, Tafforeau P, Sanchez S (2018) Wing bone geometry reveals active flight in *Archaeopteryx*. Nature Commun 9:923. https://doi.org/10.1038/s41467-018-03296-8

Wada H (2010) The development of birdsong. Nat Educ Knowledge 3(10):86. https://www.nature.com/scitable/knowledge/library/the-development-of-birdsong-16133266/

Wade ASI, Barov B, Burfield IJ, Gregory RD, Norris K, Vorisek P, Wu T, Butler SJ (2014) A niche-based framework to assess current monitoring of European forest birds and guide indicator species' selection. PLoS ONE 9(5):e97217. https://doi.org/10.1371/journal.pone.0097217. eCollection 2014

Wadley L, Hodgskiss T, Grant M (2009) Implicatios for complex cognition from the hafting of tolls with compound adhesives in the Middle Stone Age, South Africa. Proc Natl Acad Sci USA 106: 9590–9594

Wake DB, Vredenburg VT (2008) Colloquium paper: are we in the midst of the sixth mass extinction? A view from the world of amphibians. Proc Natl Acad Sci USA 115:11466–11473

Wallraff HG, Kiepenheuer J, Neumann MF, Streng A (1995) Homing experiments with starlings deprived of the sense of smell. Condor 97:20–26

Walsh PT, Hansellc M, Borellod WD, Healy SD (2013) Are elaborate bird nests built using simple rules? Avian Biol Res 6:157–162

Wang J, Hong M, Long J, Yin Y, Xie J (2022) Differences in intestinal microflora of birds among different ecological types. Front Ecol Evol 10:920869. https://doi.org/10.3389/fevo.2022.920869

Wang L, Cheung JTM, Pu F, Li D, Zhang M, Fan Y (2011) Why do woodpeckers resist head impact injury: a biomechanical investigation. PLoS ONE 6:e26490. https://doi.org/10.1371/journal.pone.0026490

Wang L, Niu X, Ni Y, Xu P, Liu X, Lu S, Ming Z, Fan Y (2013) Effect of microstructure of spongy bone in different parts of woodpecker's skull on resistance to impact injury. J Nanomater 2013: 924564. https://doi.org/10.1155/2013/924564

Wang M, Zheng X, O'Connor JK, Lloyd GT, Wang X, Wang Y, Zhang X, Zhou Z (2015) The oldest record of Ornithuromorpha from the Early Cretaceous of China. Nat Commun 6:6987. https://doi.org/10.1038/ncomms7987

Wang M, Zhou Z (2016) A new adult specimen of the basalmost ornithuromorph bird *Archaeorhynchus spathula* (Aves: Ornithuromorpha) and its implications for early avian ontogeny. J System Paleontol 15:1–18

Wang M, Zhou Z (2017) The evolution of birds with implications from new fossil evidences. In: Maina JN (ed) The biology of the avian respiratory system: evolution, development, structure and function. Springer, Berlin, pp 1–26

Wang X, O'Connor JK, Maina JN, Pan Y, Wang M, Wang Y, Zheng X, Zhou Z (2018) A specimen of *Archaeorhynchus* preserving significant soft tissue including the first probable occurrence of fossilized lungs in an archosaur. Proc Nat Acad Sci USA 115:11555–11560

Wang YS, Li M, Tell LA, Baynes RE, Davis JL, Vickroy TW, Riviere JE, Zhoumeng L (2020) Physiological parameter values for physiologically based pharmacokinetic models in food-producing animals. Part II: chicken and turkey. J Vet Pharmacol Ther 44:423–455

Ward MP, Benson TJ, Deppe J, Zenzal TJ Jr, Diehl RH, Celis-Murillo A, Bolus R, Moore FR (2018) Estimating apparent survival of songbirds crossing the Gulf of Mexico during autumn migration. Proc R Soc (Lond) B 285:20181747. https://doi.org/10.1098/rspb.2018.1747

Ward S, Bishop CM, Woakes AJ, Butler PJ (2002) Heart rate and the rate of oxygen consumption of flying and walking barnacle geese (*Branta leucopsis*) and bar-headed geese (*Anser indicus*). J Exp Biol 205:3347–3356

Ward S, Moller U, Rayner JMV, Jackson DM, Bilo D, Nachtigall W, Speakman JR (2001) Metabolic power, mechanical power and efficiency during wind tunnel flight by the European starling *Sturnus vulgaris*. J Exp Biol 204:3311–3322

Warner RW (1972a) The anatomy of the syrinx in passerine birds. J Zool (Lond) 168:381–393

Warner RW (1972b) The syrinx in family Columbidea. J Zool (Lond) 166:385–390

Warrick D, Hedrick T, Fernández J, Tobalske B, Biewener A (2012) Hummingbird flight. Curr Biol 22:R472–R477

Warrick DR, Tobalske BW, Powers DL (2005) Aerodynamics of the hovering hummingbird. Nature 435:1094–1097

Wassenbergh S, Mielke M, Böhmer C, Shadwick RE, Abourachid A (2022) Woodpeckers minimize cranial absorption of shocks. Curr Biol 32:P3189–P3194

Wassnetzov W (1932) Über die Morphologie der Schwimmblase. Zool Jahrb Abteilung Anat Ontog Tiere 56:1–36

Watson AS, Plog S, Culleton BJ, Gilman PA, LeBlanc SA, Whiteley PM, Claramunt S, Kennett DJ (2015) Early procurement of scarlet macaws and the emergence of social complexity in Chaco Canyon, NM. Proc Natl Acad Sci USA 112:8238–8243

Weber RE (2007) High-altitude adaptations in vertebrate hemoglobins. Respir Physiol Neurobiol 158:132–142

Weber RE, Jessen TH, Malte H, Tame J (1993) Mutant hemoglobins (alpha 119-Ala and beta 55-Ser): functions related to high-altitude respiration in geese. J Appl Physiol 75:2646–2655

Wedel MJ (2001) The evolution of vertebral pneumaticity in the sauropoda. MSc Thesis, University of oklahoma.

Wedel MJ (2003) Vertebral pneumaticity, air sacs, and the physiology of sauropod dinosaurs. Paleobiology 29:243–255

Wedel MJ (2006) Origin of postcranial skeletal pneumaticity in dinosaurs. Integr Zool 2:80–85

Wedel MJ (2009) Evidence for bird-like air sacs in saurischian dinosaurs. J Exp Zool 311A:611–628

Weeks BC, Willard DE, Zimova M, Ellis AA, Witynski ML, Hennen M, Winger BM (2019) Shared morphological consequences of global warming in North American migratory birds. Ecol Lett 23:316–325

Weidensaul S (1999) Living on the wind: across the hemisphere with migratory birds. North Point Press, New York

Weimerskirch H, Guionnet T, Martin J, Shaffer SA, Costa DP (2000) Fast and fuel efficient? Optimal use of wind by flying albatrosses. Proc Biol Sci 267:1869–1874

Weimerskirch H, Delord K, Guitteaud A, Phillips RA, Pinet P (2015) Extreme variation in migration strategies between and within wandering albatross populations during their sabbatical year, and their fitness consequences. Sci Rep 5:8853. https://doi.org/10.1038/srep08853

Weir JT, Schluter D (2008) Calibrating the avian molecular clock. Mol Ecol 17:2321–2328

Weis-Fogh T (1972) Energetics of hovering flight in hummingbirds and in *Drosophila*. J Exp Biol 56:79–104

Wells DJ (1993) Muscle performance in hovering hummingbirds. J Exp Biol 178:39–57

Welty JC (1982) The life of birds, 3rd edn. Saunders College Publ, Philadelphia (PA)

Welty JC, Baptista L (1988) The life of birds, 4th edn. WB Saunders and Co, London

Wesson PS (1980) Does gravity change with time? Phys Today 33(7):32. https://doi.org/10.1063/1.2914162

West B, Zhou BX (1988) Did chickens go north? New evidence for domestication. J Archaeol Sci 15:515–533

Wester P (2014) Feeding on the wing: hovering in nectar-drinking Old World birds—more common than expected. Emu 114:171–183

Whelan CJ, Wenny DG, Marquis RJ (2008) Ecosystem services provided by birds. Ann NY Acad Sci 1134:25–60

Whiteley JD, Pritchard JS, Slater PJB (1990) Strategies of mussel dropping by carrion crows *Corvus c. corone*. Bird Study 37:12–17

Whiteman L (2000) The high life. http://audubonmagazine.org/birds/birds0011.html – accessed on 24-04-22.

Wiemann J, Fabbri M, Yang T, Stein K, Sander PM, Norell MA, Briggs DEG (2018) Fossilization transforms vertebrate hard tissue proteins into N-heterocyclic polymers. Nat Commun 9:4741. https://doi.org/10.1038/s41467-018-07013-3

Wilkins MR, Ghofrani HA, Weissmann N, Aldashev A (2015) Pathophysiology and treatment of high-altitude pulmonary vascular disease. Circulation 131:582–590

Wilkins AS, Wrangham RW, Fitch WT (2014) The domestication syndrome in mammals: a unified explanation based on neural crest cell behavior and genetics. Genetics 197:795–808

Williams H (2006) Birdsong and singing behavior. Ann NY Acad Sci 1016:1–30

Williams H (2008) Bird song and singing behavior. In: Ziegler HP, Marler P (eds) The neuroscience of bird song. Cambridge University Press, Cambridge, pp 32–49

Williams HJ, Shepard LC, Holton MD, Alarcón PAE, Wilson RP, Lambertucci SA (2020) Physical limits of flight performance in the heaviest soaring bird. Proc Natl Acad Sci (USA) 117:17884–17890

Willmer P, Stone G, Johnston I (2005) Environmental physiology of animals, 2nd edn. Blackwell Science Ltd, Oxford (UK)

Wilson A, Lichtwark (2011) The anatomical arrangement of muscle and tendon enhances limb versatility and locomotor performance. Philos Trans R Soc (Lond) B 366:1540–1553

Wiltschko R, Wiltschko W (1995) Magnetic orientation in animals. Springer, Berlin (Germany)

Wing SL, Sues HD (1992) Mesozoic and early Cenozoic terrestrial ecosystems. In: Behrensmeyer AK, Damuth JD, DiMichele WA, Potts R, Sues HD, Wing SL (eds) Terrestrial ecosystems through time. University of Chicago Press, Chicago, pp 327–416

Wingfield JC, Krause JS, Perez JH, Chmura HE, Németh Z, Word KR, Calisi RM, Meddle SL (2015) A mechanistic approach to understanding range shifts in a changing world: what makes a pioneer? Gen Comp Endocrinol 222:44–53

Witmer LM (2009) Palaeontology: feathered dinosaurs in a tangle. Nature 461:601–602

Witton MP, Habib MB (2010) On the size and flight diversity of giant pterosaurs, the use of birds as pterosaur analogues and comments on pterosaur flightlessness. PLoS ONE 5:e13982. https://doi.org/10.1371/journal.pone.0013982

Wood BM, Pontzer H, Raichlen DA, Marlowe FW (2014) Mutualism and manipulation in hadza-honeyguide interactions. Evol Behav 35:540–546

Wood GA (1984) Tool use by the palm cockatoo, *Probosciger aterrimus*, during display. Corella 4: 94–95

Wood GA (1988) Further field observations of the palm cockatoo *Probosciger aterrimus* in the Cape York Peninsula, Queensland. Corella 12:48–52

World Bank (2020) Agricultural land (% of land area). The World Bank Group, Washington (DC)

Worthy TH (2001) A giant flightless pigeon gen. et sp. nov. and a new species of Ducula (Aves: Columbidae), from Quaternary deposits in FIJI. J R Soc N Z 31:763–794

Wright ADG, Northwood KS, Obispo NE (2009) Rumen-like methanogens identified from the crop of the folivorous South Ameican bird, the hoatzin (*Opisthocomus hoatzin*). ISME J 3:1120–1126

Wright D, Rubin C, Schutz K, Kerje S, Kindmark A, Brandström H, Andersson L, Pizzari T, Jensen P (2012) Onset of sexual maturity in female chickens is genetically linked to loci associated with fecundity and a sexual ornament. Reprod Domest Anim 47:31–36

Wright NA, Steadman DW, Witt CC (2016) predictable evolution toward flightlessness in volant island birds. Proc Natl Acad Sci (USA) 113:4765–4770

Wu L, Sun Y, Li J, Li Y, Wu Y, Li D (2015) A phylogeny of the Passerida (Aves: Passeriformes) based on mitochondrial 12S ribosomal RNA gene. Avian Res 6:1. https://doi.org/10.1186/s40657-015-0010-5

Wu W, Gao YD, Jianga DC, Leid J, Rena JL, Liaoe WB, Deng C, Wanga Z, Hillisg DM, Zhang YP, Lia JT (2022) Genomic adaptations for arboreal locomotion in Asian flying treefrogs. Proc Natl Acad Sci USA 119(13):e2116342119. https://doi.org/10.1073/pnas.211634211

Wuebbles DJ, Fahey DW, Hibbard KA, DeAngelo B, Doherty S, Hayhoe K, Horton R, Kossin JP, Taylor PC, Waple AM et al (2017) Executive summary. In: Wuebbles DJ, Fahey DW, Hibbard KA, Dokken DJ, Stewart BC, Maycock TK (eds) Climate Science Special Report: Fourth National Climate Assessment, Vol. I. U.S. Global Change Research Program, Washington, DC, USA, pp 12–34. https://doi.org/10.7930/J0DJ5CTG

Xiang H, Gao J, Yu B, Zhou H, Cai D, Zhang Y, Chen X, Wang X, Hofreiter M, Zhao X (2014) Early Holocene chicken domestication in northern China. Proc Natl Acad Sci (USA) 111: 17564–17569

Xu W, Wang L (2022) Editorial: the physiological and molecular response of aquatic animals to environmental stresses. Front Physiol 13:987004. https://doi.org/10.3389/fphys.2022.987004

Xu X (2006) Dinosaurs from China and the evolution of major avian characters. Integr Zool 1:4–11

Xu X, You H, Du K, Han F (2011) An *Archaeopteryx*-like theropod from China and the origin of Avialae. Nature 475:465–470

Xu X, Zheng X, Sullivan C, Wang X, Xing L, Wang Y, Zhang X, O'Connor JK, Zhang F, Pan Y (2015) A bizarre Jurassic maniraptoran theropod with preserved evidence of membranous wings. Nature 521:70–73

Xu X, Zhou Z, Dudley R, Mackem S, Chuong CM, Erickson GM, Varricchio DJ (2014) An integrative approach to understanding bird origins. Science 346:1253293. https://doi.org/10.1126/science.1253293

Yalden DW (2008) What size was *Archeopteryx*? Zool J Linn Soc 82:177–188

Yalden DW, Morris PA (1975) The lives of bats. The New York Times Book Company, New York

Yang S, He H, Jin F, Zhang F, Wu Y, Yu Z, Li Q, Wang M, O'Connor JK, Deng C, Zhu R, Zhou Z (2020) The appearance and duration of the Jehol Biota: constraint from SIMS U-Pb zircon dating for the Huajiying formation in northern China. Proc Natl Acad Sci USA 117:14299–14305

Yang Z, Jiang B, McNamara ME, Kearns SL, Pittman M, Kaye TG, Orr PJ, Xu X, Benton MJ (2019) Pterosaur integumentary structures with complex feather-like branching. Nat Ecol Evol 3:24–30

Yang L, Tan GY, Fu YQ, Feng JH, Zhang MH (2010) Effects of acute heat stress and subsequent stress removal on function of hepatic mitochondrial respiration, ros production and lipid peroxidation in broiler chickens. Comp Biochem Physiol Part C Toxicol Pharmacol 151:204–208

Yang X, Zhang T, Qin D, Kang S, Qin X (2011) Characteristics and changes in air temperature and glacier's response on the north slope of Mt. Qomolangma (Mt. Everest). Arct Antarct Alp Res 43:147–160

Yasuhiko S (2013) Terms of ornithomancy in Hittie. Tokyo University Linguistic Papers 33:219–238

Yeaton IJ, Ross SD, Baumgardner GA, Socha JJ (2020) Undulation enables gliding in flying snakes. Nat Phys 16:974–982

Yates AM, Wedel MJ, Bonnan MF (2012) The early evolution of postcranial skeletal pneumaticity in sauropodomorph dinosaurs. Acta Pal Pol 57:85–100

York YM, Scadeng M, McCracken KG, Milsom WK (2018) Respiratory mechanics and morphology of Tibetan and Andean high-altitude geese with divergent life histories. J Exp Biol 221: jeb170738. https://doi.org/10.1242/jeb.170738

Young GC (2010) Placoderms (armored fish): dominant vertebrates of the Devonian Period. Annu Rev Earth Planet Sci 38:523–550

Yu JJ, Non AL, Heinrich EC, Gu W, Alcock J, Moya EA, Lawrence ES, Tift MS, O'Brien KA, Storz JF et al (2022) Time domains of hypoxia responses and -omics insights. Front Physiol 13: 885295. https://doi.org/10.3389/fphys.2022.885295

Zaccone G, Lauriano ER, Capillo G, Kuciel M (2018) Air-breathing fish: air-breathing organs and control of respiration: nerves and neurotransmitters in the air-breathing organs and the skin. Acta Histochem 120:630–641

Zach R (1978) Selection and dropping of whelks by northwestern crows. Behaviour 67:134–138

Zach R (1979) Shell dropping: decision-making and optimal foraging in northwestern crows. Behaviour 68:106–117

Zeigler HP, Marler P (eds) (2008) Neuroscience of birdsong. Cambridge University Press, Cambridge (UK)

Zhang F, Kearns SL, Orr PJ, Benton MJ, Zhou Z, Johnson D, Xu X, Wang X (2010) Fossilized melanosomes and the colour of Cretaceous dinosaurs and birds. Nature 463:1075–1078

Zhang F, Zhou Z, Dyke GJ (2006) Feathers and 'feather-like' integumentary structures in Liaoning birds and dinosaurs. Geol J 41:395–404

Zhang G, Li C, Li Q, Li B, Larkin DM, Lee C, Storz JF, Antunes A, Greenwold MJ, Meredith RW et al (2014) Comparative genomics reveals insights into avian genome evolution and adaptation. Science 346:1311–1320

Zhang L, Zhang P, Li Q, Gaur U, Liu Y, Zhu Q, Zhao X, Wang Y, Yin H, Hu Y, Liu A, Li D (2017) Genetic evidence from mitochondrial DNA corroborates the origin of Tibetan chickens. PLOS ONE 12:e0172945. https://doi.org/10.1371/journal.pone.0172945

Zhao P, Li S, He Z, Zhao F, Wang J, Liu X, Li M, Hu J, Zhao Z, Luo Y (2022) Physiology and proteomic basis of lung adaptation to high-altitude hypoxia in tibetan sheep. Animals 12:2134. https://doi.org/10.3390/ani12162134

Zheng X, O'Connor J, Huchzermeyer F, Wang X, Wang Y, Wang M, Zhou Z (2013) Preservation of ovarian follicles reveals early evolution of avian reproductive behaviour. Nature 495:507–511

Zheng X, O'Connor JK, Huchzermeyer F, Wang X, Wang Y, Zhang X, Zhouet Z (2014) New specimens of *Yanornis* indicate a piscivorous diet and modern alimentary canal. PLoS ONE 9(4):e95036. https://doi.org/10.1371/journal.pone.0095036

Zheng X, O'Connor JK, Wang X, Pan Y, Wang Y, Wang M, Zhou Z (2017) Exceptional preservation of soft tissue in a new specimen of Eoconfuciusornis and its biological implications. Natnl Sci Rev 4:441–452

Zheng W, Wang Z, Collins JE, Andrews RM, Stemple D, Gong Z (2011) Comparative transcriptome analyses indicate molecular homology of zebrafish swim-bladder and mammalian lung. PLoS ONE 6:e24019. https://doi.org/10.1371/journal.pone.0024019

Zhou Z (2004) The origin and early evolution of birds: discoveries, disputes and perspectives from fossil evidence. Naturwissenschaften 91:455–471

Zhou Z (2014) The Jehol Biota, an Early Cretaceous terrestrial Lagerstätte: new discoveries and implications. Nat Sci Rev 1:543–559

Zhou ZH, Zhang FC (2006) Mesozoic birds of China—a synoptic review. Vertebr Palasiat 44:74–98

Zhou S, Zhou Z, O'Connor JK (2013) Anatomy of the basal ornithuromorph bird *Archaeorhynchus spathula* from the Early Cretaceous of Liaoning, China. J Vert Paleontol 33:141–152

Zhu Z, Zhang W, Wu C (2014) Energy conversion in woodpecker on successive peckings and its role on anti-shock protection of the brain. Sci China Technol Sci **57**:1269–1275

Zhu ZD, Ma GJ, Wu CW, Chen S (2012) Numerical study of the impact response of woodpecker's head. AIP Adv 2:042173. https://doi.org/10.1063/1.4770305

Zimmer C (1996) From teeth to beak. Discover 17:44–50

Zinßmeister D, Troupin D, Sapir N (2022) Autumn migrating passerines at a desert edge: do birds depart for migration after reaching a threshold fuel load or vary it according to the rate of fuel deposition? Front Ecol Evol 10:874923. https://doi.org/10.3389/fevo.2022.874923

Zollinger SA, Dorado-Correa A, Goymann W, Forstmeier W, Knief U, Bastidas-Urrutia AM, Brumm H (2019) Traffic noise exposure depresses plasma corticosterone and delays offspring growth in breeding zebra finches. Conserv Physiol 7(1):coz056. https://doi.org/10.1093/conphys/coz056. eCollection 2019

Chapter 2
Development of the Avian Respiratory System

Abstract Comprehensive systematic account of the development of the avian respiratory system, specifically that of the domestic fowl, *Gallus gallus* variant *domesticus*, is presented. Factors and conditions that determine and drive the intricate morphogenetic processes are presented and the most important morphological changes are specified. Ensuing developmental processes such as branching morphogenesis and the role that molecular factors (morphogenetic cues) play in elaboration of the complex morphology of the avian respiratory system is specified. Very early in its development, the lung is firmly affixed to the vertebrae and the ribs on the dorsolateral aspects and the air sacs form as blister-like outgrowths from the cranial, the ventral, and the caudal edges of the organ. At the end of the incubation period, i.e., at hatching (day 21), for the precocial domestic fowl, the respiratory system is well-developed for gas exchange: the air- and the blood capillaries are well-formed, the blood-gas barrier is very thin, and the respiratory surface area is large.

2.1 Developmental Biology: The less Well-Known and Explored Aspects

Embryonic development starts with fertilization that forms a zygote, a single cell that results from the fusion of gametes (i.e., the ovum or egg cell and the sperm) that divides into many cells that differentiate, grow, and migrate in certain ways to self-organize into structurally coherent and functionally well-coordinated assemblages that form diverse specialized tissues, organs, and organ systems (Ten Have-Opbroek 1981, 1991; Merkus and Ten Have-Opbroek 1996; Gilbert et al. 1996; Wolf et al. 2018; Gilbert and Sarkar 2000; Gilbert 2000; Gjorevski and Nelson 2010a; Beloussov 2010; Chanet and Martin 2014; Kindberg et al. 2020; Baral et al. 2022). Cells utilize intrinsic (genetic) and extrinsic signals by which they interact with the surrounding milieu to determine and establish their developmental trajectories and destinies (Callahan et al. 1997; West-Eberhard 2005; Wake 2008; Ungerer et al. 2008). In the long run, except for stem cells that retain their embryonic pluripotent replicative capacities (Lanza and Atala 2006; Allman 2006; Slack

© The Author(s), under exclusive license to Springer Nature Switzerland AG 2023

J. N. Maina, *Current Perspectives on the Functional Design of the Avian Respiratory System*, Zoological Monographs 8, https://doi.org/10.1007/978-3-031-35180-8_2

2021; Baral et al. 2022), most of the cells of the body differentiate into or adopt definitive distinctiveness (Lewontin 2000; Dietmar et al. 2013). Epigenetics, a term conceived by Waddington (1957) to mean a combination of mechanisms and phenomena that can induce variations in the phenotype of a cell or an organism without affecting the structure of its DNA is a decisive player in many biological processes (Baral et al. 2022). In organs and structures that exhibit well-defined three-dimensional (3D) physical configurations such as lungs where airways and blood vessels form by branching morphogenesis (Ten Have-Opbroek 1981; Hogan 1999; Wongtrakool et al. 2003; Lu et al. 2006; Yates et al. 2010) and secretory organs such as mammary and salivary glands and kidneys (Shah et al. 2004; Gomes et al. 2015), a complex morphology originates from a hard-wired genetic inventory of intricately orchestrated spatiotemporal expression of molecular cues (factors) as well as from existing mechanical forces that instruct cell differentiation processes such as proliferation, migration, death and apoptosis, i.e., programmed cell death (Hogan et al. 1997; Alsberg et al. 2004; Mattiske et al. 2006; Gjorevski and Nelson 2010a; Hsu et al. 2011; Maina 2012, 2013; Herriges and Morrisey 2014; Xu et al. 2017; Fernandes-Silva et al. 2017a).

Presently, the process of morphogenesis is one of the most vibrant, interesting, challenging, and important areas of medical sciences (Goodwin 1985; Beloussov and Grabovsky 2006; Funk et al. 2009; Farge 2013; Beloussov and Volodyaev 2013; Tyler 2014; Sagai et al. 2017; Whitsett et al. 2019; Hallou and Brunet 2020). For cells, tissues, and organs to function optimally, proper regulation of their growth, shape, size, and organization is necessary (Alsberg et al. 2004; Whitsett et al. 2019). Well before the now very active discipline of molecular biology was well-established, Alan Turing (1912–1954) (Turing 1952) advanced the "chemical kinetic theory" that stipulated that fast diffusing inhibitory and excitatory molecules which he termed "morphogens," spread through embryonic cells, setting up cellular patterning. It is now well-known that in both plants and animals, molecular cues (signals) and cytokines disseminate within tissues and cells, directing and establishing cell uniqueness while directing cell differentiation and creating spatial cell patterning (Rogers and Schier 2011; Kicheva et al. 2012; Guo et al. 2014; Green and Sharpe 2015). Cytokines are molecular factors that are involved in the regulation of the physiological processes that provide homeostasis (Hegazi and Abdel-Rahman 2015): they initiate and control movement, differentiation, growth, deaths, and replacement of cells (Toews 2001). For the mouse lung, Notch signaling regulates the balance between ciliated and secretory cells and their fates during airway development (Tsao et al. 2009). The concept of "morphospace" (Beloussov 2001; McGhee 2006; Mitteroecker and Huttegger 2009; Bizzarri et al. 2011; Levin 2012; Tyler 2014; Höckel 2015) has been advanced as means of mathematically analyzing and defining the particular morphologies that are practical in groups of organisms/animals occupying certain habitats and practicing particular lifestyles.

In his classic book entitled *On Growth and Form*, by mathematical and physical reasoning, D'Arcy Wentworth Thompson (1860–1948) (Thompson 1917) advanced elegant concepts on the mechanisms and processes by which tissues develop, grow, and pattern into complex assemblages that execute particular functions: he argued that the remarkable diversity in the morphologies of the extant animal life is not a stochastic consequence of evolution by natural selection but is rather buttressed by overarching "transformative laws of growth" under immutable mechanobiological physical forces (Arthur 2006; Ball 2013; Antony et al. 2016; Abzhanov 2017). Criticizing Charles Darwin's (1809–1882) (Darwin 1859) "theory of evolution of species by natural selection" as lacking in explaining the origin of new species since it totally ignored, overlooked, or de-emphasized any alternative evolutionary "experiments" that nature might have pursued during the development, growth, and advancement of animal life. Thompson (1917) advocated the notion that in the course of its development, life undeviatingly subscribed to physical laws and mechanics. While Thompson's work has been criticized for being largely descriptive, i.e., lacking in detailed mechanistic and morphometric facts and experimental data and therefore failing to explain multiplicative processes (Bookstein 1977; Arthur 2006), recently, application of robust and combinations of investigative methods such as experimental embryology, stem cell biology, microscopy, imaging, developmental genetics, molecular biology, and mathematical and computational sciences has provided important answers. Thompson's (1917) concept on the key drivers of development and growth has now been endorsed by among others the following investigators: (a) Boden (2008) stated that during the evolution of life, the effect of natural selection on its progress was secondary to that from physical influences; (b) Abzhanov (2017) contended that the functional designs of the biological entities conform with the principles of growth and development and are subservient to natural physical forces; (c) Gould (1971) contended that the internal growth parameters that regulate form and shape of biological structures can best be explained by changes in their morphospaces and by principles of physics and mathematics; (d) Briscoe and Kicheva (2017) acknowledged that physical principles constrain biological form and consequently hamper and regulate evolutionary progresses; Wolf et al. (2018) remarked that "biological entities obey the fundamental physical laws"; and e) Cockell (2017) noted that the remarkable diversity of life on Earth is regulated by universal principles that exist both at the macro- and the microworlds and went on to suggest that "physical laws certainly do limit the engineering solutions to life's problems."

Thompson (1942, p. 16) defined form as a "diagram of forces in which the external and internal forces are resolved in a state of stasis." We opine that if the laws of physics inflexibly specified morphological outcomes, life would either not have evolved at all and if it materialized, only one type would have continued: changes such as transformation from unicellular to multicellular forms would not have materialized. Katsnelson et al. (2018) remarked that "biological evolution in no way ignores the laws of physics, but the emerging biological phenomena initiate the further development of physics itself." The concept of "transformations by growth" (Thompson 1917, 1942) has now synergized with other scientific disciplines such as morphology, molecular biology, genetics, evolutionary biology, applied mathematics, and synthetic biology (Klingenberg and Zaklan 2000; Weston 2003; Klingenberg 2010; Mallarino and Abzhanov 2012; Yuri et al. 2018; Katsnelson et al. 2018). Fractal geometry is a branch of mathematics that seeks to explain the development of complex iterated patterns in natural structures (Mandelbrot 1977,

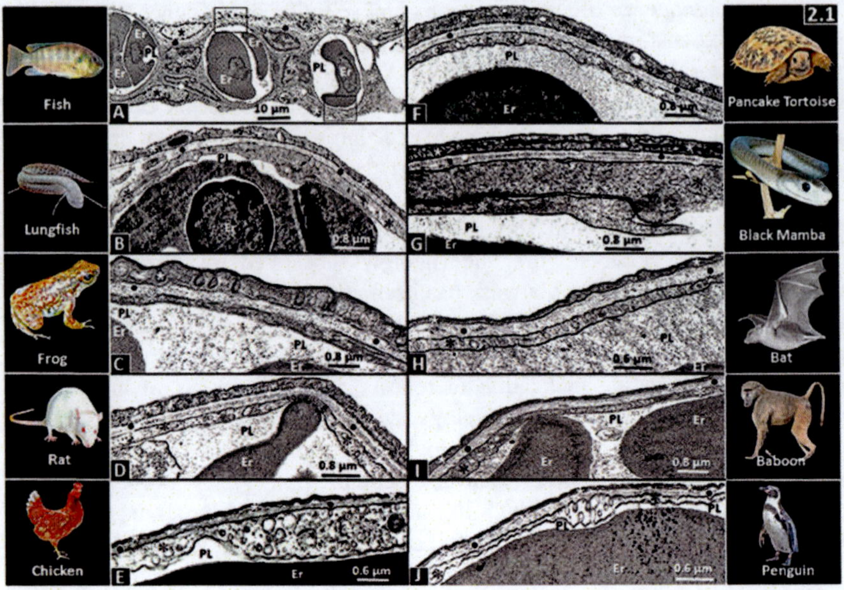

Plate 2.1 The highly conserved trilaminar or tripartite design (morphology) of the water-blood barrier of fish gills (**a**) and the blood-gas barriers of lungs of air-breathing vertebrates (**b-j**). The parts of the tissue barriers essentially comprise an epithelial cell (star, ★), a basement membrane (dot, •), an endothelial cell (asterisk, *). *PL* plasma layer; *Er* erythrocyte

1978, 1989; Weibel 2005; Di Ieva 2016; Essey and Maina 2020) and the growth and development of optimized biological structures (Weibel 1991, 1994; West et al. 1991; Brown et al. 2002; Losa 2002, 2009; Glenny 2011; Essey and Maina 2020).

For gas exchangers, the evolved gross morphological differences (Plate 1.6) could be attributed to variations of factors such as phylogenetic levels of development, lifestyles pursued, and habitat occupied (Maina 1998, 2000, 2002a, b, c, d, 2014; Hsia et al. 2013; Perry et al. 2019), common, i.e., shared, structural features such as the tripartite design of the water and blood gas (air) barriers (Plate 2.1, b-j) (Maina and West 2005) and increasing of the respiratory surface area by internal subdivision of lungs have happened (Plate 1.7). In part, the design of gas exchangers was influenced by the physical laws of gas diffusion such as the Graham's and Fick's Laws (Maxwell 1860; Petrucci et al. 2002). For example, depending on the respiratory fluid medium utilized (water or air), according to Fick's Law (Fick 1855), the diffusing capacity of a gas exchanger for O_2 correlates inversely with the thicknesses of the water and the blood gas (tissue) barriers and directly with the respiratory surface area (Weibel 2000). Morphogenesis comprises self-organizing developmental changes that are driven by physical forces that are created and conveyed across different levels of the structural organization under genetic and molecular instructions that work through multiple feedback loops (Coen 1999; Suzuki et al. 2012; Keller 2012; Heisenberg and Bellaïche 2013; Antony et al. 2016; Delile et al. 2017;

Chan et al. 2017). The forces that control cell and tissue growth and development are fundamentally generated by molecular motors and transmitted across cytoskeletal elements and adhesion molecules within and between cells (Ingber 2006; Parsons et al. 2010; Fletcher and Mullins 2010; Sanghvi-Shah and Weber 2017; Kanoldt et al. 2019). The importance of physical forces in the instruction of morphogenetic processes has given rise to new fields of developmental biology such as "mechanobiology," "physical biology," and "mechanotrasduction biology" (Wozniak and Chen 2009; Keller 2012; Mammoto et al. 2013). Mechanical forces lead to changes in cell size, shape, number, location, and gene expression (Ralston and Shaw 2008; Chan et al. 2017) while interactions between tissue mechanics and biochemical signaling orchestrate tissue patterning (Heisenberg and Bellaïche 2013).

Biological structures contend with constantly changing external and internal environmental states. Albeit the existing dynamic conditions, for physiological processes to occur optimally, at cellular level, homeostasis has to be maintained (Lim et al. 2006). A biomechanical environment is an important epigenetic determinant of the form, function, and fate of cells (Ingber and Folkman 1989; Alsberg et al. 2004; Müller and Newman 2005; Mammoto and Ingber 2010; Gjorevski and Nelson 2010a; Fletcher and Mullins 2010; Navis and Nelson 2016). Cells experience forces that originate from inside (endogenous forces), i.e., from deformable cytoskeleton which is a highly dynamic system that comprises an intricate network of filamentous polymers and regulatory proteins, and those that originate externally, i.e., exogenous or applied forces (Fletcher and Mullins 2010; Heisenberg and Bellaïche 2013). Cells constantly sense and assess their physiological conditions and accordingly attune their internal and external states and transduce biochemical signals into a range of cellular responses (Stamenović and Wang 1985; Janmey and McCulloch 2007). The forces that a cell is exposed to strengthen, soften, and re-organize cytoskeletal filaments (Katsumi et al. 2002; Ingber 2005; Discher et al. 2005; Gardel et al. 2006; Mammoto and Ingber 2010). More than 150 proteins contain binding domains for the protein actin that polymerizes to form important cytoskeletal filaments in cells (Dos Remedios et al. 2003). In various ways, mechanosensory and mechanotransductive events occur at the periphery of the cell (Discher et al. 2005; Giannone and Sheetz 2006; Farge 2011; Petridou et al. 2017; Roca-Cusachs et al. 2017; Chen et al. 2018; Hallou and Brunet 2020). From embryonic development to maturity, through mechanotransductive forces, stem cells and their immediate microenvironment crosstalk to regulate cell shape, behavior, and fate, ultimately determining outcomes such as cell patterning (Lecuit et al. 2011; Clément et al. 2012a, b; Vining and Mooney 2017; Uhler and Shivashankar 2017; Chan et al. 2017; Romani et al. 2021). The nucleus is an important mechanosensor (Kirby and Lammerding 2018) and integrin extracellular matrix receptors that form connections between the interior of a cell and the external microenvironment allow the molecular factors (integrins) to biomechanically detect, react to epithelial-mesenchymal interactions and interact with the extracellular membrane/matrix (Shannon et al. 1998; Weaver et al. 2000; Chinoy 2003; Turcatel et al. 2013; Antony et al. 2016; Spurlin et al. 2019; Kechagia et al. 2019).

In higher chordates, at different stages of embryogenesis, change of epithelial into mesenchymal cells is a defining process for the development of organ-specific structures (Hay 1995, 2005; Shook and Keller 2003; Lee et al. 2006; Kalluri and Weinberg 2009). Mediated by growth factors (GFs) and cytokines that serve as paracrine signals that control gene responses which lead to cell division and differentiation, the epithelial-mesenchymal cell transitions are phenotypically instructive (Ornitz and Itoh 2001; Antony et al. 2016; Witsch et al. 2010; Maina 2012; Nicolas et al. 2020). For organs and structures that develop a tubular morphology such as the gastrointestinal system, the neural tube of the spinal cord and those that possess branching tubular systems such as blood vessels and airways of vertebrate lungs and in ducts of secretory organs such as glands, the formation of a lumen entails intricately coordinated biomechanical and biochemical cellular interactions that control movements and organizations of cells around an open space and elongation along the hollow core (Hogan and Kolodziej 2002; Lubarsky and Krasnow 2003; Girosi et al. 2006; Yates et al. 2010; Sigurbjörnsdóttir et al. 2014; Navis and Nelson 2016). Tenascin, a multifunctional extracellular matrix glycoprotein (Ekblom and Aufderheide 1989; Chiquet-Ehrismann and Tucker 2011), is contained in the mesenchyme (Koch et al. 1991) where it promotes the lengthening of a developing bud or sprout (Abbott et al. 1991).

Albeit the important role(s) that the use of the wheel has contributed to human technological developments, it is perplexing that true (= active) locomotory wheels and propellers have not evolved in biological systems at the macro (gross) level of organization (Full et al. 1993; Gould 1981; LaBarbera 1983; Diamond 1983; Walker 1991; Meyer and Halbeisen 2006; Scholtz 2008). Limitations imposed by natural selection may have hindered the development of functional wheels during the progression toward complexity (Gould 1981; Walker 1991; Meyer and Halbeisen 2006). In order to freely rotate around an axis, functional wheels need a measure of independence from the main structure, i.e., the frame or the body. For the reason for proper control of processes and mechanisms, different parts of the organism/animal have to be interconnected, e.g., by the nervous system for coordination of function (s) and the cardiovascular system supplies O_2 and nutrients and removes the end-products of metabolism such as CO_2, extreme physical independence of the moving (locomotory) part is not very much tenable in the design of living entities. Furthermore, functional (locomotory) wheels may not have developed in the natural world because for them to be effective, they have to interact physically with a firm (hard) surface/substrate that provides grip (resistance) to create forward push or thrust. During the early evolution of animals when invertebrate animal life existed in abundance, the surrounding materials and surfaces were largely soft and/or fluidly and were therefore unsuitable for the development of a wheel-driven type of propulsion. The statement "Let us beware of saying there are laws in nature. There are only necessities," that is attributed to the philosopher Friedrich Wilhelm Nietzsche (1844–1900) (Spinks 2003), very much speaks to the natural world. Interestingly, a freely-rotating structure that is employed for digestion and not forward motion has been reported in certain bivalves and gastropods (Bailey and Worboys 1960; Kristensen 1972; Bernard 1973; Judd 1979; Owen 1980). In abundance, molecular wheels have been reported in living cells (Amendola 2001; Schliwa 2003; Oster and Wang 2003; Kolomeisky 2015; Zhou et al. 2020). Cilia and flagella are good examples of structures that work as biological wheels: attached to a basal body a rotary motor that is powered by the flow of ions down an electrochemical gradient across the cytoplasmic membrane into the cell generates constant torque (twisting force that creates propulsion and movements that push surrounding material(s)) (LaBarbera 1983; Yorimitsu and Homma 2001; Berg 2003; Xing et al. 2006; Minamino et al. 2008; Beeby et al. 2016; Shum 2019): the bacterial flagellar motor utilizes ion flow to drive rotary motion at speeds that reach 100,000 rounds per minute (Kojima and Blair 2004).

Interestingly, biological systems appear to conform with the premise that "the whole is more than the sum of its parts," a supposition that without explicit evidence has been associated with the philosopher Aristotle (384–322 BC) (Michael 1975). The viewpoint has been accepted by among others Maurovich-Horvat (2017) and Halatek et al. (2018), questioned by Scaltsas (1990), and characterized as "worthless" by Itkonen (2017). Organisms and animals develop, exist, and function in accordance with the Laws of Thermodynamics (Prigogine and Stengers 1984; Styler 2008; Schreiber and Gimbel 2010; Baverstock and Rönkkö 2014), especially the second law that for natural bodies speaks to the empirical relationships between temperature, heat dissipation, thermodynamic work, and entropy (Dolev and Elitzur 1998; Atkins 2007; Garcia et al. 2011; Katsnelson et al. 2018). Individual animals, large populations, and ecosystems are thermodynamically open spaces where energy freely flows in and out (Toussaint and Schneider 1998; Pascal et al. 2013). In extant animals, thermodynamic gradient in an oxygenic environment was a decisive state for the development of complexity (Goldstein and Inge 1993; Koch and Britton 2008): at all levels of biological design, aerobic metabolism supports sophisticated structures and physiologies (Koch and Britton 2008). Molecular cell biological and genetic studies have elucidated the mechanisms and the processes that regulate and control cell development and function (Levin 2012; Whitsett et al. 2019) and have contributed importantly to both understanding of the phylogenetic diversity of the extant animal life (Goodwin 1994) and establishment of structural-functional correlations in the make-ups of biological structures (Radinsky 1987; Padian 1995; Alexander 1996; Weibel et al. 1991, 1998; Weibel 2000; Schwenk 2000; Lister 2014; Vizcaíno and Bargo 2021; Cassini et al. 2021). Radinsky (1987) observed that "form-function correlation involves looking for the behaviors or functions that are correlated with a particular anatomical form in living species."

Early embryonic development occurs under hypoxic and even anoxic environmental conditions (Pabon et al. 1989; Genbacev et al. 1997; Harvey 2007; Carroll et al. 2021). For mammals in particular, where direct studies have been made, the blastocyst travels down to the uterus from the fallopian tubes along a diminishing O_2 gradient (Yedwab et al. 1976; Fischer and Bavister 1993; Li and Winuthayanon 2017; De Munck et al. 2019). In general, mammalian embryos can tolerate anoxia for as long as 30–60 min (Webster and Abela 2007). The low O_2 tension in the oviducts may adaptively decrease the production of reactive O_2 species and consequently protect the delicate embryonic cells and tissues from injurious molecular factors (Catt and Henman 2000). Although prenatal hypoxia commonly results in negative outcomes on the fetus, resulting in fatalities and certain birth defects (Webster and Abela 2007), molecular O_2 considerably supports normal embryonic development by properly regulating and determining the fates of stem cells (Panchision 2009; Prado-Lopez et al. 2010; Stamati et al. 2011; Liu et al. 2017; Fathollahipour et al. 2018; Podkalicka et al. 2020; Covello et al. 2021). From embryonic to adult stages of organogenesis, the levels of O_2 to which the cells are exposed are narrowly regulated within a very narrow range, largely by the hypoxia-inducible factors (HIFs) that detect and control presence of O_2 in various tissue spaces (Simon and Keith 2008; Mazumdar et al. 2010; Braunschweig et al. 2015; Liu et al. 2017; Podkalicka et al. 2020). During the early stages of embryogenesis, of the processes that influence cell proliferation, low O_2 concentration is the most important factor for development (Mohyeldin et al. 2010; Eliasson and Jönsson 2010): the kidney develops under a hypoxic environment (Liu et al. 2017). Showing that too much O_2 may affect development, for the human embryo, during the first trimester, O_2 supply to the tissues is highly regulated (Webster and Abela 2007). For the avian embryo, Carroll et al. (2021) observed the following: (a) a hypoxic phase with an O_2 concentration of <5% occurs between days 1 and 2 of incubation, a time when HIF-α expression starts, and (b) low O_2 concentrations are critical to proper development of blood vessels. The observations show that the hypoxic conditions under which life started have been conserved in some of the later developmental processes of animal life.

2.2 Developments of Gas Exchangers

The vertebrate lungs originate from the ventral aspect of the primitive foregut (pharynx) (Zorn and Wells 2009; Goss et al. 2009; Green et al. 2011; Maina 2012; Bellairs and Osmond 2014; White et al. 2016; Tatsumi et al. 2016). Well before the various organs that originate along the anteroposterior axis of the developing foregut tubes such as the thyroid gland, the lung, and the stomach become conspicuous, transcription factors or markers of such organs are expressed in select groups of endodermal cells that are set aside very early. The reserved cells proliferate and invade the surrounding splanchnic mesenchyme. Consistent with the development of other organs and tissues of the body, the establishment of the lung occurs through interactions between different signaling and sensing molecules that control cell replication, migration, polarity, differentiation, survival, death, and function (Wittenberg 1993; Hogan et al. 1997; Hogan 1999; Hogan and Kolodziej 2002; Alberts et al. 2002; Davé et al. 2008; Nowotschin and Hadjantonakis 2010; O'Farrell 2015). While some of the transcription factors are specific to particular cell types, e.g., mesenchymal or epithelium ones, others are expressed in different tissue spaces where the pattern of distribution possibly changes during development (Maeda et al. 2007). Well-coordinated morphogenetic processes lead to the formation of flat layers of epithelial cells, pits (depressions = invaginations = inward folds), projections (protrusions = evaginations = outward folds), and tubular configurations (Lewis 1947; Ettensohn 1985; Pilot and Lecuit 2005; Lecuit and Lenne 2007; Keller and Shook 2011; Gritsun et al. 2012; Davidson 2012; Pearl et al. 2017; Blanchard 2018; Zhou et al. 2019). During the development of tissues and organs, cell association is a well-recognized process by which groups or sheets of cells, commonly epithelial ones, aggregate to form complex structures (Sadler 2010; Ray and Niswander 2012; Chan et al. 2021). Cell polarity is a particularly important feature in proper development of many organs, especially the "luminized" ones (Martín-Belmonte and Rodríguez-Fraticelli 2009; St Johnston and Sanson 2011; Willenborg and Prekeris 2011; Román-Fernández and Bryant 2016; Jewett and Prekeris 2018).

For air-breathing vertebrates, the lungs are important sites of gas exchange and acid-base balance (Weibel 1984; Maina 1998; Jürgens and Gros 2002; Levitzky 2013). They display common morphological features of which the most important ones are (a) large respiratory surface area, (b) large pulmonary capillary blood volume, and (c) thin blood-gas (tissue) barrier (Maina 1998, 2000, 2011; Hsia et al. 2016; Gillich et al. 2020): the morphogenetic processes of the lung are directed toward achievement these prototypical features. Morphologically, lungs range from single-chambered (saccular), i.e., rather smooth-surfaced, forms in the lower vertebrates (lungfish, amphibians, and less-derived reptiles) to the highly structurally complex and functionally efficient multichambered (multicameral) ones of the higher vertebrates (e.g. birds and mammals) (Duncker 1978a, 1981, 2004; Perry and Duncker 1978, 1980; Perry 1998; Perry and Sander 2004; Kardong 2014) (Plate 1.6). Highly conserved genes and molecular factors that are involved in lung development have now been identified in mammals (Warburton et al. 1998, 2000, 2001, 2003; Zeng et al. 1998; Metzger and Krasnow 1999; Ambros 2001; Bartel and Chen 2004; Kumar et al. 2005; Cardoso and Lü 2006; Metzger et al. 2008; Gontan et al. 2008; Goss et al. 2009; Nasarre et al. 2010; Whitsett et al. 2011; Machado 2012; Schittny 2017; Fernandes-Silva et al. 2017b; Warburton 2021) and birds (Maina and Madan 2003; Maina et al. 2003; Moura et al. 2011, 2014, 2015, 2016; Maina 2012; Moura and Correia-Pinto 2017). Imprudently, the findings on the mammalian lung have been extrapolated to other vertebrate lungs, especially those of birds. Ostensibly, presently, no meaningful molecular developmental data exist on the morphogenesis of the amphibian and the reptilian lungs and those of air-breathing fish.

Among the air-breathing vertebrates, the avian respiratory system is structurally the most complex (King 1966, 1975; King and Molony 1971; Duncker 1971; King and King 1979; King and McLelland 1984, 1989; Seller 1987; McLelland 1989; Maina 2005, 2006, 2009; Maina et al. 2010a, b; Lawson et al. 2021; Mostafa et al. 2022) (Plate 2.2, a) and functionally the most efficient (Scheid 1979; Fedde 1980; Scheid and Piiper 1989; Powell 2015). Although its morphology has been investigated over many years (Coitier 1573; Harvey 1651; Huxley 1882; Selenka 1866; Ross 1899; Müller 1908; Juillet 1911), the molecular aspects of the development of the avian respiratory system have received much attention (Maina 2012, 2013; Moura and Correia-Pinto 2017). Some of the investigations include those by Locy and Larsell (1916a, b), Delphia (1961), Kloek and Casler (1972), Jones and Radnor

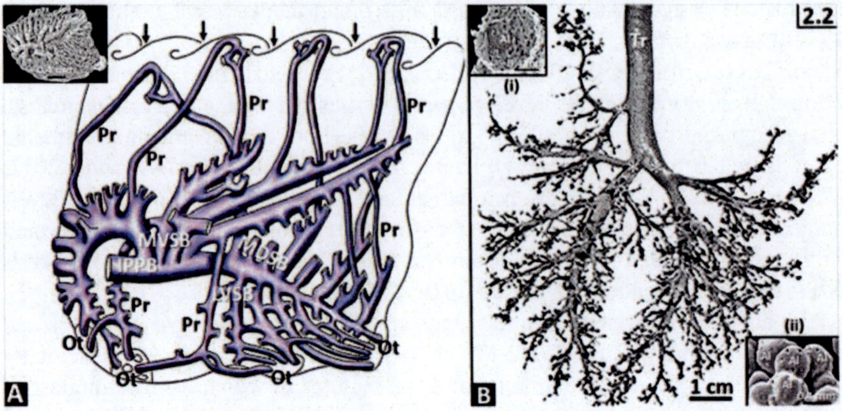

Plate 2.2 (**a**) Schematic illustration (**a**) and latex rubber cast preparation (**inset**) of the airway system of the avian lung. The parabronchi (Pr) (tertiary bronchi) interconnect the medioventral secondary bronchi (MVSB), the mediodorsal secondary bronchi (MDSB) and the lateral ventral secondary bronchi (LVSB): they form continuous hoop-like structures. The laterodorsal secondary bronchi are excluded to simplify the illustration. IPPB, intrapulmonary primary bronchus. *Ot* ostia; arrows, costovertebral impressions. (**b**) Latex rubber cast preparation of the airway system of the lung of the Pig (*Sus scrofa*). *Tr* trachea; *, principal bronchi; ▯, tracheal bronchus. **Insets–(i), (ii):** The airway system of the mammalian lung terminates blindly in alveoli (Al)

(1972a, b), Duncker (1978b), Maina (2003a, b, 2004a, b, 2006, 2012, 2015a), Maina et al. (2003), Seymour et al. (2004), Runciman et al. (2005), Makanya et al. (2007), Makanya and Djonov (2008, 2009), Miura et al. (2009), Moura et al. (2011), Gleghorn et al. (2012), Moura and Correia-Pinto (2017), Palmer and Nelson (2020) and Mostafa et al. (2022): aspects on the development of the avian respiratory system are missing in many existing publications (Seller 1987; King 1966, 1975; Duncker 1971; King and King 1979; King and McLelland 1984, 1989; McLelland 1989; Powell 2015). Among the large number of extant species, only a few such as the Domestic Fowl (*Gallus gallus* variant *domesticus*), the House Sparrow (*Passer domesticus*), the Rock Dove (*Columba livia*), Parrots (Family: *Psittaciformes)*, the Guinea Fowl (*Numida meleagris*), the Budgerigar (*Mellopsittaccus undulatas*), the Turkey *(Meleagris gallopavo*), the Common Quail (*Coturnix coturnix*), and the Muscovy Duck (*Cairina moschata*) have been meaningfully scientifically studied (Stern 2004, 2005; Bateson and Feenders 2010; Kain et al. 2014; St Johnston 2015; Erdoğan et al. 2015; Wittig and Münsterberg 2016; Burggren et al. 2016; Lansford and Rugonyi 2020).

The process of the development of the airways of the mammalian lung by branching morphogenesis is similar to that of tubular structures (ducts) in organs such as the kidneys; the pancreas; and the mammary, the prostrate, and the salivary glands (Bedford et al. 1996; Duan et al. 1999; Driskell et al. 2004, 2007; Lu et al. 2006; Reidy and Rosenblum 2009; Malhotra et al. 2014; Lynch et al. 2016; Anderson et al. 2017). For such anatomical structures, a large surface area is required for their functions such as secretion, exchange, and absorption of substances/

materials. Different cellular phenotypes are spread along the tracheobronchial system and the alveolar region of the mammalian lung (Sorokin 1970; Perelman et al. 1981; Crapo et al. 1982; Weibel 1984; McQualter and Bertoncello 2012; Hogan et al. 2014; Feng et al. 2015; Whitsett et al. 2019). For example, mucous secretory and ciliated cells are situated in the upper airways to trap and remove harmful materials/substances and particulate matter while surfactant secreting granular pneumocytes (type-II cells) and macrophages are found on the respiratory surface. Between 40 and 60 different types of cells, that are among others classified into epithelial, endothelial, mesenchymal, and immune (protective) categories, exist in various parts of the adult mammalian lung (Sorokin 1970, 1977; Weibel 1984; Bowden 1987; Warburton et al. 1998; Rawlins and Hogan 2006; Franks et al. 2008; Morrisey and Hogan 2010; Kotton and Morrisey 2014; El Agha and Bellusci 2014; Yuan et al. 2018; Cunniff et al. 2020): as many as 58 so-called "molecular" cell types have been reported in the human lung (Travaglini et al. 2020). The remarkable assortment of cell types in the lung (that derive from ectodermal, mesenchymal, and endodermal regions) explains the complex gene transcription processes that occur during the development of the organ and its definitive structural complexity at maturity (Stone et al. 1992; Maeda et al. 2007; Franks et al. 2008; Price et al. 2010). From its particularly low cell turnover rate, the vertebrate lung is an exceptionally stable organ (Akram et al. 2016): less than 5% of the epithelial cells of the organ divide (Boers et al. 1996). In the mature human lung, compared to other organs, with a mitotic rate of ~3 cells per 1000 cells, the average cell turnover in the lung is very low (Adamson 1985). The rate of cell turnover in the larger airways of the mammalian lung is faster than that in the terminal parts, i.e., bronchioles and alveoli (Bowden (1983). The turnover time for epithelial cells in the small bronchi and the respiratory bronchioles in the mouse lung is 7–10 days while in the large bronchi, the superficial (epithelial) cells have a shorter lifespan of 3 days (Spencer and Shorter 1962). Cell turnover in the lung generally correlates with replacement of the injured and aged cells (Bowden 1983). Like for other organs such as the liver and the pancreas, after injury, the pulmonary airway epithelium has an exceptional capacity for repair, with total restoration of the original complex structure possible (Erjefalt et al. 1995; Erjefält and Persson 1997; Yahaya 2012; Hogan et al. 2014; Akram et al. 2016).

The vertebrate lungs generally develop from the median part of the floor (surface) of the embryonic (primitive) foregut as ventral invaginations (outpouchings) of the endodermal epithelial cell lining into the underlying mesenchyme (the splanchnic mesoderm) to form lung primordium (anlage) (Rosenquist 1970; Maina 2003a, b; Roth-Kleiner and Post 2003; White et al. 2016; Tata and Rajagopal 2017; Schittny 2017; Kina et al. 2021). The mesenchymal space contains an assortment of progenitor and differentiated cells and is a reservoir of important growth factors, signaling networks, and physical forces that vitally drive lung development (Warburton et al. 2005, 2010; Morrisey and Hogan 2010; El Agha and Bellusci 2014; McCulley et al. 2015; Antony et al. 2016). The committed group of endodermal cells that form lungs are set apart very early, at least before the development of the gastrointestinal system and the thyroid gland (Cardoso 2000; Cardoso and Lü 2006; Hackett et al. 1996;

Moura and Correia-Pinto 2017). The earliest molecular markers of lung development that are specific to the respiratory endoderm of the primitive foregut are the transcription factors TBx-1 (thyroid transcription factor-1), Tbx-4, Tbx-5, HNF-3β (hepatocyte nuclear factor), and NKX2.1 (Homeobox-2.1) (Lazzaro et al. 1991; Hackett et al. 1996; Zeng et al. 1998; Minoo et al. 1999; Sakiyama et al. 2003; Serls et al. 2005; Funk et al. 2020a, b): they are expressed before the lung buds become morphologically discernable (Maeda et al. 2007; Boggaram 2009; Yang et al. 2012; Pogoriler and Husain 2014; Hogan et al. 2014; Phelps et al. 2018). TTF-1 is also an important transcription factor for branching morphogenesis (Yang et al. 2012; Pogoriler and Husain 2014). The commitment of the cells that form the airways and the rest of the lung occurs well before the formation of the trachea and the lung buds (Perl et al. 2002; Whitsett et al. 2011; Phelps et al. 2018). The epithelial cells secrete an extracellular membrane or basement membrane that physically separates the cells from the mesenchymal space that originates from the lateral plate mesoderm. The various components of the lung that comprise connective tissue and smooth muscle tissue elements (that are associated with the airways and the blood vessels), tracheal cartilages, pericytes, lymphatics, and the pleural covering of the lung are mesodermal in origin (Cardoso and Lü 2006; Herriges and Morrisey 2014; Whitsett et al. 2019). The presence of lung mesoderm is vital to proper lung development (Rudnick 1933; Alescio and Cassini 1962; Herriges and Morrisey 2014).

Development of efficient air-breathing organs (gas exchangers) was a decisive occurrence for the transition of animal life from aquatic to terrestrial existence (Dejours 1988; Randall et al. 1981; Little 1990; Maina 1998): it occurred in the Palaeozoic (Devonian 419–359 million years ago) (Cupello et al. 2022). The transition caused the development of important morphological and physiological changes/modifications/specializations that concerned the acquirement of lungs. The lung is thought to have started as a single organ that changed into paired organs (with the evolution of tetrapods) with a large surface area and high compliance and volume for breathing air on land (Cupello et al. 2022). The development and growth of the pulmonary airway system involve complexly spatiotemporarily well-coordinated expression of molecular growth factors by which an intricate branched morphology is fashioned (Warburton et al. 1998, 2000, 2003; Schittny and Burri 2004; Kumar et al. 2005; Cardoso and Lü 2006; Morrisey and Hogan 2010; Turcatel et al. 2013; Moura et al. 2014; Herriges and Morrisey 2014; Fernandes-Silva et al. 2020; Hill 2021). Cross-talk or cell-to-cell signaling especially between epithelial and mesenchymal cells which consequences in activation of transcriptional factors is decisive to correct lung development. For the mammalian lung, the airways undergo iterative (stereotypical) branching to form the so-called "respiratory tree" while the mesoderm contributes to formation of parts of the lung such as the pulmonary vasculature (Hogan 1999; Rubin et al. 2004; Plosa et al. 2004; Alsberg et al. 2004; Cardoso and Lü 2006; Yates et al. 2010; Whitsett et al. 2019): inadequate development of the lung leads to embryonic lethality (Berrocal et al. 2004; Cullinane et al. 1992; DeMello 2004; Groenman et al. 2005). Experimental replacement of tracheal mesenchyme of an embryonic lung with that from the periphery of the lung causes ectopic branching of tracheal epithelium (Wessells 1970; Minoo and King 1994). Epithelium that is separated from its mesenchyme and cultured in matrigel medium containing acidic fibroblast growth factor (aFGF) undergoes branching morphogenesis that resembles the one that occurs during normal lung development (Nogawa and Ito 1995): the conducting airways of the lung form before the blood vessels (Maina 2003a, b, 2004a, b, 2006; Schittny and Burri 2004; Schittny 2017).

By means of the process of angiogenesis, the pulmonary vasculature forms by extension of blood vessels from the aortic arches and the left atrium of the heart while in the lung mesenchyme, blood vessels form in situ by the process of vasculogenesis (Wood et al. 1997; DeMello et al. 1997; Gebb and Shannon 2000; Cardoso and Lü 2006): before the end of embryonic life, the locally formed blood vessels connect to the larger (systemic) ones. In the development of the mammalian (Lammert et al. 2001; Matsumoto et al. 2001; Pauling and Vu 2004) and the avian lungs (Maina 2003a, b; Makanya et al. 2007, 2011a, b; Makanya 2017), the airways greatly impact on the development of the blood vessels. The complexity and patterning of the airway and vascular systems of the lung originate from events that include the expression of mesenchymal-derived signaling molecules and their receptors (that are located on epithelium) in interaction with the cell-extracellular matrix (ECM) (McGowan 1992; Plosa et al. 2004; Cardoso and Lü 2006; Maeda et al. 2007; Morrisey and Hogan 2010).

For the air-breathing vertebrates, during development, the respiratory surfaces of lungs come to be lined by a surface active substance (surfactant), the blood-gas barrier becomes thinner and the volume of the pulmonary capillary blood increases (Hislop and Reid 1972; Gehr et al. 1978; Weibel 1984; Maina 1994, 1998, 2000; Daniels and Orgeig 2003; Schittny and Burri 2004; Sano and Kuroki 2005). A very thin aqueous film that underlies the surfactant dissolves, concentrates, and increases the partial pressure gradient of O_2, impelling it across the blood-gas barrier (by diffusion) (Scarpelli 1988; Bastacky et al. 1995; Notter 2000; Baoukina et al. 2007; Zuo et al. 2006). With phosphatidylcholine (PC) being the main lipid species, a pulmonary surfactant that comprises 90% lipids and 10% protein, is a complex surface-active membrane-based lipid-protein system (Daniels and Orgeig 2003): it stabilizes the terminal respiratory units of the lung and its immunomodulatory properties neutralize and destroy pathogenic microorganisms and injurious particulates, thereby protecting the lung in particular and the body in general (Chroneos et al. 2010; Glasser and Mallampalli 2012; Han and Mallampalli 2015; Cañadas et al. 2020). As well, the surfactant prevents the exudation of blood plasma onto the respiratory surface (Levine and Mellins 1975; Günther et al. 2001; Ainsworth and Milligan 2002; Nicod 2005). Compton and Goeringer (1981) observed that at variance with the mammalian lung, for the chicks of the Domestic Fowl (*Gallus gallus* variant *domesticus*), neither of the two pathways that are involved in the synthesis of phosphatidylcholine (PC) in lungs, namely, the choline incorporation pathway (I) and the phosphatidylethanolamine (PE) methylation pathway (II), occur during the incubation of the eggs or after hatching. Unlike the mammalian lung, PC, especially its disaturated part (DSPC), constitutes the main phospholipid of the avian surfactant and the production of DSPC is significantly increased before internal pipping occurs (Hylka and Doneen 1982). For a developing chicken lung, in a process termed "heterokairy," which was defined by Spicer and Burggren (2003) as "changes in the timing and/or onset of a particular physiological system during development within a given species," Blacker et al. (2004) observed that hypoxia varies the start and the rate of the development of the surfactant, conceivably by means of endogenous glucocorticoids so as to develop individuals able to stay alive after premature hatching.

2.3 Branching Morphogenesis

Branching morphogenesis is the process by which biological structures such as the airways of the lung, the collecting ducts (tubules) of the kidney, and the secretory ducts of organs such as the mammary, the salivary, and the pancreas glands develop: it entails stereotypical (formulaic) bifurcation of the airways under signaling molecules that regulate epithelial-mesenchymal crosstalk and cellular translocation, patterning and reorganization during development (Metzger and Krasnow 1999; Desai and Cardoso 2002; Antony et al. 2016; Maeda et al. 2007; Metzger et al. 2008; Yates et al. 2010; Maina 2012, 2013; Cellière et al. 2012; Rockich et al. 2013; Hines and Sun 2014; Moura and Correia-Pinto 2017). Biomechanical forces that derive from contractile events in the surrounding mesenchyme, specifically from smooth muscle and myoepithelial cells and enlargement of developing structures, significantly determine the growth and the development of the lung (Ferguson et al. 2005; Antony et al. 2016; Metzger et al. 2008; Morrisey and Hogan 2010; McCulley et al. 2015; Spurlin and Nelson 2017; Schittny 2017; Neumann et al. 2018): mesenchymal cells differentiate and form various structures such as the tracheal and the bronchial cartilages as well as parts of blood vessels (Morrisey and Hogan 2010; McCulley et al. 2015).

In nature, branching morphogenesis is a common structural feature (Lu and Werb 2008; Iber and Menshykau 2013; Hannezo and Simons 2019; Uçar et al. 2021). It is well-displayed in trees, where a stem fashions a series of branches of which the terminal ones support leaves that capture sunlight for photosynthesis. In animal life, some organs or their parts develop by branching morphogenesis (Lu et al. 2006; Miura 2008, 2015; Warburton 2008; Horowitz and Simons 2008, 2009; Yates et al. 2010; Andrew and Ewald 2010; Ochoa-Espinosa and Affolter 2012; Manivannan and Nelson 2012; Clément et al. 2012b; Maina 2013; Lang et al. 2018, 2021). Among these are (a) the airway (bronchial) (Plate 2.12, a, b) and vascular systems of the mammalian lung (Plate 2.12, b, c, e, f), (b) the collecting and filtration ducts of the mammalian kidneys, (c) ducts of glandular structures such as mammary and salivary glands and the pancreas, and (d) the tracheal (respiratory) system of insects (Maina 1989; Ghabrial et al. 2003; Fraga and Guttentag 2012; Greaves 2012; Short and Smyth 2016; Essey and Maina 2020). Commonly termed "respiratory tree," the airway system of the human lung (Plates 2.2, b and 2.12, d) is a well-known example of a structure that develops by branching morphogenesis: it displays an inverted tree-

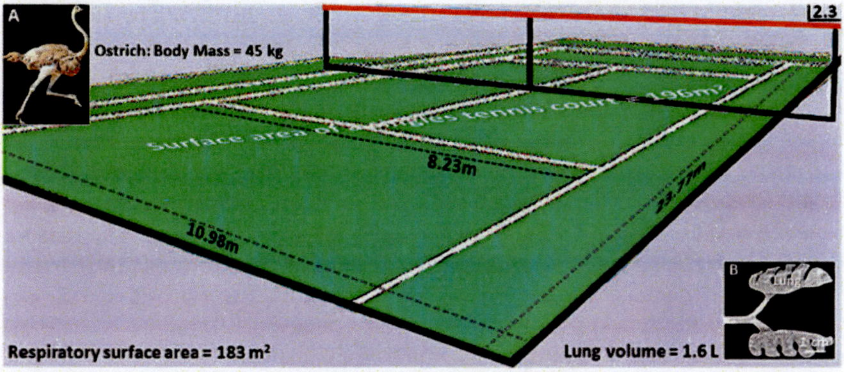

Plate 2.3 (**a**) The respiratory surface area of an Ostrich (*Struthio camelus*) lung. The surface area of the air capillaries (183 m^2) is equivalent to that of the surface of a singles tennis court (196 m^2). (**b**) The extensive surface area is contained in a total lung volume of 1.6 L. The large respiratory surface area is achieved by intense internal subdivision of the exchange tissue of the lung into extremely small terminal respiratory units

like morphology. The adult human lung comprises ~17 million airway branches that terminate in about half a billion (500 million) alveoli and the arterial and the venous "trees" (systems) give off more than 250,000 arterioles and venules (Ochs et al. 2004; Jones et al. 2021). Branching morphogenesis is a space optimization process (Lu et al. 2006; Glenny 2011; Ochoa-Espinosa and Affolter 2012; Spurlin and Nelson 2017; Glenny et al. 2020; Essey and Maina 2020): large surface-to-volume ratio is important in gas exchange, secreting, and absorbing processes. For example, for a human being (75 kg body mass), a respiratory surface area of 142 m^2 is contained in only 4.5 L of lung volume (Gehr et al. 1978) while for the lung of an immature 45 kg body mass Ostrich (*Struthio camelus*) (Plate 2.3, a), the respiratory surface area of 183 m^2, which is equivalent to that of a singles tennis court, is contained in a volume of the lung of 1.6 L (Maina and Nathaniel 2001) (Plate 2.3, b).

Important advances have been made in the identification of the various molecular signaling factors that are involved in the development of the mammalian lung (Cardoso 2000; Warburton et al. 2000, 2010; Cardoso and Lü 2006; Gleghorn et al. 2012; Clément et al. 2012a, b; Cellière et al. 2012; Swarr and Morrisey 2015; Moura and Correia-Pinto 2017). The airways and the blood vessels form by branching morphogenesis, a process by which branching points, successive epithelial budding, and elongation of the tubular parts (sections) are specified by well-orchestrated expression of molecular cues (factors) (Iber 2021). The evolution of the transcriptional systems and mechanisms that control the development of various functional parts of the lung and the coordination of the intricate processes largely remains equivocal (Maeda et al. 2007; Lang et al. 2018). In spite of this, it is clear that the molecular factors involved in the development of branched biological structures have been extensively and for a long time conserved (Le Roith et al. 1993; Peters et al. 1994; Perl and Whitsett 1999; De Miguel et al. 2016; Warburton and Bellusci 2004; Maina 2012, 2013; Whitsett et al. 2011; Warburton 2021). Biological structures/features that develop by such processes have maintained constant (fixed) morphology over a long evolutionary time and have therefore been designated "Bauplans" or "Blue Prints" (Orgeig et al. 2007; Sander et al. 2011; Evans et al. 2012; Edelman et al. 2016; Maina 2011, 2012, 2017; Scanes et al. 2022; Cavin and Alvarez 2022). It is interesting that the while the fundamental "instruments" that are involved in "sculpting" the lung (i.e., the molecular signaling pathways and transcription factors) are essentially similar in many organs and tissues (Compton and Goeringer 1981; Ghabrial et al. 2003; Ferguson et al. 2005; Maina 2012, 2013; Moura and Correia-Pinto 2017; Raś et al. 2018; Hayashi and Kondo 2018), among the animal species, the "products," "outcomes," "results," or "developments" are in certain ways structurally and functionally different.

In birds, the respiratory system (the lung-air sac system) is totally separated into a gas exchanger (lung) and ventilators (air sacs) (King 1966; Duncker 1971; Scheid 1979; McLelland 1989; Maina 2005, 2012; Harvey and Ben-Tal 2016; Makanya 2017) (Plate 1.1, d): in contrast, for the mammalian (bronchioalveolar) lung (Plate 2.4, a, b), the contractile (structural) elements of the lung are incorporated in the gas exchanger (the lung) (Comroe 1974; Weibel 1984). For insects, the trachea delivers O_2 directly to the tissue cells (Weis-Fogh 1964a, b, 1967). The airways of the mammalian lung form by dichotomous branching (Plate 2.4, c) (Weibel 1984; Kitaoka et al. 1996; Kovar et al. 2002; Cardoso and Lü 2006; Hyde et al. 2007; Schittny and Burri 2004, 2008; Schittny and Mund 2008; Loscertales et al. 2008; Mund et al. 2008; Yates et al. 2010; Schittny 2017; Goodwin and Nelson 2020) and terminate blindly in alveoli (Cellière et al. 2012; Lang et al. 2018) (Plates 2.2, b-inset and 2.4, a, b) while for the avian one (Plates 2.2, a and 2.4, d, e), where the airways develop by monopodial budding (Plate 2.4, f), a continuous, i.e., a closed hoop, airway system forms (Maina 2003a, b, 2005, 2012; Loscertales et al. 2008; Gleghorn et al. 2012; Kim et al. 2013; Miura 2015; Tzou et al. 2016; Fujii et al. 2021) (Plate 2.2, a).

The development of the mammalian lung has been well-investigated (Ten Have-Opbroek 1981, 1991; Kitaoka et al. 1996; Burri 1997, 1999; Kovar et al. 2002; Hislop 2002; Schittny and Burri 2004, 2008; Morrisey et al. 2013; Tschanz et al. 2014; Hines and Sun 2014; Fraga and Guttentag 2012; Swarr and Morrisey 2015; Woods and Schittny 2016; Schittny 2017). Regrettably, without being selective, observations on mammals have been used as morphogenetic reference points of other vertebrate lungs, especially those of birds (Moura and Correia-Pinto 2017). For the mammalian lung, as many as six developmental stages, namely, embryonic, pseudoglandular, canalicular, saccular, alveolar, and vascular maturation have been identified (Wessells 1970; Spooner and Wessells 1970; Ten Have-Opbroek 1981, 1991; Roth-Kleiner and Post 2003; Schittny and Burri 2004, 2008; Schittny 2017) while for the avian respiratory system, the development is a noticeably continuous process (Locy and Larsell 1916a, b; Awkati 1939; Duncker 1978b; Maina 2003a, b, 2004a, b; Palmer and Nelson 2017, 2020): there are no clear developmental stages/phases. In precocial species of birds, where chicks display independence soon after hatching, the lungs are very well-developed at internal pipping, when the chick takes

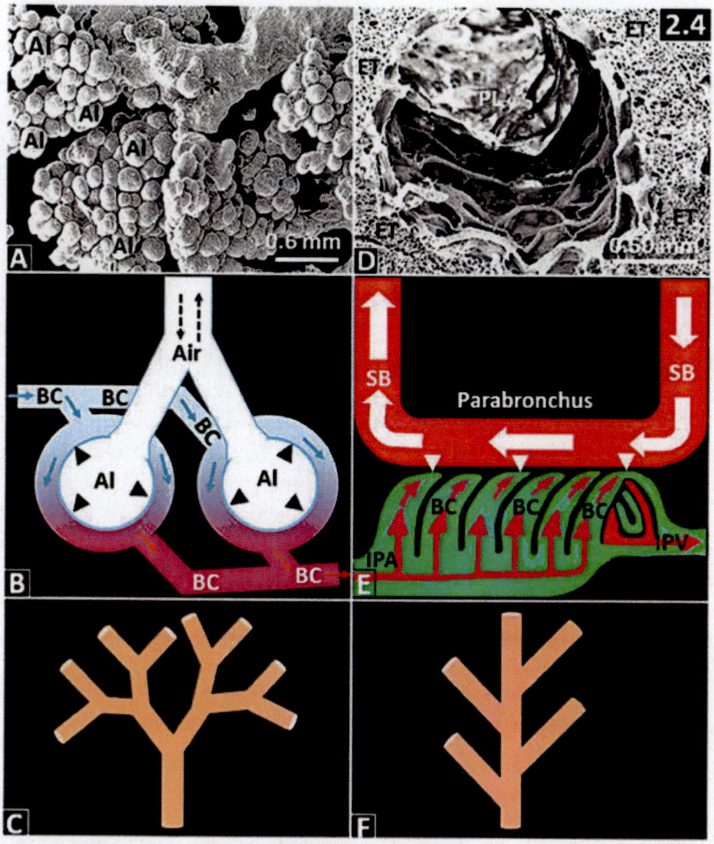

Plate 2.4 (**a**) Latex cast preparation of the lung of the Lesser Bushbaby (*Galago senegalensis*) showing branching of the respiratory bronchi (Asterisk, *) that give rise to acini which terminate in alveoli (Al). (**b**) Schematic illustration of alveoli (Al) which are supplied with blood by blood capillaries (BC). The alveoli are tidally ventilated with air (double dashed arrows). From the alveoli, oxygen diffuses across the blood-gas barrier into the capillary blood (arrow heads). (**c**) Regular dichotomous branching of airways. (**d**) Parabronchus of the lung of the Domestic Fowl (*Gallus gallus* variant *domesticus*) of which a parabronchial lumen (PL) is surrounded by exchange tissue (ET). (**e**) Crosscurrent gas exchange system in the avian lung that is formed by perpendicular (orthogonal) disposition between the directions of air flow through the parabronchial lumen (large white arrows) and that of blood across the exchange tissue (pink arrows). SB, secondary bronchus; BC, blood capillaries; IPA (interparabronchial artery)–venous blood entering the exchange tissue; IPV (interparabronchial vein)–arterial blood leaving the parabronchus; arrow heads (▶), diffusion of oxygen across the blood-gas barrier. (**f**) For the avian lung, the airways develop by monopodial alternating branching

its first breath (Nice 1962; Duncker 1978b; McNichols and McNabb 1988; Ricklefs and Starck 1998; Christensen 2009; Lewallen and Burggren 2015). In the Domestic Fowl (*Gallus gallus* variant *domesticus*) (a precocial species), 2–3 days before hatching occurs, the lungs and the air sacs are well-developed (Duncker 1978b;

Maina 2003a, b, 2004a, b). Runciman et al. (2005) observed that for altricial chicks of the Australian Pelican (*Pelicanus conspicillatus*) (a group of birds where chicks are "underdeveloped" at hatching), the thickness of the blood-gas barrier decreased before hatching, remained constant in the juveniles and thinned further toward maturity. For the same species, from juvenile to adulthood, Seymour et al. (2004) observed that the surface area of the blood-gas barrier markedly increased while the harmonic mean thickness of the blood-gas barrier did not change. For the human lung, alveolarization continues until early adulthood, a process that parallels microvascular maturation (Schittny et al. 2008). In chickens, Tona et al. (2022) observed that depending on the period and the stage of exposure, incubation conditions importantly influence embryonic development and consequently post-hatching growth. From changes in the metabolic profiles of chicks during the early stages of lung development, Fernandes-Silva et al. (2017b) noted that acetate and lactate utilization were instructive metabolic biomarkers, and with subsequent morphogenetic changes, glycolytic utilization increased and Krebs cycle metabolism changed into lactate production. In the lung of the Domestic Fowl (*Gallus gallus* variant *domesticus*), the air capillaries (the terminal respiratory units of the avian lung) which are analogous to the alveoli of the mammalian (bronchioalveolar) lung (Plate 2.4, a, b), are well-developed and the blood-gas barrier is particularly thin at the end of the 21 days of incubation (Maina 2004a, b). During the early stages of development, for both the mammalian and the avian lungs, the airways direct the growth of the blood vessels (Hislop 2002; Perl et al. 2003; Maina 2003a, b, 2004a, b, 2005; Cellière et al. 2012).

2.4 Development of the Avian Lung

The development of the lungs of the Domestic Fowl (*Gallus gallus* variant *domesticus*) has among the avian species been best studied. Its primordia first appear as ridge-like outgrowths (anlage) (that comprise masses of mesenchymal cells) on the ventral aspect of the primitive foregut on the third day of incubation (Locy and Larsell 1916a, b; Hamburger and Hamilton 1951, 1992; Duncker 1978b; Maina 2003a, b; Bellairs and Osmond 2014; Moura and Correia-Pinto 2017) (Plate 2.5, a, b). For the white leghorn breed of chicken, after studying the movement of ^{3}H thymidine labeled cells of prelung cells that were isolated from the medium-streak at the fourth somite stage of embryogenesis and transplanted onto the epiblast, Rosenquist (1970) made the following observations: (a) at the medium-streak stage, the special endodermal cells that were predestined to form the lung were located in the anterior one-third of the primitive streak; (b) the prelung cells reached their definitive location (at the ventrolateral wall of the gut) between the tenth and the 16th somite stage; and (c) at ~22nd somite stage, the inaugural lung endodermal cells began to invade the underlying splanchnic mesoderm. From expression and localization of molecular growth factors such as the Wnt-5a gene, in the chick embryo, the site where the avian lung forms can be identified as early as 24 h from the onset of

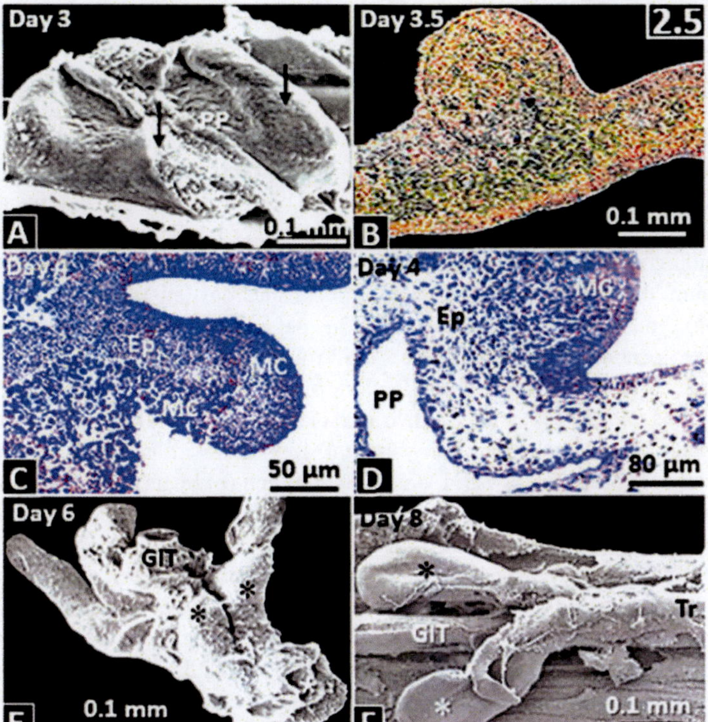

Plate 2.5 (**a**) Early development of the lung of the Domestic Fowl (*Gallus gallus* variant *domesticus*). The anlage start on day 3 of incubation as paired swellings on the floor of the primitive pharynx which at day 3.5 of incubation (**b**) comprise a homogenous group of mesenchymal cells (dashed circle). By day 4 of incubation, the cells that constitute the swellings have differentiated into epithelial cells (Ep) that are surrounded by mesenchymal ones (MC) (**c, d**). PP, primitive pharynx. The lung buds (asterisks, *) are conspicuous from day 6 of incubation (**e-g**). At day 10 of incubation, the lungs (stars, ▯) have engaged and inserted into the ribs (dots, •) and the vertebrae on

incubation (Loscertales et al. 2008; Moura et al. 2014; Maina 2012, 2013, 2015a). In birds, the process of "staging" expresses the morphogenetic features that occur during embryogenesis, i.e., from the first cleavage divisions up to hatching (Romanoff 1960; Hamburger and Hamilton 1951, 1992; Martinsen 2005; Sellier et al. 2006; Ainsworth et al. 2010; Vergara and Canto-Soler 2012; Bellairs and Osmond 2014; Hill 2022). For the Domestic Fowl (*Gallus gallus* variant *domesticus*), Maina (2003a, b) made the following observations: (a) between days 3 and 5 of incubation, the epithelial cells that are destined to form the lung multiply and spread into the surrounding mesenchymal cell space (Plate 2.5, c, d), (b) between days 6 and 8 of incubation, the lung buds have divided into conspicuous left and right lungs (Plate 2.5, e-g), (c) at day 10 of incubation, the developing lungs had reached their definitive locations in the body (i.e., the "roof" on the craniodorsal part of the coelomic cavity), developed a wedge-like shape and become deeply implanted into the ribs and the vertebrae on their dorsolateral aspects (Plate 2.5, h), (d) from about day 12 of incubation, the vertebral and costal impressions (sulci) were conspicuously deep (Plates 2.5, i, j and 2.6, a-c) and (e) by day 15 of incubation, a large part of the volume of the lung was held (contained) between the ribs and the vertebrae and the parabronchi are well-developed (Plate 2.6, c). The mechanism by which the avian lungs engage and insert into the boney structures (ribs and vertebrae) is unclear (Maina 2003a): it has not been identified whether it is the ribs and the vertebrae that slot into the substance of the developing lung or it is the lung that slots in the vertebrae and the ribs or if the boney structures and the lungs participate in the process of attachment. Interestingly, poorly developed depressions (sulci) form on the dorsolateral aspects of embryonic chicken lung explants, i.e., lungs grown (cultured) in vivo in a growth or nutrient medium such as matrigel, where ribs and vertebrae are nonexistent (Maina 2003a): it suggests that the formation of the costovertebral sulci (impressions) may to an unknown extent be genetically determined. The connection of the avian lung to the body wall and the horizontal septum renders them practically rigid (Jones et al. 1985; Makanya et al. 2021) and causes them to be strong and non-compliant (Maina 2007a, b; West et al. 2006). In contrast, the mammalian lung, which is compliant inflates and deflates with ventilatory actions (Bennett and Tenney 1982; Leith 1983; Harris 2005; Maina et al. 2010a, b): the avian lung is a passive organ that is ventilated continuously and unidirectionally in a caudocranial direction by the bellows-like activity of a cranial and a caudal group of air sacs (Scheid 1979; Fedde 1980). Between the respiratory cycles, in the duck, the lung changes in volume by a mere ~1.4% (Jones et al. 1985). Recently, in the adult Domestic Ducks (*Cairina moschata*), Makanya et al. (2021) determined that insignificant volume changes occur from "slight morphological adaptation rather than a change in the archetypical design of the avian lung parenchyma." The terminal gas exchange units of the avian lung, i.e., the air capillaries

---

Plate 2.5 (continued) their dorsolateral aspects (**h**). From day 11 of incubation (**i, j**), the costal sulci (impressions) are deep and conspicuous (arrows, ↓). *GIT* gastrointestinal system

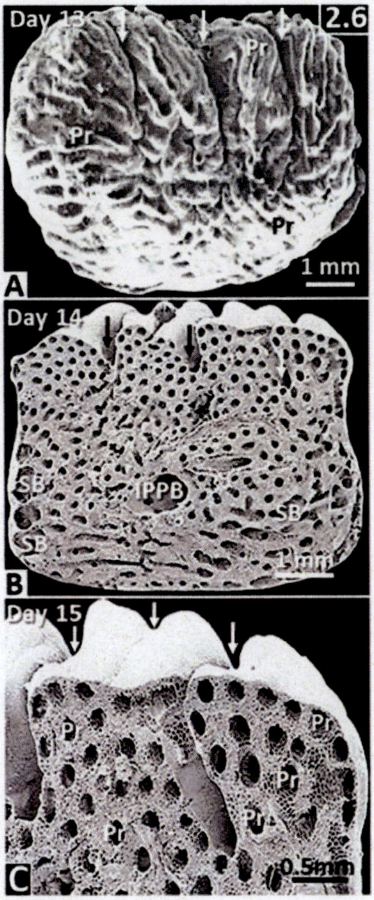

Plate 2.6 (a-c) Late development of the lung of the Domestic Fowl (*Gallus gallus* variant *domesticus*) showing costovertebral impressions. From day 13 of incubation, the im pressions (arrows, ↓) and the airways are well-developed. *IPPB* intrapulmonary primary bronchus; *SB* secondary bronchi; *Pr* parabronchi

and the blood capillaries, are particularly strong (Maina 2007a, b, 2015b; West et al. 2006, 2007, 2010; Watson et al. 2008).

2.4.1 Development of the Airway (Bronchial) System

For the Domestic Fowl (*Gallus gallus* variant *domesticus*), which has an incubation period of ~21 days, the intrapulmonary primary bronchus is the first airway to develop from the second day of incubation (Plate 2.7, a, b). As it extends in a proximodistal direction, in quick succession, the airway gives rise to secondary bronchi from its luminal aspect from day 3 of incubation (Plate 2.7, c-h). Involved in the growth process, mesenchymal cells cluster on the apical ends leading to the developing airways (Plate 2.7, i). The secondary bronchi "grow" by budding

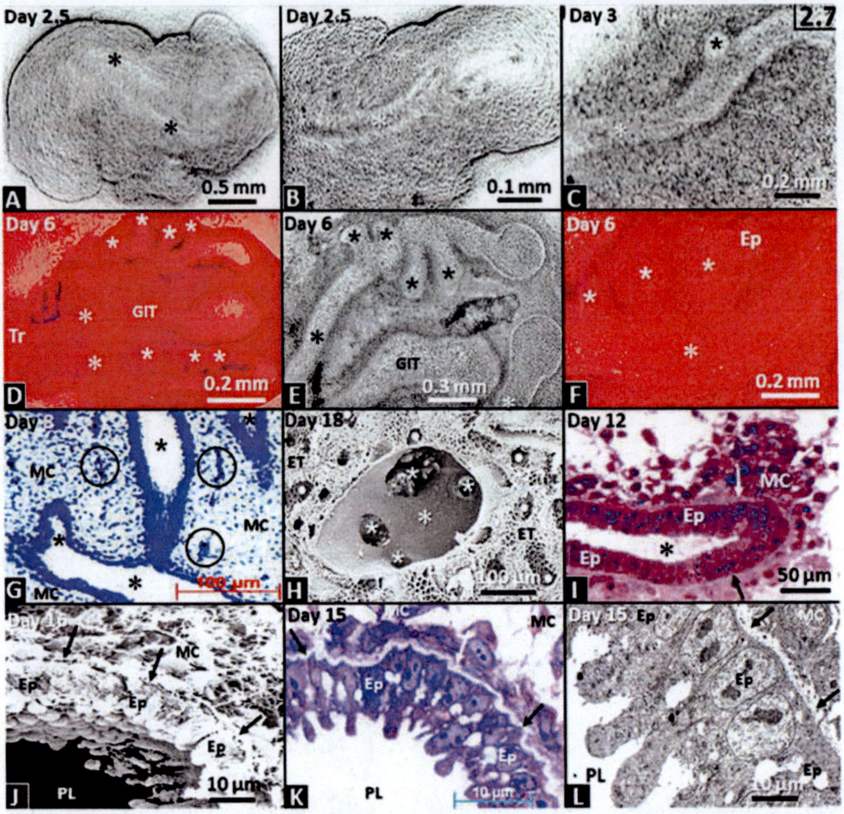

Plate 2.7 Development of the airways in the lung of the Domestic Fowl (*Gallus gallus* variant *domesticus*) cultured in matrigel growth medium (**a-f**). The intrapulmonary primary bronchus (asterisk, *) starts to develop from day 2.5 of incubation (**a, b**) (asterisk, *) and the first secondary bronchus appears from day 3 (**c**) (star, ☆). **At** day 6 of incubation, various sets of secondary bronchi have formed (**d-g**) from the intrapulmonary primary bronchus (stars, ☆). The intrapulmonary primary bronchus (**g-l**) is lined by a pseudostratified columnar epithelium (**i-l**). The airways extend by accretion of epithelial and mesenchymal cells at their growing tips (**I**). *GIT* gastrointestinal system (**d, e**); dot (•), parabronchi (**h**); *Tr* trachea (**d**)**;** *MC* mesenchymal cells; circle (O), developing blood vessel (**g**); *Ep* epithelium (**I-L**); arrows (←), basement lamina (**i-l**); *PL* parabronchial lumen (**j-l**); *ET* exchange tissue (**h**)

(sprouting), a process that investigators like Abbott et al. (1991), Maina (2003a, b, 2005), Gleghorn et al. (2012), Kim et al. (2013), Martin and Goldstein (2014), Tzou et al. (2016), Sawyer et al. (2010) and Makanya (2017) have characterized as "apical constriction" (Plate 2.7, c-g). By day 15 of incubation, the primary bronchus and the secondary bronchi (Plate 2.7, g, h) are lined by a pseudostratified columnar epithelium (Plate 2.7, j- l). Parabronchi, i.e., tertiary bronchi (Plate 2.7, h) originate from the secondary bronchi by a process similar to that by which secondary bronchi develop from the intrapulmonary primary bronchus, i.e., budding, sprouting, or

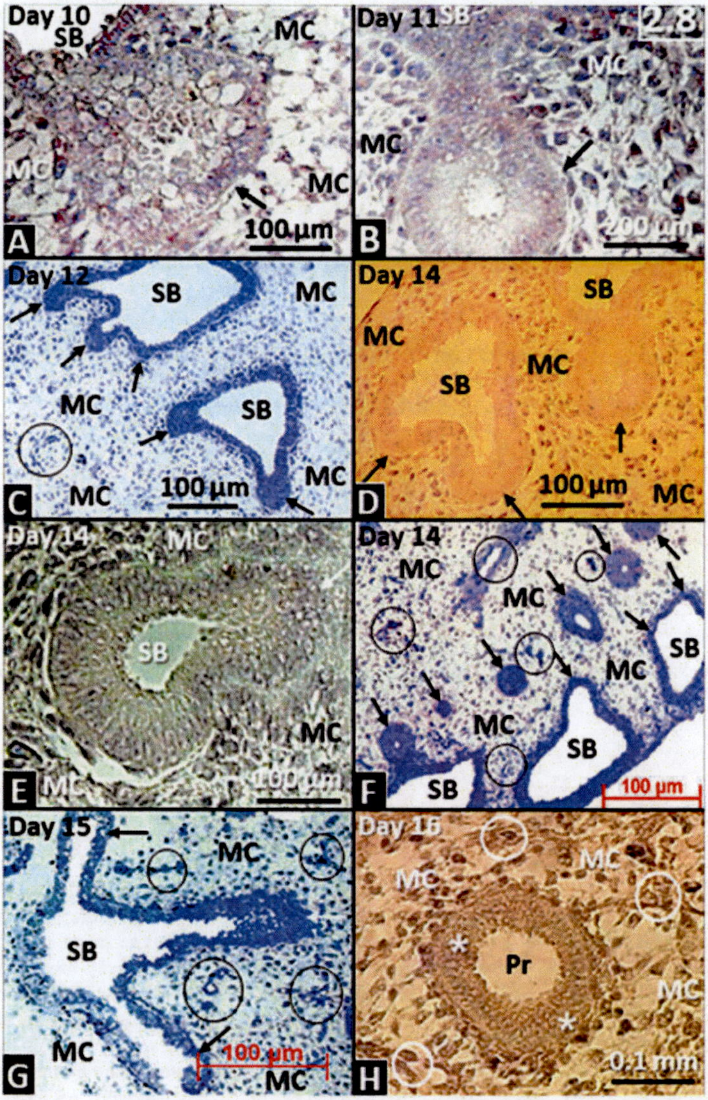

Plate 2.8 Development of the parabronchi (tertiary bronchi) (arrows, ↓) of the developing lung of the Domestic Fowl (*Gallus gallus* variant *domesticus*) by budding from the secondary bronchi (SB) (**a-g**). The developing parabronchi are lined by a pseudostrstified columnar epithelium (stars,) (**h**). *MC* mesenchymal cells; circles, developing blood vessels

apical constriction (Plate 2.8, a-g): the parabronchi are lined by a tall, pseudostratified columnar epithelium (Plate 2.8, h).

For the Ostrich (*Struthio camelus*), which has an incubation period of 42 days, the airways are well-developed at day 32 of incubation (Plate 2.9, a-c). The

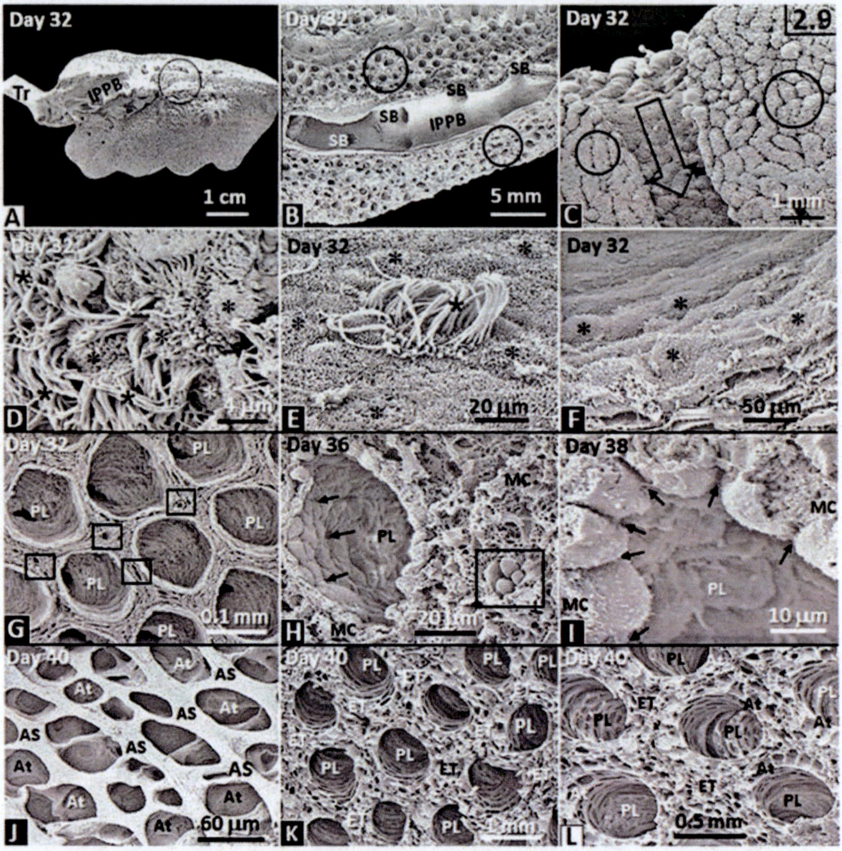

Plate 2.9 Development of the lung of the Ostrich (*Struthio camelus*). At day 32 of incubation, the airways are well-developed (circles) (**a-c**) and the lung is attached to the ribs and the vertebrae (large open arrow) (**C**). Ciliated cells (stars, ▯) and microvillous covered mucous secreting cells (asterisk, *) are found in the epithelial lining of the lumina of the intrapulmonary primary bronchus (IPPB) and the secondary bronchi (SB) (**D-F**). Initially, lined by flat or smooth epithelium (**g**), from day 36, evaginations (atria) project into the surrounding mesenchymal tissue space (arrows) (**h, i**). The atria (At) are well-developed at day 40 of incubation (**J**) and the mesenchymal cell space has converted into exchange tissue (ET) (**k, l**). Tr trachea; encircled areas, secondary bronchi (**b, c**); PL parabronchial lumen; MC mesenchymal cells; boxed areas (**g, h**), blood vessels; AS atrial septa

intrapulmonary primary bronchus, the secondary bronchi, and the parabronchi are conspicuous and the lungs are firmly attached to the vertebrae and the ribs, respectively, on the dorsal and lateral aspects (Plate 2.9, c). The intrapulmonary primary bronchus and the secondary bronchi are lined by an epithelium that contains ciliated epithelial cells among which are interspersed goblet cells (Plate 2.9, d, e). From the part that connects to the secondary bronchus, progressively, the parabronchial epithelium changes from a tall pseudostratified, a columnar to a cuboidal one (Plate 2.9, f-h). From day 36 of incubation, evaginations, i.e., outward projecting spaces, develop from the parabronchial lumina (Plate 2.9, h, i): the spaces form the early stages of the development of the atria which from day 40 of incubation are cell-developed (Plate 2.9, j). The air capillaries exist in the gas exchange tissue of the parabronchi from day 40 of incubation (Plate 2.9, k, l). The conversion of the loose mesenchymal tissue into gas exchange tissue occurs in preparation for air-breathing at pipping and hatching. Amaral-Silva et al. (2019) observed that exposure of developing embryo to hypoxia during the time the parabronchi are developing leads to morphological remodeling of the lung which is characterized by increased volume density and respiratory surface area of structures which are involved in gas exchange at the expense of the air conducting passageways in chicks up to 10 days old. Islam et al. (2022) observed that exposure of broiler chickens to Dexamethasone (Decadron®: a corticosteroid–empirical formula $C_{22}H_{29}FO_5$) affected the development of the lung in the following important ways: (a) the weight, the length, and the width of the lung decreased significantly; (b) vascular and parabronchial lumen congestion occurred; (c) there was lack of conspicuous interparabronchial septa; (d) pathological changes of the lung parenchyma occurred and; (e) significant reduction in the diameters of the parabronchi and the thickness of the parabronchial exchange tissue happened. Mostafa et al. (2022) showed that the epithelial developmental changes that occurred in the lung of the Japanese Quail (*Coturnix coturnix japonica*) from the second to the tenth embryonic days were driven by epithelial-mesenchyme interactions.

The thickness of the blood-gas barrier of the developing Ostrich (*Struthio camelus*) lung is 2.21 μm at day 39 of incubation (Makanya et al. 2012). For the Domestic Fowl (*Gallus gallus* variant *domesticus*), from day 9 of incubation, parabronchi start to form by clustering of mesenchymal cells (Plate 2.10, a) and by day 10, the cells are surrounded by a conspicuous basement membrane (Plate 2.10, b). The cell clusters develop a centrally located lumen at day 11 of incubation (Plate 2.10, c, d) and the epithelial cells are columnar from day 12 (Plate 2.10, e-g). As the luminal diameters widen, the epithelial cells become rather cuboidal in shape (Plate 2.10, h). From day 14, blood vessels start to develop from mesenchymal cells (Plate 2.10, g, h). The processes of the development of the atria, the infundibula, and the air capillaries of the lung of the Domestic Fowl are fundamentally similar to those of the lung of the Ostrich, which has an incubation time of 42 days (Plate 2.11): with the openings surrounded by interatrial septa, the atria start forming as evaginations from the parabronchial lumina (Plate 2.11, a-i) and the air capillaries originate from the terminal part of the infundibula (Plate 2.11, j-l). In embryonic white leghorn chicken lungs, from day 14 of incubation, Jones and Radnor (1972a, b) made out the development of atrial evaginations from the parabronchial lumen into surrounding mesenchymal tissue. The type-II cells, i.e., the granular or the cuboidal pneumocytes of the developing lungs of the Domestic Fowl are located on the surfaces of the atria and infundibula (Plate 2.11, d, j, k): they are totally lacking in the air capillaries (Plate 2.11, l). Aspects of the development of the lung of the embryonic white leghorn chicken and the type-II cells that secrete surfactant (Pattle 1978; Nicholas 1996; Bernhard et al. 2001; Knudsen and Ochs 2018) were studied by Jones and Radnor (1972a, b) from day 13 of incubation to hatching (day 21): at day 13, the

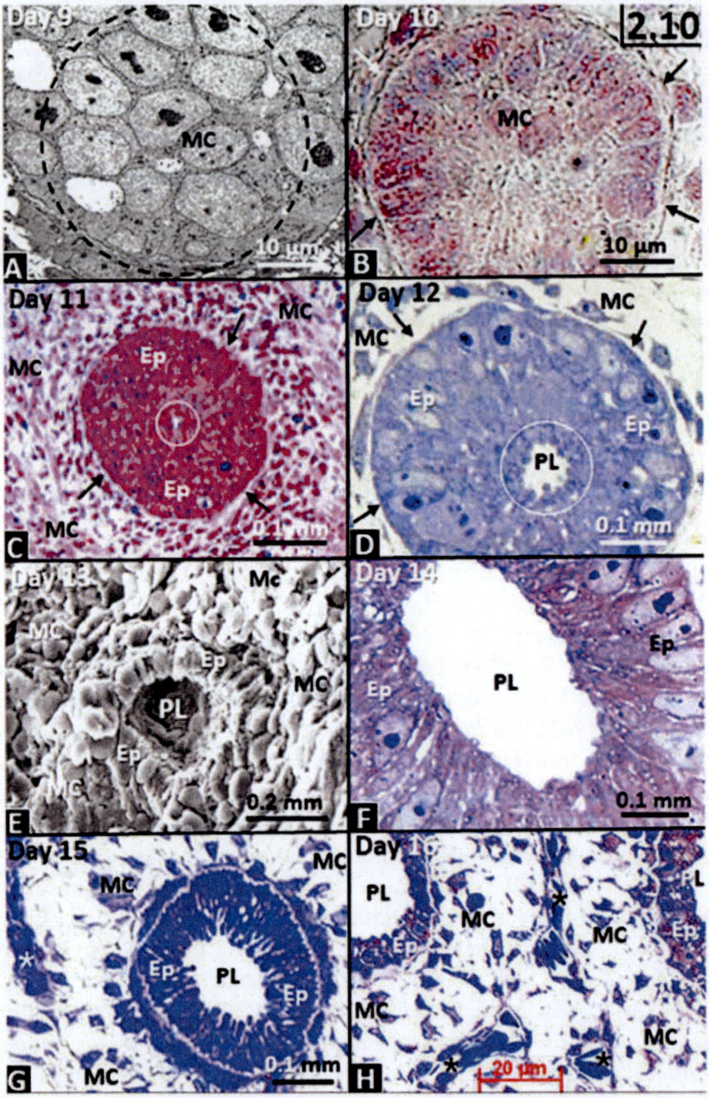

Plate 2.10 Early development of the lung of the parabronchi of the lung of the Domestic Fowl (*Gallus gallus* variant *domesticus*). Clusters of committed mesenchymal cells (MC) (dashed circle) (**a**) start to form from day 9 of incubation. By day 10 of incubation, the groups of cells become encapsulated by a common basement membrane (arrows) (**b-d**) and at day 11 of incubation, a lumen begins to form at the center of the cell clusters (**c, d**) (circles). After widening of the lumen, the mesenchymal cells come to form epithelial cells (Ep) (**c-g**). The parabronchial lumen (PL) is lined by a columnar epithelium (**e–g**). By day 15 of incubation, blood vessels (stars, ⊞) have formed in the mesenchymal tissue space, between the parabronchi (**g, h**)

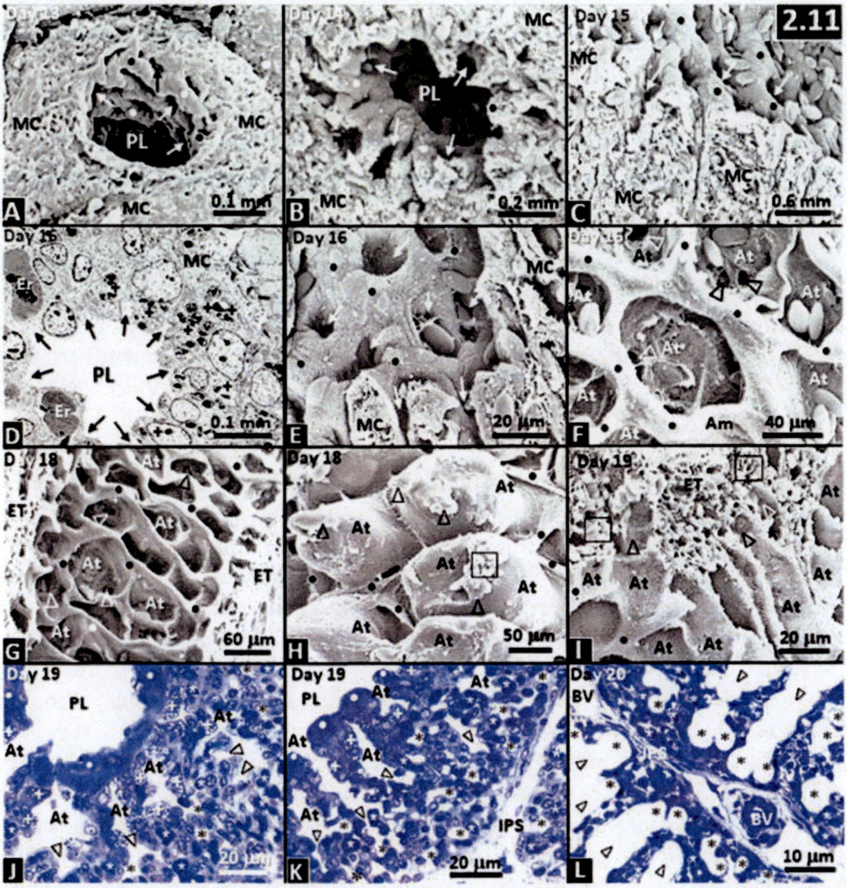

Plate 2.11 Late development of the parabronchi of the lung of the Domestic Fowl (*Gallus gallus* variant *domesticus*). From day 13 of incubation, evaginations from the parabronchial lumina (PL) form atria (arrows, ←) (**a-k**): the infundibula start to develop from day 16 (**f**) and air capillaries from day 18 (**h-l**). Type-2 pneumocytes exist on the surfaces of the atria and infundibula from day 16 (**d, j, k**). Dots (•), interatrial folds/septa (**a-c, e-k**); *MC* mesenchymal cells (**a-e**); *PL* parabronchial lumen (**a, b, j, k**); stars, ▯, erythrocytes (**j-l**); crosses (+), type-2 cells (**d, j, k**); *At* atria; *ET* exchange tissue (**g, i**); arrow heads (Δ), infundibula (**f-l**); boxed areas (**h, i**), air capillaries; asterisks (*), air capillaries (**j-l**); *BV* interparabronchial blood vessel (**i**); *IPS* interparabronchial septum (**k, l**)

airways comprised simple hollow tubes that were lined by a columnar epithelium that was surrounded by mesenchymal cells and by day 16, type-II cells existed on the surfaces of the atria and the infundibulae. In the chicks of the Domestic Fowl (*Gallus gallus* variant *domesticus*), Maina (2003a, b) noted that the type-II cells existed on atria and infundibula from day 16 of incubation (Plate 2.11, d, j, k): containing osmiophilic lamellated bodies in their cytoplasm (Plate 2.11, d), the cells are absent from the very narrow air capillaries (Plate 2.11, l). In the avian lung, the surfactant must spread out from the type-II cells of the atria and the infundibula to cover the blood-gas barrier of the air and the blood capillaries (Plate 2.1, e, j). In the avian lung, the means by which surfactant is recycled is currently unclear, particularly given that macrophages are lacking on the surface of the air capillaries. Bódi et al. (2016) observed that in the avian lung, a surfactant may be secreted in two directions: (a) into the parabronchus, atrium, and infundibulum, where it forms a trilaminar substance that performs a respiratory role of lowering surface tension, and (b) into the basolateral aspects, i.e., into the intercellular spaces of type-II cells (pneumocytes) where it contributes to the innate immune responses of the lung.

The development of the avian lung correlates with factors such as the body mass at hatchling, ambient temperature, prevailing O_2 tension, and incubation period (Vleck and Kenagy 1980; Whittow 1980; Stock et al. 1983; Stock and Metcalfe 1984; Hoyt 1987; Xu and Mortola 1988; Tazawa et al. 1989, 1992; Vleck and Hoyt 1991). Between precocial species (**nidifugous group**) of birds (where chicks can live independently soon after hatching) and altricial ones (**nidicolous group**) (where hatchlings require parental care during early stages of life), important differences exist in the modes of embryonic development, growth, and metabolism (Nice 1962; Vleck et al. 1979, 1980; Vleck and Vleck 1987; Duncker 1978b; Starck and Ricklefs 1998; Ricklefs and Starck 1998; Gefen and Ar 2001; Pearson et al. 2002; Frank 2003; Dietz et al. 2015; Begazo 2019). The precocial species of birds include Mallard Ducks (*Anas platyrhynchos*), Turkeys (*Melleagris gallopavo*), Ostriches (*Struthio camelus*), and Herring Gulls (*Larus argentatus*) while species such as Rock Doves (*Columba livia*) and the groups of birds such as penguins (Family: Sphenisdae), the passerines (Order: Passeriformes), hummingbirds (Family: Trochlidae), swallows (Family: Hirundinidae), woodpeckers (Family: Picidae), and raptors (Order: Falconformes) are altricial (Begazo 2019; Emmet 2022). Classification of birds into precocial and altricial taxa is not a exact given that some species lie in the middle. The extremely precocial megapode birds (Family: Megapodiidae; Order: Galliformes), the so-called "incubator birds" or "mound-builders," have been designated "superprecocial birds": for some species of the taxon, the chicks hatch with full-flight feathers and can fly on the same day they exit the eggs (Starck and Ricklefs 1998). In birds, precociality is supposed to be ancestral to atriciality (Elzanowski and Brett-Surman 1995). Bucher (1987) observed that variations in the incubation periods could have evolved from natural selection acting on adult birds, a process that arose during the development of eggs with lower energy content and by natural selection supporting changes in adult bird basal metabolic rate. In altricial species of birds, at hatching, the parabronchi are small, the gas exchange mantle is thin and the atria are shallow while the parabronchi continue to develop after hatching, a process completed just before the chicks can fly.

For the precocial species, the respiratory system is structurally and functionally well-developed for gas exchange at hatching: it permits the chicks to move about, feed and escape from predators (Pearson et al. 2002; Seymour et al. 2004, 2008; Runciman et al. 2005; Maina 2003a, b, 2006, 2009, 2015a). On account of altricial species having shorter incubation periods compared with the precocial ones, Vleck and Vleck (1987) suggested that the evolution of the group of birds from the ancestral precocial ones might have happened by reduction of energy commitment in egg production, reduction of the egg size and increase of water content in the egg. For birds, the incubation periods of eggs appear to increase linearly with egg-size to the power of one-quarter (Kooijman 1986): it may explain why groups of birds such as the kiwi (Family: Apterygida) and petrels (Family: Procellariidae), species that lay relatively larger eggs for their body masses, have to care for them over a longer period of time compared with larger birds that lay eggs of equivalent size. In mammalian biology, human beings are the best-known altricial animals!

2.4.2 Development of the Pulmonary Blood Vessels

During embryonic development, the cardiovascular system is one of the earliest organ systems to form and become functional (Pansky 1982; Burggren 1991; Copp 1995; Halliday et al. 1998; Moorman et al. 2003; McGrath et al. 2003; Wittig and Münsterberg 2016; Crivellato and Ribatti 2006). For the human embryo, the heart and circulatory system are operational by the end of the third week of life, a time when the placenta cannot meet the O_2 demands and the nutrient requirements of the developing embryo (Takada et al. 2017; Tan and Lewandowski 2020). Pulsating (beating) heart tube and connection of the yolk sac vasculature to the embryonic vasculature marks the origin of circulatory function (McGrath et al. 2003). In the mammalian lung, the mechanisms and processes by which airways and blood vessels develop are similar (DeMello et al. 1997; Weibel 1984) (Plate 2.12) and close patterning between the pulmonary artery (and its branches) and the bronchial (airway) system, maximizes the respiratory surface area and makes the blood-gas (tissue) barrier thin.

For the lung of the Domestic Fowl (*Gallus gallus* variant *domesticus*), the earliest indication of the onset of in situ blood vessel development is that of transformation of pluripotent mesenchymal cells into stellate angioblastic cells gather in the mesenchymal cell space and hemangiogenetic cells that start to accumulate hemoglobin (Maina 2003a, b) (Plate 2.13, a-d). The developmental processes and changes are perceptible from as early as the fourth day of incubation (Maina 2003a, 2004a, 2005). The angioblastic cells are characterized by conspicuous filopodial extensions (Plate 2.13), a feature that specifies their ambulatory nature: the mesenchymal cells that convert directly into erythroblasts do not exhibit such morphological feature (Plates 2.13 and 2.14, a-h). The free angioblastic cells appear to be chemotactically attracted by the erythroblasts and come to surround them, forming blood vessels by the vasculogenetic process (Plate 2.13, e-i): the angioblastic cells contribute to the formation of the walls of the blood vessels (Plate 2.13, g-i). The mesenchymal cells that convert into erythroblasts appear to induce proximate mesenchymal cells to convert into similar cells which instantly start to accumulate hemoglobin (Plate 2.14, c, d): a succession of such erythroblastic cells forms a blood vessel which ultimately comes to surround an air space (Plate 2.14, d). In the developing lung of the

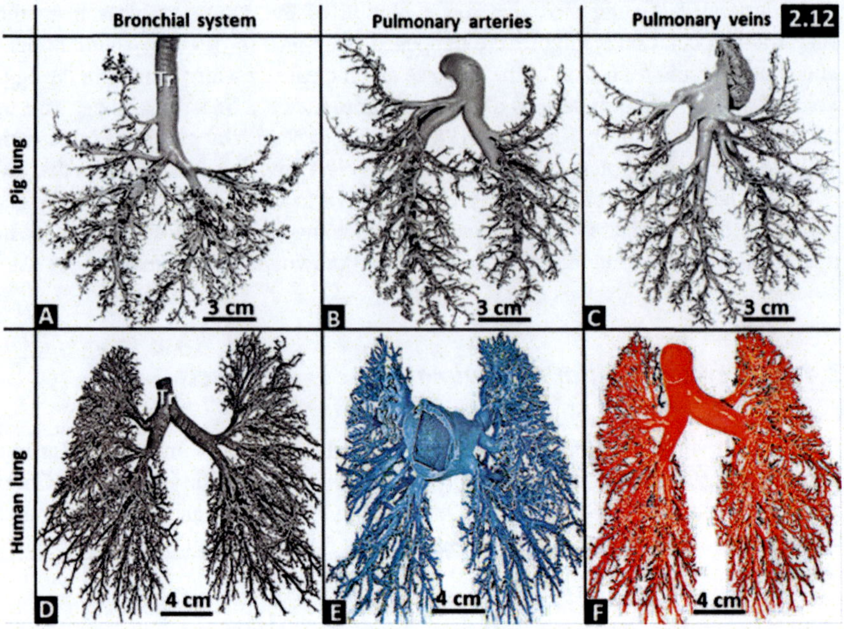

Plate 2.12 Branched morphologies of the airway (bronchial), the pulmonary artery and the pulmonary vein systems of the mammalian lung. The pulmonary artery tracts the bronchial system more closely than the pulmonary vein that runs intersegmentary. The particular closeness between the airway and pulmonary artery provides large respiratory surface area while it reduces the thickness of the blood-gas barrier. *Tr* trachea

Domestic Fowl (*Gallus gallus* variant *domesticus*), like for the erythroblasts (Plate 2.14, e, f), granulocytes also derive directly from "fixed" mesenchymal cells (Plate 2.14, e, f). In rarer cases, angioblastic cells also convert directly into granulocytes (Plate 2.14, g). In the process of transforming the mesenchymal cell space into gas exchange tissue, cells undergo marked apoptosis (Plate 2.14, f–inset). Interestingly, the erythroblastic cells undergo mitotic division in the mesenchymal cell space (Plate 2.14, h). Near the time of hatching (i.e., from day 18 of incubation), for the ling of the Domestic Fowl, white and red blood cells are found in the pulmonary blood vessels (Plate 2.14, i).

The classification of embryonic angiogenetic, erythroblastic, and vasoformative cells and the cells they derive from and differentiate into has been an area of intense research over the past times but is still uncertain and contentious (Akeson et al. 2000; Downs 2003; Urbich and Dimmeler 2004; Schatteman and Awad 2004; Schatteman 2004; Dzierzak and Speck 2008; Lim et al. 2013; Dzierzak and Philipsen 2013). Early in time, His (1900) postulated that in a developing embryo, angiogenetic cells derive from extraembryonic yolk sac precursor cells and that the primordial blood vessels grow and proliferate throughout the body by programmed budding, branching, and dissemination. Later, the suggestion that angiogenetic cells are of

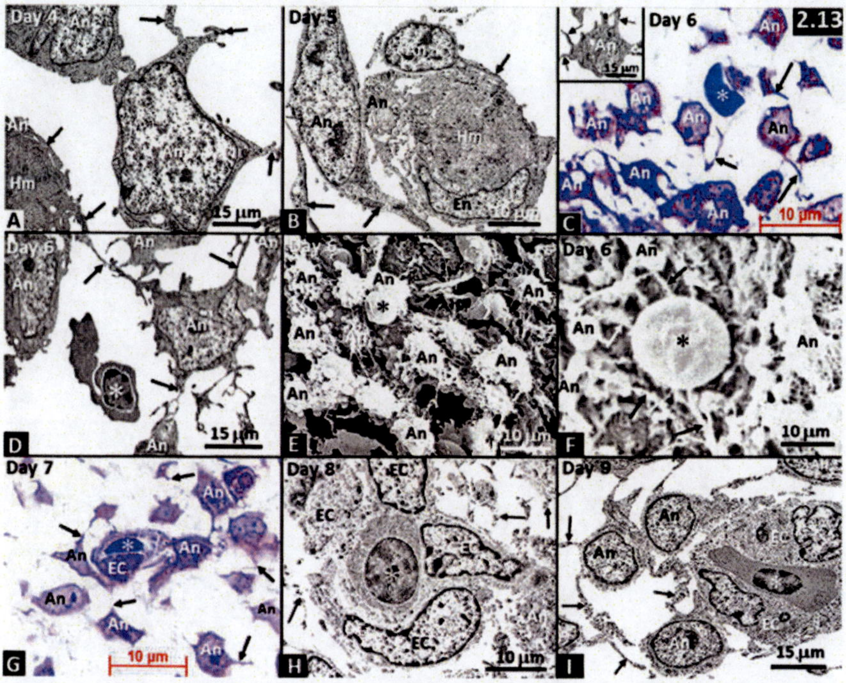

Plate 2.13 Early stages of development of blood vessels in the the lung of the Domestic Fowl (*Gallus gallus* variant *domesticus*). The process starts with differentiation of mesenchymal cells into angioblastic cells (An) and some of them into hemangioblasts (Hm) that accumulate hemoglobin in their cytoplasm **(a-c)** and ultimately become nucleated **(d, h, i)**. The hemangioblasts appear to attract angioblastic cells that come to surround them **(a, b, e-i)**, forming blood vessel walls **(g-i)**. The first angioblstic cells that contact the hemangioblasts form the endothelial cells (EC) of the formative blood vessels. With well-developed filopodial extension (arrows, ←) **(a-d, g-i)**, the angioblastic cells appear to be motile in the mesenchymal cell space

mesodermal or are of intraembryonic origin was endorsed by investigators such as Hahn (1909), Evans (1909), Miller and McWhorter (1914), Reagan (1916) and Ferguson et al. (2005). His (1900) used the term " angioblast" to define the embryonic mesenchymal cells that belonged to the endothelial cell lineage; Lillie (1918) applied the term (angioblast) to the layer of mesoderm in the area opaca of the chick embryo that originated from blood islands and; Bremer (1912) and Sabin (1917, 1920) proposed that the term should describe the type of cell that developmentally lies between the mesodermal and the endothelial cells. The embryonic origin of the vasculoendothelial and the hematogenetic cells, particularly erythrocytes, is distinctly uncertain (Pèault et al. 1983; Pardanaud et al. 1989, 1996; Mills et al. 1999; Lacaud et al. 2001, 2004; Wang et al. 2004; Davidson and Zon 2004; Jaffredo et al. 2005; Wilting and Becker 2006; Kauts et al. 2016; Lis et al. 2017; Canu and Ruhrberg 2021). The " hemangioblast" is deemed to be a multipotent precursor cell that gives rise to both endothelial and blood cells (erythroblasts)

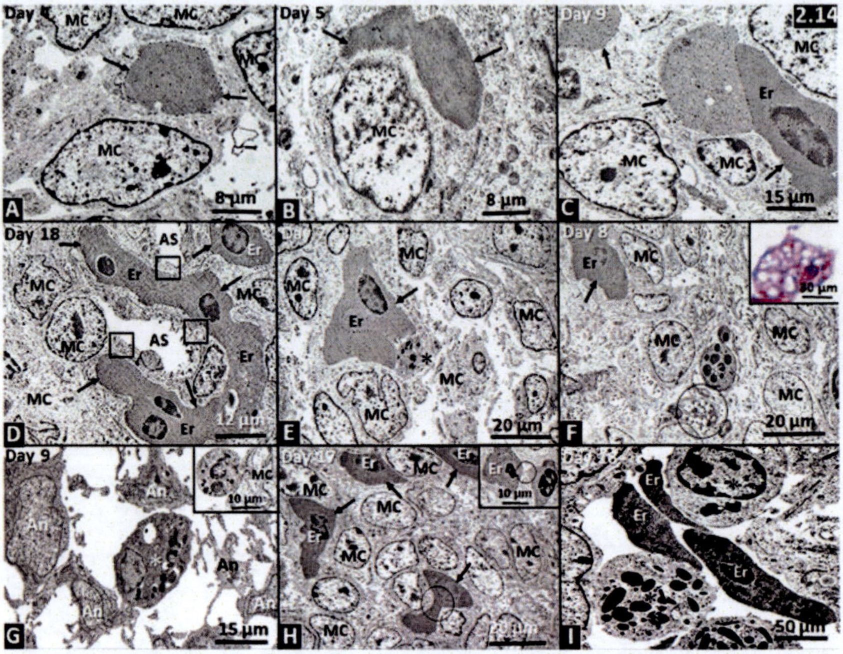

Plate 2.14 Development of blood cells and blood vessels in the embryonic lung of the Domestic Fowl (*Gallus gallus* variant *domesticus*). Committed mesenchymal cells (MC) (hemangioblasts) accumulate hemoglobin in their cytoplasm (arrows) (**a-f, h**). Such cells induce proximate cells to accumulate hemoglobin (**c, d**) and nucleate to into erythrocytes (Er) (**c-e, d**). Erythrocytes and lecukocytes develop from mesenchymal cells (**e-g**). Erythrocytes divide mitotically (**H**–encircled area and inset). During later stages of embryonic development, erythrocytes and leukocytes appear in the blood vessels (**i**). Boxed area (**D**), blood-gas barrier; AS (**d**), air spaces that form air capillaries; asterisks (*) (**e-g, i**), granular leukocytes; encircled area and inset (**f**), a vesiculated mesenchymal cell undergoing apoptosis

(Risau and Flamme 1995; Risau 1997; Lacaud et al. 2001; Eichmann et al. 2002; Jaffredo et al. 2005; Cao and Yao 2011). Presumed to be a hemangioblast, Lacaud et al. (2001) identified a precursor cell with the capacity of differentiating into both primitive and definitive hematopoietic cells, together with cells with endothelial cell properties. Stockard (1915) and Ogawa et al. (2001) proposed that although cells such as endothelial cells, erythrocytes, and leukocytes originate from mesenchymal cells, they are essentially polyphyletic in origin. Local tissue conditions drive the process of vasculogenesis (Davis and Bayless 2003; Mani et al. 2009; Kaipainen and Bielenberg 2010). By producing various molecular (growth) factors/cues, Lacaud et al. (2001) and Nicosia and Villaschi (1999) observed that interactions between endothelial cells, pericytes, smooth muscle cells, and fibroblasts can regulate angiogenesis. A definitive "vascular stem cell," i.e., an immediate precursor cell to all vascular endothelial and/or mural cells has so far not been identified (Chaudhury et al. 2011). Without excluding the existence of an absolute hemangioblast, the current view is that the direct precursor for the definitive hematopoietic stem cells is an endothelial cell with hematogenetic, i.e., blood-forming, capacity (Lengerke and Daley 2005; Jaffredo et al. 2005). For the avian lung, endothelial and hematopoietic cells derive from a common mesenchymal cell precursor, the hemangioblast (Asahara et al. 1997; Choi et al. 1998).

In birds, the cellular mechanisms that drive vasculogenetic and angiogenetic processes vary greatly (Coffin and Poole 1988; Poole and Coffin 1989, 1991; Pardanaud et al. 1989; Poole 1994) and are regulated by different molecular factors (cues) (Yancopoulous et al. 2000; Ferguson et al. 2005). According to Noden (1989), Risau and Flamme (1995), Risau (1997), DeMello et al. (1997), and Schittny (2017), the first stage of the development of blood vessels, entails the differentiation of endothelial cells that form blood vessels by the process of vasculogenesis that is accompanied by migration of angioblasts while the second one (angiogenesis)s entails the formation of new blood vessels by extension and/or remodeling of existing ones by sprouting and intussusceptive microvascular growths (Hudlicka and Tyler 1986; Hudlická et al. 1998; Larrivée and Karsan 2000; Ferguson et al. 2005; Demir et al. 2007; Ribatti and Pezzella 2021; Laschke et al. 2022). Given that gas exchange in the lung only commences when a chick is first exposed to air at internal pipping, the erythrocytic cells that are formed early in the embryonic lung only serve the purpose of initiating and regulating local blood vessel development (vasculogenesis).

2.4.3 Development of the Air and the Blood Capillaries

The air and the blood capillaries are the terminal respiratory units of the avian lung (McLelland 1989; Maina 2005). The lung of the Domestic Fowl (*Gallus gallus* variant *domesticus*) develops from day 15 of incubation (Maina 2003a, 2004b, 2005) (Plate 2.15, a, b): the process occurs by complex reorganization of mesenchymal cells and programmed cell death (apoptosis) (Plate 2.15, c-f). Programmed cell death plays an important part in the sculpturing of embryonic tissues and organs (Kerr et al. 1972; King and Cidlowski 1998; Abud 2004; Ray and Mehendale 2005; Penaloza et al. 2006, 2008; Elmore 2007; Krysko et al. 2008; Brown et al. 2014; Zakeri et al. 2015; Escobar et al. 2015; D'Arcy 2019; Çıkla-Süzgün and Küçükgüzel 2019; Li et al. 2020). Apoptosis is an autonomous process by which excess, misplaced, redundant, abnormal, and nonfunctional cells are eliminated without inciting an inflammatory reaction: it contrasts necrotic cell death that stems from direct environmental damages. Driven by certain biochemical pathways that cause specific morphological changes, apoptosis is an active process that requires energy to regulate cellular changes (Le Roith and Karas 2003; Choi and Wang 2006; Elmore 2007; Xu et al. 2019; Li et al. 2020): activation of cysteine proteases (caspases) is integral to the process of apoptosis (Saraste and Pulkki 2000; Boatright and Salvesen 2003; Strzyz 2007).

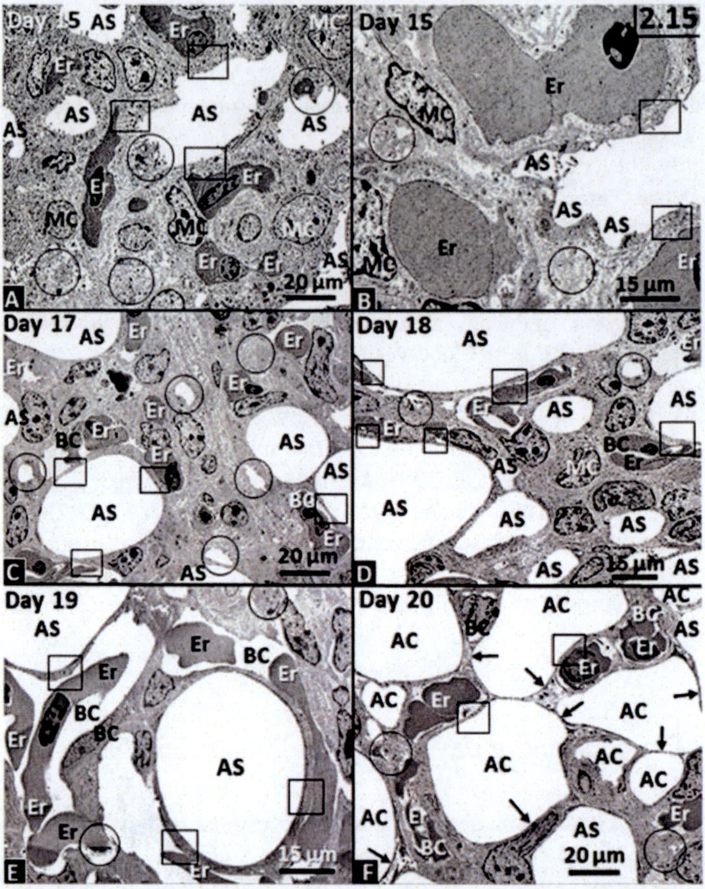

Plate 2.15 Development of air and blood capillaries of the lung of the Domestic Fowl (*Gallus gallus* variant *domesticus*). The air spaces (AS) (**a-e**) that ultimately for air capillaries (AC) (**f**) develop by massive transformation, displacement and apoptosis of mesenchymal cells (MC) (circles, O) (**a-f**). The dying cells create air spaces (**a-d**) and vascular channels (**e, f**). The tissue barrier between the air spaces (**a-e**)/ air capillaries (**f**) and the blood capillaries (BC) that contain erythrocytes (Er) (**c-f**) is thin (boxed areas, □) (**c-f**). Close to pipping and hatching, the air capillaries (AC) and the blood capillaries are well-developed and the blood capillaries are interconnected by epithelial cell retinaculae (arrows, ←) (**f**)

In the avian lung, broad cellular apoptotic processes are involved in the formation of the air and the blood capillaries in the existing mesenchymal tissue spaces (Plate 2.15). For the mammalian lung, the process of apoptosis also plays an important role in the development of the alveoli and the blood capillaries (Schittny et al. 1998; Schittny and Burri 2008; Bruce et al. 1999; Stiles et al. 2001; Wongtrakool and Roman 2008). Schittny et al. (1998) observed that maturation of the mammalian lung is marked by a reduction of the number of fibroblasts and type-II epithelial cells during the third postnatal week of embryogenesis while Schittny and Burri et al. (2004) noted that albeit a 25% increase of lung volume, for the rat lung, thinning of the interalveolar septa resulted in a reduction of the absolute mass of the interstitial tissue. For the rat lung, the absolute number of fibroblasts was reduced by 10 to 20% and the epithelial cells by more than 10% during its development (Kaufman et al. 1974). For the lung of the Domestic Fowl (*Gallus gallus* variant *domesticus*), as development progresses, massive cellular reorganization in the exchange tissue occurs with displacement and apoptosis of cells, leading to sites where blood capillaries interconnect across epithelial-epithelial cell extensions (epithelial laminae) and in rarer cases connecting directly (Plate 2.15, f). Close to hatching, the blood-gas barrier of the chicken's lung is very thin (Maina 2004b) (Plate 2.15, f): it develops by a complex process that entails displacement, transformation, differentiation and programmed cell death (apoptosis) of various types of cells, especially those intervening between the epithelial and the endothelial cells that respectively line the surface of the air capillaries and the blood lumina of the capillaries (Plate 2.16, a-d). For the blood-gas barrier, after the epithelial and the endothelial cells contact, a basement membrane is deposited in the intervening space (Plate 2.16, e, f). At hatching (day 21), for the lung of the Domestic Fowl, the structure of the blood-gas barrier comprises an epithelial cell, an endothelial cell, and an intervening basement lamina, a tripartite structure (Plate 2.16, g, h).

2.5 Development of the Air Sacs

At various times and stages of chick embryonic development, the air sacs develop as blister-like outgrowths from the cranial, the ventral, and the caudal borders of the wedge-shaped lung (Plate 2.17): roughly, the cervical and the clavicular air sacs originate from the cranial border of the lung, the cranial thoracic and the cardiothoracic ones from the ventral border and the abdominal air sacs from the caudal one. The places where air sacs join the lung are called ostia (McLelland 1989): the connections occur across the intrapulmonary primary bronchus, secondary bronchi, and rarely by means of parabronchi. For the Domestic Fowl (*Gallus gallus* variant *domesticus*), in quick succession, the primordia of the air sacs form between days 8 and 13 (Maina 2003a) (Plate 2.17, a-h). The abdominal air sacs pierce the horizontal septum and extend into the coelomic cavity (Plate 2.17, i), where they come to lie between the various visceral structures such as stomach and loops of intestines. At maturity, in some species of birds such as the Ostrich (*Struthio camelus*), the abdominal air sacs outstretch out of the coelomic cavity and come to lie subcutaneously (Bezuidenhout et al. 1999): there, they are susceptible to trauma and conditions such as airsacculitis (Fletcher 1980; Ficken and Barnes 1989). With marked interspecific variations, the air sacs enter (pneumatized) proximate bones (Duncker 1971; Wedel 2003; King and McLelland 1984; McLelland 1989; Lambertz et al. 2018). For the chicks of the Domestic Fowl, the abdominal air

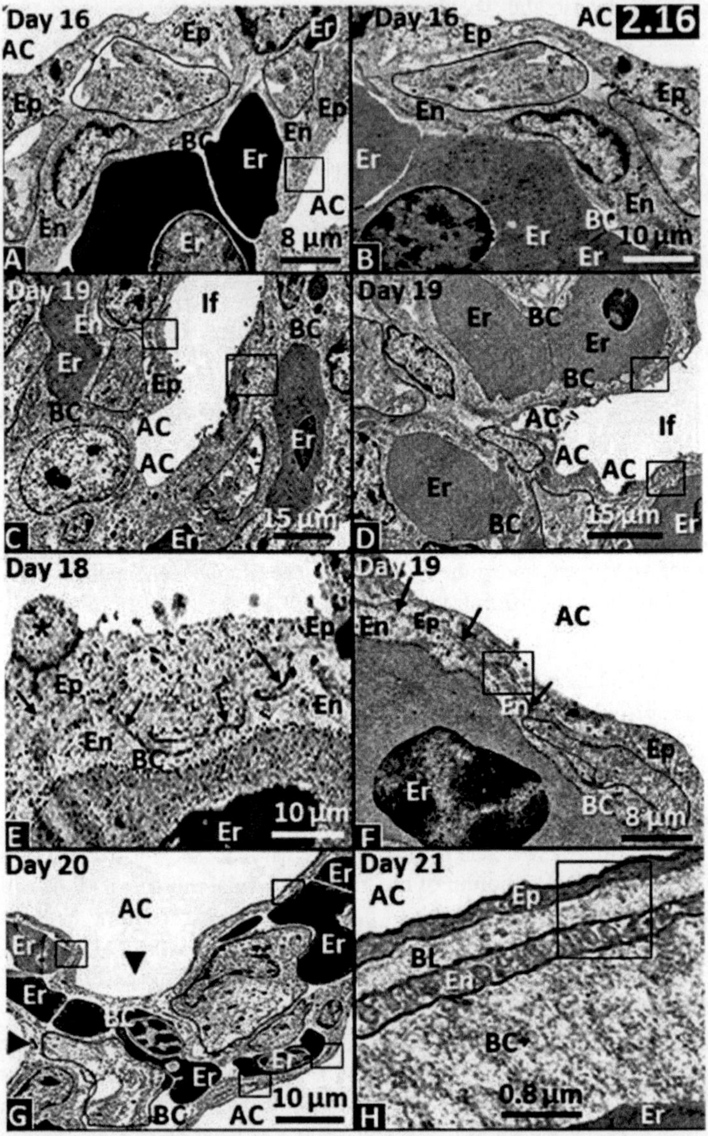

Plate 2.16 Development of the blood-gas barrier of the lung of the Domestic Fowl (*Gallus gallus* variant *domesticus*): it develops where air spaces and later air capillaries come to lie next to each other. The intervening cells/structures undergo apoptosis (encircled areas) (**a-d**), bringing the apithelial cell (Ep) lining the air capillary (AC) close to the endothelial cells (En) lining the blood capillary (BC) (**a-h**), forming thin blood-gas barrier forms (boxed areas) (**a, c, d, f-h**). When epithelial and endothelial cells contact, basement membrane is secreted (arrows, ←) (**e, f, h**). The air capillaries (AC) extend as mesenchymal cells undego apoptosis (**G**) (arrow head, ▶) (**g**) and with further apoptosis of mesenchymal cells, the blood-gas barrier becomes thinner (boxed areas) (**a-d, f**). Close to internal pipping (i.e., breathing air), the blood-gas barrier is very thin (**G**) and a tripartite structure which comprises an epithelial cell (Ep), a basement lamina (BL) and an endothelial cell (En) exists (**h**). *Er* erythrocyte; *BC* blood capillary

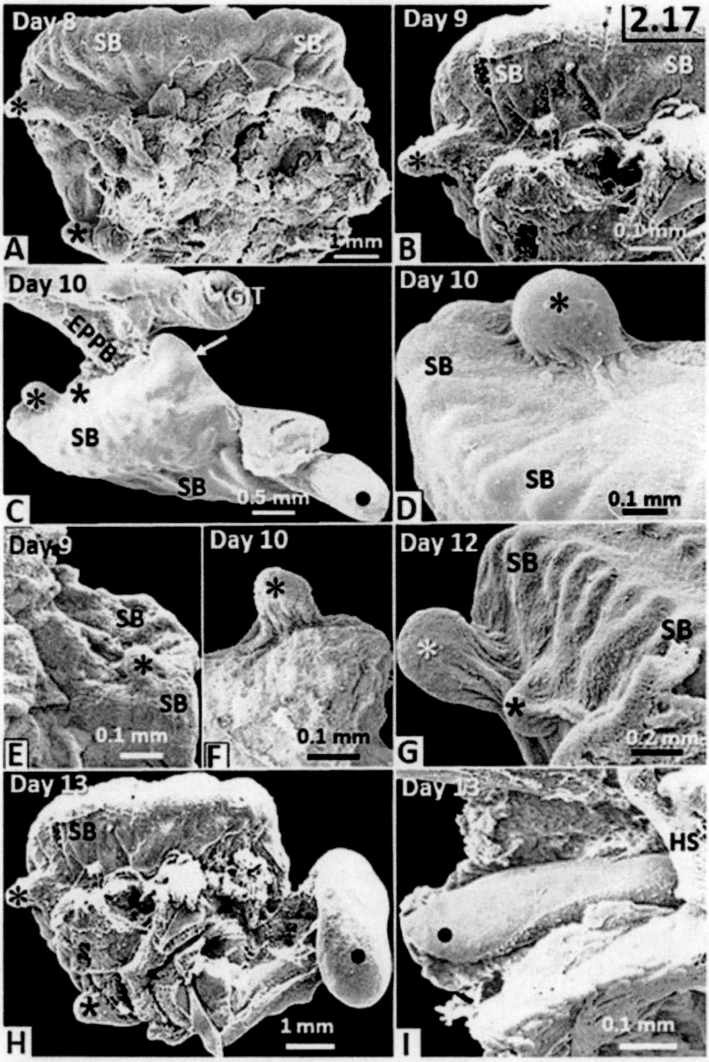

Plate 2.17 Development of the air sacs of the lung of the Domestic Fowl (*Gallus gallus* variant *domesticus*). The air sacs form on the cranial, ventral, and caudal sedges of the lung at different places and times from day 8 of incubation (**a-h**). Asterisk (*), cervical air sac; star (⦷), interclavicular air sac; arrow (←), cranial thoracic air sac; dot (•), abdominal air sac. The abdomical air sacs are the only ones that perforate the horizontal septum (HS) to enter the coelomic cavity (**I**). *SB* secondary bronchi

sacs are the largest at hatching (day 21 of incubation) and in maturity. In hummingbirds (Trochilids), the abdominal air sacs are the smallest (McLelland 1989).

2.6 Molecular and Genetic Aspects of the Development of the Avian Respiratory System

Cross-talk between epithelial and mesenchymal cells importantly determines the development of the respiratory system (Sakiyama et al. 2003; Liu et al. 2003; Miura 2008; Miura et al. 2009; Moura et al. 2014; Moura and Correia-Pinto 2017): it entails drastic reorganization of cells into highly complex functional structures (Roman 1997). While the structure of the mature avian respiratory system has been well-studied, especially that of the Domestic Fowl (*Gallus gallus* variant *domesticus*) (Duncker 1971; Abdalla and King 1975, 1976a, b, 1977; West et al. 1977; Maina 1982, 1988; Abdalla 1989; Maina et al. 2021), the molecular mechanisms and processes that are involved in its development are less understood (Sakiyama et al. 2000; Maina et al. 2003; Makanya et al. 2013; Maina 2012, 2013; Moura and Correia-Pinto 2017). In particular, for the mammalian lung, it has been shown that the development of the lung is very much driven by intricate transactions between various molecular signaling pathways and cell-to-cell transduction events (Metzger and Krasnow 1999; Lebeche et al. 1999; Warburton et al. 2000; Cardoso 2000, 2001; Hyatt et al. 2002; Cardoso and Lü 2006; Metzger et al. 2008). Among others, the molecular factors that directly or indirectly drive the process of lung development are the fibroblast growth factors (FGFs), WNTs (wingless and Int-1 genes), bone morphogenic proteins (BMPs), sonic hedgehogs (SHH), retinoic acid, notch, transforming growth factor β (TGF-β), Sox9 [SRY-related (HMG (high mobility group)] transcription factors, and VEGF (Vascular Endothelial Growth Factor). Interestingly, while the molecular signaling factors that are involved in the development of the vertebrate respiratory systems are basically similar (Metzger and Krasnow 1999; Schepers et al. 2002; Maina et al. 2003; Cardoso and Lü 2006; Warburton et al. 2010; Sylva et al. 2011; Moura et al. 2011; Maina 2012; Moura and Correia-Pinto 2017), the structures/organs that develop are morphologically different (Plates 1.6; 2.2 and 2.4). While the differences may largely stem from phylogenetic differences, variations in spatiotemporal expressions of the molecular (growth) factors may meaningfully explain the morphological differences among organs/structures that exist between animal taxa. The following are some of the important occurrences: (a) in the mouse lung, cross-talk between canonical WNT and FGF signaling occurs, with FGFr-2 (a FGF-10 cognate receptor) expression being markedly reduced in the lung airway epithelium when β-catenin is lacking (Shu et al. 2002); (b) WNT-5a regulates SHH and FGF-10 signaling during lung development (Li et al. 2005, 2010); (c) during neural development, SHH acts together with other morphogens, especially the WNTs (Wilson and Stoeckli 2012); (d) Bridges et al. (2014) showed that sterol regulatory element-binding proteins (SREBP) signaling is integrated with the WNT/b-catenin and glucocorticoid receptor signaling pathways during perinatal lung maturation; (e) Ulloa and Martí (2010) observed that antagonistic interactions between canonical WNT (promoting dorsal identities) and SHH pathways (inducing the ventral ones) characterize the dorsoventral patterning of the developing central nervous system; f) together with FGF-10 and BMP-4, WNT-5a acts upstream of SHH and fibronectin levels that are controlled by the WNT ligand and indirectly by its regulation of SHH (Sakiyama et al. 2000; Moura et al. 2014); and g) an epithelial-mesenchymal signaling pathway involving SHH-FOXF-1 (Forkhead Domain-Containing Transcription Factors) and TBX-4 (T-Box Transcription Factors) FGF10 and TBX-2 pathways are a vital transcriptional network that drives early lung morphogenesis (Karolak et al. 2021). Bellusci et al. (1997), Lebeche et al. (1999). Weaver et al. (2000) showed that: (a) FGF-10 and FGF-7, molecular factors that are vital in the patterning and the growth of the lung buds, are in various ways controlled by FGF-1, FGF-2, and SHH; (b) the FGFs regulate expression of *SHH*, BMP-4, and other *FGF* family members; and (c) suggesting that high levels of SHH downregulate FGF-10, in transgenic mouse lungs, overexpression of SHH in the endoderm leads to reduction in transcription of FGF-10. In the chicken (*Gallus gallus* variant *domesticus*) lung, Sakiyama et al. (2000) observed that the development of the lung and the air sacs, structures that are structurally very different, can be explained by the fact that HOXb genes, BMP-2, BMP-4, WNT-5a, and WNT-11 specify regional variations in morphogenetic- and cytodifferential mechanisms in the various parts of the developing respiratory tract: the expression domains of HOXb-6 to HOXb-9 genes correspond with the morphological sectors of the air sacs, along a proximodistal direction.

Some of the notable molecular investigations that have been done on the development of the avian respiratory system are the following: Goldin and Opperman (1980) and Hacohen et al. (1998) studied the stimulation of DNA synthesis in embryonic chick lung and the trachea by expression of the epidermal growth factor (EGF); Chen et al. (1986) examined the expression and the distribution of the cell-to-cell adhesion molecules, fibronectin and laminin in embryonic chick lung cells; using lectin probes and cationic dyes, Gallanger (1986) studied the development of the basement membrane and showed that the transcription factors, hepatocyte nuclear factor 3β (HNF-3β) and TTF-1 control the differentiation of epithelial cells and the expression of genes during early stages of development of the chick lung buds from the fourth day of incubation; Zeng et al. (1998) and Muraoka et al. (2000) showed correlation between the expression of the nuclear factor-kappa-β and epithelial growth and the branching of developing airways in embryonic chick lung; using tissue recombination experiments, Sakiyama et al. (2003) noted that for the avian respiratory system, the development of the airways and the air sacs stem from region-specific mesenchymal properties and that HOXb genes are expressed around the ventral-distal apices of the lung; Stabellini et al. (2001) examined the roles of polyamines and TGF-β1 on the branching of the airways; Maina et al. (2003) observed that bFGF-2 (basic FGF-2) is expressed and remains upregulated in the epithelial and mesenchymal cells of the developing chick lung through the incubation period (Plates 2.18, 2.19, and 2.20); like for the mammalian lung, where the fate of the cells of the primitive pharynx is characterized by endodermal cell expression of TTF-1 or NKX-2.1 that convey craniocaudal and dorsoventral organ-specific information to the gut endoderm (Lazzaro et al. 1991; Minoo et al. 1999), for the chick lung, endodermal expression of NKX-2.1 recognizes the respiratory system committed cells as early as the HH14 stage of embryonic development (Sakiyama

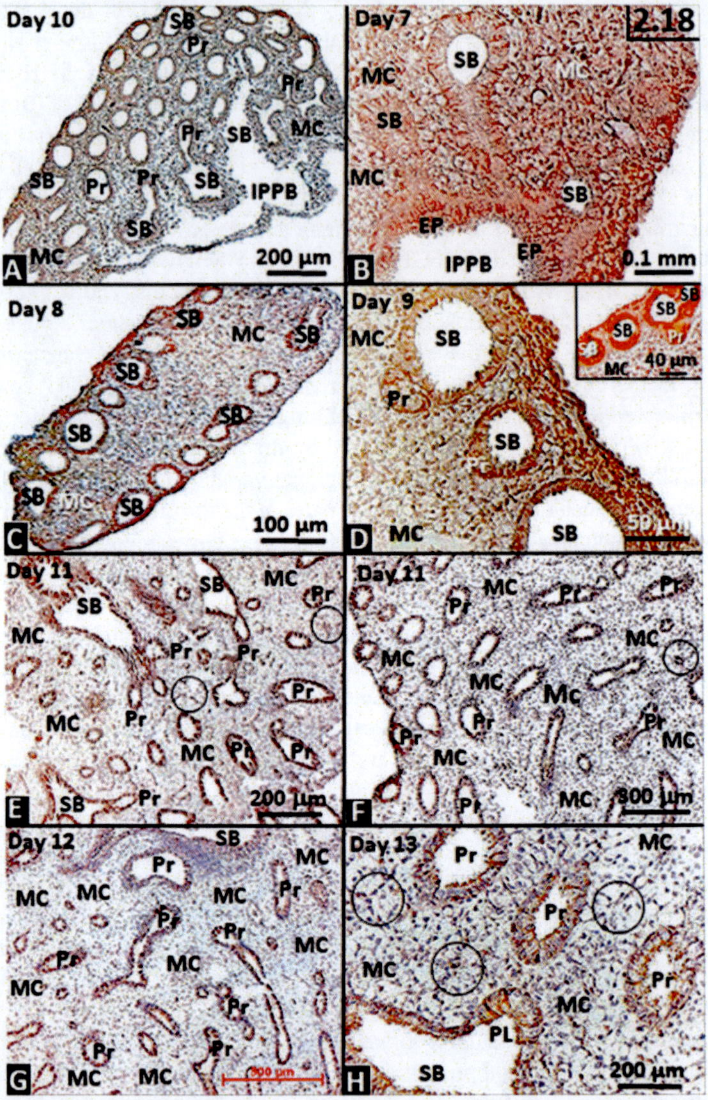

Plate 2.18 Expression of basic fibroblast factor-2 (bFGF-2) in the developing lung of the Domestic Fowl (*Gallus gallus* variant *domesticus*). The intrapulmonary primary bronchus (IPPB), which is the first airway to form, gives rise to secondary bronchi (SB) that in turn originate parabronchi (tertiay bronchi) (Pr) (**a-d**). The parabronchi, that constitute the bulk of the volume of the lung, continue to develop by budding (sprouting) as development continues (**e-h**). *MC* mesenchymal cells; circles (O), developing blood vessels

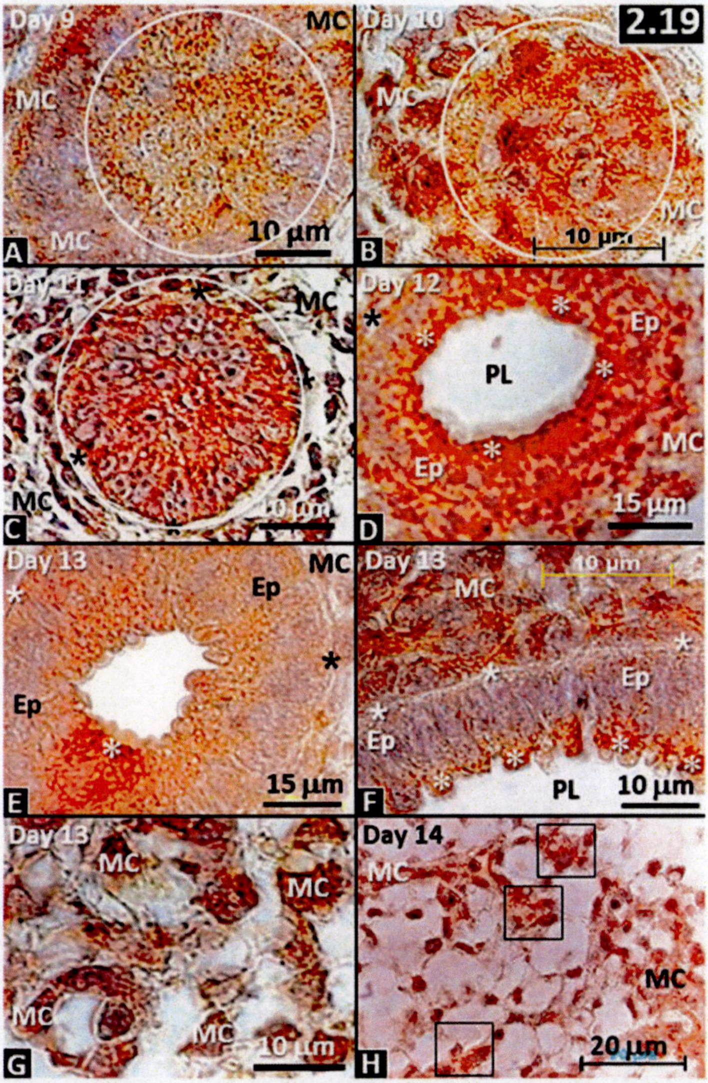

Plate 2.19 Early expression of basic fibroblast factor-2 (bFGF-2) in the developing parabronchi and mesenchymal tissue space cells of the Domestic Fowl (*Gallus gallus* variant *domesticus*). The parabronchi (tertiary bronchi) develop from clusters of mesenchymal cells (encircled cells, **a-c**) that highly express bFGF-2. Subsequently, the parabronchial lumina are lined by a columnar epithelium (**d-f**) of which bFGF-2 is overexpressed at the apical ends (asterisks, *). On the basal aspect, the epithelial cells cells attach to a basement membrane (star, ⌷) (**e, f**). The mesenchymal cell (MC) highly express bFGF-2 (**f-h**). Blood vessel forming cells (angioblats and hemangioblasts) highly express bFGF-2 (rectangle, □ enclosed cells) (**h**)

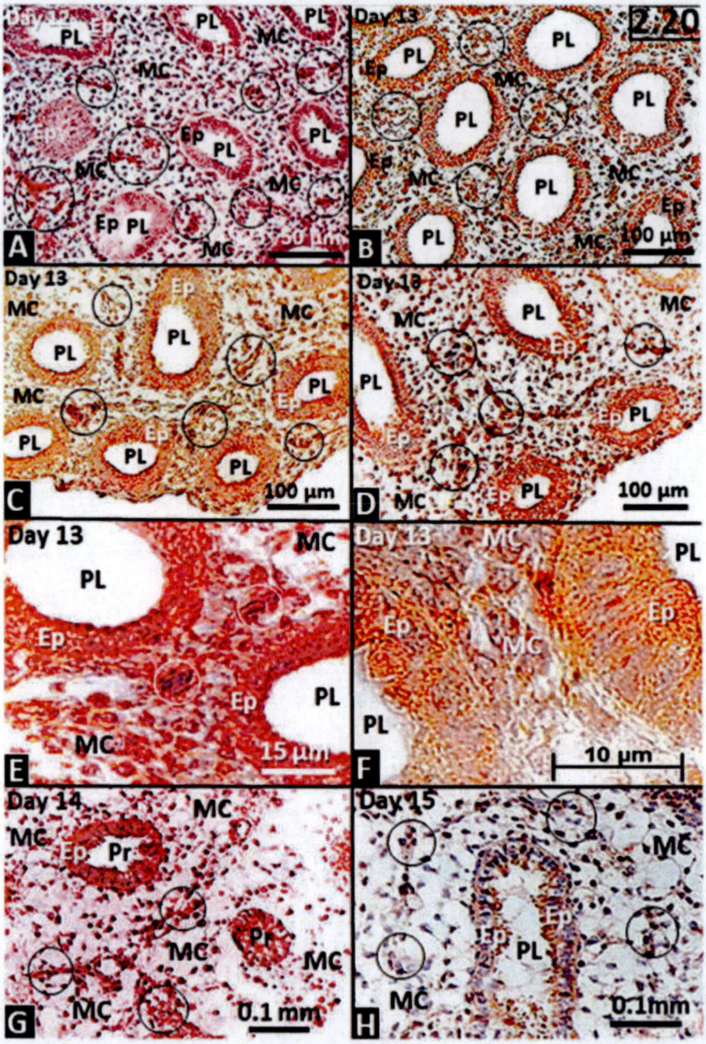

Plate 2.20 Late expression of basic fibroblast factor-2 (bFGF-2) in the parabronchi and associated blood vessels of the embryonic lung of the Domestic Fowl (*Gallus gallus* variant *domesticus*). bFBF-2 is expressed by the mesenchymal cells (MC), the epithelial cells (Ep) lining the parabronchial lumen (PL) and mesenchymal cells forming blood vessels (circles, O) (**a-e, g, h**)

et al. 2003); Miura et al. (2009) reported that the development of air sacs (which were termed "cyst-like" structures) on the ventral part of the lung stems from differences in the diffusion coefficients of morphogens such as FGF-10 between the dorsal and ventral parts of the lung, a process determined by differences of cell densities in the two parts of the lung and conjectured that heparin sulphate proteoglycan (HSPG) that traps FGF-10 is more strongly expressed in the dorsal part, slowing down its diffusion; on comparing findings and interpretations of the molecular aspects of the development of the mammalian (Cardoso 2000, 2001; Cardoso and Lü 2006; Warburton 2021; Warburton et al. 2010)—and the avian (Moura and Correia-Pinto 2017) lungs and the tracheal system of insects (Ghabrial et al. 2003), Maina (2012) drew attention to the conservation of the molecular signaling (growth) factors across phylogenetically very different animal taxa.

The most important and best-known molecular (growth) factors are succinctly appraised below.

2.6.1 Fibroblast Growth Factor (FGF) Signaling

Fibroblast growth factors (FGFs) are a family of growth factors (GFs) that exist in many multicellular animal species (Ornitz and Itoh 2001; Itoh 2004; Ornitz and Sannes 2006; Mario and Daniele 2010; Ornitz and Itoh 2015): the GFs are involved in various important biological processes that include organogenesis, homeostasis, cellular repair and metabolism (Bikfalvi et al. 1997; Lebeche et al. 1999; McKeehan et al. 1998; Szebenyi and Fallon 1999; Warburton et al. 2000; Kumar et al. 2005; Herriges and Morrisey 2014). During lung development, the FGFs are important mediators of mesenchymal-epithelial cells interactions (Min et al. 1998; Shannon and Hyatt 2004). Twenty-two FGFs, that are categorized into seven different subfamilies, i.e., FGF-1, FGF-4, FGF-7, FGF-8, FGF-9, FGF-11, and FGF-15/19 occur (Klagsbrun 1989; Baird and Klagsbrun 1991; Baird and Walicke 1989; Weinstein et al. 1998; Suzuki et al. 2000; Ornitz and Itoh 2001): they have been identified from the similarities of their biochemical function, molecular sequence, and evolutionary relationships.

The fibroblast growth factor 10 (FGF-10) is an important morphogenetic cue of lung development and repair (Ramasamy et al. 2007; Wu et al. 2018; Jones et al. 2021). It is a member of the FGF-7 subfamily of secreted growth factors that are largely secreted by the mesenchyme and act on the epithelium via the FGF receptors 2b (FGFR-2b) and FGFR-1b (Ohuchi et al. 2000; Ornitz and Itoh 2001; Yuan et al. 2018; Jones et al. 2021). For the mouse lung, FGF-10-expressing cells constitute a population of mesenchymal progenitor cells in embryonic and postnatal lungs (El Agha et al. 2014). The importance of FGF-10 in the development of the lung is evident in the fact that complete failure of lung development and embryonic lethality occurs in *FGF-10* deficient mice (Min et al. 1998; Xu et al. 1998; Sekine et al. 1999). In the vertebrate lung, paracrine signaling that is mediated by FGF-10 and FGF-R2IIIb receptors is critical to the development of the lung (Perl et al. 2003). Targeted deletion of FGF-10 causes total pulmonary agenesis (Min et al. 1998). During lung development, FGF-10 is expressed very early in the distal mesenchymal areas where bifurcation occurs (Bellusci et al. 1997; Park et al. 1998; Cardoso 2000, 2001; Cardoso and Lü 2006; Ramasamy et al. 2007; El Agha and Bellusci 2014; Lü et al. 2005). Branching morphogenesis is considerably dependent on FGF10 signaling (Bellusci et al. 1997; Lang et al. 2018) and in the mouse lung, the FGF-10-SHH signaling loop is crucial to proper developmental process (Morrisey and Hogan

2010). FGF-10 is expressed in the distal mesenchymal cells through FGFR-2 in the epithelium, driving SHH expression by cells at the branching tips (Goodwin and Nelson 2021). FGF-10 knockout embryos display lung agenesis as well as other morphogenic and organogenic abnormalities (Min et al. 1998; Sekine et al. 1999; Suzuki et al. 2000; Sakaue et al. 2002; Jaskoll et al. 2005).

In the embryonic chick lung, FGF-10, which is expressed in the mesenchyme from the very early stages of development, plays a proliferating role by stimulating distal epithelial cell growth by activating epithelial FGFR-2 (Moura and Correia-Pinto 2017). In the mouse lung, blocking FGFR-2, the dominant negative FGF receptor, terminates airway branching and epithelial cell differentiation (Peters et al. 1994). The FGFs are important molecular agents of chick lung branching morphogenesis (Maina and Madan 2003; Maina et al. 2003; Moura et al. 2011): inhibition of their signaling pathways impairs the branching process (Moura et al. 2011). Differences in FGF-10 diffusion rates between the ventral and the dorsal parts of the chick lung mesenchyme appear to drive ventral cyst (air sac) development and branching of the airways on the dorsal part of the lung (Miura et al. 2009). Moura et al. (2011) showed that inhibition of FGF signaling by SU5402, that is an FGF-10 receptor antagonist, causes the formation of secondary bronchi with cystic shape. Compared with lungs that develop under normal conditions (Plate 2.21, a, b), Maina (2012) observed that cystic airways developed in explanted chick lungs grown in matrigel (a nutrient medium) under high concentration of FGF-10 (Plate 2.21, c-g) or when a bead soaked in a high concentration of the growth factor (FGF-10) is placed close to the developing lung (Maina 2012) (Plate 2.21, h). In mice, overexpression of FGF-7 after application of the human Sftpc promoter causes severe pulmonary malformations that include bronchial airway enlargement, cystic lung lesions, and impaired branching morphogenesis, features that result in embryonic lethality (Simonet et al. 1995). For the mouse lung, Weinstein et al. (1998) observed that reciprocal function of FGFR-3 and FGFR-4 accelerated the formation of alveoli during postnatal lung development. FGF-2 is a pluripotent, greatly conserved GF member of the FGF family and is mitogenic in a broad range of mesoderm and ectoderm-derived cells (Gospodarowicz 1991): it (FGF-2) is also a potent angiogenic (Abraham et al. 1986; Folkman and Klagsbrun 1987; Klein et al. 1993; Bikfalvi et al. 1997) and neurotrophic factor (Gospodarowicz 1991). In a human epithelial cell line (VA 10), FGFR-2 and sprouty-2 were expressed at the growing tips of the branching structures, and the process was inhibited by the FGFR-small molecule inhibitor SU5402 (Franzdóttir et al. 2010). In addition to its involvement in branching morphogenesis (Arman et al. 1999), FGF-2 induces expression of VEGF by vascular endothelial cells by means of autocrine and paracrine mechanisms (Seghezzi et al. 1998). Interestingly, mice genetically deficient in FGF-2 (FGF-2 knockouts) show no defects that relate to compromised angiogenesis (Ortega et al. 1998; Zhou et al. 1998). Liberti et al. (2021) reported that FGFr-2 determines type-II alveolar pneumocyte cell fate and maintains a balance between their proliferation and differentiation during lung alveolar regeneration.

For the developing lung of the Domestic Fowl (*Gallus gallus* variant *domesticus*), the basic FGF-2 (FGF-2b), which is involved in the growth and the development of

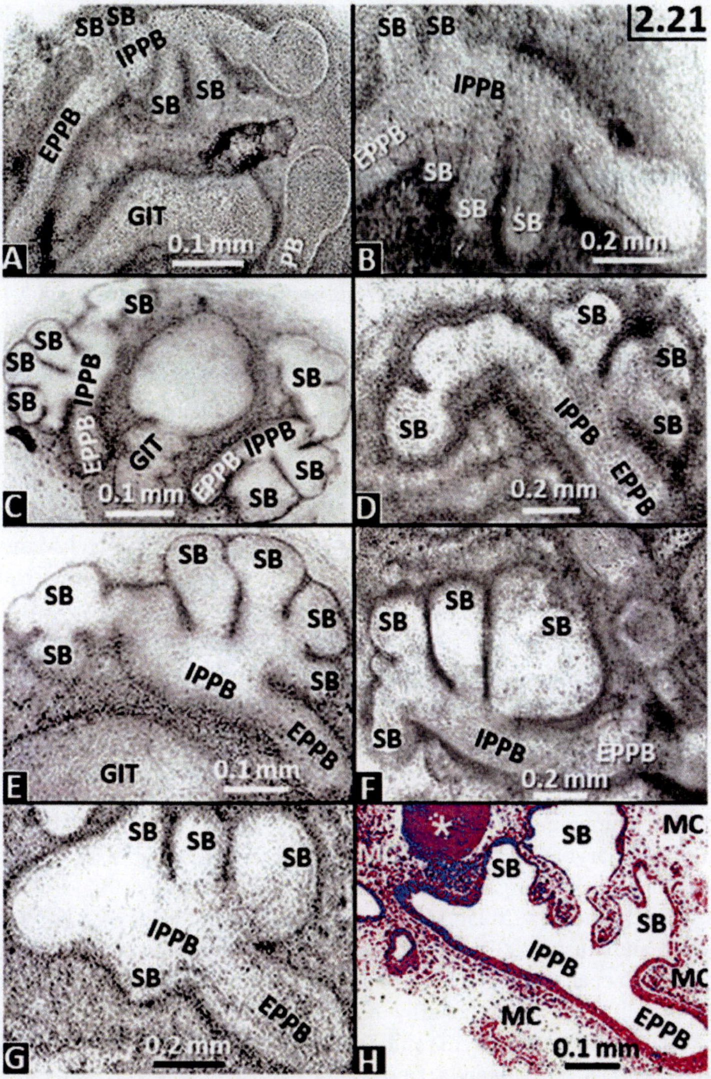

Plate 2.21 Effect of high concentration of fibroblast growth factor-10 (FGF-10) on early development of the lung of the Domestic Fowl (*Gallus gallus* variant *domesticus*). **(a, b)** Normal development of the airways in lungs cultured in matrigel, a nutrient growth medium. IPPB, intrapulmonary primary bronchus; SB, secondary bronchi; EPPB, intrapulmonary primary bronchus; GIT, gastrointestinal system. **(c-g)** Cystic airways in lungs grown under high concentration of FGF-10 and after implantation of a bead (star, ⌷) soaked in FGF-10 close to a cultured lung **(h)**. *MC* mesenchymal cells

airways and blood vessels, is extensively expressed in the mesenchymal cells from as early as day 3.5 of incubation and the upregulation is maintained throughout the incubation period (Maina and Madan 2003; Maina et al. 2003) (Plates 2.18, 2.19 and 2.20). FGF-2b is expressed in the epithelia of the airways, namely the primary bronchus, the secondary bronchi, and the parabronchi as well as in the mesenchymal cells (Plates 2.18, 2.19, and 2.20). The parabronchi develop from the ninth day of incubation and originate as a loose assemblage of mesenchymal cells that express FGF-2b (Plate 2.19, a-c): a lumen develops at the center of the clustering of cells and a basement membrane onto which the cells are attached forms at the periphery. FGF-2b is highly expressed at the apical ends of the epithelial cells (Plate 2.19, d-f). The epithelial, the mesenchymal, and the committed angioblastic cells, i.e., those cells that are involved in development of blood vessels, highly express bFGF-2 (Plates 2.19, g, h and 2.20). For embryonic and fetal rat lungs, Han et al. (1992) observed that FGF-2b is expressed in the epithelial cells of the airways, the basement membranes, and the extracellular matrix. Jones et al. (2021) termed FGFR-2b a "master conductor" of lung branching. Beyea et al. (2005) determined an unnamed 15-kDa protein (growth factor) was expressed early in mesenchymal and epithelial cells of the developing chick lungs. Maina and Madan (2003) and Maina et al. (2003) observed that for the developing chick lung, FGF2b is directly involved in vasculogenesis (Plates 2.19, h and 2.20, a, c-h).

2.6.2 *WNT Genes Signaling*

WNTligands comprise a large family of secreted, hydrophobic, glycoproteins that bind to cell membranes or extracellular matrix and are importantly involved in regulation of embryonic and adult developmental processes and tissue homeostasis in all adult multicellular organisms/animals (Nusse and Varmus 1992; Huelsken and Behrens 2002; Widelitz 2005; Mikels and Nusse 2006; Zhang et al. 2008; van Amerongen and Nusse 2009; Willert and Nusse 2012; Moura et al. 2014; Clevers et al. 2014; Loh et al. 2016; Ramakrishnan and Cadigan 2017; Aros et al. 2021). For the human being, 19 evolutionarily conserved WNT glycoproteins have been reported by Morrisey et al. (2013) and Kotton and Morrisey 2014), and in chicken, there are 18 known WNT ligands (Geetha-Loganathan et al. 2009). After binding to different receptors of proximate cells, WNTs activate intracellular signaling pathways that initiate changes in gene transcription. From their capacity of mediating cell-to-cell interactions and determining cell planar polarity and resolving cell fates, WNT signaling is considerably involved in embryogenesis and in many developmental abnormalities and pathologies, even disorders such as cancers (Wodarz and Nusse 1998; Pfeifer and Polakis 2000; Cohen et al. 2009; Okubo and Hogan 2004; Li et al. 2010; Polakis 2012; Holland et al. 2013).

From different intracellular processes that are activated after WNT-FZD binding, WNT signaling has been divided into canonical (WNT/β-catenin pathway) and noncanonical pathways (Dale 1998; Zhou et al. 2012). For canonical WNT signaling pathway, WNT binding to Frizzled (FZD) and LRP5/6 co-receptors causes β-catenin protein stabilization and entry into the nucleus, where it influences transcription of target genes and WNT-frizzled interactions activate intracellular pathways and cause β-catenin accumulation or increase in intracellular Ca^{2+} concentration (Cadigan and Nusse 1997; Miller et al. 1999; Kuhl et al. 2000; Mikels and Nusse 2006; Baarsma et al. 2013). The relatively less well-known, noncanonical pathway signaling requires binding of specific WNTs to specific FZDs that results in intracytoplasmic release of Ca^{2+} and FAK pathway activation independent of β-catenin transcription function (Cohen et al. 2002): the signaling pathway is mediated by intracellular calcium ion and c-Jun N-terminal kinase (JNK) (Semenov et al. 2007). By regulating endothelial cell growth and function, WNT signaling is involved in angiogenesis (Cheng et al. 2003). There are several soluble protein antagonists (modulators) of WNT signaling that include secreted frizzled-related protein (sFRP), WNT inhibitory factor (WIF), and Dickkopf (DKK) (Moon et al. 1997a, b; Hsieh et al. 1999; Fedi et al. 1999; Baarsma et al. 2013). By regulating epithelial cell differentiation and proliferation, in both mammalian and avian lungs, WNT signaling has been shown to be critical to proper lung development (Okubo and Hogan 2004; Logan and Nusse 2004; Hans 2006; Moura et al. 2014). With WNT2/2b signaling being vital to specifying the lung endodermal progenitor cells within the anterior foregut (Goss et al. 2009), embryos lacking expression WNT-2b display a total lack of lung development (agenesis) (Okubo and Hogan 2004). WNT-7b null mice succumb to respiratory failure (Shu et al. 2002). Various WNT signaling molecular factors such as WNT-1, WNT-*2b*, WNT-*3a*, WNT-*5a*, WNT-*7b*, WNT-*8b*, WNT-*9a*, *lRP-5,* IRP-*6*, SFRP-1, *DKK-1, β-catenin*, and AXIN-2 are expressed from the early stages of the development of a chick (Maina 2012, 2013; Moura et al. 2014; Moura and Correia-Pinto 2017) (Plate 2.22, a), in various parts of the developing lung (Plate 2.22, b-h) and the air sacs (Plate 2.23) at different instances of development. Expression of *WNT-5a* and *WNT-11* in the developing chick respiratory tract was investigated by Sakiyama et al. (2000). For the mouse lung, WNT-7b gene expression in the lung epithelium is regulated in a combinatorial manner by TTF-1, GATA-6, and Foxa-2 (Weidenfeld et al. 2002). Noncanonical WNT signaling appears to be more commonly involved in chick lung development, especially during the mid-developmental stage (Moura and Correia-Pinto 2017). In regulating pulmonary distal airway and vascular development, WNT-5a, which is a mainly noncanonical WNT ligand, acts through the ROR-2 receptor (Loscertales et al. 2008). In the mouse lung, WNT-5a is involved in alveologenesis (Li et al. 2020) and WNT-5a null mice have a defective pulmonary phenotype (Li et al. 2002). In chicks, a complex combination of expression of WNT signaling molecules occurs during the development of the lung and the various air sacs (Maina 2012) (Plate 2.23).

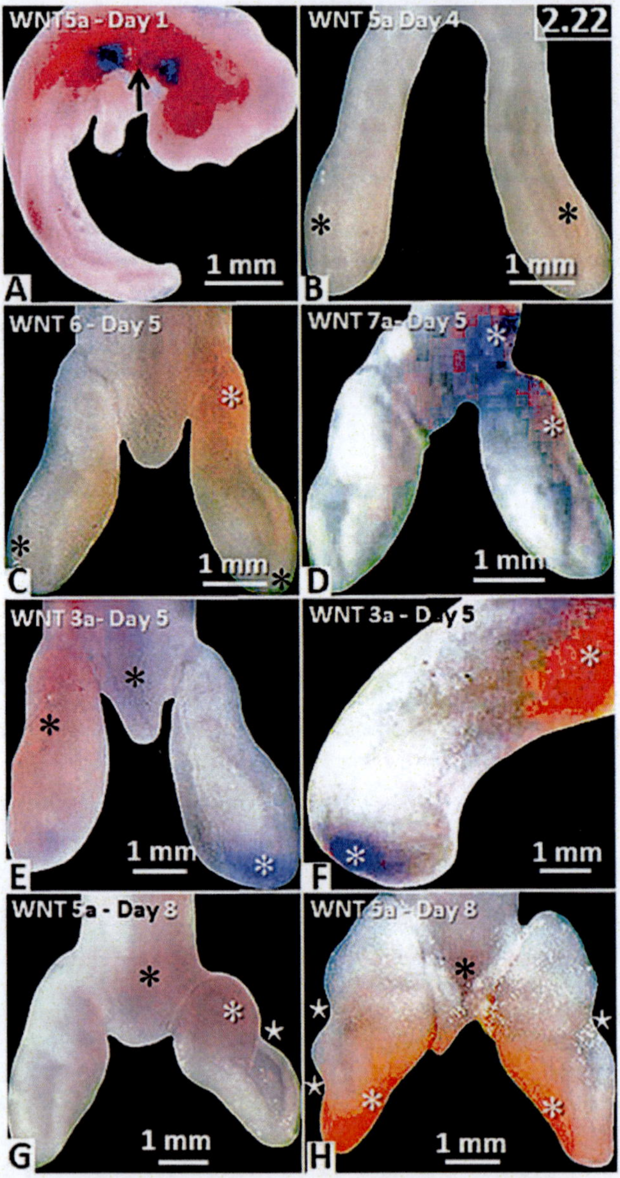

Plate 2.22 **(a-h)** Expression of various WNT molecular (growth) factors in a chick embryo **(a)** and various parts of cultured developing lungs of the Domestic Fowl (*Gallus gallus* variant *domesticus*) at different times of incubation (asterisks, *) **(b-h)**. Arrow **(a)**, ←, cells destined to form the lung. Stars (⃞) **(g, h)**, depressions on the surface of the developing lung

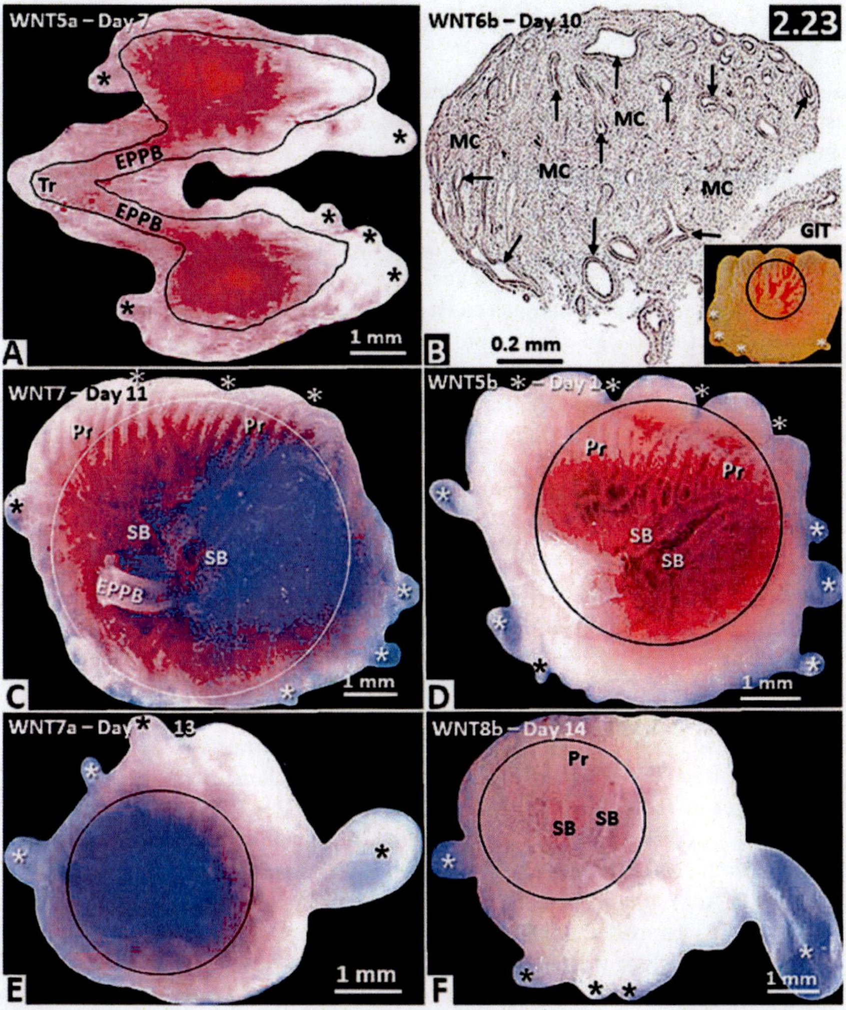

Plate 2.23 Expression of various WNT molecular (growth) factors in different parts of the developing lung and air sacs of the Domestic Fowl (*Gallus gallus* variant *domesticus*) at different times of incubation (encloded areas). At day 10 of incubation, mesenchymal cells (MC) and epithelial cells lining the airways (arrows, ↓) express WNT-6b (**b**). The air sacs develop at different times and places as blister-like outgrowths from the cranial, the ventral, and caudal edges of the embryonic lung (stars, ⨅). *Tr* trachea, *EPPB* intraprulmonary primary bronchus; *SB* secondary bronchi; *Pr* parabronchi (tertiary bronchi). Asterisks (*), costovertebral sulci; *GIT* gastrointestinal system

2.6.3 Sonic Hedgehog SHH) Signaling

Sonic Hedgehog (SHH) is an important morphogen that is involved in various aspects of organogenesis by regulating cellular processes such as growth and survival, differentiation, migration, and cell polarity (Sagai et al. 2009; Ingham and McMahon 2001; Varjosalo and Taipale 2008; Choudhry et al. 2014; Ramsbottom and Pownall 2016; Skoda et al. 2018): it is crucial to proper embryonic development. The complex canonical SHH signal transduction depends on the presence of transmembrane receptors such as Patched-1 (PTCH-1) and smoothened (SMO) and accessory proteins such as Hedgehog-interacting protein-1 and intracellular effector proteins like glioblastoma (GLI) zinc finger transcription factors that are the true drivers of the pathway (Cross and Bury 2004; Fernandes-Silva et al. 2017a). Pulmonary morphogenesis is greatly dependent on SHH and on its molecular interactions with other signaling pathways (Pepicelli et al. 1998; Fernandes-Silva et al. 2017a). In the chick lung, SHH is directly involved in lung specification, primary lung bud formation, and branching morphogenesis (Fernandes-Silva et al. 2017b): it is expressed in the trachea of the chick embryonic lung as early as day 6 of embryonic development (Davey et al. 2014) and in the distal and proximal epithelial regions at day 10 (Loscertales et al. 2008). SHH, PTCH-1, SMO, GLI-1, and HHIP (a hedgehog-interacting protein) are expressed very early in the development of the embryonic chick lung (Moura et al. 2014). In chick embryos with a recessive mutant trait dubbed *talpid*^{3}, of which the embryos die by about the 18th day of development (Harris et al. 2006), SHH signaling is lacking (Davey et al. 2014): the lungs are hypoplastic and have serious deficiencies of both the epithelial and the mesenchymal parts. In contrast to the mammalian lung, for the chick lung, SHH is not a downstream target of FGF signaling (Moura et al. 2016). Regarding SHH expression during lung morphogenesis, bird lungs appear to possess species-specific properties/features (Moura and Correia-Pinto 2017) where in particular secondary bronchi and parabronchi form in a rather non-dichotomous branching manner (Plate 2.4, F) in contrast to the mammalian lung where the airway system generally forms by regular dichotomous branching (Plates 2.2, b, 2.4, c and 2.12, a, d): between the mammalian and the avian lungs, the processes and mechanisms of airway bifurcation lead to totally different morphologies (Plate 2.2, a, b).

2.6.4 Vascular Endothelial Growth Factor (VEGF)

Vascular embryogenesis is a complex process which entails organization of cells into various shapes that ultimately fashions tubular structures (Galambos and DeMello 2007; Goldie et al. 2008; Gao and Raj 2010; Cellière et al. 2012; Udan et al. 2013) (Plate 2.24). Laminin and heparan sulfate proteoglycan drive epithelial cell polarization and lumen formation (Schuger et al. 1996). The following processes are entailed in the development of blood vessels: (a) differentiation of mesenchymal

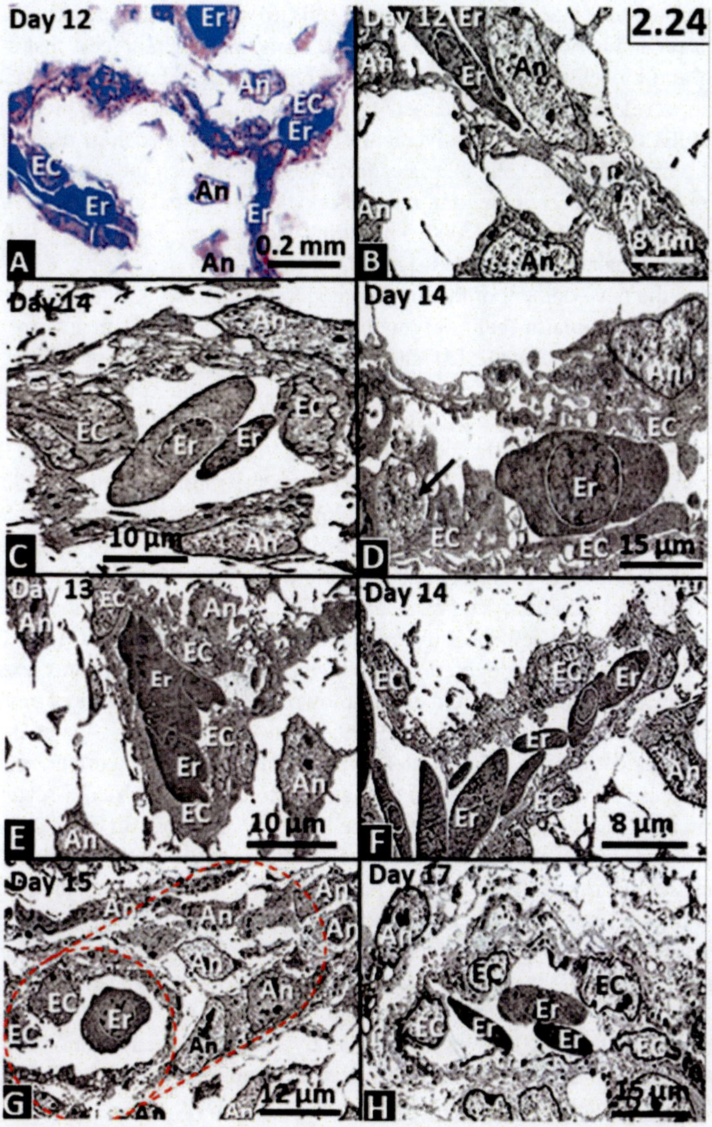

Plate 2.24 Development of the tubular configuration of blood vessels in the developing lung of the Domestic Fowl (*Gallus gallus* variant *domesticus*) at different times of incubation. Angioblasts (An) cluster around hemangioblasts that transform into erythrocytes (Er) (**a-f**). The angioblastic cells that come to lie next to the erythrocytes form endothelial cells (EC) of the developing blood vessels while the outermost ones form the outer part of the vessel wall. The blood vessels lengthen by addition of angioblasts on the leading part of the developing blood vessel (**e, f**). From day 15 of incubation (**g**), the blood vessels have adopted a three-dimensional configuration and the walls of the blood vessels are well-developed by day 17 (**h**). Arrow (↓), an endothelial cell undergoing apoptosis (**d**)

(mesodermal) cells into vascular progenitor cells, (b) formation of primitive vascular cell clusters and networks, (c) modification of blood vessels, a process that includes pruning and branching, (d) specialization of blood vessels into arteries, veins and lymphatics, (e) securing blood vessels to the surrounding tissues by production of extracellular matrix and; (f) involvement of supporting cells such as mesenchyme-derived peri-endothelial cells that later form pericytes which are incorporated into the formation of the endothelium and preservation of vascular structural integrity (Ferrara 2000a, b; Carmeliet 2000; Jain 2003; Herbert and Stainier 2011; Cellière et al. 2012; Ardalani et al. 2014; Patan 2021).

During the development of the lung of the Domestic Fowl (*Gallus gallus* variant *domesticus*), angioblasts (cells originating from morphological transformation of mesenchymal cells) (Plate 2.13) and erythroblasts (cells that stem from direct transformation of mesenchymal cells into hematogenetic cells that accumulate hemoglobin in their cytoplasm) (Plate 2.24) are involved in de novo vasculogenesis. Vasculoendothelial growth factors (VEGFs), that are members of the VEGF gene family and endothelial-cell-specific mitogens, are the most important regulators of blood vessel formation: they are both vasculogenetic and angiogenetic (Ferrara et al. 1992, 1996, 2003; Ferrara and Davis-Smyth 1997; Neufeld et al. 1999; Ferrara and Gerber 1999; Larrivée and Karsan 2000; Ferrara 2000a, b; Patan 2004; Jośko et al. 2000; Cellière et al. 2012; Yun et al. 2016; Gao et al. 2016). Together with their designated endothelial cell-specific tyrosine kinase receptors VEGFR-1, VEGFR-2, and VEGFR-3 and the angiopoietin/tie and the ephrin-B/EpH-B systems, expressed in angioblastic cells, VEGFs are the best-known molecular regulators of embryonic blood vessel formation (Petrova et al. 1999; Patan 2000; Nimmagadda et al. 2004). Lethality after deletion of a single allele draws attention to the importance of VEGF in the process (Carmeliet et al. 1996). Signaling through FLK-1, VEGF-A stimulates differentiation and proliferation of endothelial cells, inducting the formation of the earliest blood vessels and their precursors (Shalaby et al. 1995; Ng et al. 2001). Lewallen and Burggren (2015) investigated the effect of hypoxia and hyperoxia on the developing lung of the Domestic Fowl (chicken) (*Gallus gallus* variant *domesticus*) and observed that although rather malleable, pulmonary gross morphological elaboration in the embryo is not significantly affected by O_2incubation level: pulmonary VEGF concentration decreased by ~30% on day 18 of incubation compared with that on D18 in normoxia, hypoxia increased pulmonary VEGF concentration by ~35–75% on both D16 and D18 and unexpectedly, VEGF expression was also equally increased by hyperoxia only on D18.

The vertebrate lung, of which the primary role is gas exchange, has a large respiratory surface area, a thin blood-gas (tissue) barrier, and the organ is intensely vascularized (Weibel 1984; Sahasrabudhe et al. 2000; Maina 2005; Stewart and Morrisey 2016; Schittny 2017). Because the volume of the capillary bed is about equal to stroke volume, the whole pulmonary capillary blood volume is replaced with each heartbeat (Shimoda 2006). Coordination of the development of the pulmonary vasculature and that of the airway (bronchial) system ensures that airway and vascular systems closely pattern (Risau and Flamme 1995; Risau 1997; Maina 2003a, b; Kallapur and Kotecha 2016; Schittny 2017) (Plate 2.12), generating thin blood-gas barrier and large respiratory surface area (Djonov and Burri 2006; Hislop 2002; Mitzner and Wagner 2004; Cellière et al. 2012; Peng and Morrisey 2013; Gao et al. 2016).

Vascular development is controlled by a battery of signaling pathways that interact with a broad range of other inputs and outputs (McGrath et al. 2003). The various molecular (factors) cues that drive the process are predominantly the VEGF, acidic fibroblast growth factor 2 (aFGF-2), basic fibroblast growth factor-2 (bFGF-2), transforming growth factor-alpha (TGF-α), TGF-β, hepatocyte growth factor/scatter factor (HGF/SF), tumor necrosis factor-alpha (TNF-α), angiogenin and interleukin-8 (IL-8) (Hudlická et al. 1998; Seghezzi et al. 1998; Tallquist et al. 1999; Poole et al. 2001; Tomanek et al. 2002; Galambos and DeMello 2007). Members of the VEGF (Ferrara 1999, 2000a, b; D'Angio and Maniscalco 2002) and the angiopoietin and empirin families (Dumont et al. 1994; Hall et al. 2002) are the most important molecular drivers of the development of pulmonary vasculature. Blood vessels form by two processes, namely vasculogenesis and angiogenesis (Vailhé et al. 2001; Patan 2004; Demir et al. 2007; Kolte et al. 2016; Grant and Coultas 2019): vasculogenesis is defined as de novo development of blood vessels from angioblasts or endothelial precursor cells (mesodermal derived hemangioblasts) that move and differentiate in response to local signals (growth factors and extracellular matrix) to form a tubular (cylindrical) shape (Plate 2.24) while angiogenesis is the establishment of new blood vessels from pre-existing ones by processes such as endothelial sprouting, remodeling, intussusceptive microvascular growth, and distal angiogenesis (Burri et al. 2004; Parera et al. 2005; Makanya et al. 2005, 2006, 2007; De Spiegelaere et al. 2012; Peng and Morrisey 2013; Patan 2000, 2004). Sprouting entails endothelial cell migration, proliferation, and tube formation while intussusceptive microvascular growth occurs by splitting of blood vessels by formation and insertion of tissue folds and pillars of interstitial tissue across vessel lumina (Djonov et al. 2000; Augustin 2001; Kurz et al. 2003; Djonov and Makanya 2005; Díaz-Flores et al. 2020). In the avian kidney, sprouting precedes intussusceptive angiogenesis that leads to the development of the vascular system (Makanya et al. 2005). For the embryonic chick lung, sprouting angiogenesis contributes to the formation of the basic pulmonary vasculature while intussusceptive angiogenesis is involved in expansion and maturation of the system (Makanya et al. 2007). The developmental processes are regulated by interactions between various angiogenic molecular factors of which VEGF and bFGF play the most important roles (Conway et al. 2001; Nanka and Grim 2009; Ucuzian et al. 2010; Niu and Chen 2010). In both fetal and newborn rats, reduced pulmonary vascular development by VEGF inhibitors leads to defective alveolar development (Jakkula et al. 2000). In embryos of wild-type FVB and Tie2-LacZ mice, angiogenesis is the most important process by which pulmonary blood vessels form (Parera et al. 2005).

Among the various organs of the body, the lung displays the greatest expression of VEGF (Voelkel et al. 2006). The epithelial cells of the airways synthesize VEGF during the development of the organ and secrete it into the subendothelial matrix while pulmonary endothelial cells produce relevant receptors (Acarregui et al. 1999). In the chick embryo, vasculogenesis is the main process by which the pulmonary vasculature develops (Veikkola and Alitalo 1999; Veikkola et al. 2000; Anderson-Berry et al. 2005). During the development of the chick lung, Makanya et al. (2007) observed the following: (a) moderate levels bFGF-2 and VEGF-A were expressed and the levels were maintained during the sprouting phase of development, a time when PDGF-B expression was negligible; (b) expression of three growth factors, namely bFGF-2, VEGF and PDGF-B, was elevated during the intussusceptive phase of blood vessel development; and (c) VEGF was largely expressed in the epithelial cells while bFGF-2 was restricted to the mesenchymal compartment. In the avian lung, by means of well-coordinated interactive processes such as sprouting and intussusceptive angiogenesis, a structurally complex and exceptionally functionally efficient respiratory system is formed (Scheid 1979; McLelland 1989; Maina 2005; Makanya et al. 2007).

Branching morphogenesis occurs through the reorganization of epithelial cells into a complex three-dimensional (3D) stereotypically bifurcated tubular system (Affolter et al. 2003; Gjorevski and Nelson 2010a, b; Varner and Nelson 2014) (Plate 2.4, c, f). While the configuration of a tubular structure may look deceptively simple, in organs such as the lung, where the airways and the blood vessels display branched tubular morphology, the complex mechanisms by which epithelial and endothelial cells organize, move and interconnect to form a cylindrical shape with a lumen (Plate 2.24, c-h) is instructed by the intricate interaction of genetic and molecular programs that involve expressions of stimulatory and inhibitory growth factors as well as physical forces (Taichman et al. 2002; Gerritsen et al. 2003; Van Tuyl et al. 2004; Gerhardt and Betsholtz 2005; Gerhardt 2008; Boric and Figueroa 2022). Associated with epithelial and endothelial cells, the formation of a lumen and subsequent configuration of tubular shape is probably the highest mark of biological (angiogenetic) bioengineering (Ucuzian et al. 2010). In in vitro 3-D gel preparations, Gerritsen et al. (2003), Alsberg et al. (2004) and Van Tuyl et al. (2004) observed that as many as 1000 different genes are expressed or upregulated during endothelial tubulogenesis. During the formation of blood vessels and airways, the cells that are located at the extremity of the airway and the vascular sprouts use their long filopodia to detect and negotiate the external mesenchymal environment while those within the sprouting trunk replicate and create a lumen (Gerhardt and Betsholtz 2005). Balance between the mechanical properties that are generated by fibrin and the fibrinolytic actions of each type of cells in the presence of endothelial cells may regulate the formation of capillary-like structures (Vailhé et al. 1998). The migration of cell tips depends on existing concentration gradients of VEGF-A and activation of VEGFR-2 at the tips of the cell filopodia (Gerhardt and Betsholtz 2005).

References

Abbott LA, Lester SM, Erickson CA (1991) Changes in mesenchymal cell-shape, matrix collagen and tenascin accompany bud formation in the early chick lung. Anat Embryol 183:299–311

Abdalla MA (1989) The blood supply to the lung. In: King AS, McLelland J (eds) Form and function in birds, vol 4. Academic Press, London, pp 281–306

Abdalla MA, King AS (1975) The functional anatomy of the pulmonary circulation of the domestic fowl. Respir Physiol 23:267–290

Abdalla MA, King AS (1976a) Pulmonary arteriovenous anastomoses in the avian lung: do they exist? Respir Physiol 27:187–191

Abdalla MA, King AS (1976b) The functional anatomy of the bronchial circulation of the domestic fowl. J Anat 121:537–550

Abdalla MA, King AS (1977) The avian bronchial arteries: species variations. J Anat 123:697–704

Abraham JA, Mergia A, Wang JL, Tumolo A, Friedman KA, Hyerild D, Gospodarowicz D, Fiddes JC (1986) Nucleotide sequence of a bovine cDNA clone encoding the angiogenetic protein, basic fibroblast growth factor. Science 233:545–548

Abud HE (2004) Shaping developing tissues by apoptosis. Cell Death Differ 11:797–799

Abzhanov A (2017) The old and new faces of morphology: the legacy of D'Arcy Thompson's 'theory of transformations' and 'laws of growth'. Development 144:4284–4297

Acarregui MJ, Penisten ST, Goss KL, Ramirez K, Snyder JM (1999) Vascular endothelial growth factor gene expression in human fetal lung *in vitro*. Am J Respir Cell Mol Biol 20:14–23

Adamson IYR (1985) Cellular kinetics of the lung. In: Witschi HP, Brain JD (eds) Toxicology of inhaled materials. Springer-Verlag, New York, pp 289–317

Affolter M, Bellusci S, Itoh N, Shilo B, Thiery JP, Werb Z (2003) Tube or not tube: remodeling epithelial tissues by branching morphogenesis. Dev Cell 4:11–18

Ainsworth SB, Milligan DWA (2002) Surfactant therapy for respiratory distress syndrome in premature neonates. Am J Respir Med 1:417–433

Ainsworth SJ, Stanley RL, Evans DJR (2010) Developmental stages of the Japanese quail. J Anat 216:3–15

Akeson AL, Wetzel B, Thompson FY, Brooks SK, Paradis H, Gendron RL, Greenberg JM (2000) Embryonic vasculogenesis by endothelial precursor cells derived from lung mesenchyme. Anat Rec 217:11–23

Akram KM, Patel N, Spiteri MA, Forsyth NR (2016) Lung regeneration: endogenous and exogenous stem cell mediated therapeutic approaches. Int J Mol Sci 17:128. https://doi.org/10.3390/ijms17010128

Alberts B, Johnson A, Lewis J, Raff M, Roberts K, Walter P (2002) Molecular biology of the cell, 4th edn. Garland Science, New York

Alescio T, Cassini A (1962) Induction in vitro of tracheal buds by pulmonary mesenchyme grafted on tracheal epithelium. J Exp Zool 150:83–94

Alexander RMN (1996) Optima for animals. Princeton University Press, Princeton (NJ)

Allman T (2006) Stem cells. Lucent Books, Detroit

Alsberg E, Moore K, Huang S, Polte T, Inger DE (2004) The mechanical and cytoskeletal basis of lung morphogenesis. In: Massaro DJ, Massaro GC, Chambon P (eds) Lung development and regeneration. Marcel Dekker Inc., New York, pp 247–274

Amaral-Silva L, Lambertz M, Fernando JZ, Wilfried K, Luciane HG, Kênia CB (2019) Parabronchial remodeling in chicks in response to embryonic hypoxia. J Exp Biol 222: jeb197970. https://doi.org/10.1242/jeb.197970

Ambros V (2001) MicroRNAs: tiny regulators with great potential. Cell 107:823–826

Amendola V (ed) (2001) Molecular machines and motors. Springer, Berlin

Anderson-Berry A, O'Brien EA, Bleyl SB, Lawson A, Gundersen N, Ryssman D, Sweeley J, Dahl MJ, Drake CJ, Schoenwolf GC et al (2005) Vasculogenesis drives pulmonary vascular growth in the developing chick embryo. Dev Dyn 233:145–153

Anderson PJ, Lynch TJ, Engelhardt JF (2017) Multipotent myoepithelial progenitor cells are born early during airway submucosal gland development. Am J Respir Cell Mol Biol 56:716–726

Andrew DJ, Ewald AJ (2010) Morphogenesis of epithelial tubes: insights into tube formation, elongation, and elaboration. Dev Biol 341:34–55

Antony N, McDougall AR, Mantamadiotis T, Cole TJ, Bird AD (2016) Creb1 regulates late stage mammalian lung development via respiratory epithelial and mesenchymal-independent mechanisms. Sci Rep 6:25569. https://doi.org/10.1038/srep25569

Ardalani H, Assadi AH, Murphy WL (2014) Structure, function, and development of blood vessels: lessons for tissue engineering. In: Cai W (ed) Engineering in translational medicine. Springer, London, pp 155–182

Arman E, Haffner-Krausz R, Gorivodsky M, Lonai P (1999) FGF-2 is required for limb outgrowth and lung branching morphogenesis. Proc Natl Acad Sci U S A 96:11895–11899

Aros CJ, Pantoja CJ, Gomperts BN (2021) Wnt signaling in lung development, regeneration, and disease progression. Commun Biol 4:601. https://doi.org/10.1038/s42003-021-02118-w

Arthur W (2006) D'Arcy Thompson and the theory of transformations. Nat Rev Genet 7:401–406

Asahara T, Murohara T, Sullivan A, Silver M, van der Zee R, Li T, Witzenbichler B, Schatterman G, Isner JM (1997) Isolation of putative progenitor endothelial cells for angiogenesis. Science 275:964–967

Atkins P (2007) Four laws that drive the universe. Oxford University Press, Oxford

Augustin HG (2001) Tubes, branches, and pillars: the many ways of forming a new vasculature. Circ Res 89:645–647

Awkati HA (1939) The development of the avian lung. University of Edinburgh. PhD Thesis

Baarsma HA, Königshoff M, Gosens R (2013) The WNT signaling pathway from ligand secretion to gene transcription: molecular mechanisms and pharmacological targets. Pharmacol Ther 138: 66–83

Bailey K, Worboys BD (1960) The lamellibranch crystalline style. Biochem J 76:487–491

Baird A, Klagsbrun M (1991) The fibroblast growth factor family: an overview. Ann N Y Acad Sci 638:xi–xii

Baird A, Walicke PA (1989) Fibroblast growth factors. Brit Med Bull 45:438–452

Ball P (2013) In retrospect: on growth and form. Nature 494:32–33

Baoukina S, Monticelli L, Amrein M, Tieleman DP (2007) The molecular mechanism of monolayer-bilayer transformations of lung surfactant from molecular dynamics simulations. Biophys J 93:3775–3782

Baral I, Varghese PC, Dutta D (2022) Epigenetics as "conductor" in "orchestra" of pluripotent states. Cell Tiss Res 390:141–172

Bartel DP, Chen CZ (2004) Micomanagers of gene expression: the potentially widespread influence of metazoan microRNAs. Nat Rev Genet 5:396–400

Bastacky J, Lee CY, Goerke J, Koushafar H, Yager D, Kenaga L, Speed TP, Chen Y, Clements JA (1995) Alveolar lining layer is thin and continuous: low-temperature scanning electron microscopy of rat lung. J Appl Physiol 79:1615–1628

Bateson M, Feenders G (2010) The use of passerine bird species in laboratory research: implications of basic biology for husbandry and welfare. ILAR J 51:394–408

Baverstock K, Rönkkö M (2014) The evolutionary origin of form and function. J Physiol 592: 2261–2265

Bedford FK, Christopherson K, Nachtigal MW, Shen WH, Julius DJ, Ingraham HA (1996) Molecular biology of pituitary development and disease. Horm Res 45:19–21

Beeby M, Ribardo DA, Brennan CA, Rubyc EG, Jensend GJ, Hendrixson DR (2016) Diverse high-torque bacterial flagellar motors assemble wider stator rings using a conserved protein scaffold. Proc Natl Acad Sci U S A 113:E1917–E1926

Begazo A (2019) Altricial or precocial young birds: know the difference. https://avianreport.com/altricial-precocial-birds/#. Accessed on 09–07–22

Bellairs R, Osmond M (2014) Atlas of chick development, 3rd edn. Academic Press, London

Bellusci S, Grindley J, Emoto H, Itoh N, Hogan BL (1997) Fibroblast growth factor-10 (FGF-10) and branching morphogenesis in the embryonic mouse lung. Development 124:4867–4878

Beloussov LV (2001) Morphogenetic fields: outlining the alternatives and enlarging the context. Riv Biol 94:219–235

Beloussov LV (2010) The primacy of organic form. Riv Biol 103:13–18

Beloussov LV, Grabovsky VI (2006) Morphomechanics: goals, basic experiments and models. Int J Dev Biol 50:81–92
Beloussov LV, Volodyaev IV (2013) From molecular machines to macroscopic fields: an accent to characteristic times. Eur J Biophys 1:6–15
Bennett FM, Tenney SM (1982) Comparative mechanics of mammalian respiratory system. Respir Physiol 49:131–140
Berg HC (2003) The rotary motor of bacterial flagella. Annu Rev Biochem 72:19–54
Bernard FR (1973) Crystalline style formation and function in the oyster *Crassostrea gigas* (Thunberg 1795). Ophelia 12:159–170
Bernhard W, Gebert A, Vieten G, Rau G, Hohlfeld JM, Posttle AD, Freihorst J (2001) Pulmonary surfactant in birds, coping with surface tension in a tubular lung. Am J Physiol Regul Integr Comp Physiol 281:R327–R337
Berrocal T, Madrid C, Novo S, Gutierrez J, Arjonilla A, GomezLeon N (2004) Congenital anomalies of the tracheobronchial tree, lung, and mediastinum: embryology, radiology, and pathology. Radiographics 24:e17. https://doi.org/10.1148/rg.e17
Beyea JA, Olson DM, Vandergriend RA, Harvey S (2005) Expression of growth hormone and its receptor in the lungs of embryonic chicks. Cell Tissue Res 322:379–392
Bezuidenhout AJ, Groenewald HB, Soley JT (1999) An anatomical study of the respiratory air sacs in ostriches. Onderst J Vet Res 66:317–325
Bikfalvi A, Klein S, Guiseppe P, Rifkin D (1997) Biological roles of fibroblast growth factor-2. Endocr Rev 18:26–45
Bizzarri M, Cucina A, Biava PM, Proietti S, D'Anselmi F, Dinicola S, Pasqualato A, Lisi E (2011) Embryonic morphogenetic field induces phenotypic reversion in cancer cells. Curr Pharm Biotechnol 12:243–253
Blanchard G (2018) A 3D cell shape that enables tube formation. Nature 561:182–183
Blacker HA, Orgeig S, Daniels CB (2004) Hypoxic control of the development of the surfactant system in the chicken: evidence for physiological heterokairy. Am J Physiol Regul Integr Comp Physiol 287:R403–R410
Boatright KM, Salvesen GS (2003) Mechanisms of caspase activation. Curr Opin Cell Biol 15:725–731
Boden MA (2008) Mind as machine: a history of cognitive science. Oxford University Press, Oxford (UK)
Bódi I, Kocsis K, Benyeda Z, Fejsz N, Molnár D, Nagy N, Olah I (2016) Dual secretion locations on type-II cells in the avian lung suggest local as well as general roles of surfactant. J Morphol 277: 1062–1071
Boers JE, Den Brok JLM, Koudstaal J, Arends JW, Thunnissen FBJM (1996) Number and proliferation of neuroendocrine cells in normal human airway epithelium. Amer J Respir Crit Care Med 154:758–763
Boggaram V (2009) Thyroid transcription factor-1 (TTF-1/Nkx2.1/TITF1) gene regulation in the lung. Clin Sci (Lond) 116:27–35
Bookstein FL (1977) The study of shape transformation after D'Arcy Thompson. Math Biosci 34: 177–219
Boric MP, Figueroa XF (2022) Editorial: Cell communication in vascular biology, V. II. Front Physiol 13:903056. https://doi.org/10.3389/fphys.2022.903056
Bowden DH (1983) Cell turnover in the lung. Am Rev Respir Dis 128:S46–S48
Bowden DH (1987) Macrophages, dust, and pulmonary diseases. Exp Lung Res 12:89–107
Braunschweig L, Meyer AK, Wagenführ L, Storch A (2015) Oxygen regulates proliferation of neural stem cells through Wnt/β-catenin signalling. Mol Cell Neurosci 67:84–92
Bremer JL (1912) The development of the aorta and aortic arches in rabbits. Am J Anat 13:111–128
Bridges JP, Schehr A, Wang Y, Huo L, Besnard V, Machiko I, Whitsett J, Xu Y (2014) Epithelial SCAP/INSIG/SREBP signaling regulates multiple biological processes duringperinatal lung maturation. PLoS One 9:e91376. https://doi.org/10.1371/journal.pone.0091376

Briscoe J, Kicheva A (2017) The physics of development 100 years after D'Arcy Thompson's 'on growth and form'. Mech Dev 145:26–31

Brown JH, Gupta VK, Li BL, Milne BT, Restrepo C, West GB (2002) The fractal nature of nature: power laws, ecological complexity and biodiversity. Philos Trans R Soc Lond Ser B Biol Sci 357:619–626

Brown DA, Yang N, Ray SD, Wexler P (2014) Apoptosis. In: Wexler P (ed) Encyclopedia of toxicology, 3rd edtn. Academic Press, Oxford, pp 287–294

Bruce MC, Honaker CE, Ross RJ (1999) Lung fibroblasts undergo apoptosis following alveolization. Am J Respir Cell Mol Biol 20:228–236

Bucher TL (1987) Patterns in the mass-independent energetics of avian development. J Exp Zool Suppl 1:139–150

Burggren WW (1991) Ontogeny of cardiovascular and respiratory physiology in lower vertebrates. Annu Rev Phhsiol 53:107–135

Burggren WW, Santin JF, Antich MR (2016) Cardio-respiratory development in bird embryos: new insights from a venerable animal model. R Bras Zootec 45. https://doi.org/10.1590/s1806-92902016001100010

Burri PH (1997) Structural aspects of pre- and postnatal development and growth of the lung. In: McDonald J (ed) Growth and development of the lung. Marcel Dekker, New York, pp 1–35

Burri PH (1999) Lung development and pulmonary angiogenesis. In: Gaultier C, Bourbon J, Post M (eds) Lung disease. Oxford University Press, New York, pp 122–151

Burri PH, Hlushchuk R, Djonov V (2004) Intussusceptive angiogenesis: its emergence, its characteristics, and its significance. Dev Dyn 231:474–488

Cadigan KM, Nusse R (1997) Wnt signaling: a common term in animal development. Genes Dev 11:3286–3305

Callahan HS, Pigliucci M, Schlichting CD (1997) Developmental phenotypic plasticity: where ecology and evolution meet molecular biology. BioEssays 19:519–525

Cañadas O, Olmeda B, Alonso A, Pérez-Gil J (2020) Lipid-protein and protein–protein interactions in the pulmonary surfactant system and their role in lung homeostasis. Int J Mol Sci 21:3708. https://doi.org/10.3390/ijms21103708

Canu G, Ruhrberg C (2021) First blood: the endothelial origins of hematopoietic progenitors. Angiogenesis. https://doi.org/10.1007/s10456-021-09783-9

Cao N, Yao ZX (2011) The hemangioblast: from concept to authentication. Anat Rec 294:580–588

Cardoso WN (2000) Lung morphogenesis revisted: old facts, current ideas. Dev Dyn 219:121–130

Cardoso WV (2001) Molecular regulation of lung development. Annu Rev Physiol 63:471–494

Cardoso WV, Lü J (2006) Regulation of early lung morphogenesis: questions, facts and controversies. Development 133:1611–1624

Carmeliet P (2000) Mechanisms of angiogenesis and arteriogenesis. Nat Med 6:389–395

Carmeliet P, Ferrara V, Breier G, Pollefeyt S, Kieckens L, Gertsenstein M, Fahrig M, Vandenhoeck A, Harpal K, Eberhardt C et al (1996) Abnormal blood vessel development and lethality in embryos lacking a single vascular endothelial growth factor allele. Nature 380:435–439

Carroll C, Engström N, Nilsson PF, Haxen ER, Mohlin S, Berg P, Glud RN, Hammarlund EU (2021) Hypoxia generated by avian embryo growth induces the HIF-α response and critical vascularization. Front Ecol Evol 9:675800. https://doi.org/10.3389/fevo.2021.675800

Cassini GH, Toledo N, Vizcaíno SF (2021) Form-function correlation paradigm in mammalogy. A tribute to Leonard B. Radinsky (1937-1985). J Mammal Evol 28:1–5

Catt JW, Henman M (2000) Toxic effects of oxygen on human embryo development. Hum Reprod 15:199–206

Cavin L, Alvarez N (2022) Why coelacanths are almost "living fossils"? Front Ecol Evol 10: 896111. https://doi.org/10.3389/fevo.2022.896111

Cellière G, Menshykau D, Iber D (2012) Simulations demonstrate a simple network to be sufficient to control branch point selection, smooth muscle and vasculature formation during lung branching morphogenesis. Bio Open 1:775–788

Chan BHC, Moosajee M, Rainger J (2021) Closing the gap: mechanisms of epithelial fusion during optic fissure closure. Front Cell Dev 8:620774. https://doi.org/10.3389/fcell.2020.620774

Chan CJ, Heisenberg CP, Hiiragi T (2017) Coordination of morphogenesis and cell-fate specification in development. Curr Biol 27:R1024–R1035

Chanet S, Martin AC (2014) Mechanical force sensing in tissues. Prog Mol Biol Transl Sci 126: 317–352

Chaudhury H, Goldie LC, Hirschi KK (2011) Vascular precursor cells. Genes. Cancer 2:1081–1084

Chen T, Saw TB, Mège RM, Ladoux B (2018) Mechanical forces in monolayers. J Cell Sci 131: jcs218156. https://doi.org/10.1242/jcs.218156

Chen WT, Chen JM, Mueller SC (1986) Coupled expression and colocalization of 140K cell adhesion molecules, fibronectin, and laminin during morphogenesis and cytodifferentiation of chick lung cells. J Cell Biol 103:1073–1090

Cheng CW, Smith SK, Charnock-Jones DS (2003) Wnt-1 signaling inhibits human umbilical vein endothelial cell proliferation and alters cell morphology. Exp Cell Res 291:415–425

Chinoy MR (2003) Lung growth and development. Front Biosci 8:392–415

Chiquet-Ehrismann R, Tucker RP (2011) Tenascins and the importance of adhesion modulation. Cold Spring Harb Perspect Biol 3:a004960. https://doi.org/10.1101/cshperspect.a004960

Choi AMK, Wang X (2006) Apoptosis. In: Lurent GJ, Shapiro SD (eds) Encyclopedia of respiratory medicine. Academic Press, New York, pp 134–140

Choi K, Kennedy M, Kazarov A, Papadimitriou JC, Keller G (1998) A common precursor for hematopoietic and endothelial cells. Development 125:725–732

Choudhry Z, Rikani AA, Choudhry AM, Tariq S, Zakaria F, Asghar MW, Sarfraz MK, Haider K, Shafiq AA, Mobassarah NJ (2014) Sonic hedgehog signalling pathway: a complex network. Ann Neurosci 21:28–31

Christensen VL (2009) Development during the first seven days post-hatching. Avian Biol Res 2: 27–33

Chroneos ZC, Sever-Chroneos Z, Shepherd VL (2010) Pulmonary surfactant: an immunological perspective. Cell Physiol Biochem 25:13–26

Çıkla-Süzgün P, Küçükgüzel ŞG (2019) Recent advances in apoptosis: the role of hydrazones. Mini Rev Med Chem 19:1427–1442

Clément R, Blanc P, Mauroy B, Sapin V, Douady S (2012a) Shape self-regulation in early lung morphogenesis. PLoS One 7:e36925. https://doi.org/10.1371/journal.pone.0036925

Clément R, Douady S, Mauroy B (2012b) Branching geometry induced by lung self-regulated growth. Phys Biol 9:066006. https://doi.org/10.1088/1478-3975/9/6/066006

Clevers H, Loh KM, Nusse R (2014) Stem cell signaling. An integral program for tissue renewal and regeneration: Wnt signaling and stem cell control. Science 346:1248012. https://doi.org/10.1126/science.1248012

Cockell CS (2017) The laws of life. Phys Today 70(3):42. https://doi.org/10.1063/PT.3.3493

Coen E (1999) The art of genes: how organisms make themselves. Oxford University Press, Oxfrord (UK)

Coffin JD, Poole TJ (1988) Embryonic vascular development: immunohistochemical identification of the origin and subsequent morphogenensis of the major vessel primordia in the quail embryos. Development 102:735–748

Cohen ED, Ihida-Stansbury K, Lu MM, Panettieri RA, Jones PL, Morrisey EE (2009) Wnt signaling regulates smooth muscle precursor development in the mouse lung via a tenascin C/PDGFR pathway. J Clin Invest 119:2538–2549

Cohen ED, Mariol MC, Wallace RM, Weyers J, Kamberov YG, Pradel J, Wilder EL (2002) DWnt4 regulates cell movement and focal adhesion kinase during *drosophila* ovarian morphogenesis. Dev Cell 2:437–448

Coitier V (1573) Anatomie avium. In: Externum et internarum praecipalium humani corporis partium tabulae atque anatomicae exercitationes observationesque varieae. Nuremberg, Germany, pp 130–133

Compton SK, Goeringer GC (1981) Development in the chick embryo. I. Phosphatidylcholine synthesis in the developing chick lung. Pediatr Res 15:866–869

Comroe JH (1974) Physiology of respiration: an introductory text. Year Book Medical Publishers, Inc, Chicago (IL)

Conway EM, Collen D, Carmeliet P (2001) Molecular mechanisms of blood vessel growth. Cardiovasc Res 49:507–521

Copp AJ (1995) Death before birth: clues from gene knockouts and mutations. Trends Genet 11:87–93

Covello KL, Kehler J, Yu H, Gordan JD, Arsham AM, Hu CJ, Labosky PA, Simon MC, Keith B (2021) HIF-2α regulates Oct-4: effects of hypoxia on stem cell function, embryonic development, and tumor growth. Genes Dev 20:557–570

Crapo JD, Barry BE, Gehr P, Bachofen M, Weibel ER (1982) Cell number and cell characteristics of the normal human lung. Am Rev Resp Dis 126:332–337

Crivellato E, Ribatti D (2006) Aristotle: the first student of angiogenesis. Leukemia 20:1209–1210

Cross SS, Bury JP (2004) The hedgehog signalling pathways in human pathology. Curr Diagnostic Pathol 10:157–168

Cullinane C, Cox PN, Silver MM (1992) Persistent pulmonary hypertension of the newborn due to alveolar capillary dysplasia. Pediatr Pathol 12:499–514

Cunniff B, Druso JE, van der Velden JL (2020) Lung organoids: advances in generation and 3D-visualization. Histochem Cell Biol 155:301–308

Cupello C, Hirasawa T, Tatsumi N, Yabumoto Y, Gueriau P, Isogai S, Matsumoto R, Saruwatari T, King A, Hoshino M et al (2022) Lung evolution in vertebrates and the water-to-land transition. eLife 11:e77156. https://doi.org/10.7554/eLife.77156

Dale TC (1998) Signal transduction by the Wnt family of ligands. Biochem J 329:209–223

D'Angio CT, Maniscalco WM (2002) The role of vascular growth factors in hyperoxia-induced injury to the developing lung. Front Biosci 7:1609–1623

D'Arcy MS (2019) Cell death: a review of the major forms of apoptosis, necrosis and autophagy. Cell Biol Int 43:582–592

Daniels CB, Orgeig S (2003) Pulmonary surfactant: the key to the evolution of air breathing. News Physiol Sci 18:151–157

Darwin C (1859) On the origin of species by means of natural selection, or the preservation of favoured races in the struggle for life, 2nd edn. D. Appleton and Company, New York

Davé V, Wert SE, Tanner T, Thitoff AR, Loudy DE, Whitsett JA (2008) Conditional deletion of pten causes bronchiolar hyperplasia. Am J Respir Cell Mol Biol 38:337–345

Davey MG, McTeir L, Barrie AM, Freem LJ, Stephen LA (2014) Loss of cilia causes embryonic lung hypoplasia, liver fibrosis, and cholestasis in the talpid3 ciliopathy mutant. Organogenesis 10:177–185

Davidson LA (2012) Epithelial machines that shape the embryo. Trends Cell Biol 22:82–87

Davidson AJ, Zon LI (2004) The "definitive" (and "primitive") guide to zebrafish hematopoiesis. Oncogene 23:7233–7246

Davis GE, Bayless KJ (2003) An integrin and rho GTPase-dependent pinocytotic vacuole mechanism controls capillary lumen formation in collagen and fibrin matrices. Microcirculation 10:27–44

Dejours P (1988) Respiration in water and air: adaptation, regulation and evolution. Elsevier, Amsterdam

Delile J, Herrmann M, Peyriéras N, Doursat R (2017) A cell-based computational model of early embryogenesis coupling mechanical behaviour and gene regulation. Nat Commun 8:13929. https://doi.org/10.1038/ncomms13929

Delphia JM (1961) Early development of the secondary bronchi in the house sparrow *Passer domesticus* (Linnaeus). Amer Midland Natur 65:44–59

DeMello DE (2004) Pulmonary pathology. Semin Neonatol 9:311–329

DeMello DE, Sawyer D, Galvin N, Reid LM (1997) Early fetal development of lung vasculature. Am J Respir Cell Mol Biol 16:568–581

De Miguel C, Linsler F, Casanova J, Franco-Marro X (2016) Genetic basis for the evolution of organ morphomenesis: the case of Spalt and cut in development of insect trachea. Development 143:3615–3622

Demir R, Seval Y, Huppertz B (2007) Vasculogenesis and angiogenesis in the early human placenta. Acta Histochem 109:257–265

De Munck N, Janssens R, Segers I, Tournaye H, Van de Velde H, Verheyen G (2019) Influence of ultra-low oxygen (2%) tension on *in-vitro* human embryo development. Hum Reprod 34:228–234

Desai TJ, Cardoso WV (2002) Growth factors in lung development and disease: friends or foe. Respir Res 3:2. https://doi.org/10.1186/rr169

De Spiegelaere W, Casteleyn C, Van den Broeck W, Plendl J, Bahramsoltani M, Simoens P, Djonov V, Cornillie P (2012) Intussusceptive angiogenesis: a biologically relevant form of angiogenesis. J Vasc Res 49:390–404

Diamond J (1983) Transport mechanisms: the biology of the wheel. Nature 302:572–573

Díaz-Flores L, Gutiérrez R, Gayoso S, García MP, González-Gómez M, Díaz-Flores L, Sánchez R, Carrasco JL, Madrid JF (2020) Intussusceptive angiogenesis and its counterpart intussusceptive lymphangiogenesis. Histol Histopathol 35:1083–1103

Dietmar K, Clayron DF, Robinson GE, Albertson C, Carey HV, Cummings ME, Dewar K, Edwards SV, Hofman HA, Gross LJ et al (2013) New frontiers for organismal biology. Bioscience 63:464–471

Dietz MW, van Kampen M, van Griensven MJM, van Mourik S (2015) Energy budgets of avian embryos: the paradox of the plateau phase in egg metabolic rate. Physiol Zool 71:147–156

Di Ieva A (2016) The fractal geometry of the brain: an overview. In: Di Ieva A (ed) The fractal geometry of the brain. Springer Series in Computational Neuroscience, Springer, New York, pp 3–12

Discher DE, Janmey P, Wang Y (2005) Tissue cells fell and respond to the stiffness of their substrate. Science 310:1139–1143

Djonov V, Burri PH (2006) Vascular development: Vasculogenesis and angiogenesis. In: Shepro D (ed) Microvascular research: biology and pathology. Elsevier, Philadelphia, pp 91–96

Djonov V, Makanya AN (2005) New insights into intussusceptive angiogenesis. EXS 94:17–33

Djonov V, Schmid M, Tschanz SA, Burri PH (2000) Intussusceptive angiogenesis: its role in embryonic vascular network formation. Circ Res 86:286–292

Dolev S, Elitzur AC (1998) Biology and thermodynamics: seemingly-opposite phenomena in search of a unified paradigm. Einstein Quart J Biol Med 15:1–16

Dos Remedios CG, Chhabra D, Kekic M, Dedova IV, Tsubakihara M, Berry DA, Nosworthy NJ (2003) Actin binding proteins: regulation of cytoskeletal microfilaments. Physiol Rev 83:433–473

Downs KM (2003) Florence Sabin and the mechanism of blood vessel lumenization during vasculogenesis. Microcirculation 10:5–25

Driskell RR, Goodheart M, Neff T, Liu X, Luo M, Moothart C, Sigmund CD, Hosokawa R, Chai Y, Engelhardt JF (2007) Wnt3a regulates Lef-1 expression during airway submucosal gland morphogenesis. Dev Biol 305:90–102

Driskell RR, Liu X, Luo M, Filali M, Zhou W, Abbott D, Cheng N, Moothart C, Sigmund CD, Engelhardt JF (2004) Wnt-responsive element controls Lef-1 promoter expression during submucosal gland morphogenesis. Am J Physiol Lung Cell Mol Physiol 287:L752–L763

Duan D, Yue Y, Zhou W, Labed B, Ritchie TC, Grosschedl R, Engelhardt JF (1999) Submucosal gland development in the airway is controlled by lymphoid enhancer binding factor 1 (LEF1). Development 126:4441–4453

Dumont DJ, Gradwohl G, Fong GH, Puri MC, Gertsenstein M, Auerbach A, Breitman ML (1994) Dominant-negative and targeted null mutations in the endothelial receptor tyrosine kinase, tek, reveal a critical role in vasculogenesis of the embryo. Genes Dev 8:1897–1909

Duncker HR (1971) The lung-air sac system of birds. A contribution to the functional anatomy of the respiratory apparatus. Ergeb Anat Entwicklung 45:1–171

Duncker H-R (1978a) General morphological principles of amniotic lungs. In: Piiper J (ed) Respiratory function in birds, adult and embryonic. Springer, Berlin, pp 2–15

Duncker HR (1978b) Development of the avian respiratory and circulatory systems. In: Piiper J (ed) Respiratory function in birds, adult and embryonic. Springer-Verlag, Berlin-Heidelberg, pp 260–273

Duncker HR (1981) Evolution of the structural and functional principles of vertebrate lungs. Verh Anat Ges 75:129–303

Duncker HR (2004) Vertebrate lungs: structure, topography and mechanics. A comparative perspective of the progressive integration of respiratory system, locomotor apparatus and ontogenetic development. Respir Physiol Neurobiol 144:111–124

Dzierzak E, Philipsen S (2013) Erythropoiesis: development and differentiation. Cold Spring Harb Perspect Med 3(4):a011601. https://doi.org/10.1101/cshperspect.a011601

Dzierzak E, Speck NA (2008) Of lineage and legacy: the development of mammalian hematopoietic stem cells. Nat Immunol 9:129–136

Edelman DB, McMenamin M, Sheesley P, Pivar S (2016) Origin of the vertebrate body plan via mechanically biased conservation of regular geometrical patterns in the structure of the blastula. Prog Biophys Mol Biol 121:212–244

Eichmann A, Pardanaud L, Yuan L, Moyon D (2002) Vasculogenesis and the search for the hemangioblast. J Hematother Stem Cell Res 11:207–214

Ekblom P, Aufderheide E (1989) Stimulation of tenascin expression in mesenchyme by epithelial-mesenchymal interactions. Int J Dev Biol 33:71–79

El Agha E, Bellusci S (2014) Walking along the fibroblast growth factor 10 route: a key pathway to understand the control and regulation of epithelial and mesenchymal cell-lineage formation during lung development and repair after injury. Scientifica 2014:538379. https://doi.org/10.1155/2014/538379

El Agha E, Herold S, Al Alam D, Quantius J, MacKenzie B, Carraro G, Moiseenko A, Chao CM, Minoo P, Seeger W, Bellusci S (2014) FgF10-positive cells represent a progenitor cell population during lung development and postnatally. Development 141:296–306

Eliasson P, Jönsson JI (2010) The hematopoietic stem cell niche: low in oxygen but a nice place to be. J Cell Physiol 222:17–22

Elmore S (2007) Apoptosis: a review of programmed cell death. Toxicol Pathol 35:495–516

Elzanowski A, Brett-Surman MK (1995) Avian premaxilla and tarsometatarsus from the uppermost cretaceous of Montana. Auk 112:762–767

Emmet M (2022) Birds: precocial and altricial. https://southafrica.co.za/birds-precocial-and-altricial. html. Accessed on 09–07–22

Erdoğan S, Sağsöz H, Paulsen F (2015) Functional anatomy of the syrinx of the chukar partridge (Galliformes: *Alectoris chukar*) as a model for phonation research. Anat Rec 298:602–617

Erjefalt JS, Erjefalt I, Sundler F, Persson CGA (1995) *In vivo* restitution of airway epithelium. Cell Tiss Res 281:305–315

Erjefält JS, Persson CGA (1997) Airway epithelial repair: breathtakingly quick and multipotentially pathogenic. Thorax 52:1010–1012

Escobar LM, Echeverría OM, Vázquez-Nin GH (2015) Necrosis as programmed cell death. In: Ntuli TM (ed) Cell death–autophagy, apoptosis and necrosis. IntechOpen, Rijeka (Croatia), pp 419–434. https://doi.org/10.5772/61483

Essey M, Maina JN (2020) Fractal analysis of concurrently prepared latex rubber casts of the bronchial and vascular systems of the human lung. Open Biol 10:190249. https://doi.org/10.1098/rsob.190249

Ettensohn CA (1985) Mechanisms of epithelial invagination. Q Rev Biol 60:289–307

Evans AR, Jones D, Boyer AG, Brown JH, Costa DP, Ernest SM, Fitzgerald EMG, Fortelius M, Gittleman JL, Hamilton MJ et al (2012) The maximum rate of mammal evolution. Proc Natl Acad Sci U S A 109:4187–4190

Evans HE (1909) On the development of the aortae, cardinal veins, and other blood vessels of vertebrate embryos from capillaries. Anat Rec 3:498–518

Farge E (2011) Mechanotransduction in development. Curr Top Dev Biol 95:243–265
Farge E (2013) Mechano-sensing in embryonic biochemical and morphologic patterning: evolutionary perspectives in the emergence of primary organisms. Biol Theory 8:232–244
Fathollahipour S, Patil PS, Leipzig ND (2018) Oxygen regulation in development: lessons from embryogenesis towards tissue engineering. Cells Tissues Organs 205:350–371
Fedde MR (1980) The structure and gas flow pattern in the avian lung. Poult Sci 59:2642–2653
Fedi P, Bafico A, Nieto-Soria A, Burgess WH, Miki T, Bottaro DP, Kraus MH, Aaronson SA (1999) Isolation and biochemical characterization of the human Dkk-1 homologue, a novel inhibitor of mammalian Wnt signaling. J Biol Chem 274:19465–19472
Feng L, Jinxi H, Jun W, Cho WC, Xiaoming L (2015) Diversity of epithelial stem cell types in adult lung. Stem Cell Intern 2015:728307. https://doi.org/10.1155/2015/728307
Ferguson JE, Kelley RW, Patterson C (2005) Mechanisms of endothelial differentiation in embryonic vasculogenesis. Arterioscl Thromosis Vasc Biol 25:2246–2554
Fernandes-Silva H, Alves MG, Correia-Pinto J, Oliveira PF, Moura RS (2017a) Lung branching morphogenesis in the chicken model is accompanied by temporal metabolic changes. Porto Biomed J 2:222–223
Fernandes-Silva H, Araújo-Silva H, Correia-Pinto J, Moura RS (2020) Retinoic acid: a key regulator of lung development. Biomol Ther 10:152. https://doi.org/10.3390/biom10010152
Fernandes-Silva H, Correia-Pinto J, Moura RA (2017b) Canonical sonic hedgehog signaling in early lung development. J Dev Biol J Dev Biol 5:3. https://doi.org/10.3390/jdb5010003
Ferrara N (1999) Molecular and biological properties of vascular endothelial growth factor. J Mol Med 77:527–543
Ferrara N (2000a) The role of vascular endothelial growth factor in angiogenesis. In: Ware JA, Simons M (eds) Angiogenesis in health and disease: basic mechanisms and clinical applications. Mercel Dekker, New York, pp 47–73
Ferrara N (2000b) Vascular endothelial growth factor and the regulation of angiogenesis. Recent Prog Horm Res 55:15–35
Ferrara N, Carver-Moore K, Chen H, Dowd M, Lu L, O'Shea KS, Powell-Braxton L, Hillan KJ, Moore MW (1996) Heterozygous embryonic lethality induced by targeted inactivation of the VEGF gene. Nature 380:439–442
Ferrara N, Davis-Smyth T (1997) The biology of vascular endothelial growth factor. Endocr Rev 18:4–25
Ferrara N, Gerber HP (1999) The vascular endothelial growth factor family. In: Ware JA, Simons M (eds) Angiogenesis and cardiovascular disease. Oxford University Press, New York, pp 101–127
Ferrara N, Gerber HP, LeCouter J (2003) The biology of VEGF and its receptors. Nat Med 9:669–676
Ferrara N, Houck K, Jakeman L, Leung DW (1992) Molecular and biological properties of the vascular endothelial growth factor family of proteins. Endocr Rev 13:18–32
Fick A (1855) Ueber Diffusion. Annalen der Physik (in German) 94:59–86
Ficken MD, Barnes HJ (1989) Acute air sacculitis in turkeys inoculated with *Pasteurella multocida*. Vet Pathol 26:231–237
Fischer B, Bavister BD (1993) Oxygen tension in the oviduct and uterus of rhesus monkeys, hamsters and rabbits. J Reprod Fert 99:673–679
Fletcher DA, Mullins RD (2010) Cell mechanics and the cytoskeleton. Nature 463:485–492
Fletcher OJ (1980) Pathology of the avian respiratory system. Poult Sci 59:2666–2679
Folkman J, Klagsbrun M (1987) Angiogenic factors. Science 235:442–447
Fraga MV, Guttentag S (2012) Lung development: embryology, growth, maturation, and developmental biology. In: Gleason CA, Devaskar SU (eds) Avery's diseases of the newborn, 9th edn. Elsevier BV, New York, pp 571–583
Frank GB (2003) Ornithology, 2nd edn. WH Freeman and Company, New York
Franks TJ, Colby TV, Travis WD, Tuder RM, Reynolds HY, Brody AR, Cardoso WV, Crystal RG, Drake CJ, Engelhardt J et al (2008) Resident cellular components of the human lung: current

knowledge and goals for research on cell phenotyping and function. Proc Am Thorac Soc 15: 763–766

Franzdóttir SR, Axelsson IT, Arason AJ, Baldursson Ó, Gudjonsson T, Magnusson MK (2010) Airway branching morphogenesis in three dimensional culture. Respir Res 11:1–10

Fujii S, Muranaka T, Matsubayashi J, Yamada S, Yoneyama A, Takakuwa T (2021) Bronchial tree of the human embryo: categorization of the branching mode as monopodial and dipodial. PLoS One 15:e0245558. https://doi.org/10.1371/journal.pone.0245558

Full R, Earis K, Wong M, Caldwell R (1993) Locomotion like a wheel? Nature 365:495. https://doi.org/10.1038/365495A0

Funk EC, Breen C, Sanketi BD, Kurpios N, McCune A (2020a) Changes in Nkx2.1, Sox2, Bmp4, and Bmp16 expression underlying the lung-to-gas bladder evolutionary transition in ray-finned fishes. Evol Dev 22:384–402

Funk E, Lencer E, McCune A (2020b) Dorsoventral inversion of the air-filled organ (lungs, gas bladder) in vertebrates: RNAsequencing of laser capture microdissected embryonic tissue. J Exp Zool B Mol Dev Evol 334:325–338

Funk RH, Monsees T, Ozkucur N (2009) Electromagnetic effects–from cell biology to medicine. Prog Histochem Cytochem 43:177–264

Galambos C, DeMello DE (2007) Molecular mechanisms of pulmonary vascular development. Pediatric Dev Pathol 10:1–18

Gallanger BC (1986) Branching morphogenesis in the avian lung: Electronmicroscopic studies using cataionic dyes. J Embryol Exp Morphol 94:189–205

Gao Y, Cornfield DN, Stenmark KR, Thébaud B, Abman SH, Raj JU (2016) Unique aspects of the developing lung circulation: structural development and regulation of vasomotor tone. Pulm Circ 6:407–425

Gao Y, Raj JU (2010) Regulation of the pulmonary circulation in the fetus and newborn. Physiol Rev 90:1291–1335

Garcia HG, Kondev J, Orme N, Theriot JA, Phillips R (2011) Thermodynamics of biological processes. Methods Enzymol 492:27–59

Gardel ML, Nakamura F, Hartwig JH, Crocker JC, Stossel TP, Weitz DA (2006) Prestressed F-actin networks cross-linked by hinged filamins replicate mechanical properties of cells. Proc Natl Acad Sci U S A 103:1762–1767

Gebb SA, Shannon JM (2000) Tissue interactions mediate early events in pulmonary vasculogenesis. Dev Dyn 217:159–169

Geetha-Loganathan P, Nimmagadda S, Antoni L, Fu K, Whiting CJ, Francis-West P, Richman JM (2009) Expression of WNT signalling pathway genes during chicken craniofacial development. Dev Dyn 238:1150–1165

Gefen E, Ar A (2001) Gas exchange and energy metabolism on the ostrich (*Struthio camelus*) embryo. Comp Biochem Pysiol A: Mol Integr Physiol 130:689–699

Gehr P, Bachofen M, Weibel ER (1978) The normal human lung: ultrastructure and morphometric estimation of diffusion capacity. Respir Physiol 32:121–140

Genbacev O, Zhou Y, Ludlow JW, Fisher SJ (1997) Regulation of human placental development by oxygen tension. Science 277:1669–1672

Gerritsen ME, Soriano R, Yang S, Zlot C, Ingle G, Toy K, Williams PM (2003) Branching out: a molecular fingerprint of endothelial differentiation into tube-like structures generated by affymetrix oligonucleotide arrays. Microcirculation 10:63–81

Gerhardt H (2008) VEGF and endothelial guidance in angiogenic sprouting. Organogenesis 4:241–246

Gerhardt H, Betsholtz C (2005) How do endothelia cells orientate? EXS 94:3–15

Ghabrial A, Luschnig S, Metzstein MM, Krasnow MA (2003) Branching morphogenesis of the *drosophila* tracheal system. Annu Rev Cell Dev Biol 19:623–647

Giannone G, Sheetz MP (2006) Substrate rigidity and force define form through tyrosine phosphatase and kinase pathways. Trends Cell Biol 16:213–223

Girosi D, Bellodi S, Sabatini F, Rossi GA (2006) The lung and the gut: common origins, close links. Paediatr Resp Rev 7S:S235–S239

Gjorevski N, Nelson CM (2010a) The mechanics of development: models and methods for tissue morphogenesis. Birth Defects Res C Embryo Today 90:193–202

Gjorevski N, Nelson CM (2010b) Endogenous patterns of mechanical stress are required for branching morphogenesis. Integr Biol 2:424–434

Gilbert SF (2000) Developmental biology, 6th ed Sunderland (MA), Sinauer Associates: 2000. Morphogenesis and Cell Adhesion. https://www.ncbi.nlm.nih.gov/books/NBK10021/

Gilbert SF, Opitz JM, Raff RA (1996) Resynthesizing evolutionary and developmental biology. Dev Biol 173:357–372

Gilbert SF, Sarkar S (2000) Embracing complexity: organism for the 21st century. Dev Dyn 219:1–9

Gillich A, Zhang F, Farmer CG, Travaglini KJ, Tan SY, Gu M, Zhou B, Feinstein JA, Krasnow MA, Metzger RJ (2020) Capillary cell-type specialization in the alveolus. Nature 586:785–789

Glasser JR, Mallampalli RK (2012) Surfactant and its role in the pathobiology of pulmonary infection. Microbes Infect 14:17–25

Gleghorn JP, Kwak J, Pavlovich AL, Nelson CM (2012) Inhibitory morphogens and monopodial branching of the embryonic chicken lung. Dev Dyn 241:852–862

Glenny RW (2011) Emergence of matched airway and vascular trees from fractal rules. J Appl Physiol 110:1119–1129

Glenny RW, Krueger M, Bauer C, Beichel RR (2020) The fractal geometry of bronchial trees differs by strain in mice. J Appl Physiol 128:362–367

Goldie LC, Nix MK, Hirschi KK (2008) Embryonic vasculogenesis and hematopoietic specification. Organogenesis 4:257–263

Goldin GV, Opperman LA (1980) Induction of supernumerary tracheal buds and the stimulation of DNA synthesis in the embryonic chick lung and trachea by epidermal growth factor. J Embryol Exp Morphol 60:235–243

Goldstein M, Inge F (1993) The refrigerator and the universe: understanding the laws of energy. Harvard University Press, Harvard (MA)

Gontan C, Munck A, Vermeij M, Grosveld F, Tibboel D, Rottier R (2008) Sox2 is important for two crucial processes in lung development: branching morphogenesis and epithelial cell differentiation. Dev Biol 317:296–309

Goodwin BC (1985) The causes of morphogenesis. BioEssays 3:32–36

Goodwin BC (1994) How the leopard changed its spots: the evolution of complexity. Charles Scribner's Sons, New York

Goodwin K, Nelson CM (2020) Branching morphogenesis. Development 147:dev184499. https://doi.org/10.1242/dev.184499

Goodwin K, Nelson CM (2021) Uncovering cellular networks in branching morphogenesis using single-cell transcriptomics. Curr Topics Dev Biol 143:239–280. https://doi.org/10.1016/bs.ctdb.2020.09.004

Gomes AM, Bhat R, Correia AL, Mott JD, Ilan N, Vlodavsky I, Pavão MS, Bissell M (2015) Mammary branching morphogenesis requires reciprocal signaling by heparanase and MMP-14. J Cell Biochem 116:1668–1679

Gospodarowicz D (1991) Biological activities of fibroblast growth factors. In: Baird A, Klagsbrun M (eds) The fibroblast growth factor family. New York Academy of Science, New York, pp 1–8

Goss AM, Tian Y, Tsukiyama T, Cohen ED, Zhou D, Lu MM, Yamaguchi TP, Morrisey EE (2009) Wnt2/2b and β-catenin signaling are necessary and sufficient to specify lung progenitors in the foregut. Dev Cell 17:290–298

Gould J (1971) D'Arcy Thompson and the science of form. New Lit Hist 2:229–258

Gould SJ (1981) Kingdoms without wheels. Natur. History 90:42–48

Grant ZL, Coultas L (2019) Growth factor signaling pathways in vascular development and disease. Growth Factors 37:53–67

Greaves P (2012) Histopathology of preclinical toxicity studies: interpretation and relevance in drug safety studies, 4th edn Mammary gland. Academic Press, Boston, pp 69–97

Green JBA, Sharpe J (2015) Positional information and reaction-diffusion: two big ideas in developmental biology combine. Development 142:1203–1211

Green M, Chen A, Nostro MC, d'Souza SL, Schaniel C, Lemischka IR, Gonon-Evans V, Keller G, Snoeck HW (2011) Generation of anterior foregut endoderm from human embryonic and induced pluripotent stem cells. Nat Biotechnol 29:267–272

Groenman F, Unger S, Post M (2005) The molecular basis for abnormal human lung development. Biol Neonate 87:164–177

Gritsun TA, le Feber J, Rutten WLC (2012) Growth dynamics explain the development of spatiotemporal burst activity of young cultured neuronal networks in detail. PLoS One 7: e43352. https://doi.org/10.1371/journal.pone.0043352

Guo Y, Sun M, Garfinkel A, Zhao X (2014) Mechanisms of side branching and tip splitting in a model of branching morphogenesis. PLoS One 9:e102718. https://doi.org/10.1371/journal.pone.0102718

Günther A, Ruppert C, Schmidt R, Markart P, Grimminger F, Walmrath D, Seeger W (2001) Surfactant alteration and replacement in acute respiratory distress syndrome. Respir Res 2:353–364

Hackett BP, Binge CD, Gitlin JD (1996) Mechanisms of gene expression and cell fate determination in the developing pulmonary epithelium. Annu Rev Physiol 58:51–71

Hacohen N, Kramer S, Sutherland D, Hiromi Y, Krasnow M (1998) *Sprouty* encodes a novel antagonist of FGF signaling that patterns apical branching of the *drosophila* airways. Cell 92: 253–263

Hahn H (1909) Experimental studies on the development of the blood and the first blood vessels in the chick. Arch Entwicklungsmechanik Organ 4:140–143

Halatek J, Brauns F, Frey E (2018) Self-organization principles of intracellular pattern formation. Phil Trans R Soc (Lond) B 373:20170107. https://doi.org/10.1098/rstb.2017.0107

Hall SM, Hislop AA, Harworth SG (2002) Origin, differentiation, and maturation of human pulmonary veins. Am J Respir Cell Mol Biol 26:333–340

Halliday A, Hunt BJ, Poston L, Schachter M (eds) (1998) An introduction to vascular biology: from physiology to pathophysiology. Cambridge University Press, Cambridge, pp 3–19

Hallou A, Brunet T (2020) On growth and force: mechanical forces in development. Development 147:dev187302. https://doi.org/10.1242/dev.187302

Hamburger V, Hamilton HL (1951) A series of normal stages in the development of the chick embryo. J Morph 88:49–92

Hamburger V, Hamilton HL (1992) A series of normal stages in the development of the chick embryo. Dev Dyn 195:231–272

Han RNN, Liu J, Tanswell AK, Post M (1992) Expression of basic fibroblast growth factor and receptor: immunolocalization studies in developing rat fetal lung. Pediatr Res 31:435–440

Han S, Mallampalli RK (2015) The role of surfactant in lung disease and host defense against pulmonary infections. Ann Amer Thoracic Soc 12:765–774

Hannezo E, Simons BD (2019) Multiscale dynamics of branching morphogenesis. Curr Opin Cell Biol 60:99–105

Hans C (2006) Wnt/β-catenin signaling in development and disease. Cell 127:469–480

Harris MP, Hasso SM, Ferguson MWJ, Fallon JF (2006) The development of archosaurian first-generation teeth in a chicken mutant. Curr Biol 16(4):371–377

Harris RS (2005) Pressure-volume curves of the respiratory system. Respir Care 50:78–98

Harvey AJ (2007) The role of oxygen in ruminant preimplantation embryo development and metabolism. Anim Reprod Sci 98:113–128

Harvey EP, Ben-Tal A (2016) Robust unidirectional airflow through avian lungs: new insights from a piecewise linear mathematical model. PLoS Comput Biol 12:e1004637. https://doi.org/10.1371/journal.pcbi.1004637

Harvey W (1651) Exercitatio 3. Exercitationes de generatione animalium. Elzevir, Jansson and Ravesteyn, Amsterdam

Hay ED (1995) An overview of epithelia-mesenchymal transformation. Acta Anat 154:8–20

Hay ED (2005) The mesenchymal cell, its role in the embryo, and the remarkable signaling mechanisms that create it. Dev Dyn 233:706–720

Hayashi S, Kondo T (2018) Development and function of *drosophila* tracheal system. Genetics 209:367–380

Hegazi AG, Abdel-Rahman EH (2015) Cytokines. Austin Publishing Group. Innovative Immunology, pp. 1–38. www.austinpublishinggroup.com/ebooks

Heisenberg CP, Bellaïche Y (2013) Forces in tissue morphogenesis and patterning. Cell 153:948–962

Herbert SP, Stainier DY (2011) Molecular control of endothelial cell behaviour during blood vessel morphogenesis. Nat Rev Mol Cell Biol 12:551–564

Herriges M, Morrisey EE (2014) Lung development: orchestrating the generation and regeneration of a complex organ. Development 141:502–513

Hill MA (2021) Embryology: respiratory system development. Retrieved from https://embryology.med.unsw.edu.au/emberyology/index.php/Respiratory_System_Development on 21-3-2021

Hill MA (2022) Embryology: chicken development. Retrieved from https://embryology.med.unsw.edu.au/embryology/index.php/Chicken_Development on 01-05-22

Hines EA, Sun X (2014) Tissue crosstalk in lung development. J Cell Biochem 115:1469–1477

His W (1900) Lecithoblast und Angioblast der Wirbeltiere. Abhandl Math-Phys Gen Wiss 26:171–328

Hislop AA (2002) Airway and blood vessel interaction during lung development. J Anat 201:325–334

Hislop A, Reid L (1972) Intra-pulmonary arterial development during fetal life-branching pattern and structure. J Anat 113:35–48

Höckel M (2015) Morphogenetic fileds of embryonic development in locoregional cancer spread. Lancet Oncol 16:148–151

Hogan BLM (1999) Morphogenesis Cell 96:225–233

Hogan BL, Barkauskas CE, Chapman HA, Epstein JA, Jain R, Hsia CW, Niklason L, Calle E, Le A, Randell SH et al (2014) Repair and regeneration of the respiratory system: complexity, plasticity and mechanisms of lung stem cell function. Cell Stem Cell 15:123–138

Hogan BLM, Grindley J, Bellusci S, Dunn NR, Emoto H, Itoh N (1997) Branching morphogenesis of the lung: new models for classical problems. Cold Spring Harbour Symp Quart Biol 62:249–256

Hogan BL, Kolodziej PA (2002) Organogenesis: molecular mechanisms of tubulogenesis. Nat Rev Genet 3:513–523

Holland JD, Klaus A, Garratt AN, Birchmeier W (2013) Wnt signaling in stem and cancer stem cells. Curr Opin Cell Biol 25:254–264

Horowitz A, Simons M (2008) Branching morphogenesis. Circ Res 103:784–795

Horowitz A, Simons M (2009) Branching morphogenesis. Circ Res 104:e21. https://doi.org/10.1161/CIRCRESAHA.108.191494

Hoyt DF (1987) A new model of avian embryonic metabolism. J Exp Zool Suppl 1:127–138

Hsia CC, Hyde DM, Weibel ER (2016) Lung structure and the intrinsic challenges of gas exchange. Compr Physiol 6:827–895

Hsia CC, Schmitz A, Lambertz M, Perry SF, Maina JN (2013) Evolution of air breathing: oxygen homeostasis and the transitions from water to land and sky. Compr Physiol 3:849–915

Hsieh JC, Kodjabachian L, Rebbert ML, Rattner A, Smallwood PM, Samos CH, Nusse R, Dawid IB, Nathans J (1999) A new secreted protein that binds to Wnt proteins and inhibits their activities. Nature 398:431–436

Hsu YC, Osinski J, Campbell CE, Litwack ED, Wang D, Liu S, Bachurski CJ, Gronostajski RM (2011) Mesenchymal nuclear factor I B regulates cell proliferation and epithelial differentiation during lung maturation. Dev Biol 354:242–252

Hudlická O, Brown MD, Egginton S (1998) Angiogenesis: basic concepts and methodology. In: Halliday A, Hunt BJ, Poston L, Schachter M (eds) An introduction to vascular biology: from physiology to pathophysiology. Cambridge University Press, Cambridge, pp 3–19

Hudlicka O, Tyler KR (1986) Angiogenesis: the growth of the vascular system. Academic Press, London

Huelsken J, Behrens J (2002) The Wnt signalling pathway. J Cell Sci 115:3977–3978

Huxley TH (1882) On the respiratory organs of *apteryx*. Proc Zool Soc (Lond) 1882:560–569

Hyatt BA, Shangguan X, Shannon JM (2002) BMP-4 modulates fibroblast growth factor-mediated induction of proximal and distal lung differentiation in mouse embryonic tracheal epithelium in mesenchyme free culture. Dev Dyn 225:153–165

Hyde DM, Blozis SA, Avdalovic MV, Putney LF, Dettorre R, Quesenberry NJ, Singh P, Tyler NK (2007) Alveoli increase in number but not size from birth to adulthood in rhesus monkeys. Am J Physiol Lung Cell Mol Physiol 293:L570–L579

Hylka VW, Doneen BA (1982) Lung phospholipids in the emberyonic and immature chicken: changes in lipid composition biosynthesis during maturation of the surfactant system. J Exp Zool 220:71–80

Iber D (2021) The control of lung branching morphogenesis. Curr Top Dev Biol 143:205–237

Iber D, Menshykau D (2013) The control of branching morphogenesis. Open Biol 3:130088. https://doi.org/10.1098/rsob.130088

Ingber DE (2005) Mechanical control of tissue growth: function follows form. Proc Natl Acad Sci U S A 102:11571–11572

Ingber DE (2006) Cellular mechano-transduction: putting all the pieces together again. FASEB J 20:811–827

Ingber DE, Folkman J (1989) Mechanochemical switching between growth and differentiation during fibroblast growth factor-stimulated angiogenesis *in vitro*: role of extracellular membrane. J Cell Biol 109:317–330

Ingham PW, McMahon AP (2001) Hedgehog signaling in animal development: paradigms and principles. Genes Dev 15:3059–3087

Islam R, Sultana N, Alam R, Akter A, Anisuzzaman MD, Haque Z, Khan MD, ABN ALI (2022) Dose and time-dependent effects of glucocorticoid: a morphologic and morphometric study in the broiler lung. Turkish J Vet Anim Sci 46(4). https://doi.org/10.55730/1300-0128.4233

Itkonen E (2017) The whole is greater than the sum of its parts–true, false or meaningless. J Semiotics 7:21–50

Itoh N (2004) Fibroblast growth factor (FGF). In: Luciano M (ed) Encyclopedia of endocrine diseases. Elseier, New York, pp 67–71

Jaffredo T, Nottingham W, Liddiard K, Bollerot K, Pouget C, de Bruijn M (2005) From hemangioblast to hematopoietic stem cell: an endothelial connection. Exp Hematol 33:1029–1040

Jakkula M, Le Cras TD, Gebb S, Hirth KP, Tuder RM, Voelkel NF, Abman SH (2000) Inhibition of angiogenesis decreases alveolarization in the developing rat lung. Am J Physiol Lung Cell Mol Physiol 279:L600–L607

Jain RK (2003) Molecular regulation of vessel maturation. Nat Med 9:685–693

Janmey PA, McCulloch CA (2007) Cell mechanics: integrating cell responses to mechanical stimuli. Annu Rev Biomed Eng 9:1–34

Jaskoll T, Abichaker G, Witcher D, Sala FG, Bellusci S, Hajihosseini MK, Melnick M (2005) FGF10/FGFR2b signaling plays essential roles during in vivo embryonic submandibular salivary gland morphogenesis. BMC Dev Biol 5:11. https://doi.org/10.1186/1471-213X-5-11

Jewett CE, Prekeris R (2018) Insane in the apical membrane: trafficking events mediating apicobasal epithelial polarity during tube morphogenesis. Traffic. https://doi.org/10.1111/tra.12579

Jones AW, Radnor CJP (1972a) The development of the chick tertiary bronchus. I. General development and the mode of production of the osmiophilic inclusion body. J Anat 113:303–324

Jones AW, Radnor CJP (1972b) The development of the chick tertiary bronchus. II. The origin of the surface lining system. J Anat 113:325–340

Jones JH, Effmann EL, Schmidt-Nielsen K (1985) Lung volume changes during respiration in ducks. Respir Physiol 59:15–25

Jones MR, Chong L, Belleusci S (2021) Fgf10/Fgfr2b signaling orchestrates the symphony of molecular, cellular, and physical processes required for harmonious airway branching morphogenesis. Front Cell Dev Biol 8:620667. https://doi.org/10.3389/fcell.2020.620667

Jośko J, Gwóźdź B, Jedrzejowska-Szypułka H, Hendryk S (2000) Vascular endothelial growth factor (VEGF) and its effect on angiogenesis. Med Sci Monit 5:1047–1052

Judd W (1979) The secretions and fine structure of bivalve crystalline style sacs. Ophelia 18:205–233

Juillet A (1911) Recherches anatomiques, embryologiques, histologiques et comparatives sur le poumon des oiseaux. Arch Zool Exp Gen 9:207–371

Jürgens KD, Gros G (2002) Phylogeny of gas exchange systems. Anasthesiol Intensivmed Notfallmed Schmerzther 37:185–198

Kain KH, Miller JW, Jones-Paris CR, Thomason RT, Lewis JD, Bader DM, Barnett JV, Zijlstra A (2014) The chick embryo as an expanding experimental model for cancer and cardiovascular research. Dev Dyn 243:216–228

Kaipainen A, Bielenberg DR (2010) Hemangiogenesis versus lymphangiogenesis. In: Dartt DA (ed) Encyclopedia of the eye. Academic Press, Oxford, pp 227–232

Kallapur S, Kotecha S (2016) Perinatal modifiers of lung structure and function. In: Jobe A, Whitsett J, Abman S (eds) Fetal and neonatal lung development: clinical correlates and technologies for the future. Cambridge University Press, Cambridge, pp 187–204

Kalluri R, Weinberg RA (2009) The basics of epithelial-mesenchymal transition. J Clin Invest 119: 1420–1428

Kanoldt V, Fischera L, Grashoff C (2019) Unforgettable force–crosstalk and memory of mechanosensitive structures. Biol Chem 400:687–698

Kardong KV (2014) Vertebrates–comparative anatomy, function, evolution, 6th edn. McGraw-Hill, New York

Karolak JA, Gambin T, Szafranski P, Stankiewicz P (2021) Potential interactions between the TBX4-FGF10 and SHH-FOXF1 signaling during human lung development revealed using ChIP-seq. Respir Res 22:26. https://doi.org/10.1186/s12931-021-01617-y

Katsnelson MI, Wolf YI, Koonin EV (2018) Towards physical principles of biological evolution. Phys Sci 93:043001. https://doi.org/10.1101/182378

Katsumi A, Milanini J, Kiosses WB, del Pozo MA, Kaunas R, Chien S, Hahn KM, Schwartz MA (2002) Effects of cell tension on the small GTPase Rac. J Cell Biol 158:153–164

Kaufman SL, Burri PH, Weibel ER (1974) The postnatal growth of the rat lung. II. Autoradiography. Anat Rec 180:63–76

Kechagia JZ, Ivaska J, Roca-Cusachs P (2019) Integrins as biomechanical sensors of the microenvironment. Nat Rev Mol Cell Biol 20:457–473

Keller R (2012) Physical biology returns to morphogenesis. Science 338:201–203

Keller R, Shook D (2011) The bending of cell sheets–from folding to rolling. BMC Biol 9:90. https://doi.org/10.1186/1741-7007-9-90

Kerr JFR, Wyllie AH, Currie AR (1972) Apoptosis: a basic biological phenomenon with wide-ranging implications in tissue kinetics. Br J Cancer 26:239–257

Kicheva A, Cohen M, Briscoe J (2012) Developmental pattern formation: insights from physics and biology. Science 338:210–212

Kim HY, Varner VD, Nelson CM (2013) Apical constriction initiates new bud formation during monopodial branching of the embryonic chicken lung. Development 140:3146–3155

Kina YP, Khadim A, Seeger W, El Agha E (2021) The lung vasculature: a driver or passenger in lung branching morphogenesis? Front Cell Dev Biol 8:623868. https://doi.org/10.3389/fcell.2020.623868

Kindberg A, Hu JK, Bush JO (2020) Forced to communicate: integration of mechanical and biochemical signaling in morphogenesis. Curr Opin Cell Biol 66:59–68
King AS (1966) Structural and functional aspects of the avian lung and its air sacs. Intern Rev Gen Exp Zool 2:171–267
King AS (1975) Aves respiratory system. In: Getty R (ed) Sisson and Grossman's the anatomy of the domestic animals, vol 2, 5th edn. Saunders, Philadelphia, pp 1011–1075
King AS, King DZ (1979) Avian morphology: general principles. In: King AS, McLelland J (eds) Form and function in birds, vol 1. Academic Press, London, pp 1–38
King AS, McLelland J (1984) Birds: their structure and function, 2nd edn. Bailliéle Tindall, London
King AS, McLelland J (eds) (1989) Form and function in birds, vol 4. Academic Press, London
King AS, Molony V (1971) The anatomy of respiration. In: Bell DF, Freeman BM (eds) Physiology and biochemistry of the domestic fowl, vol 1. Academic Press, London, pp 347–384
King KL, Cidlowski JA (1998) Cell cycle regulation and apoptosis. Annu Rev Physiol 60:601–617
Kirby TJ, Lammerding J (2018) Emerging views of the nucleus as a cellular mechanosensor. Nat Cell Biol 20:373–381
Kitaoka H, Burri PH, Weibel ER (1996) Development of the human fetal airway tree: analysis of the numerical density of airway endtips. Anat Rec 244:207–213
Klagsbrun M (1989) The fibroblast growth factor family: structural and biological properties. Prog Growth Factor Res 1:207–235
Klein S, Giancotti FG, Presta M, Albelda SM, Buck CA, Rifkin DB (1993) Basic fibroblast growth factor modulates integrin expression in microvascular endothelial cells. Mol Biol Cell 4:973–982
Klingenberg CP (2010) Evolution and development of shape: integrating quantitative approaches. Nat Rev Genet 11:623–635
Klingenberg CP, Zaklan SD (2000) Morphological integration between developmental compartments in the *drosophila* wing. Evolution 54:1273–1285
Kloek GP, Casler C (1972) The lung and air-sac system of the common grackle. Auk 89:817–825
Knudsen L, Ochs M (2018) The micromechanics of lung alveoli: structure and function of surfactant and tissue components. Histochem Cell Biol 150:661–676
Koch LG, Britton SL (2008) Aerobic metabolism underlies complexity and capacity. J Physiol 586: 83–95
Koch M, Wehrle-Haller B, Baumgartner S, Spring J, Brubacher D, Chiquet M (1991) Epithelial synthesis of tenascin at tips of growing bronchi and graded accumulation in basement membrane and mesenchyme. Exp Cell Res 194:297–300
Kojima S, Blair DF (2004) The bacterial flagellar motor: structure and function of a complex molecular machine. Int Rev Cytol 233:93–134
Kooijman SA (1986) What the hen can tell about her eggs: egg development on the basis of energy budgets. J Math Biol 23:163–185
Kolomeisky AB (2015) Motor proteins and molecular motors. CRC Press, Boca Raton (FL)
Kolte D, McClung JA, Aronow WS (2016) Vasculogenesis and angiogenesis. In: Aronow WS, McClung JA (eds) Translational research in coronary artery disease. Academic Press, Boston, pp 49–65
Kotton DN, Morrisey EE (2014) Lung regeneration: mechanisms, applications and emerging stem cell populations. Nature Med 20:822–832
Kovar J, Sly PD, Willet KE (2002) Postnatal alveolar development of the rabbit. J Appl Physiol 93: 629–635
Kauts ML, Vink CS, Dzierzak E (2016) Hematopoietic (stem) cell development–how divergent are the roads taken? FEBS Lett 590:3975–3986
Kristensen JH (1972) Structure and function of crystalline styles of bivalves. Ophelia 10:91–108
Krysko DV, Berghe TV, D'Herde K, Vandenabeele K (2008) Apoptosis and necrosis: detection, discrimination and phagocytosis. Methods 44:205–221
Kuhl M, Sheldahl LC, Park M, Miller JR, Moon RT (2000) The Wnt/Ca2+ pathway: a new vertebrate Wnt signaling pathway takes shape. Trends Genet 16:279–283

Kumar VH, Lakshminrusimha S, El Abiad MT, Chess PR, Ryan RM (2005) Growth factors in lung development. Adv Clin Chem 40:261–316

Kurz H, Burri PH, Djonov VG (2003) Angiogenesis and vascular remodeling by intussusception: from form to function. News Physiol Sci 18:65–70

LaBarbera M (1983) Why the wheels won't go. Amer Natur 121:395–408

Lacaud C, Robertson S, Pallis J, Kennedy M, Keller G (2001) Regulation of hemangioblast development. Ann N Y Acad Sci 938:96–107

Lacaud G, Keller G, Kouskoff V (2004) Tracking mesoderm formation and specification to the hemangioblast in vitro. Trends Cardiovasc Med 14:314–317

Lambertz M, Bertozzo F, Sander PM (2018) Bone histological correlates for air sacs and their implications for understanding the origin of the dinosaurian respiratory system. Biol Lett 14: 20170514. https://doi.org/10.1098/rsbl.2017.0514

Lammert E, Cleaver O, Melton D (2001) Induction of pancreatic differentiation by signals from blood vessels. Science 294:564–567

Lang C, Conrad L, Iber D (2021) Organ-specific branching morphogenesis. Front Cell Dev Biol 9: 671402. https://doi.org/10.3389/fcell.2021.671402

Lang C, Conrad L, Michos O (2018) Mathematical approaches of branching morphogenesis. Front Genet 9:673. https://doi.org/10.3389/fgene.2018.00673

Lansford R, Rugonyi S (2020) Follow me! A tale of avian heart development with comparisons to mammal heart development. J Cardiovasc Dev Dis 7:8. https://doi.org/10.3390/jcdd7010008

Lanza R, Atala A (eds) (2006) Essentials of stem cell biology. Elsevier, Amsterdam

Larrivée B, Karsan A (2000) Signaling pathways induced by vascular endothelial growth factor. Int J Mol Med 5:447–545

Laschke MW, Gu Y, Menger MD (2022) Replacement in angiogenesis research: studying mechanisms of blood vessel development by animalfree *in vitro*, *in vivo* and in silico approaches. Front Physiol 13:981161. https://doi.org/10.3389/fphys.2022.981161

Lawson AB, Hedrick BP, Echols S, Schachner ER (2021) Anatomy, variation, and asymmetry of the bronchial tree in the African grey parrot (*Psittacus erithacus*). J Morph 2021:1–19. https://doi.org/10.1002/jmor.21340

Lazzaro D, Price M, de Felice M, Di Lauro R (1991) The transcription factor TTF-1 is expressed at the onset of thyroid and lung morphogenesis and in restricted regions of the fetal brain. Development 113:1093–1104

Lebeche M, Marpel S, Cardoso WV (1999) Fibroblast growth factor interactions in the developing lung. Mech Dev 86:125–136

Lecuit T, Lenne PF (2007) Cell surface mechanics and the control of cell shape, tissue patterns and morphogenesis. Nat Rev Mol Cell Biol 8:633–644

Lecuit T, Lenne PF, Munro E (2011) Force generation, transmission, and integration during cell and tissue morphogenesis. Annu Rev Cell Dev Biol 27:157–184

Lee JM, Dedhar S, Kalluri R, Thompson EW (2006) The epithelial-mesenchymal transition: new insights in signaling, development, and disease. J Cell Biol 172:973–981

Leith DE (1983) Comparative mammalian respiratory mechanics. Am Rev Respir Dis 128:S77–S82

Lengerke C, Daley GQ (2005) Patterning definitive hematopoietic stem cells from embryonic stem cells. Exp Hematol 33:971–979

Le Roith D, Karas M (2003) Apoptosis. In: Henry HL, Normann AW (eds) Encyclopedia of hormones. Academic Press, New York, pp 153–156

Le Roith D, Kavsan VM, Koval AP, Roberts CT (1993) Phylogeny of the insulin-like growth factors (IGFs) and receptors: a molecular approach. Mol Reprod Dev 35:332–338

Levin M (2012) Morphogenetic fields in embryogenesis, regeneration, and cancer: non-local control of complex patterning. Bio Systems 109:243–261

Levine OR, Mellins RB (1975) Liquid balance in the lung and pulmonary edema. In: Scarpelli D (ed) Pulmonary physiology of the fetus, newborn and child. Lea and Febiger, Philadelphia, pp 239–258

Levitzky MG (2013) Pulmonary physiology, 8th edn. McGraw Hill Education, New Yolk

Lewallen MA, Burggren WW (2015) Chronic hypoxia and hyperoxia modifies morphology and VEGF concentration of the lungs of the developing chicken (*Gallus gallus* variant *domesticus*). Respir Physiol Neurobiol 219:85–94

Lewis WH (1947) Mechanics of invagination. Anat Rec 97:139–156

Lewontin R (2000) The triple helix: gene, organism, and environment. Harvard University Press, Harvard (MA)

Li C, Hu L, Xiao J, Chen H, Li JT, Bellusci S, Delanghe S, Minoo P (2005) Wnt5a regulates shh and Fgf10 signaling during lung development. Dev Biol 287:86–97

Li C, Smith SM, Peinado N, Gao F, Li W, Lee MK, Zhou B, Bellusci S, Pryhuber GS, Ho HH, Borok Z, Minoo P (2010) WNT5a-ROR signaling is essential for alveologenesis. Cell 7:384. https://doi.org/10.3390/cells9020384

Li C, Xiao J, Hormi K, Borok Z, Minoo P (2002) Wnt5a participates in distal lung morphogenesis. Dev Biol 248:68–81

Li M, Liao L, Tian W (2020) Extracellular vesicles derived from apoptotic cells: an essential link between death and regeneration. Front Cell Dev Biol 2(8):573511. https://doi.org/10.3389/fcell.2020.573511

Li S, Winuthayanon W (2017) Oviduct: roles in fertilization and early embryo development. J Endocrinol 232:R1–R26

Liberti DC, Kremp MM, Liberti WA, Penkala IJ, Li S, Zhou S, Morrisey EE (2021) Alveolar epithelial cell fate is maintained in a spatially restricted manner to promote lung regeneration after acute injury. Cell Rep 35:109092. https://doi.org/10.1016/j.celrep.2021.10909

Lillie FR (1918) The development of the chick, 2nd edn. Holt, New Yolk

Lim CT, Zhou EH, Quek ST (2006) Mechanical models for living cells–a review. J Biomech 39: 195–216

Lim WF, Inoue-Yokoo T, Tan KS, Lai MI, Sugiyama D (2013) Hematopoietic cell differentiation from embryonic and induced pluripotent stem cells. Stem Cell Res Ther 4(3):71. https://doi.org/10.1186/scrt222

Lis R, Karrasch CC, Poulos MG, Kunar B, Redmond D, Duran JGB, Badwe CR, Schachterle W, Ginsberg M, Xiang J et al (2017) Conversion of adult endothelium to immunocompetent haematopoietic stem cells. Nature 545:439–445

Lister AM (2014) Behavioural leads in evolution: evidence from the fossil record. Biol J Linn Soc 112:315–331

Little C (1990) The terrestrial invasion: an ecophysiological approach to the origins of land animals: an ecophysiological approach to the origins of land animals. Cambridge University Press, Cambridge

Liu Y, Jiang H, Crawford HC, Hogan BLM (2003) Role for ETS domain transcription factors Pea3/erm in mouse lung development. Dev Biol 261:10–24

Liu S, Song N, He J, Yu X, Guo J, Jiao X, Ding X, Teng J (2017) Effect of hypoxia on the differentiation and the self-renewal of metanephrogenic mesenchymal stem cells. Stem Cells Internat 2017:7168687. https://doi.org/10.1155/2017/7168687

Locy WA, Larsell O (1916a) The embryology of the bird's lung based on observations of the domestic fowl. Part II Am J Anat 20:1–44

Locy WA, Larsell O (1916b) The embryology of the bird's lung based on observations of the bronchial tree. Part I Am J Anat 19:447–504

Logan CY, Nusse R (2004) The Wnt signaling pathway in development and disease. Annu Rev Cell Dev Biol 20:781–810

Loh KM, van Amerongen R, Nusse R (2016) Generating cellular diversity and spatial form: Wnt signaling and the evolution of multicellular animals. Dev Cell 38:643–655

Losa GA (2002) Fractal morphometry of cell complexity. Riv Biol 95:239–258

Losa GA (2009) The fractal geometry of life. Riv Biol 102:29–59

Loscertales M, Mikels AJ, Hu JKH, Donahoe PK, Roberts DJ (2008) Chick pulmonary Wnt5a directs airway and vascular tubulogenesis. Development 135:1365–1376

Lü J, Izvolsky KI, Qian J, Cardoso WV (2005) Identification of FGF10 targets in the embryonic lung epithelium during bud morphogenesis. J Biol Chem 280:4834–4841

Lu P, Sternlicht MD, Werb Z (2006) Comparative mechanisms of branching morphogenesis in diverse systems. J Mammary Gland Biol Neoplasia 11:213–228

Lu P, Werb Z (2008) Patterning mechanisms of branched organs. Science 322:1506–1509

Lubarsky B, Krasnow MA (2003) Tube morphogenesis: making and shaping biological tubes. Cell 112:19–28

Lynch TJ, Anderson PJ, Xie W, Crooke AK, Liu X, Tyler SR, Luo M, Kusner DM, Zhang Y, Neff T et al (2016) Wnt signaling regulates airway epithelial stem cells in adult murine submucosal glands. Stem Cells 34:2758–2771

Machado RD (2012) The molecular genetics and cellular mechanisms underlying pulmonary arterial hypertension. Scientifica 2012:106576. https://doi.org/10.6064/2012/106576

Maeda Y, Davé V, Whitsett JA (2007) Transcriptional control of lung morphogenesis. Physiol Rev 87:219–244

Maina JN (1982) A scanning electron microscopic study of the air- and blood capillaries of the lung of the domestic fowl (*Gallus domesticus*). Experientia 35:614–616

Maina JN (1988) Scanning electron microscopic study of the spatial organization of the air- and blood conducting components of the avian lung *(Gallus gallus domesticus)*. Anat Rec 222:145–153

Maina JN (1989) A scanning and transmission electron microscopic study of the tracheal air-sac system in a grasshopper (*Chrotogonus senegalensis*, Kraus)- (orthoptera: Acrididae: Pygomorphinae). Anat Rec 223:393–405

Maina JN (1994) Comparative respiratory morphology and morphometry: the functional design of the respiratory systems. In: Gilles R (ed) Advances in comparative and environmental physiology. Springer-Verlag, Berlin, pp 111–232

Maina JN (1998) The gas exchangers: structure, function, and evolution of the respiratory processes. Springer-Verlag, Heidelberg

Maina JN (2000) Comparative respiratory morphology: themes and principles in the design and construction of the gas exchangers. Anat Rec 261:25–44

Maina JN (2002a) Functional morphology of the vertebrate respiratory systems. Science Publishers Inc, Enfield (NH)

Maina JN (2002b) Fundamental structural aspects in the bioengineering of the gas exchangers: comparative perspectives. Adv Anat Embryol Cell Biol 163:1–112

Maina JN (2002c) Some recent advances of the study and understanding of the functional design of the avian lung: morphological and morphometric perspectives. Biol Rev 77:97–152

Maina JN (2002d) Structure, function and evolution of the gas exchangers: comparative perspectives. J Anat 201:281–304

Maina JN (2003a) A systematic study of the development of the airway (bronchial) system of the avian lung from days 3 to 26 of embryogenesis: a transmission electron microscopic study on the domestic fowl, *Gallus gallus* variant *domesticus*. Tissue Cell 35:375–391

Maina JN (2003b) Developmental dynamics of the bronchial (airway)- and air sac systems of the avian respiratory system from days 3 to 26 of life: a scanning electron microscopic study of the domestic fowl, *Gallus gallus* variant *domesticus*. Anat Embryol 207:119–134

Maina JN (2004a) A systematic study of hematopoiesis, vasculogenesis, and angiogensis in the developing avian lung, *Gallus gallus* variant *domesticus*. Tissue Cell 36:307–322

Maina JN (2004b) Morphogenesis of the laminated tripartite cytoarchitectural design of the blood-gas barrier of the avian lung: a systematic electron microscopic study of the domestic fowl, *Gallus gallus* variant *domesticus*. Tissue Cell 36:129–139

Maina JN (2005) The lung-air sac system of birds: development, structure, and function. Springer-Verlag, Berlin

Maina JN (2006) Development, structure, and function of a novel respiratory organ, the lung-air sac system of birds: to go where no other vertebrate has gone. Biol Rev 81:545–579

Maina JN (2007a) Minutialization at its extreme best! The underpinnings of the remarkable strenths of the air and the blood capillaries of the avian lung: a conundrum. Respir Physionl Neurobiol 159:141–145

Maina JN (2007b) Tensegrity principle explains the mechanical strength of the avian lung. Respir Physiol Neurobiol 155:1–10

Maina JN (2009) Functional morphology of the avian respiratory system, the lung-air sac system: efficiency built on complexity. Ostrich (J Afric Ornithol) 79:117–132

Maina JN (2011) Bioengineering aspects in the design of gas exchangers: comparative evolutionary, morphological, functional, and molecular perspectives. Springer-Verlag, Berlin

Maina JN (2012) Comparative molecular developmental aspects of the mammalian- and the avian lungs, and the insectan tracheal system by branching morphogenesis: recent advances and future directions. Front Zool 9:16. https://doi.org/10.1186/1742-9994-9-16

Maina JN (2013) Molecular biological aspects of lung development: current status of understanding and future research directions. In: Berhardt LV (ed) Advances in medicine and biology, vol 60. Nova Biomedical, New York, pp 49–87

Maina JN (2014) Comparative respiratory physiology: the fundamental mechanisms and the functional designs of the gas exchangers. Open Access Animal Physiol 6:53–66

Maina JN (2015a) The design of the avian respiratory system: development, morphology and function. J Ornithol 156:41–63

Maina JN (2015b) Structural and biomechanical properties of the exchange tissue of the avian lung. Anat Rec 298:1673–1688

Maina JN (2017) Critical appraisal of some factors pertinent to the functional designs of the gas exchangers. Cell Tiss Res 367:747–767

Maina JN, Jimoh SA, Hosie M (2010a) Implicit mechanistic role of the collagen, smooth muscle, and elastic tissue components in strengthening the air- and blood capillaries of the avian lung. J Anat 217:597–608

Maina JN, Madan AK (2003) Occurrence and distribution of fibroblast growth factor-2 (FGF-2) in the early development of the avian lung. FASEB J 17(779):468.10A

Maina JN, Madan AK, Alison B (2003) Expression of fibroblast growth factor-2 (FGF-2) in early stages (days 3-11) of the development of the avian lung, *Gallus gallus* variant *domesticus*. J Anat 203:505–512

Maina JN, Nathaniel C (2001) A qualitative and quantitative study of the lung of an ostrich, *Struthio camelus*. J Exp Biol 204:2313–2330

Maina JN, Ramonisi Y, Mashiteng R, Mokae L, Woodward JD (2021) 3D computer reconstruction of the airway and the vascular systems of the lung of the domestic fowl, *Gallus gallus* variant *domesticus*. J Appl Math Comput 5:89–104

Maina JN, West JB (2005) Thin and strong! The bioengineering dilemma in the structural and functional design of the blood-gas barrier: comparative and evolutionary perspectives. Physiol Rev 85:811–844

Maina JN, West JB, Orgeig S, Foot NJ, Daniels CB, Kiama SG, Gehr P, Mühlfeld C, Blank F, Müller L et al (2010b) Recent advances into understanding some aspects of the structure and function of the mammalian- and avian lungs. Physiol Biochem Zool 83:792–807

Makanya AN (2017) Development of the airways and the vasculature in lungs of birds. In: Maina JN (ed) The biology of the avian respiratory system: evolution, development, structure and function. Springer, Cham (Switzerland), pp 147–178

Makanya A, Anagnostopoulou A, Djonov V (2013) Development and remodeling of the vertebrate blood-gas barrier. BioMed Res Internat 2013:101597. https://doi.org/10.1155/2013/101597

Makanya AN, Darawish YE, Kavoi BM, Djonov V (2011a) Spatial and functional relationships between air conduits and blood capillaries in the pulmonary gas exchange tissue of adult and developing chickens. Microsc Res Intern 74:159–169

Makanya AN, Djonov V (2008) Developmental and spatial organization of the air conduits in the lung of the domestic fowl *Gallus gallus* variant *domesticus*. Microsc Res Tech 71:689–702

Makanya AN, Djonov V (2009) Parabronchial angioarchitecture in developing and adult chickens. J Appl Physiol 106:1959–1969

Makanya AN, Hlushchuk R, Baum O, Velinov N, Ochs M, Djonov V (2007) Microvascular endowment in the developing chicken embryo lung. Am J Physiol Lung Cell Mol Physiol 292:L1136–L1146

Makanya AN, Hlushchuk R, Djonov V (2011b) The pulmonary blood-gas barrier in the avian embryo: inauguration, development and refinement. Respir Physiol Neurobiol 178:30–88

Makanya AN, Hlushchuk R, Duncker HR, Draeger A, Djonov V (2006) Epithelial transformations in the establishment of the blood-gas barrier in the developing chick embryo lung. Dev Dyn 235: 68–81

Makanya AN, Kavoi BM, Kihurani DO (2021) Slight volume chages in the do not imply a fundamental change in the structure of the parenchyma. Anat Histol Embryol 50:169–174

Makanya AN, Koller T, Hlushchuk R, Djonov V (2012) Pre-hatch lung development in the ostrich. Respir Physiol Neurobiol 180:183–192

Makanya AN, Stauffer D, Ribatti D, Burri PH, Djonov V (2005) Microvascular growth, development, and remodeling in the embryonic avian kidney: the interplay between sprouting and intussusceptive angiogenic mechanisms. Microsc Res Tech 66:275–288

Malhotra GK, Zhao X, Edwards E, Kopp JL, Naramura M, Sander M, Band H, Band V (2014) The role of Sox9 in mouse mammary gland development and maintenance of mammary stem and luminal progenitor cells. BMC Dev Biol 14:47. https://doi.org/10.1186/s12861-014-0047-4

Mallarino R, Abzhanov A (2012) Paths less travelled: evo-devo approaches to investigating animal morphological evolution. Annu Rev Cell Dev Biol 28:743–763

Mammoto T, Ingber DE (2010) Mechanical control of tissue and organ development. Development 137:1407–1420

Mammoto T, Mammoto A, Ingber DE (2013) Mechanobiology and developmental control. Annu Rev Cell Dev Biol 29:27–61

Mandelbrot BB (1977) Fractals: form, chance, and dimension. Freeman, San Francisco

Mandelbrot BB (1978) The fractal geometry of trees and other natural phenomena. In: Miles RE, Serra J (eds) Geometrical probability and biological structures, Lecture notes in biomathematics, vol 23. Springer Verlag, Heidelberg, pp 235–250

Mandelbrot BB (1989) Fractal geometry: what is it, and what does it do? Proc R Soc Lond A 433:3–16

Mani VT, Jeffrey DM, Michael YJ, Richard GL, Jay PD (2009) A three-dimensional model of vasculogenesis. Biomaterials 30:1098–1112

Manivannan S, Nelson CM (2012) Dynamics of branched tissue assembly. Stem Cell Res Ther 3: 42. https://doi.org/10.1186/scrt133

Mario M, Daniele M (2010) Encyclopedia of movement disorders. In: Kompoliti K, Verhagen ML (eds) Encyclopedia of movement disorders. Academic Press, Oxford, pp 91–95

Martin AC, Goldstein B (2014) Apical constriction: themes and variations on a cellular mechanism driving morphogenesis. Development 141:1987–1898

Martín-Belmonte F, Rodríguez-Fraticelli AE (2009) Acquisition of membrane polarity in epithelial tube formation patterns, signaling pathways, molecular mechanisms, and disease. Int Rev Cell Mol Biol 274:129–182

Martinsen BJ (2005) Reference guide to the stages of chick heart embryology. Dev Dyn 233:1217–1237

Matsumoto K, Yoshitomi H, Rossant J, Zaret KS (2001) Liver organogenesis promoted by endothelial cells prior to vascular function. Science 294:559–563

Mattiske D, Sommer P, Kidson SH, Hogan BLM (2006) The role of the forkhead transcription factor, Foxc1, in the development of the mouse lacrimal gland. Dev Dyn 235:1074–1080

Maurovich-Horvat P (2017) The whole is more than the sum of its parts–Aristotle. Cardiovasc Imaging 18:294–295

Maxwell JC (1860) Illustrations of the dynamical theory of gases, part II. On the process of diffusion of two or more kinds of moving particles among one another. Philos Mag. (4th series) 20:21–37

Mazumdar J, O'Brien WT, Johnson RS, LaManna JC, Chavez JC, Klein PS, Simon MC (2010) O_2 regulates stem cells through Wnt/β-catenin signalling. Nat Cell Biol 12:1007–1013

McCulley D, Wienhold M, Sun X (2015) The pulmonary mesenchyme directs lung development. Curr Opin Genet Dev 32:98–105

McGhee GR (2006) The geometry of evolution: adaptive landscapes and theoretical morphospaces. Cambridge University Press, Cambridge (UK)

McGowan SE (1992) Extracellular matrix and the regulation of lung development and repair. FASEB J 6:2895–2904

McGrath KE, Koniski AD, Malik J, Palis J (2003) Circulation is established in a stepwise pattern in the mammalian embryo. Blood 101:1669–1676

McKeehan WL, Wang F, Kan M (1998) The heparan sulfate-fibroblast growth factor family: diversity of structure and function. Prog Nucleic Acid Res Mol Biol 59:135–176

McLelland J (1989) Anatomy of the lungs and air sacs. In: King AS, McLelland J (eds) Form and function of the avian lung, vol 4. Academic Press, London, pp 221–279

McNichols MJ, McNabb FMA (1988) Development of thyroid function and its pituitary control in embryonic and hatchling precocial Japanese quail and altricial ring doves. Gen Comp Endocrin 69:109–118

McQualter JL, Bertoncello I (2012) Concise review: deconstructing the lung to reveal its regenerative potential. Stem Cells 30:811–816

Merkus PJFM, Ten Have-Opbroek AAW (1996) Human lung growth: a review. Pediatr Pulmonol 21:583–397

Metzger RJ, Krasnow MA (1999) Genetic control of branching morphogenesis. Science 284:1635–1639

Metzger RJ, Klein OD, Martin GR, Krasnow MA (2008) The branching programme of mouse lung development. Nature 453:745–750

Meyer VR, Halbeisen M (2006) Warum gibt es in der Natur keine Räder? Biologie in Unserer Zeit 36:120–123

Mohyeldin A, Garzón-Muvdi T, Quiñones-Hinojosa A (2010) Oxygen in stem cell biology: a critical component of the stem cell niche. Cell Stem Cell 6:150–161

Michael F (1975) Stoic vs peripatetic syllogistic. Arch History Philos 56:99–124

Mikels A, Nusse R (2006) Wnts as ligands: processing, secretion and reception. Oncogene 25: 7461–7468

Miller AM, McWhorter JE (1914) Experiments on the development of blood vessels in the area pellucida and embryonic body of the chick. Anat Rec 8:203–227

Miller JR, Hocking AM, Brown JD, Moon RT (1999) Mechanism and function of signal transudation by the Wnt/β-catenin and Wnt/Ca^{2+} pathways. Oncogene 18:7860–7872

Mills KR, Kruep D, Saha MS (1999) Elucidating the origins of the vascular system: a fate map of the vascular endothelial and red blood cell lineages in *Xenopus laevis*. Dev Biol 209:352–568

Min H, Danilenko DM, Scully SA, Bolon B, Ring BD, Tarpley JE, DeRose M, Simmonett WS (1998) FGF-10 is required for both limb and lung development and exhibits striking functional similarities to *drosophila* branchless. Genes Dev 12:3156–3161

Minamino T, Imada K, Namba K (2008) Molecular motors of the bacterial flagella. Curr Opin Struct Biol 18:693–701

Minoo P, King RJ (1994) Epithelial-mesenchymal interactions in lung development. Annu Rev Physiol 56:13–45

Minoo P, Su G, Drum H, Bringas P, Kimura S (1999) Defects in tracheoesophangeal and lung orphogenesis in Nkx2.1 (−/−) mouse embryos. Dev Biol 209:60–71

Mitteroecker P, Huttegger SM (2009) The concept of morphospaces in evolutionary and developmental biology: mathematics and metaphors. Biol Theory 4:54–67

Mitzner W, Wagner EM (2004) Vascular remodeling in the circulations of the lung. J Appl Physiol 97:1999–2004

Miura T (2008) Modeling lung branching morphogenesis. Curr Top Dev Biol 81:291–310

Miura T (2015) Models of lung branching morphogenesis. J Biochem 157:121–127

Miura T, Hartmann D, Kinboshi M, Komada M, Ishibashi M, Shiota K (2009) The cyst-branch difference in developing chick lung results from a different morphogen diffusion coefficient. Mech Dev 126:160–172

Moon RT, Brown JD, Torres M (1997a) Wnts modulate cell fate and behaviour during vertebrate development. Trends Genet 14:452–162

Moon RT, Brown JD, Yang-Snyder JA, Miller JR (1997b) Structurally related receptors and antagonists compete for secreted Wnt ligands. Cell 88:725–728

Moorman A, Webb S, Brown NA, Lamers W, Anderson RH (2003) Development of the heart: (1) formation of the cardiac chambers and arterial trunks. Heart 89:806–814

Morrisey EE, Cardoso WV, Lane RH, Rabinovitch M, Abman SH, Ai X, Albertine KH, Bland RD, Chapman HA, Checkley W et al (2013) Molecular determinants of lung development. Ann Am Thorac Soc 10:S12–S16

Morrisey EE, Hogan BL (2010) Preparing for the first breath: genetic and cellular mechanisms in lung development. Dev Cell 18:8–23

Mostafa H, Hussein MT, Elnaeim M (2022) Developmental events in the lung of the Japanese quail (*Coturnix coturnix japonica*): morphological, histochemical and electron-microscopic studies. Microsc Res Tech 2022:1–16. https://doi.org/10.1002/jemt.24225

Moura RS, Carvalho-Correia E, daMota P, Correia-Pinto J (2014) Canonical Wnt signaling activity in early stages of chick lung development. PLoS One 9:e112388. https://doi.org/10.1371/journal.pone.0112388

Moura RS, Correia-Pinto J (2017) Molecular aspects of avian lung development. In: Maina JN (ed) The biology of the avian respiratory system: evolution, development, structure and function. Springer, Cham, (Switzerland), pp 129–146

Moura RS, Coutinho-Borges JP, Pacheco AP, daMota PO, Correia-Pinto J (2011) FGF Signaling pathway in the developing chick lung: expression and inhibition studies. PLoS One 6:e17660. https://doi.org/10.1371/journal.pone.0017660

Moura RS, Silva-Gonçalves C, Vaz-Cunha P, Correia-Pinto P (2016) Expression analysis of shh signaling members in early stages of chick lung development. Histochem Cell Biol 146:457–466

Moura RS, Vaz-Cunha P, Silva-Gonçalves C, Correia-Pinto J (2015) Characterization of miRNA processing machinery in the embryonic chick lung. Cell Tissue Res 362:569–575

Müller B (1908) The air sacs of the pigeon. Smithson misc Colls 50:365–414

Müller GB, Newman SA (2005) The innovation triad: an EvoDevo agenda. J Exp Zool 304B:487–503

Mund SI, Stampanoni M, Schittny JC (2008) Developmental alveolarization of the mouse lung. Dev Dyn 237:2108–2116

Muraoka RS, Bushdid PB, Brantley DM, Yull FE, Kerr LD (2000) Mesenchymal expression of nuclear factor-kappaβ inhibits epithelial growth and branching in the embryonic chick lung. Dev Biol 225:322–338

Nanka O, Grim M (2009) Formation of the vascular bed: a review of its molecular mechanisms and therapeutic implications. Cas Lek Cesk 148:158–163

Nasarre P, Potiron V, Drabkin H, Roche J (2010) Guidance molecules in lung cancer. Cell Adhesion and Migration 4:130–145

Navis A, Nelson CM (2016) Pulling together: tissue-generated forces that drive lumen morphogenesis. Semin Cell Dev Biol 55:139–147

Neufeld G, Cohen T, Gengrinovitch S, Poltorak Z (1999) Vascular endothelial growth factor (VEGF) and its receptors. FASEB J 13:9–22

Neumann NM, Perrone MC, Veldhuis JH, Huebner RJ, Zhan H, Devreotes PN, Brodland GW, Ewald AJ (2018) Coordination of receptor tyrosine kinase signaling and interfacial tension dynamics drives radial intercalation and tube elongation. Dev Cell 45:67–82

Ng YS, Rohan R, Sunday ME, Demello DE, D'Amore PA (2001) Dufferential expression of VEGF isoforms in mouse during development and in the adult. Dev Dyn 220:112–121

Nice MM (1962) Development of behavior in precocial birds. Trans Linn Soc New York 8:1–211

Nicholas TE (1996) Pulmonary surfactant: no mere paint on the alveolar wall. Respirology 1:247–257

Nicod LP (2005) Lung defences: an overview. Eur Respir Rev 14(95):45–50

Nicolas J, Magli S, Rabbachin L, Sampaolesi S, Nicotra F, Russo L (2020) 3D extracellular matrix mimics: fundamental concepts and role of materials chemistry to influence stem cell fate. Biomacromolecules 21:1968–1994

Nicosia RF, Villaschi S (1999) Autoregulation of angiogenesis by cells of the vascular wall. Int Rev Cytol 185:1–43

Nimmagadda S, Loganathan PG, Wilting J, Christ B, Huang R (2004) Expression pattern of VEGFR-2 (Quek1) during quail development. Anat Embryol 208:219–224

Niu G, Chen X (2010) Vascular endothelial growth factor as an anti-angiogenic target for cancer therapy. Curr Drug Targets 11:1000–1017

Noden DM (1989) Embryonic origins and assembly of blood vessels. Am Rev Respir Dis 140: 1097–1103

Nogawa H, Ito T (1995) Branching morphogenesis of embryonic mouse lung epithelium in mesenchyme-free culture. Dev Dent 121:1015–1022

Notter RH (2000) Lung surfactant, basic science and clinical application. Marcel Dekker, New York

Nowotschin S, Hadjantonakis AK (2010) Cellular dynamics in the early mouse embryo: from axis formation to gastrulation. Curr Opin Genet Dev 20:420–427. https://doi.org/10.1016/j.gde.2010.05.008

Nusse R, Varmus HE (1992) Wnt genes. Cell 69:1073–1087

Ochoa-Espinosa A, Affolter M (2012) Branching morphogenesis: from cells to organs and back. Cold Spring Harb Perspect Biol 4:a008243. https://doi.org/10.1101/cshperspect.a008243

Ochs M, Nyengaard JR, Jung A, Knudsen L, Voigt M, Wahlers T, Richter J, Gundersen HJ (2004) The number of alveoli in the human lung. Am J Respir Crit Care Med 169:120–124

O'Farrell PH (2015) Growing an embryo from a single cell: a hurdle in animal life. Cold Spring Harbour Perspect Biol 7:a019042. https://doi.org/10.1101/cshperspect.a019042

Ogawa M, Fraser S, Fujimoto T, Endoh M, Nishikawa S, Nishikawa SI (2001) Origin of hematopoietic progenitors during embryogenesis. Int Rev Immunol 20:21–44

Ohuchi H, Hori Y, Yamasaki M, Harada H, Sekine K, Kato S, Itoh N (2000) FGF-10 acts as a major ligand for FGF receptor 2 IIIb in mouse multi-organ development. Biochem Biophys Res Commun 277:643–649

Okubo T, Hogan BLM (2004) Hyperactive Wnt signaling changes the developmental potential of embryonic lung endoderm. J Biol 3:11–31

Orgeig S, Bernhard W, Biswas SC, Daniels CB, Hall SB, Hetz SK, Lang CJ, Maina JN, Panda AK, Perez-Gill J et al (2007) The anatomy, physical, and physiology of gas exchange surfaces: is there a universal function for pulmonary surfactant in animal respiratory structures? Integr Comp Biol 47:610–627

Ornitz DM, Itoh N (2001) Fibroblast growth factors. Genome Biol 2:REVIEWS3005. https://doi.org/10.1186/gb-2001-2-3-reviews3005

Ornitz DM, Itoh N (2015) The fibroblast growth factor signaling pathway. Dev Biol 4:215–266

Ornitz D, Sannes P (2006) Fibroblast growth factors. In: Lourent GJ, Shapiro SD (eds) Encyclopedia of respiratory medicine. Academic Press, Oxford, pp 210–213

Ortega S, Ittmann M, Tsang SH, Ehrlich M, Basilico C (1998) Neuronal defects and delayed wound healing in mice lacking fibroblast growth factor 2. Proc Natl Acad Sci U S A 95:5672–5677

Oster G, Wang H (2003) Rotary protein motors. Trends Cell Biol 13:114–121

Owen J (1980) Filter-feeding: feeding strategy. University of Chicago Press, Chicago (IL)

Pabon JE, Findley WE, Gibbons WE (1989) The toxic effect of short exposures to the atmospheric oxygen concentration on early mouse embryonic development. Fertil Steril 51:896–900

Padian K (1995) Form and function: the evolution of a dialectic. In: Thomason JJ (ed) Functional morphology and vertebrate paleontology. Cambridge University Press, Cambridge, pp 264–277

Palmer MA, Nelson CM (2017) Epithelial tube fusion as a mechanism for the development of complex lumen-containing organs. Trends Dev Biol 10:57–69

Palmer MA, Nelson CM (2020) Fusion of airways during avian lung development constitutes a novel mechanism for the formation of continuous lumena in multicellular epithelia. Dev Dyn 249:1318–1333

Panchision DM (2009) The role of oxygen in regulating neural stem cells in development and disease. J Cell Physiol 220:562–568

Pansky B (1982) Review of medical embryology. McGraw-Hill, New York

Pardanaud L, Luton D, Prigent M, Bourcheix LM, Catala M, Dieterlen-Lièvre F (1996) Two distinct endothelial lineages in ontogeny, one of them related to hemopoiesis. Development 122:1363–1371

Pardanaud L, Yasiine F, Dieterlen-Lièvre F (1989) Relationship between vasculogenesis, angiogenesis, and haemopiesis during avian ontogeny. Development 105:473–485

Parera MC, van Dooren M, van Kempen M, de Krijger R, Grosveld F, Tibboel D, Rottier R (2005) Distal angiogenesis: a new concept for lung vascular morphogenesis. Am J Physiol Lung Cell Mol Physiol 288:L141–L149

Park WY, Miranda B, Lebeche D, Hashimoto G, Cardoso WV (1998) FGF-10 is a chemotactic factor for distal epithelial buds during lung development. Dev Biol 201:125–134

Parsons JT, Horwitz AR, Schwartz MA (2010) Cell adhesion: integrating cytoskeletal dynamics and cellular tension. Nature Rev Mol Cell Biol 11:633–643

Pascal R, Pross A, Sutherland JD (2013) Towards an evolutionary theory of the origin of life based on kinetics and thermodynamics. Open Biol 3:130156. https://doi.org/10.1098/rsob.130156

Patan S (2000) Vasculogenesis and angiogenesis as mechanisms of vascular network formation, growth and remodeling. J Neuro-Oncol 50:1–15

Patan S (2004) Vasculogenesis and angiogenesis. Cancer Treat Res 117:3–32

Patan S (2021) Peri-endothelial cells, the versatile players in the alternating circular events of vascular growth and differentiation, a switch guided by lycat and cloche, fine tuning angiogenic responsiveness beyond linear hypoxia stimulation. Acad Letts 2418. https://doi.org/10.20935/AL2418

Pattle RE (1978) Lung surfactant and lung lining in birds. In: Piiper J (ed) Respiratory function in birds, adult and embryonic. Springer-Verlag, Berlin, pp 23–32

Pauling MH, Vu TH (2004) Mechanisms and regulation of lung vascular development. Curr Top Dev Biol 64:73–99

Pèault B, Thiery JP, Le Douarin NM (1983) A surface marker for the hemopoietic and endothelial cell lineage in the quail species defined by a monoclonal antibody. Proc Natl Acad Sci U S A 80: 2976–2980

Pearl EJ, Li J, Green JBA (2017) Cellular systems for epithelial invagination. Phil Trans R Soc (Lond). B 372:20150526. https://doi.org/10.1098/rstb.2015.0526

Pearson JT, Seymour RS, Baudinette RV, Runciman S (2002) Respiration and energetics of embryonic development in a large altricial bird, the Australian pelican (*Pelecanus conspicillatus*). J Exp Biol 205:2925–2933

Penaloza C, Lin L, Lockshin RA, Zakeri Z (2006) Cell death in development: shaping the embryo. Histochem Cell Biol 126:149–158

Penaloza C, Orlanski S, Ye Y, Entezari-Zaher T, Javdan M, Zakeri Z (2008) Cell death in mammalian development. Curr Pharm Des 14:184–196

Peng T, Morrisey EE (2013) Development of the pulmonary vasculature: current understanding and concepts for the future. Pulm Circ 3:176–178

Pepicelli CV, Lewis P, McMahon A (1998) Sonic hedgehog regulates branching morphogenesis in the mammalian lung. Curr Biol 8:1083–1086

Perelman RH, Engle MJ, Farrell PM (1981) Perspectives on fetal lung development. Lung 159:53–80

Perl AKT, Hokuto I, Impagnatiello MA, Christofori G, Whitsetta GA (2003) Temporal effects of Sprouty on lung morphogenesis. Dev Biol 258:154–168

Perl AK, Wert SE, Nagy A, Lobe CG, Whitsett JA (2002) Early restriction of peripheral and proximal cell lineages during formation of the lung. Proc Natl Acad Sci U S A 99:10482–10487

Perl AK, Whitsett JA (1999) Molecular mechanisms controlling lung morphogenesis. Clin Genet 56:14–27

Perry SF (1998) Lungs: comparative anatomy, functional morphology, and evolution. In: Gans C, Gaunt AS (eds) Biology of the reptilia, vol 19. Society for the Study of Amphibians and Reptiles, Ithaca (NY), pp 1–92

Perry SF, Duncker HR (1978) Lung architecture, volume and static mechanics in five species of lizards. Respir Physiol 34:61–81

Perry SF, Duncker HR (1980) Interrelationship of static mechanical factors and anatomical structure in lung ventilation. J Comp Physiol 138:321–334

Perry SF, Lambertz M, Schmitz A (2019) Respiratory biology of animals: evolutionary and functional morphology. Oxford University Press, Oxford

Perry SF, Sander M (2004) Reconstructing the evolution of the respiratory apparatus in tetrapods. Respir Physiol Neurobiol 144:125–139

Peters K, Werner S, Liao X, Wert S, Whitsett J, Williams L (1994) Targeted expression of a dominant negative FGF receptor blocks branching morphogenesis and epithelial differentiation of the mouse. EMBO J 13:3296–3301

Petridou N, Spiró Z, Heisenberg CP (2017) Multiscale force sensing in development. Nat Cell Biol 19:581–588

Petrova TV, Makinen T, Alitalo K (1999) Signaling via vascular endothelial growth factor receptors. Exp Cell Res 253:117–130

Petrucci RH, Harwood WS, Herring FG (2002) General chemistry, 8th edtn. Prentice-Hall, Upper Saddle River (NJ)

Pfeifer M, Polakis P (2000) Wnt signaling in oncogenesis and embryogenesis: a look outside the nucleus. Science 287:1606–1609

Phelps CA, Lai SC, Mu D (2018) Roles of thyroid transcription factor 1 in lung cancer biology. Vitam Horm 106:517–544

Pilot F, Lecuit T (2005) Compartmentalized morphogenesis in epithelia: from cell to tissue shape. Dev Dyn 232:685–694

Plosa EJ, Young LR, Gulleman PM, Polosukhin VV, ZaynagetdinovR BJT, Im AM, van der Meer R, Gleaves LA, Bulus N et al (2004) Epithelial β1 integrin is required for lung branching morphogenesis and alveolarization. Development 141:4751–4762

Podkalicka P, Stępniewski J, Mucha O, Kachamakova-Trojanowska N, Dulak J, Łoboda A (2020) Hypoxia as a driving force of pluripotent stem cell reprogramming and differentiation to endothelial cells. Biomol Ther 10:1614. https://doi.org/10.3390/biom10121614

Pogoriler J, Husain AN (2014) Pulmonary development and pediatric lung diseases. In: McManus LM, Mitchell RN (eds) Pathobiology of human disease. Academic Press, San Diego, pp 2575–2587

Polakis P (2012) Wnt signaling in cancer. Cold Spring Harb Perspect Biol 4:a008052. https://doi.org/10.1101/cshperspect.a008052

Poole TJ (1994) Cellular and molecular biology of endothelial cell differentiation during embryonic development. In: Maragoudakis ME, Gullino PM, Lelkes PI (eds) Angiogenesis. NATO ASI series (series a: life sciences), 263. Springer, Boston (MA). https://doi.org/10.1007/978-1-4757-9188-4_2

Poole TJ, Coffin JD (1989) Vasculogenesis and angiogenensis: two distinct morphogenetic mechanisms establish embryonic vascular pattern. J Exp Zool 251:224–231

Poole TJ, Coffin JD (1991) Morphogenetic mechanisms in avian vascular development. In: Feinberg RN, Sherer GK, Auerbach R (eds) The development of the vascular system, vol 14. Karger S, Basel, pp 25–36

Poole TJ, Finkelstein EB, Cox CM (2001) The role of FGF and VEGF in angioblast induction and migration during vascular development. Dev Dyn 220:1–17

Powell FL (2015) Respiration. In: Scanes CG (ed) Sturkie's avain physiology, 6th edn. Elsevier, Amsterdam, pp 301–336

Prado-Lopez S, Conesa A, Arminan A, Martinez-Losa M, Escobedo-Lucea C, Gandia C, Tarazona S, Melguizo D, Blesa D, Montaner D et al (2010) Hypoxia promotes efficient differentiation of human embryonic stem cells to functional endothelium. Stem Cells 28:407–418

Price AP, England KA, Matson AM, Blazar BR, Panoskaltsis-Mortari A (2010) Development of a decellularized lung bioreactor system for bioengineering the lung: the matrix reloaded. Tissue Engineer A 16:2581–2591

Prigogine I, Stengers I (1984) Order out of chaos: Man's new dialogue with nature. Flamingo, London

Radinsky LB (1987) The evolution of vertebrate design. University of Chicago Press, Chicago

Ralston A and Shaw K (2008) Gene expression regulates cell differentiation. Nature Ed 1, 127. https://www.nature.com/scitable/topicpage/gene-expression-regulates-cell-differentiation-931/. Accessed on 10–07–22

Ramakrishnan AB, Cadigan KM (2017) Wnt target genes and where to find them. F1000Res 6:746. https://doi.org/10.12688/f1000research.11034.1

Ramasamy SK, Mailleux AA, Gupte VV, Mata F, Sala FG, Veltmaat JM, Del Moral PM, De Langhe S, Parsa S, Kelly LK et al (2007) FGF10 dosage is critical for the amplification of epithelial cell progenitors and for the formation of multiple mesenchymal lineages during lung development. Dev Biol 307:237–247

Ramsbottom SA, Pownall ME (2016) Regulation of hedgehog signalling inside and outside the cell. J Dev Biol 4:23. https://doi.org/10.3390/jdb4030023

Randall DJ, Burggren WW, Farrell AP, Haswell MS (1981) The evolution of air-breathing in vertebrates. Cambridge University Press, Cambridge

Raś M, Iwan D, Kamiński MJ (2018) The tracheal system in postembryonic development of holometabolous insects: a case study using mealworm beetle. J Anat 232:997–1015

Rawlins EL, Hogan BL (2006) Epithelial stem cells of the lung: privileged few or opportunities for many? Development 133:2455–2465

Ray HJ, Niswander L (2012) Mechanisms of tissue fusion during development. Development 139: 1701–1711

Ray SD, Mehendale HM (2005) Apoptosis. In: Wexler P (ed) Encyclopedia of toxicology, 2nd edn. Elsevier, New York, pp 153–167

Reagan FP (1916) A further study of the origin of blood vascular tissues in chemically treated teleost embryos, with special reference to haematopoiesis in the anterior mesenchyme and in the heart. Anat Rec 10:99–118

Reidy KJ, Rosenblum ND (2009) Cell and molecular biology of kidney development. Semin Nephrol 29:321–337

Ribatti D, Pezzella F (2021) Overview on the different patterns of tumor vascularization. Cell 10: 639. https://doi.org/10.3390/cells10030639

Ricklefs RE, Starck JM (1998) Embryonic growth and development. In: Starck JM, Ricklefs RE (eds) Avian growth and development, evolution within the altricial-precocial spectrum. Oxford University Press, New York, pp 31–58

Risau W (1997) Mechanisms of angiogenesis. Nature 386:671–674

Risau W, Flamme I (1995) Vasculogenesis. Annu Rev Cell Dev Biol 11:73–91

Roca-Cusachs P, Conte V, Trepat X (2017) Quantifying forces in cell biology. Nat Cell Biol 19: 742–751

Rockich BE, Hryca SM, Shih HP, Nagy MS, Ferguson MAH, Kopp JL, Sander M, Wellik DM, Spence JR (2013) Sox9 plays multiple roles in the lung epithelium during branching morphogenesis. Proc Natl Acad Sci U S A 110:E4456–E4464

Rogers KW, Schier AF (2011) Morphogen gradients: from generation to interpretation. Annu Rev Cell Dev Biol 27:377–407

Roman J (1997) Cellular mechanisms of lung development. In: Fishman A (ed) Pulmonary diseases and disorders. McGraw Hill, New York, pp 73–89

Román-Fernández A, Bryant DM (2016) Complex polarity: building multicellular tissues through apical membrane traffic. Traffic 17:1244–1261

Romani P, Valcarcel-Jimenez L, Frezza C, Dupont S (2021) Crosstalk between mechanotransduction and metabolism. Nat Rev Mol Cell Biol 22:22–38

Romanoff AL (1960) The avian embryo. Macmillan, New York

Rosenquist GC (1970) The origin and movement of prelung cells in the chick embryo as determined by radioautographic mapping. Embryol Exp Morph 24:497–509

Ross MJ (1899) Special structural features in the air-sacs of birds. Trans Amer Micr Soc 20:29–40

Roth-Kleiner M, Post M (2003) Genetic control of lung development. Biol Neonate 84:83–88

Rubin LP, Kovacs CS, De Paepe ME, Tsai SW, Torday JS, Kronenberg HM (2004) Arrested pulmonary alveolar cytodifferentiation and defective surfactant synthesis in mice missing the gene for parathyroid hormone-related protein. Dev Dyn 230:278–289

Rudnick D (1933) Developmental capacities of chick lung in chorioallantoic grafts. J Exp Zool 66: 125–154

Runciman S, Seymour RS, Baudinette RV, Pearson JT (2005) An allometric study of lung morphology during development in the Australian pelican, *Pelicanus conspicillatus*, from embryo to adult. J Anat 207:365–380

Sabin FR (1917) Preliminary note on the differentiation of angioblasts and the method by which they produce blood vessels, blood plasma and red blood cells as seen in the living chick. Anat Rec 13:199–204

Sabin FR (1920) Studies on the origin of blood vessels and of red blood corpuscles as seen in the living blastoderm of chicks during the second day of incubation. Contrib Embryol 36:213–261

Sadler TW (2010) The embryologic origin of ventral body wall defects. Semin Pediatr Surg 19: 209–214

Sagai T, Amano T, Maeno A, Kimura T, Nakamoto M, Takehana Y, Naruse K, Okada N, Kiyonari H, Shiroishi T (2017) Evolution of shh endoderm enhancers during morphological transition from ventral lungs to dorsal gas bladder. Nat Commun 8:14300. https://doi.org/10.1038/ncomms14300

Sagai T, Amano T, Tamura M, Mizushina Y, Sumiyama K, Shiroishi T (2009) A cluster of three long-range enhancers directs regional shh expression in the epithelial linings. Development 136: 1665–1674

Sahasrabudhe N, Gosney J, Hasleton P (2000) The normal lung: histology, embryology, development, aging and function. In: Hasleton P, Flieder D (eds) Spencer's pathology of the lung. Cambridge University Press, Cambridge, pp 1–40

Sakaue H, Konishi M, Ogawa W, Asaki T, Mori T, Yamasaki M, Takata M, Ueno H, Kato S, Kasuga M et al (2002) Requirement of fibroblast growth factor 10 in development of white adipose tissue. Genes Dev 16:908–912

Sakiyama J-I, Yamagishi A, Kuroiwa A (2000) Coordinated expression of hoxb genes and signaling molecules during development of chick respiratory tract. Dev Biol 227:12–27

Sakiyama J-I, Yamagishi A, Kuroiwa A (2003) Tbx-Fgf10 system controls lung bud formation during chicken embryonic development. Development 130:1225–1234

Sander PM, Christian A, Clauss M, Fechner R, Gee CT, Griebeler EM, Gunga HC, Hummel J, Mallison H, Perry SF et al (2011) Biology of the sauropod dinosaurs: the evolution of gigantism. Biol Rev 86:117–155

Sanghvi-Shah R, Weber GF (2017) Intermediate filaments at the junction of mechanotransduction, migration, and development. Front Cell Dev Biol 5:81. https://doi.org/10.3389/fcell.2017.00081

Sano H, Kuroki Y (2005) The lung collectins, SP-A and SP-D, modulate pulmonary innate immunity. Mol Immunol 42:279–287

Saraste A, Pulkki K (2000) Morphologic and biochemical hallmarks of apoptosis. Cardiovasc Res 45:528–537

Sawyer JM, Harrell JR, Shemer G, Sullivan-Brown J, Roh-Johnson M, Goldstein B (2010) Apical constriction: a cell shape change that can drive morphogenesis. Dev Biol 341:5–19

Scaltsas T (1990) Is a whole identical to its parts? Mind 99:583–598

Scanes CG, Witt J, Ebeling M, Schaller S, Baier V, Bone AJ, Preuss TG, Heckmann D (2022) Quantitative comparison of avian and mammalian physiologies for parameterization of physiologically based kinetic models. Front Physiol 13:858386. https://doi.org/10.3389/fphys.2022.858386

Scarpelli EM (1988) Surfactants and the lining of the lung. Johns Hopkins University Press, Baltimore

Schatteman GC (2004) Adult bone marrow-derived hemangioblasts, endothelial cell progenitors, and EPCs. Curr Top Dev Biol 64:141–180

Schatteman GC, Awad O (2004) Hemangioblasts, angioblasts, and adult endothelial cell progenitors. Anat Rec 276:13–21

Scheid P (1979) Mechanisms of gas exchange in bird lungs. Rev Physiol Biochem Pharmacol 86: 137–186

Scheid P, Piiper J (1989) Respiratory mechanics and air flow in birds. In: King AS, McLelland J (eds) Form and function in birds, vol 4. Academic Press, London, pp 369–391

Schepers GE, Teasdale RD, Koopman P (2002) Twenty pairs of sox: extent, homology, and nomenclature of the mouse and human sox transcription factor gene families. Dev Cell 3: 167–170

Schittny J (2017) Development of the lung. Cell Tissue Res 367:427–444

Schittny J, Burri PH (2004) Morphogenesis of the mammalian lung: aspects of structure and extracellular matrix. In: Massaro DJ, Massaro GC, Chambon P (eds) Lung development and regeneration. Marcel Dekker Inc, New York, pp 275–316

Schittny JC, Burri PH (2008) Development and growth of the lung. In: Fishman AP, Elias JA, Fishman JA, Grippi MA, Kaiser LR, Senior RM (eds) Fishman's pulmonary diseases and disorders, vol 1. McGraw-Hill, New York, pp 91–114

Schittny JC, Mund SI (2008) A re-examination of the maturation of the alveolar septa revealed that microvascular maturation takes place in parallel to alveolarization. Am J Respir Crit Care Med 177:A317

Schittny JC, Mund SI, Stampanoni M (2008) Evidence and structural mechanism for late lung alveolarization. Am J Physiol Lung Cell Mol Physiol 294:L246–L254

Schittny J, Djonov V, Fine A, Burri PH (1998) Programmed cell death contributes postnatal lung development. Am J Respir Cell Mol Biol 18:786–793

Schliwa M (ed) (2003) Molecular motors. Wiley, Darmstadt

Scholtz G (2008) Scarab beetles at the interface of wheel invention in nature and culture? Contr Zool 77:139–148

Schreiber A, Gimbel S (2010) Evolution and the second law of thermodynamics: effectively communicating to non-technicians. Evo Edu Outreach 3:99–106

Schwenk K (2000) Feeding. Form, function and evolution in tetrapod vertebrates. Academic Press, San Diego

Seghezzi G, Patel S, Ren CJ, Gualandris A, Pintucci G, Robbins ES, Shapiro RL, Galloway AC, Rifkin DB, Mignatti P (1998) Fibroblast growth factor-2 (FGF-2) induces vascular endothelial growth factor (VEGF) expression in the endothelial cells of forming capillaries: an autocrine mechanism contributing to angiogenesis. J Cell Biol 141:1659–1673

Sekine K, Ohuchi H, Fujiwara M, Yamasa K, Yoshizawa T, Sato T, Yagishita N, Matsui D, Koga Y, Itoh N et al (1999) FGF-10 is essential for limb and lung formation. Nat Gent 21:138–141

Selenka E (1866) Beitrag zur Entwickelungsgeschichte der Luffsache des Huhn. Zeitsehr f wiss Zo 51:178–182

Seller TJ (ed) (1987) Bird respiration, vol I and II. CRC Press, Boca Raton (FL)

Sellier N, Brillard JP, Dupuy V, Bakst MR (2006) Comparative staging of embryo development in chicken, Turkey, duck, goose, Guinea fowl, and japanese quail assessed from five hours after fertilization through seventy-two hours of incubation. J Appl Poult Res 15:219–228

Semenov MV, Habas R, MacDonald BT (2007) Noncanonical Wnt signaling pathways. Cell 131: 1378. https://doi.org/10.1016/j.cell.2007.12.011

Serls AE, Doherty S, Parvatiyar P, Wells JM, Deutsch GH (2005) Different thresholds of fibroblast growth factors pattern the ventral foregut into liver and lung. Development 132:35–47

Seymour RS, Runciman S, Baudinette RV (2008) Development of maximum metabolic rate and pulmonary diffusing capacity in the superprecocial Australian brush Turkey *Alectura lathami*: an allometric and morphometric study. Comp Biochem Physiol A Mol Integr Physiol 150:169–175

Seymour RS, Runciman S, Baudinette RV, Pearson JT (2004) Developmental allometry of pulmonary structure and function in the altricial Australian pelican *Pelecanus conspicillatus*. J Exp Biol 207:2663–2669

Shah MM, Sampogna RV, Sakurai H, Bush KT, Nigam SK (2004) Branching morphogenesis and kidney disease Development 131:1449–1462

Shalaby F, Rossant J, Yamaguchi TP, Gertsenstein M, Wu XF, Breitman M, Schuh AC (1995) Failure of blood-Island formation and vasculogenesis in Flk-1-deficient mice. Nature 376:62–66

Shannon JM, Hyatt BA (2004) Epithelial-mesenchymal interactions in the developing lung. Annu Rev Physiol 66:625–645

Shannon JM, Nielsen LD, Gebb SA, Randell SH (1998) Mesenchyme specifies epithelial differentiation in reciprocal recombinants of embryonic lung and trachea. Dev Dyn 212:482–494

Shook D, Keller R (2003) Mechanisms, mechanics and function of epithelial-mesenchymal transitions in early development. Mech Dev 120:1351–1383

Short KM, Smyth IM (2016) Quantification of developmental branching morphogenesis. In: Little MH (ed) Kidney development, disease, repair and regeneration. Elsevier Inc., New York, pp 57–65

Shimoda LA (2006) Pulmonary circulation. In: Laurent GJ, Shapiro SD (eds) Encyclopedia of medicine. Elsevier Ltd, New York, pp 537–544

Shu W, Jiang YQ, Lu MM, Morrisey EE (2002) Wnt7b regulates mesenchymal proliferation and vascular development in the lung. Development 129:4831–4842

Schuger L, Skubitz AP, Gilbride K, Mandel R, He L (1996) Laminin and heparan sulfate proteoglycan mediate epithelial cell polarization in organotypic cultures of embryonic lung cells: evidence implicating involvement of the inner globular region of laminin beta 1 chain and the heparan sulfate groups of heparan sulfate proteoglycan. Dev Biol 179:264–273

Shum H (2019) Microswimmer propulsion by two steadily rotating helical flagella. Micromachines 10(1):65. https://doi.org/10.3390/mi10010065

Sigurbjörnsdóttir S, Mathew R, Leptin M (2014) Molecular mechanisms of de novo lumen formation. Nat Rev Mol Cell Biol 15:665–676

Simon MC, Keith B (2008) The role of oxygen availability in embryonic development and stem cell function. Nat Rev Mol Cell Biol 9:285–296

Simonet WS, DeRose ML, Bucay N, Nguyen HQ, Wert SE, Zhou L, Ulich TR, Thomason A, Danilenko DM, Whitsett JA (1995) Pulmonary malformation in transgenic mice expressing human keratinocyte growth factor in the lung. Proc Natl Acad Sci U S A 92:12461–12465

Skoda AM, Simovic D, Karin V, Kardum V, Vranic S, Serman L (2018) The role of the hedgehog signaling pathway in cancer: a comprehensive review. Bosnian J Basic Med Sci 18:8–20

Slack J (2021) Stem cells: a very short introduction, 2nd edn. Oxford University Press, Oxford (UK)

Sorokin SP (1970) The cells of the lungs. In: Nettesheim P, Hanna MG, Deatherage JW (eds) Morphology of experimental respiratory carcinogenesis. Proceedings of a Biology Division, Oak Ridge National Laboratory Conference; Gatlinburg (TN), pp 3–43

Sorokin SP (1977) The respiratory system. In: Weiss L, Greep RD (eds) Histology. McGraw Hill Book Company, New York, pp 765–782

Spencer H, Shorter RG (1962) Cell turnover in pulmonary tissues. Nature 194:880. https://doi.org/10.1038/194880a0

Spicer JI, Burggren WW (2003) Development of physiological regulatory systems: altering the timing of crucial events. Zoology 106:91–99

Spinks L (2003) Friedrich Nietzsch. Taylor and Francis, Boca Raton

Spooner BS, Wessells NK (1970) Mammalian lung development: interactions in primordium formation and bronchial morphogenesis. J Exp Zool 175:445–454

Spurlin JW, Nelson CM (2017) Building branched tissue structures: from single cell guidance to coordinated construction. Philos Trans R Soc Lond Ser B Biol Sci 372:20150527. https://doi.org/10.1098/rstb.2015.0527

Spurlin JW, Siedlik MJ, Nerger BA, Pang MF, Jayaraman S, Zhang R, Nelson CM (2019) Mesenchymal proteases and tissue fluidity remodel the extracellular matrix during airway epithelial branching in the embryonic avian lung. Development 146:dev175257. https://doi.org/10.1242/dev.175257

Stabellini G, Locci P, Calvitti M, Evangelisti R, Marinucci L, Bodo M, Carusio A, Canaider S, Carinci P (2001) Epithelial-mesenchymal interactions and lung branching morphogenesis, role of polyamines and transforming growth factor beta1. Eur J Histochem 45:151–162

Stamati K, Mudera V, Cheema U (2011) Evolution of oxygen utilization in multicellular organisms and implications for cell signalling in tissue engineering. J Tissue Eng 2:2041731411432365. https://doi.org/10.1177/2041731411432365

Stamenović D, Wang N (1985) Invited review: engineering approaches to cytoskeletal mechanics. J Appl Physiol 89:2085–2090

Starck JM, Ricklefs RE (1998) Avian growth and development: evolution within the altricial-precocial spectrum. Oxford University Press, New York

Stern CD (2004) The chick embryo–past, present and future as a model system in developmental biology. Mech Dev 121:1011–1013

Stern CD (2005) The chick; a great model system becomes even greater. Dev Cell 8:9–17

Stewart K, Morrisey E (2016) Early development of the mammalian lung-branching morphogenesis. In: Jobe A, Whitsett J, Abman S (eds) Fetal and neonatal lung development: clinical correlates and technologies for the future. Cambridge University Press, Cambridge, pp 22–33

Stiles AD, Chrysis D, Jorvis HW, Brighton B, Moats-Staats BM (2001) Programmed cell death in normal fetal rat lung development. Exp Lung Res 25:569–587

St Johnston D (2015) The renaissance of developmental biology. PLoS Biol 13:e1002149. https://doi.org/10.1371/journal.pbio.1002149

St Johnston D, Sanson B (2011) Epithelial polarity and morphogenesis. Curr Opin Cell Biol 23: 540–546

Stock MK, Francisco DL, Metcalfe J (1983) Organ growth in chick embryos incubated in 40% or 70% oxygen. Respir Physiol 52:1–11

Stock MK, Metcalfe J (1984) Stimulation of growth of the chick embryo by acute hyperoxia. Respir Physiol 58:351–358

Stockard CR (1915) The origin of blood and vascular endothelium in embryos without a circulation of blood and in the normal embryo. Am J Anat 18:227–325

Stone KC, Mercer RR, Gehr P, Stockstill B, Crapo JD (1992) Allometric relationships of cell numbers and size in the mammalian lung. Am J Respir Cell Mol Biol 6:235–243

Strzyz P (2007) Pulling the apoptotic trigger for necrosis. Nat Rev Mol Cell Biol 18:72. https://doi.org/10.1038/nrm.2017.1

Styler DF (2008) Entropy and evolution. Am J Phys 76:1031–1033

Suzuki M, Morita H, Ueno N (2012) Molecular mechanisms of cell shape changes that contribute to vertebrate neural tube closure. Develop Growth Differ 54:266–276

Suzuki K, Yamanishi K, Mori O, Kamikawa M, Andersen B, Kato S, Toyoda T, Yamada G (2000) Defective terminal differentiation and hypoplasia of the epidermis in mice lacking the Fgf10 gene. FEBS Lett 481:53–56

Swarr DT, Morrisey EE (2015) Lung endoderm morphogenesis: gasping for form and function. Annu Rev Cell Dev Biol 31:553–573

Sylva M, Vivian SW, Buffing AAA, van Es JH, van den Born M, van der Velden S, Gunst Q, Koolstra JH, Moorman AFM, Clevers H et al (2011) The BMP antagonist follistatin-like 1 is required for skeletal and lung organogenesis. PLoS One 6:e22616. https://doi.org/10.1371/journal.pone.0022616

Szebenyi G, Fallon JF (1999) Fibroblast growth factors as multifunctional signaling factors. Int Rev Cytology 185:45–106

Takada M, Sagawa F, Chi NC, Endo S, Kozawa S, Sato TN (2017) Re-evaluating the functional landscape of the cardiovascular system during development. Bio Open 6:1756–1770

Tallquist MD, Soriano P, Klinghoffer RA (1999) Growth factor signaling pathways in vascular development. Oncogene 18:7917–7932

Tan CM, Lewandowski AJ (2020) The transitional heart: from early embryonic and fetal development to neonatal life. Fetal Diagn Ther 47:373–386

Taichman DB, Loomes KM, Schachtner SK, Guttentag S, Vu C, Williams P, Oakley RJ, Baldwin HS (2002) Notch 1 and jagged 1 expression by the developing pulmonary vasculature. Dev Dyn 225:166–175

Tata PR, Rajagopal J (2017) Plasticity in the lung: making and breaking cell identity. Development 144:755–766

Tatsumi N, Kobayashi R, Yano T, Noda M, Fujimura K, Okada N, Okabe M (2016) Molecular developmental mechanism in polypterid fish provides insight into the origin of vertebrate lungs. Sci Rep 6:30580. https://doi.org/10.1038/srep30580

Tazawa H, Hashimoto Y, Nakazawa S, Whittow GC (1992) Metabolic responses of chicken embryos and hatchlings to altered oxygen environments. Respir Physiol 88:37–50

Tazawa H, Okuda A, Nakazawa S, Whittow GC (1989) Metabolic responses of chicken embryos to graded, prolonged alterations in ambient temperature. Comp Biochem Physiol 92:613–617

Ten Have-Opbroek AAW (1981) The development of the lung in mammals: an analysis of concepts and findings. Am J Anat 162:201–219

Ten Have-Opbroek AA (1991) Lung development in the mouse embryo. Exp Lung Res 17:111–130

Thompson D'AW (1917) On growth and form, 1st edtn. Cambridge University Press, Cambridge (UK)

Thompson DW (1942) On growth and form, 2nd edn. Cambridge University Press, Cambridge

Toews GB (2001) Cytokines and the lung. Europ Respir J 18:3s–17s. https://doi.org/10.1183/09031936.01.00266001

Tomanek RJ, Holifield JS, Reiter RS, Sandra A, Lin JJC (2002) Role of VEGF family members and receptors in coronary vessel formation. Dev Dyn 225:233–240

Tona K, Voemesse K, N'nanlé O, Oke OE, Kouame YAE, Bilalissi A, Meteyake H, Oso OM (2022) Chicken incubation conditions: role in embryo development, physiology and adaptation to the posthatch environment. Front Physiol 13:895854. https://doi.org/10.3389/fphys.2022.895854

Toussaint O, Schneider ED (1998) The thermodynamics and evolution of complexity in biological systems. Comp Biochem Physiol A Mol Integr Physiol 120:3–9

Travaglini KJ, Nabhan AN, Penland L, Sinha R, Glilich A, Sit RV, Chang S, Conley SD, Mori Y, Seita J et al (2020) A molecular cell atlas of the human lung from single-cell RNA sequencing. Nature 587:619–625

Tschanz SA, Salm LA, Roth-Kleiner M, Barre SF, Burri PH, Schittny JC (2014) Rat lungs show a biphasic formation of new alveoli during postnatal development. J Appl Physiol 117:89–95

Tsao PN, Vasconcelos M, Izvolsky KI, Qian J, Lu J, Cardoso WV (2009) Notch signaling controls the balance of ciliated and secretory cell fates in developing airways. Development 136:2297–2307

Turcatel G, Rubin N, Menke DB, Martin G, Shi W, Warburton D (2013) Lung mesenchymal expression of Sox9 plays a critical role in tracheal development. BMC Biol 11:117. http://www.biomedcentral.com/1741-7007/11/117. Accessed on 10-07-22

Turing AM (1952) The chemical basis of morphogenesis. Philos Trans R Soc Lond B 237:37–72

Tyler SEB (2014) The work surfaces of morphogenesis: the role of the morphogenetic field. Biol Theory 9:194–208

Tzou D, Spurlin JW, Pavlovich AL, Stewart CR, Gleghorn JP, Nelson CM (2016) Morphogenesis and morphometric scaling of lung airway development follows phylogeny in chicken, quail, and duck embryos. EvoDevo 7:12. https://doi.org/10.1186/s13227-016-0049-3

Uçar MC, Kamenev D, Sunadome K, Fachet D, Lallemend F, Adameyko I, Hadjab S, Hannezo E (2021) Theory of branching morphogenesis by local interactions and global guidance. Nat Commun 12:6830. https://doi.org/10.1038/s41467-021-27135-5

Ucuzian AA, Gassman AA, East AT, Greisler HP (2010) Molecular mediators of angiogenesis. J Burn Care Res 31:158–175

Udan RS, Culver JC, Dickinson ME (2013) Understanding vascular development. Wiley Interdiscip Rev Dev Biol 2:327–346

Uhler C, Shivashankar G (2017) Regulation of genome organization and gene expression by nuclear mechanotransduction. Nat Rev Mol Cell Biol 18:717–727

Ulloa F, Martí E (2010) Wnt won the war: antagonistic role of Wnt over shh controls dorso-ventral patterning of the vertebrate neural tube. Dev Dyn 239:69–76

Ungerer MC, Johnson LC, Herman MA (2008) Ecological genomics: understanding gene and genome function in the natural environment. Heredity 100:178–183

Urbich C, Dimmeler S (2004) Endothelial progenitor cells functional characterization. Trends Cardiovasc Med 14:318–322

Vailhé B, Lecomte M, Wiernsperger N, Tranqui L (1998) The formation of tubular structures by endothelial cells is under the control of fibrinolysis and mechanical factors. Angiogenesis 2: 331–344

Vailhé B, Vittet D, Feige JJ (2001) *In vitro* models of vasculogenesis and angiogenesis. Lab Investig 81:439–452

van Amerongen R, Nusse R (2009) Towards an integrated view of Wnt signaling in development. Development 136:3205–3214

Van Tuyl MV, Del Riccio V, Post M (2004) Lung branching morphogenesis: potential for regeneration of small conducting airways. In: Massaro DJ, Massaro GC, Chambon P (eds) Lung development and regeneration. Marcel Dekker Inc., New York, pp 355–393

Varjosalo M, Taipale J (2008) Hedgehog: functions and mechanisms. Genes Dev 22:2454–2472

Varner VD, Nelson CM (2014) Cellular and physical mechanisms of branching morphogenesis. Development 141:2750–2759

Veikkola T, Alitalo K (1999) VEGFs, receptors and angiogenesis. Semin Cancer Biol 9:211–220

Veikkola T, Karkkainen M, Claesson-Welsh L, Alitalo K (2000) Regulation of angiogenesis via vascular endothelial growth factor receptors. Cancer Res 60:203–212

Vergara MN, Canto-Soler MV (2012) Rediscovering the chick embryo as a model to study retinal development. Neural Dev 7:22. https://doi.org/10.1186/1749-8104-7-22

Vining K, Mooney D (2017) Mechanical forces direct stem cell behaviour in development and regeneration. Nat Rev Mol Cell Biol 18:728–742

Vizcaíno SF, Bargo MS (2021) Views on the form-function correlation and biological design. J Mammal Evol 28:15–22

Vleck CM, Hoyt DF (1991) Metabolism and energetics of reptilian and avian embryos. In: Deeming DC, MWJ F (eds) Egg incubation: its effects on embryonic development in birds and reptiles. Cambridge University Press, New York, pp 285–306

Vleck CM, Kenagy GJ (1980) Embryonic metabolism of the fork-tailed storm petrel: physiological patterns during prolonged and interrupted incubation. Physiol Zool 53:32–42

Vleck CM, Vleck D (1987) Metabolism and energetics of avian embryos. J Exp Zool Suppl 1:111–125

Vleck CM, Hoyt DF, Vleck D (1979) Metabolism of avian embryos: patterns in altricial and precocial birds. Physiol Biochem Zool 52:363–377

Vleck CM, Vleck D, Hoyt DF (1980) Patterns of metabolism and growth in avian embryos. Amer Zool 20:405–416

Voelkel NF, Vandvier RW, Tuder RM (2006) Vascular endothelial growth factor in the lung. Am J Physiol Lung Cell Mol Physiol 290:L209–L221

Waddington CH (1957) (reissued 2014) the strategy of the genes: a discussion of some aspects of theoretical biology. Routledge, Taylor and Francis Group, London

Wake MH (2008) Integrative biology: science for the 21st century. Bioscience 58:349–353

Walker I (1991) Why legs and not wheels? Acta Biotheor 39:151–155

Wang L, Li L, Shojaei F, Levac K, Cerdan C, Menendez P, Martin T, Rouleau A, Bhatia M (2004) Endothelial and hematopoietic cell fate of human embryonic stem cells originates from primitive endothelium with hemangioblastic properties. Immunity 21:31–41

Warburton D (2008) Developmental biology: order in the lung. Nature 453:733–735

Warburton D (2021) Conserved mechanisms in the formation of the airways and alveoli of the lung. Front Cell Dev Biol 9:662059. https://doi.org/10.3389/fcell.2021.662059

Warburton D, Bellusci S (2004) The molecular genetics of lung morphogenesis and injury repair. Paediatr Respir Rev Suppl 5A:S283–S287. https://doi.org/10.1016/s1526-0542(04)90052-8

Warburton D, Bellusci S, De Langhe S, Del Moral PM, Fleury V, Mailleaux A, Tefft D, Unbekandt M, Wang K, Shi W (2005) Molecular mechanisms of early lung specification and branching morphogenesis. Pediatr Res 57:26–37

Warburton D, Bellusci S, Del Moral PM, Kaartinen V, Lee MD, Shi W (2003) Growth factor signaling in lung morphogenetic centers automaticity, stereotypy and symmetry. Respir Res 19: 294–315

Warburton D, El-Hashash A, Carraro G, Tiozzo C, Sala F, Rogers O, De Langhe S, Kemp PJ, Riccardi D, Torday J et al (2010) Lung organogenesis. Curr topics. Dev Biol 90:73–158

Warburton D, Schwarz M, Tefft D, Flores-Delgado G, Anderson KD, Cardoso WV (2000) The molecular basis of lung morphogenesis. Mech Dev 92:55–81

Warburton D, Tefft D, Mailleux A, Bellusci S, Thiery JP, Zhao J, Buckley S, Shi W, Driscoll B (2001) Do lung remodeling, repair, and regeneration recapitulate respiratory ontogeny? Amer J Respir Crit Care Med 164:S59–S62

Warburton D, Wuenschell C, Flores-Delgado G, Anderson K (1998) Commitment and differentiation of lung cell lineages. Biochem Cell Biol 76:971–995

Watson RR, Fu Z, West JB (2008) Minimal distensibility of pulmonary capillaries in avian lungs compared with mammalian lungs. Respir Physiol Neurobiol 160:208–214

Weaver M, Dunn NR, Hogan BL (2000) BMP-4 and FGF-10 play opposing roles during lung bud morphogenesis. Development 127:2695–2704

Webster WS, Abela D (2007) The effect of hypoxia in development. Embryo Today Revs 81:215–228

Wedel MJ (2003) Vertebral pneumaticity, air sacs, and the physiology of sauropod dinosaurs. Paleobiology 29:243–255

Weibel ER (1984) The pathways for oxygen. Harvard University Press, Harvard (MA)

Weibel ER (1991) Fractal geometry: a design principle for living organisms. Amer J Physiol Lung Cell Mol Physiol 261:L361–L369

Weibel ER (1994) Design of biological organisms and fractal geometry. In: Nonnenmacher TF, Losa GA, Weibel ER (eds) Fractals in biology and medicine, vol 1. Birkhäuser, Basel, pp 68–85

Weibel ER (2000) Symmorphosis: on form and function in shaping life. Harvard University Press, Cambridge (MA)

Weibel ER (2005) Mandelbrot's fractals and the geometry of life: a tribute to Benoît Mandelbrot on his 80th birthday. In: Losa GA, Merlini D, Nonnenmacher TF, Weibel ER (eds) Fractals in biology and medicine. Mathematics and Biosciences in Interaction, Birkhäuser (Basel), pp 3–16

Weibel ER, Taylor CR, Bolis L (eds) (1998) Principles of animal design: the optimization and symmorphosis debate. Cambridge University Press, Cambridge (UK)

Weibel ER, Taylor CR, Hoppeler H (1991) The concept of symmorphosis: a testable hypothesis of structure-function relationship. Proc Natl Acad Sci U S A 88:10357–10361

Weidenfeld J, Shu W, Zhang L, Millar SE, Morrisey EE (2002) The Wnt7b promoter is regulated by TFT-1, GAATA6, and Foxa2 in lung epithelium. J Bio Chem 277:21061–21070

Weinstein M, Xu X, Ohyama K, Deng CX (1998) FGFR-3 and FGFR-4 function coorperatively to direct alveogenesis in the murine lung. Development 125:3615–3623

Weis-Fogh T (1964a) Functional design of the tracheal system of flying insects as compared with the avian lung. J Exp Biol 41:207–227

Weis-Fogh T (1964b) Diffusion in insect wing muscle, the most active tissue known. J Exp Biol 41: 229–256

Weis-Fogh T (1967) Respiration and tracheal ventilation in locusts and other flying insects. J Exp Biol 47:561–587

Wessells NK (1970) Mammalian lung development: interactions in formation and morphogenesis of tracheal buds. J Exp Zool 175:455–466

West-Eberhard MJ (2005) Phenotypic accommodation: adaptive innovation due to developmental plasticity. J Exp Zool 304B:610–618

West GB, Brown JH, Enquist BJ (1991) The fourth dimension of life: fractal geometry and allometric scaling of organisms. Science 284:1677–1679

West JB, Fu Z, Gu Y, Wagner HE, Carr JA, Peterson KL (2010) Pulmonary artery pressure responses to increased cardiac output in chickens with raised metabolic rate. Comp Biochem Physiol A Mol Integr Physiol 156:430–435

West JB, Watson RR, Fu Z (2006) The honeycomb-like structure of the bird lung allows a uniquely thin blood-gas barrier. Respir Physiol Neurobiol 152:115–118

West JB, Watson RR, Fu Z (2007) Major differences in the pulmonary circulation between birds and mammals. Respir Physiol Neurobiol 157:382–390

West NH, Bamford OS, Jones DR (1977) A scanning electron microscope study of the microvasculature of the avian lung. Cell Tissue Res 176:553–564

Weston EM (2003) Evolution of ontogeny in the hippopotamus skull: using allometry to dissect developmental change. Biol J Linn Soc 80:625–638

White S, Danowitz M, Solounias N (2016) Embryology and evolutionary history of the respiratory tract. Edorium J Anat Embryo 3:54–62

Whitsett JA, Haitchi HM, Maeda Y (2011) Intersection between pulmonary development and disease. Amer J Respir Crit Care Med 184:401–406

Whitsett JA, Kalin TV, Xu Y, Kalinichenko VV (2019) Building and regenerating the lung cell by cell. Physiol Rev 99:513–554

Whittow GC (1980) Physiological and ecological correlates of prolonged incubation in sea birds. Amer Zool 20:427–436

Widelitz R (2005) Wnt signaling through canonical and non-canonical pathways: recent progress. Growth Factors 23:111–116

Willenborg C, Prekeris R (2011) Apical protein transport and lumen morphogenesis in polarized epithelial cells. Biosci Rep 31:245–256

Willert K, Nusse R (2012) Wnt proteins. Cold Spring Harb Perspect Biol 4:a007864. https://doi.org/10.1101/cshperspect.a007864

Wilson NH, Stoeckli ET (2012) Sonic hedgehog regulates Wnt activity during neural circuit formation. Vitam Horm 88:173–209

Wilting J, Becker J (2006) Two endothelial cell lines derived from the somite. Anat Embryol 1:57–63

Witsch E, Sela M, Yarden Y (2010) Roles for growth factors in cancer progression. Physiology 25: 85–101

Wittenberg RW (1993) Models of self-organization in biological development. MSc Thesis,. University of Cape Town

Wittig JG, Münsterberg A (2016) The early stages of heart development: insights from chicken embryos. J Cardiovasc Dev Dis 3:12. https://doi.org/10.3390/jcdd3020012

Wodarz A, Nusse R (1998) Mechanisms of Wnt signaling in development. Annu Rev Cell Biol 14: 59–88

Wolf YI, Katsnelson MI, Koonina EV (2018) Physical foundations of biological complexity. Proc Natl Acad Sci U S A 115:E8678–E8687

Wongtrakool C, Malpel S, Gorenstein J, Sedita J, Ramirez MI, Underhill TM, Cardoso WV (2003) Down-regulation of retinoic acid receptor α signaling is required for sacculation and type-1 cell formation in the developing lung. J Biol Chem 278:46911–46918

Wongtrakool C, Roman J (2008) Apoptosis of mesenchymal cells during the pseudoglandular stage of lung development affects branching morphogenesis. Exp Lung Res 34:481–499

Wood HB, May G, Healy L, Enver T, Morriss-Kay GM (1997) CD34 expression patterns during early mouse development are related to modes of blood vessel formation and reveal additional sites of hematopoiesis. Blood 90:2300–2311

Woods JC, Schittny JC (2016) Lung structure at preterm and term birth. In: Jobe AH, Whitsett JA, Abman SH (eds) Fetal lung development–clinical correlates and future technologies. Cambridge University Press, New York, pp 126–140

Wozniak M, Chen C (2009) Mechanotransduction in development: a growing role for contractility. Nat Rev Mol Cell Biol 10:34–43

Wu J, Chu X, Chen C, Bellusci S (2018) Role of fibroblast growth factor 10 in mesenchymal cell differentiation during lung development and disease. Front Genet 9:545. https://doi.org/10.3389/fgene.2018.00545

Xing J, Bai F, Berry R, Oster G (2006) Torque-speed relationship of the bacterial flagellar motor. Proc Natl Acad Sci U S A 103:1260–1265

Xu H, Sun M, Zhao X (2017) Turing mechanism underlying a branching model for lung morphogenesis. PLoS One 12:e0174946. https://doi.org/10.1371/journal.pone.0174946

Xu LJ, Mortola JP (1988) Development of the chick embryo: effects of egg mass. Respir Physiol 74:177–186

Xu X, Lai Y, Hua ZC (2019) Apoptosis and apoptotic body: disease message and therapeutic target potentials. Biosci Rep 39:BSR20180992. https://doi.org/10.1042/BSR20180992

Xu X, Weistein M, Li C, Naski M, Cohen RI, Ornitz DM, Leder P, Deng C (1998) Fibroblast growth factor 2 (FGFR-2) mediated reciprocal regulation loop between FGF-8 and FGF-10 is essential for limb induction. Development 125:753–765

Yahaya B (2012) Understanding cellular mechanisms underlying airway epithelial repair: selecting the most appropriate animal models. Sci World J 2012:961684. https://doi.org/10.1100/2012/961684

Yancopoulous GD, Davis S, Gale NW, Rudge JS, Wiegand SJ, Holash J (2000) Vascular-specific growth factors and blood vessel formation. Nature 407:242–248

Yang L, Lin M, Ruan WJ, Dong LL, Chen EG, Wu XH, Ying KJ (2012) Nkx2-1: a novel tumor biomarker of lung cancer. J Zhejiang Univ Sci B 13:855–866

Yates LL, Schnatwinkel C, Murdoch JN, Bogani D, Formstone CJ, Townsend S, Greenfield A, Niswander LA, Dean CH (2010) The PCP genes *Celsr1* and *Vangl2* are required for normal lung branching morphogenesis. Human Mol Genet 19:2251–2267

Yedwab GA, Paz G, Homonnai TZ, David MP, Kraicer PF (1976) The temperature, pH, and partial pressure of oxygen in the cervix and uterus of women and uterus of rats during the cycle. Fertility Sterility 27:304–309

Yorimitsu T, Homma M (2001) Na+−driven flagellar motor of *Vibrio*. Biochim Biophys Acta Bioenerg 1505:83–93

Yuan T, Volckaert T, Chanda D, Thannickal VJ, De Langhe SP (2018) Fgf10 Signaling in lung development, homeostasis, disease, and repair after injury. Front Genet 9:418. https://doi.org/10.3389/fgene.2018.00418

Yun EJ, Lorizio W, Seedorf G, Abman SH, Vu TH (2016) VEGF and endothelium-derived retinoic acid regulate lung vascular and alveolar development. Am J Physiol Lung Cell Mol Physiol 310: L287–L298

Yuri IW, Katsnelson MI, Koonin EV (2018) Physical foundations of complexity. Proc Natl Acad Sci U S A 115:E8678–E8687

Zakeri Z, Penaloza CG, Smith K, Ye Y, Lockshin RA (2015) What cell death does in development. Int J Dev Biol 59:11–22

Zeng X, Yutzey KE, Whitsett JA (1998) Thyroid transcription factor-1, hepatocyte nuclear factor-3beta and surfactant protein a and B in the developing chick lung. J Anat 193:399–408

Zhang Y, Goss AM, Cohen ED, Kadzik R, Lepore JJ, MuthukumaraswamyK YJ, DeMayo FJ, Whitsett JA, Parmacek MS, Morrisey EE (2008) A Gata6-Wnt pathway required for epithelial stem cell development and airway regeneration. Nat Genet 40:862–870

Zhou J, Schor IE, Yao V, Theesfeld CL, Marco-Ferreres R, Tadych A, Furlong EEM, Troyanskaya OG (2019) Accurate genome-wide predictions of spatio-temporal gene expression during embryonic development. PLoS Genet 15:e1008382. https://doi.org/10.1371/journal.pgen.1008382

Zhou M, Sutliff RL, Paul RJ, Lorenz JN, Hoying JB, Haudenschild CC, Yin M, Coffin JD, Kong L, Kranias EG et al (1998) Fibroblast growth factor 2 control of vascular tone. Nature (Med) 4: 201–207

Zhou Q, Chen J, Luan Y, Vainikka PA, Thallmair S, Marrink SJ, Feringa BL, van Rijn P (2020) Unidirectional rotating molecular motors dynamically interact with adsorbed proteins to direct the fate of mesenchymal stem cells. Sci Adv 6:eaay2756. https://doi.org/10.1126/sciadv.aay2756

Zhou T, He X, Cheng R, Zhang B, Zhang RR, Chen Y, Takahashi Y (2012) Implication of dysregulation of the canonical wingless-type MMTV integration site (WNT) pathway in diabetic neuropathy. Diabetologia 55:255–266

Zorn AM, Wells JM (2009) Vertebrate endoderm development and organ formation. Ann Rev Cell Dev Biol 25:221–251. https://doi.org/10.1146/annurev.cellbio.042308.113344

Zuo YY, Acosta E, Policova Z, Cox PN, Hair ML, Neumann AW (2006) Effect of humidity on the stability of lung surfactant films adsorbed at air-water interfaces. Biochim Biophys Acta 1758: 1609–1620

Chapter 3
Structure of the Avian Respiratory System

Abstract The avian respiratory system is fully described from the larynx to the pneumatization of various proximate bones of the body by the different air sacs. To prominence its morphological novelty, the complexity of the lung-air sac system of birds is compared particularly with the mammalian respiratory system: together with birds, mammals are the only other endothermic-homeotherms. The evolution, the structure, and the mode of sound production (vocalization) by the syrinx are outlined. The three-tiered arrangement of the airway (bronchial) system of the avian lung is delineated, and the functional significance of the assemblage is underscored. The morphologies of the terminal respiratory units, the minuscular air- and blood capillaries, are described, and the topographical arrangement between the bronchial- and the vascular systems of the avian lung that fashion the various gas exchange designs are explained.

3.1 Fundamental Aspects of the Structural Designs of Gas Exchangers

On account that they perform the same primary function (gas exchange), the morphological diversity of the structures that have evolved for respiration is confounding (Jürgens and Gros 2002; Farmer 2006, 2010; Maina 1998, 2002a, b, c, 2011, 2017a, b, c; Perry et al. 2019). To determine how birds acquire the large amounts of O_2 needed to underwrite the enormous energetic cost of powered flight (Tucker 1971; Rayner 1982; Butler 1991, 2016; Nudds and Bryant 2000; Bishop and Butler 2015; Ton et al. 2021), the structure and function of the avian respiratory system have been investigated from the historical times to the present (Coitier 1573; Huxley 1882; Sappey 1847; King 1957, 1966; Duncker 1971; King and King 1979; McLelland 1989a; Roux 2002; Perry and Sander 2004; Farmer 2006, 2010, 2015a, b; Maina 2017a, b, c). Among the air-breathing vertebrates, regarding its morphology and physiology, the avian respiratory system rests at the pinnacle of the evolutionary ranking (Scheid 1979; Fedde 1980; Maina 1998, 2011, 2022; Schachner et al. 2021). Body temperature importantly influences all physiological aspects in the lives of animals (Huey and Stevenson 1979; Angilletta et al.

© The Author(s), under exclusive license to Springer Nature Switzerland AG 2023
J. N. Maina, *Current Perspectives on the Functional Design of the Avian Respiratory System*, Zoological Monographs 8, https://doi.org/10.1007/978-3-031-35180-8_3

2002; Mortola 2004; Angilletta 2009; Moreira et al. 2021). In tetrapods, the mechanisms and processes that maintain constant body temperature (endothermic homeothermy) have been conserved for the last 60 million years of evolutionary time (Ruben 1995; Grigg et al. 2004; Lovegrove 2012, 2017; Nespolo et al. 2011; Araújo et al. 2022). Endothermy enabled mammals and birds to occupy diverse ecological milieus (Lovegrove 2019; Benton 2021): the two endothermic-homeotherms are the most highly metabolically active groups of extant animals (Buttemer et al. 2010). Birds generally operate at relatively higher body temperatures (40–42 °C) compared with mammals (38–40 °C) (Bewig and Mitchell 2009). Presumed by some investigators to have evolved independently in birds and mammals (Ruben 1995; Poelmann et al. 2014; Padian and de Ricqlès 2020), endothermic-homeothermy (Clarke and Rothery 2008; Labocha and Hayes 2019) is a kind of homeostasis of which stable internal (body) temperature is maintained at large energy cost against fluctuating environmental temperatures (Bennett and Ruben 1979; Nespolo et al. 2011; Fristoe et al. 2015; Bonebrake et al. 2020; Osilla et al. 2022). The air-breathing vertebrates possess remarkably morphologically different respiratory structures/organs (Plates 1.6, 2.2 and 2.4), variations that may be explained by factors such as differences in phylogeny, lifestyle pursued, and environments occupied (Maina 1998). Birds and mammals respectively evolved unidirectional and tidal modes of pulmonary ventilation (Plate 2.4b, e), mechanisms and processes that appear to have diverged as far back as the Permian Period, i.e., 250–300 million years ago (Farmer 2015a). Structurally and functionally, the avian respiratory system is separated into a gas-exchanging part (lung) and a ventilatory one (air sacs) (King 1966; Duncker 1971, 1978a, b, 1979; Fedde 1980; King and King 1979; Maina 2005). While the mammalian lung is compliant (Comroe 1974; Weibel 1984; West 2000), and ventilated by the air sacs, the avian one is practically rigid (West et al. 2006; Maina 2007a, b; West 2017).

For a long time, unidirectional airflow across the avian lung (specifically across the paleopulmonic parabronchi) was believed to be unique to the avian respiratory system and that to a great extent, it explained the exceptional efficiency of the gas exchanger (Bretz and Schmidt-Nielsen 1971; Scheid 1979; Schmidt-Nielsen 1971, 1983; Maina 2005, 2017a, 2017b, 2017c). However, in the recent past, it has been shown that to an extent, a similar mode of pulmonary ventilation exists in some parts of the lungs of certain reptilian species (Farmer 2010; Schachner et al. 2013, 2014; Cieri and Farmer 2016, 2020). The presumed existence of pulmonary unidirectional airflow in a common diapsid ancestor (Farmer 2015a, b, 2017) casts doubt on the long-held view that in birds, the mode of airflow in the lung was a unique adaptive specialization for efficient gas exchange which permitted the realization of powered flight, a highly metabolically costly mode of locomotion (Tucker 1971; Rayner 1982; Butler 1991; Nudds and Bryant 2000). Furthermore, it casts doubt on whether the air sacs (which are lacking in the reptilian lung) were a requirement for the occurrence of unidirectional airflow in the avian lung. Farmer (2006) speculated that the evolution of avian air sacs may have enhanced balance and agility during flight. After observing that unidirectional pulses in airflow matched heartbeats in the lungs of the American Alligator (*Alligator mississippiensis*), Farmer (2010) suggested that unidirectional airflow evolved in the basal archosaurs or probably in earlier sauropsids to allow gas exchange during apnea by coupling airflow patterns with the heart beats.

The genomes of birds and reptiles, but not mammals, comprise a few large chromosomes and many small microchromosomes (Waters et al. 2021). In certain aspects, morphologically, the reptilian respiratory systems resemble the avian ones (Perry and Duncker 1978; Maina et al. 1989a; Maina 2005; Perry 1992, 1998, 2001; Maina et al. 1999; Perry et al. 2019; Schachner et al. 2021): at gross level, a shared structural feature is that the anteroposterior part of the lungs of some reptilian species is partitioned into gas exchange tissue (Plate 1.6f, h), a part analogous to the lung of the avian respiratory system while the posterior part which is saccular is analogous in the air sacs of the avian respiratory system (Plate 2.17) (Maina 1989a; Perry 1990, 1992). Interestingly, resembling the air sacs of the avian respiratory system, the lungs of the African Chameleon (*Chamaeleo africanus*) have saccular extensions that extend into the coelomic cavity (Patt and Patt 1969) (Plate 1.6f). The avian respiratory system appears to have evolved from a relatively structurally modest muticameral reptilian lung such as the crocodilian and alligator ones (Perry 1988; Perry et al. 1989; Sanders and Farmer 2012; Schachner et al. 2013, 2021). The unidirectional airflow in the crocodilian lung is regulated by aerodynamic valving that stems from the geometry and configuration of the airways (Farmer 2010; Schachner et al. 2013). Based on shared evolutionally features that exist between crocodilians and birds, Brocklehurst et al. (2020) speculated that a single ancestor of archosaurs that had a parabronchial lung with heterogeneously subdivided gas exchange tissue (parenchyma), unidirectional airflow across the secondary bronchi and parabronchi and a post-pulmonary septum existed. Shared pulmonary features that include separation of the thoracoabdominal cavity by an oblique septum into a cavum respiratorium and a cavum cardioabdominale exist in birds and crocodilians (Farmer 2015b). Comparisons between the bronchial morphologies of the Alligator (*Alligator mississippiensis*) and the Ostrich (*Struthio camelus*) showed various similarities that suggest that certain structural features of the bronchial tree were possibly conserved across the Archosauria and may even have existed in the ancestral archosaurian lung (Schachner et al. 2021). Noteworthy lung heterogeneity exists in, e.g., the varanid lizards (Family: Varanidae) such as the Monitor Lizard (*Varanus exanthematicus*) (Maina et al. 1989a; Schachner et al. 2014; Cieri et al. 2018) (Plate 1.6h): such lungs offer instructive models for securing insights into the structural and functional designs of the lungs of the inaugural archosaurs.

The existence of pneumatic foramina in fossilized theropod, sauropod, and pterosaur bones has been employed to reconstruct the presence of avian-type air sacs in these early animal taxa (O'Connor and Claessens 2005; O'Connor 2009; Wedel 2009; Benson et al. 2012). Schachner et al. (2009) determined that avian-like air sacs existed in basal theropods and concluded that an avian-type pulmonary system was a possible theropod trait. Postcranial pneumatization of archosaurs has focused on determining whether cranial and caudal groups of air sacs existed in the fossilized remains of the taxon (O'Connor and Claessens 2005; Wedel 2009), a trait that would suggest the existence of unidirectional ventilation in the lungs. It has interestingly been determined that aerodynamic valving (Banzett et al. 1987, 1991; Kuethe 1988; Wang et al. 1988, 1992; Brown et al. 1995; Maina and Africa 2000; Maina et al. 2009) occurs even after some of the air sacs are not working (Brackenbury et al. 1989; Brackenbury and Amaku 1990). Occurrence of unidirectional airflow in some parts of the reptilian lung (Schachner et al. 2013, 2014; Farmer and Sanders 2010; Cieri et al. 2014; Farmer 2015b), where air sacs such as those of the avian respiratory system are missing, shows that such structures (air sacs) are not required for unidirectional ventilation of a lung to occur.

Investigation of the structure of the costovertebral joints has been employed to determine the evolution of the functional design of the avian respiratory system (Schachner et al. 2011): birds and crocodilians possess bicapitate ribs that articulate with the vertebral column at two points—the rib capitulum articulates with the vertebral parapophysis and the rib tuberculum with the vertebral diapophysis (Hoffstetter and Gasc 1969). The dinosauriform, *Silesaurus*, had a rather narrow thoracic cavity which may have compressed the dorsal surface of the lung (Brocklehurst et al. 2018), in a manner similar to that of the lungs of modern birds that are deeply attached to ribs and vertebrae on the dorsolateral aspects (McLelland 1989a; Maina 2005) (Plates 1.1a, 1.6i, 2.5i, j and 2.6), making them (lungs) essentially rigid (Schachner et al. 2011; Brocklehurst et al. 2018). Showing that the structure of the avian lung has been greatly conserved, recently, Wang et al. (2018) described fossilized lungs of a specimen of ~125 million years old basal ornithuromorph bird *Archaeorhynchus*, where the putative lungs showed deep costal sulci (impressions) (Plate 1.5a–c) and airways that resembled the parabronchi of the contemporary (modern) birds and minute air spaces that looked a lot like air capillaries (Plate 1.5d–f). Fossilized materials provide the direct and instructive evidence on the evolution of biological structures: for soft tissues such as lungs, it is a very rare process. Interactions between biological and environmental conditions (especially geological ones) determine whether preservation happens (Parry et al. 2018).

Among the modern-day gas exchangers, one of the most interesting examples of structure and function similarity is that between the insectan tracheal and the avian (parabronchial) systems: insects and birds have been separated by over 300 million years of evolutionary time: the former evolved during the Early Ordovician (~479 million years ago) (Misof et al. 2014) while birds arose from theropod dinosaurs during the Jurassic Period (~165 to 150 million years ago) (Brusatte et al. 2015). The tracheal respiratory system of insects is a characteristic feature of insects (Buck 1962; Weis-Fogh 1964a, b, 1967; Wigglesworth 1972, 2008; Miller 1974; Chapman 1998; Nation 2002, 2008; Klowden 2002; Harrison et al. 2012). Simple and elegant, the diffusion-based gas exchanger makes it possible for insects to attain some of the highest known mass-specific metabolic capacities in the Animal Kingdom (Weis-Fogh 1964a; Scheid et al. 1981; Harrison 1997; Dudley 2000; Nation 2008; Niven and Scharlemann 2005; Harrison et al. 2013; Terblanche and Woods 2018). Comprising in excess of one million named species, in the Animal Kingdom, insects are the most speciose and diverse taxon (May 1988, 1992; Stork 1994, 2018; Erwin 1991; Mora et al. 2011). The tracheal respiratory system of insects, that efficiently

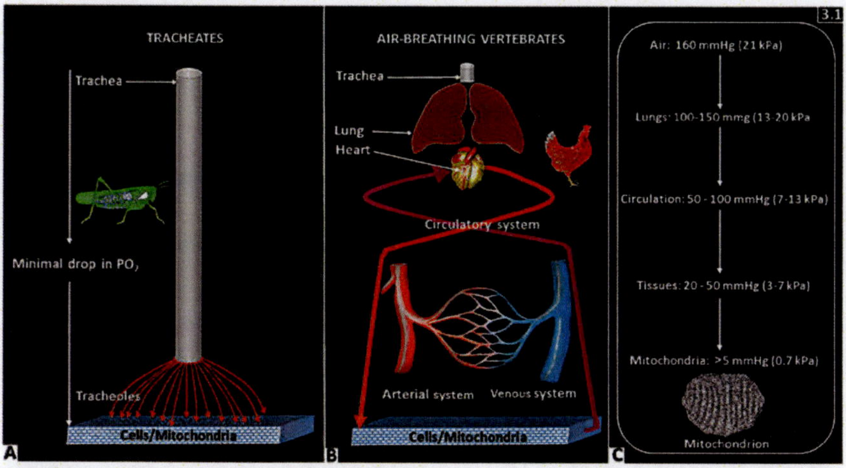

Plate 3.1 Morphological and functional differences between the insectan and the vertebrate gas exchangers. (**a**) In insects, oxygen is delivered directly from the atmosphere to the tissues/cells of the body at high partial pressure. In vertebrates, transfer of oxygen occurs across the lungs (for air-breathers) (**b**) and gills for water-breathers: it is then transported by the vascular system (in blood) to the tissues and cells of the body. (**c**): The oxygen cascade in vertebrates where the partial pressure of oxygen decreases along the transport pathway, with the partial pressure being nearly zero at mitochondrial level

supplies O_2 to the tissues, explains the evolutionary success of insects. Among animals, insects have some of the highest mass-specific metabolic rates (Dudley 1998). While in most evolved gas exchangers, especially of vertebrates, the respiratory fluid medium, water, or air (in which O_2 is dissolved) are convectively conveyed to the respiratory surface by blood vessels from where O_2 diffuses across the water-gas or the blood-gas (tissue) barriers (Plate 2.1) to reach the cells of the body, for insects (mainly the small, highly metabolically active ones), O_2 travels by diffusion from air to reach the cells of the body directly (Maina 1989b, 2012; Snyder et al. 1995; Schmitz and Perry 1999; Nation 2002; Westneat et al. 2003; Bradley 2007; Kirkton 2007; Wigglesworth 2008; Groenewald et al. 2012; Hsia et al. 2013; Harrison et al. 2013; Simelane et al. 2016; Hochgraf et al. 2018) (Plates 3.1a and 3.2a–h): a circulatory system is not involved in the respiratory process. Mitochondria comprise the O_2sink that drives the flow of O_2 across the respiratory cascade, i.e., the distance from the external respiratory fluid medium to the cells of the body (Taylor and Weibel 1981; Weibel 2000; Hoppeler and Weibel 2000; Biro 2013; Milsom et al. 2021; Dominelli et al. 2021; De Souza et al. 2021). In the same manner that the parabronchial (avian) lung is ventilated continuously and unidirectionally (Scheid 1979; Fedde 1980), in some of the species that have been investigated, ventilation of the tracheal system occurs in the same way, i.e., the airflow is continuous and unidirectional (Bailey 1954; Henderson and Prange 1995; Harrison 1997; Duncan and Byrne 2002; Nation 2008; Harrison et al. 2013; Aboelkassem and Staples 2013a, b; Webster et al. 2015; Hochgraf et al. 2018): air enters the body through

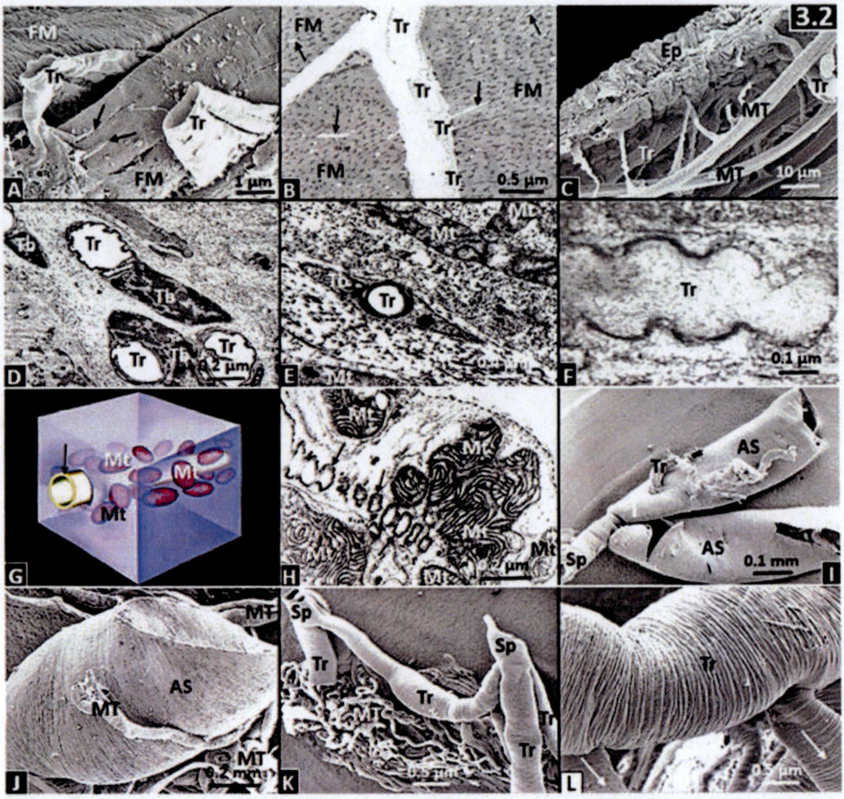

Plate 3.2 Tracheal respiratory system of a grasshopper (*Chrotogonus senegalensis*). **(a, b)** Trachea (Tr) delivers oxygen to the flight muscles (FM). Arrows (↓), tracheoles. **(c)** Trachea (Tr) delivers oxygen to the epithelial lining (Ep) of the gastrointestinal system. MT, Malpighian tubules. **(d–f)** Tracheoles (Tr) in the wall of the gastrointestinal system. Tb, tracheoblasts; Mt, mitochondria. **(g)** A schematic illustration showing a tracheole (arrow) delivering oxygen to a flight muscle cell. Mt, mitochondria. **(h)** Tracheole (arrows, ↓) delivering oxygen to a cell dashed line). Mt, mitochondria; dashed line, cell wall. **(i, j)** Air sacs (AS). SP, spiracle (i); MT, Malpighian tubules (j). **(k)** spiracular ends (Sp) of trachea (Tr). MT, Malpighian tubules. **(l)** Transverse trachea (Tr) and branches (arrows)

one set of spiracles and leaves in others (Bailey 1954; Henderson and Prange 1995; Harrison 1997; Duncan and Byrne 2002; Nation 2008; Harrison et al. 2013; Aboelkassem and Staples 2013a, b; Webster et al. 2015; Hochgraf et al. 2018). In both the insectan tracheal (respiratory system) (Plate 3.2i, j) and the avian (parabronchial) one (Plates 2.17 and 2.23), air sacs have evolved to increase the ventilatory tidal volume (Maina 1989b, 2005).

For the insectan tracheal system, O_2 reaches the cells of the body at very high partial pressure (in the very small insects practically at the atmospheric measure) while for the complex animals, the partial pressure of O_2 decreases considerably with the distance from the respiratory surface to almost zero in the mitochondria

(Wittenberg and Wittenberg 2007; Weibel 1984) (Plate 3.1c). In the larger groups of insects such as locusts and grasshoppers, rhythmic, i.e., mechanical, contractions largely of the abdominal part of the body help drive air through the tracheal system (Aboelkassem and Staples 2013a, b). Resting wasps accomplish highly efficient levels of gas exchange mostly from constant respiratory activities such as abdominal pumping and movements of the legs, the wings, and the body (Käfer et al. 2013). In very long trachea, e.g., those of the wings of insects, O_2 is conveyed by complex interactions of the hemolymphatic circulatory system and the tracheal systems: gas exchange may also occur directly by cutaneous respiration (Pass 2018). In larger species of insects such as moths, butterflies, bees, wasps, locusts, and grasshoppers, air passes through the spiracles into the air sacs (Plate 3.2i, j) and then across the trachea and the tracheoles to the tissue cells (Wigglesworth 1972; Snelling et al. 2011, 2012) (Plate 3.2k–l): the tracheoles branch and decrease in diameter until their sizes reach the mean free path of O_2 molecules which is the average distance across which a moving particle such as an atom, a molecule, or a photon travels before considerably altering its direction or energy largely from one or more collisions with other particles: for the trachea, further decrease of diameter (below the mean free path of O_2) becomes limiting to diffusion (Weis-Fogh 1964a; Pickard 1974). In highly metabolically active tissues, the tracheoles even indent cells to terminate close to mitochondria (Nation 2008) (Plate 3.2h). For insects, while allowing exceptionally high mass-specific metabolic capacities (Dudley 2000), the evolution of the tracheal system came at great cost: they had to be extremely small in size and/or acquire very thin (long) body shapes, e.g., the Stick Insect (*Morosus carausius*) (Schmitz and Perry 1999).

The evolution of the exceptionally efficient respiratory systems, namely, the tracheal system in insects and the parabronchial one in birds, should have contributed to the accomplishment of powered flight (Schmidt-Nielsen 1972; Ellington 1984; Rothe et al. 1987; Ellington et al. 1990; Maina 2000a; Dudley 2000; Morris et al. 2010; Bairlein et al. 2015; Jenni-Eiermann 2017). The very high metabolic demands that powered flight obligated may explain the structural and functional evolutionary convergence between the respiratory systems of insects and birds. It should, however, be called to attention that neither the tracheal system nor the lung-air sac system is prerequisites for powered flight: bats (Order: Chiroptera), a mammalian taxon that flies as well, if not better than most species of birds evolved powered flight while retaining the mammalian lung (Maina 1985). However, compared with those of non-flying (terrestrial or nonvolant) mammals (Gehr et al. 1981), the chiropteran lungs are spectacularly morphometrically specialized (Maina et al. 1982a, b, 1989b, 1991; Maina and King 1984).

The gas exchangers that exist in the modern animal species have been fashioned under among others factors such as phylogenetic levels of development, prevailing environmental/ecological conditions, body size and design, and lifestyle pursued (Maina 1998, 2000a, b, c, 2005, 2011, 2022; Jürgens and Gros 2002; De Souza et al. 2021). While in terms of cognitive ability the human brain is the best developed in the Animal Kingdom (Hofman 2014; Mitchell 2016: Sherwood 2018; Harris 2022), the bronchioalveolar (mammalian) lung is far from being the most functionally efficient gas exchanger (Weibel 1984; Cameron 1989; Schmidt-Nielsen 1990; Maina 1998, 2005, 2011; Mortola 2004): that position has largely been accorded to the lung-air sac system of birds (Scheid 1979; Maina 2005, 2022; Powell 1983, 2000, 2015; Powell and Hopkins 2004). Interestingly, there are no tissues or cells that are prototypical of gas exchangers, at least not in the same manner that for example neurons are to nervous tissue, osteocytes to bone, and podocytes to the kidney. While it is often submitted that the granular pneumocytes (type-II cells) that secrete surfactant (Castranova et al. 1988; Wood et al. 2000; Mason 2006; Bódi et al. 2016) are unique to the vertebrate lungs, phospholipid secreting cells exist in diverse tissues and organs such as stomachs, intestines, swim-bladders, gas-exchange mantles of pneumonate gastropods, prostate gland, female reproductive tract, lacrimal gland, mesothelial cells of the pleural linings, pericardium, peritoneum, and epithelial lining of Eustachian tubes (Daniels and Orgeig 2001, 2003; Bourbon and Chailley-Heu 2001; Luis da Cruz et al. 2009; Maina 2009). The basis of the diversity of the existing gas exchangers was captured by Jürgens and Gros (2002) as follows: "There is no such thing as an ideal gas exchange system. The system that has evolved in each species depends to an impressive extent on environmental conditions, on body build and size, on the animal's patterns of movement and on its energy consumption."

3.2 Structure of the Avian Respiratory System: The Lung-Air Sac System

The structure of the avian respiratory system has been extensively investigated by among others King (1966), Duncker (1971), King and McLelland (1984, 1989), Seller (1987), McLelland (1989a, b), and Maina (2005, 2006, 2017a, b). Succinct accounts of the various parts of the lung-air sac system are outlined below.

3.2.1 Larynx

For air-breathing vertebrates, the larynx is located at the proximal end of the trachea. In birds, the structure has been well-investigated (White 1975; Bock 1978; Zweers et al. 1981; Homberger 2017a) and the findings reviewed by King and McLelland (1984) and McLelland (1989b). Both structurally and functionally, the avian larynx differs from the mammalian one (Homberger 2017a, b). Constituting a laryngeal chamber that lacks vocal folds, its body comprises four individual cartilages, namely, the cricoid and the paired (i.e., the right and left) arytenoid cartilages that form the main part of the frame and a small procricoid cartilage (King and McLelland 1984; Homberger 1979, 2017a, b; Habib 2019) (Plate 3.3a, b): in contrast to the mammalian one, thyroid and epiglottic cartilages are lacking. In birds, the

Plate 3.3 (**a, b**) Lateral (**a**) (diagrammatic) and dorsal (**b**) (gross) views of the larynx of birds showing the cricoid (cr), arytenoid (Ar), and procricoid (Pc) cartilages. Tr, trachea; Gl, glottis (**b**). (**c**) Interlocking complete cartilages of the trachea (TC). Figures modified from King and McLelland (1984) and McLelland (1989a)

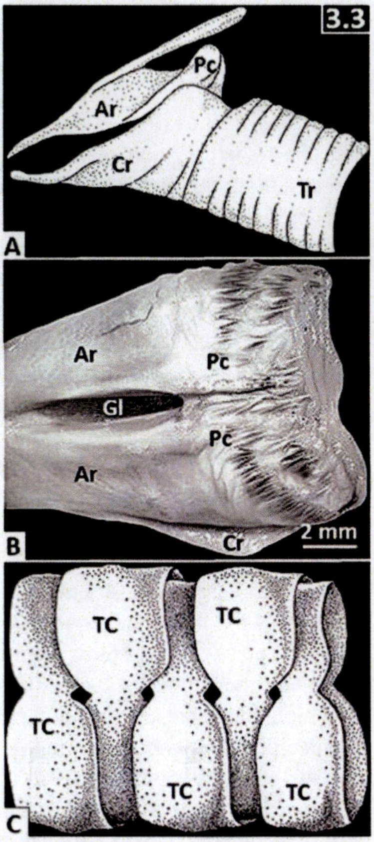

cricoid cartilage forms the ventral and the lateral walls of the larynx, and the small procricoid cartilage lies dorsally in the midline between the two arytenoids cranially and the right and left wings of the cricoid cartilage caudally: it articulates with the two cartilages. Dorsally located, the right and left arytenoid cartilages border and support the margins of the glottis (Plate 3.3a, b). The laryngeal cartilages ossify with age (Zweers et al. 1981; Hogg 1982). The glottis, which is a slit-like opening of the larynx, is lined by a pseudostratified columnar epithelium with mucous glands (McLelland 1989b): the size of the aperture is regulated by abductor (dilator) and adductor (constrictor) muscles (White and Chubb 1967; White 1975; Bock 1978; Zweers et al. 1981). While in some amphibian, reptilian, and mammalian species, the larynx serves as an organ of vocalization, in birds, a novel sound-producing organ (the syrinx), which is located at the distal end of the trachea (King 1989; Brackenbury 1989), exists. Fundamentally, the avian larynx serves as a vulvar structure which prevents foreign materials from entering the lower respiratory tract.

3.2.2 Trachea

Relative to their body sizes, birds have particularly long necks that are associated with a long trachea (Plate 1.1a). It is presumed that the long necks of birds evolved with the transformation of forelimbs into wings which were totally dedicated to powered flight rather than walking (King and King 1979; King and McLelland 1984; Van der Leeuw et al. 2001a, b; Böhmer et al. 2019; Marek et al. 2021). The long and highly flexible necks of birds acquired some of the important roles previously done by the forelimbs such as food acquisition and manipulation, nest making, feeding, preening of feathers, sexual display, and defense (Zweers et al. 1994; Van der Leeuw et al. 2001b; Hansell 2005). After investigating the beak and the neck muscles of the Rosy-Faced Lovebird (*Agapornis roseicollis*), Young et al. (2022) observed that the structures "have been co-opted to function biomechanically as a propulsive third limb during vertical climbing." Interestingly, in contrast to birds, bat wings have maintained many of the functions executed by the forelimbs of mammals include the capacity of catching, holding, and manipulating food as well as walk (though clumsy) on the ground (Vaughan and Bateman 1970; Geipel et al. 2013): the wings of birds have very limited if any, capacities of performing such functions (Gatesy and Dial 1996). Böhmer et al. (2019) observed that "the avian neck can be regarded as the functional equivalent to the arm and in combination with the beak, the head-neck system of birds even acts as a tool-bearing arm."

Among the extant tetrapods, the necks of birds are morphologically one of the most highly variable anatomical features (Terray et al. 2020). The differences largely emanate from variations in number, shape, and structure of the cervical vertebrae (Zweers et al. 1986; Bout 1997; Tambussi et al. 2012). The neck of the extinct *Andalgalornis steulleti* from the upper Miocene to Lower Pliocene (6 million years ago) of Argentina had 13 cervical vertebrae (Tambussi et al. 2012). The medium-sized patagornithine phorusrhacid bird belonged to the avian taxon designated "terrorbirds" that constituted the apex predators of their time (much of the Cenozoic): to kill their prey, it made sudden sagittal neck movements (Tambussi et al. 2012). In those species of birds that have secondarily lost flight such as the emu (Family: Dromaiidae) or the penguins (Family: Spheniscidae) (Sackton et al. 2019), the forelimbs are either vestigial or are adapted for locomotory activities such as balance and "flight" through the water.

Compared with mammals, birds have relatively more cervical (neck) vertebrae (Terray et al. 2020): they range from as few as 10 in parrots to as many as 26 in swans (Böhmer et al. 2019). Interestingly, for mammals, except for the manatees (Family: Trichechidae) and some sloths (Families: Megalonychidae and Bradypodidae) (Varela-Lasheras et al. 2011), even for the Giraffe (*Giraffa camelopardalis*) that has a particularly long neck, there are only seven neck vertebrae (Badlangana et al. 2009). Owls display unique neck and head mobility (Krings et al. 2014, 2017; Kambic et al. 2017). Without strangulation of the blood vessels of the neck, owls can rotate their heads through an angle of more than 270° in each direction on the horizontal axis and tilt their heads through one of 180° on the vertical axis (de Kok-Mercado et al. 2013): adaptively, the internal carotid and the vertebral arteries are located in distinctive foramina of the cervical vertebrae, a feature that prevents them from being strangulated during the twisting of the neck. The owls increase horizontal head rotation by unique independent processes of yawing and rolling in various parts of the neck (Krings et al. 2017). During terrestrial locomotion, long necks stabilize vision by minimizing head bobbling (Abourachid et al. 2011; Nyakatura and Andrada 2014) while during flight, the neck stabilizes the head, a process without which the animals would not be able to maintain level flight (Warrick et al. 2002; Pete et al. 2015).

In most birds, the trachea runs the length of the neck, directly connecting the larynx with the syrinx. For the Domestic Fowl (*Gallus gallus* variant *domesticus*), ~120 cartilages, that decrease in diameter proximodistally, comprise the trachea (King and McLelland 1984). The airway is supported by interlocking, generally complete signet ring-like shaped cartilages (Plate 3.3c) that are interconnected by bands of tough connective tissue (Mathey 1965). Sexual dimorphism occurs in the morphologies of the avian trachea (King and McLelland 1984; McLelland 1989b). With age, the tracheal cartilages ossify (Hogg 1982; Brown et al. 1997). The interlocking arrangement of the tracheal cartilages of the long trachea of birds (Plate 3.3c) ensures greater flexibility and strength to the airway. For animals of equivalent body mass, the length of a bird's trachea is 2.7 times longer than that of a mammal (Pandey et al. 2015). To compensate for the increased resistance in a longer airway, the diameter of a bird's trachea is 1.29 times wider compared with that of a similar-sized mammal (Hinds and Calder 1971): consequently, in birds and mammals, tracheal airflow resistance is comparable. The tracheal dimensions of birds make the tracheal dead space of the long and wide trachea four and half times greater than that of a similar-sized mammal (Hinds and Calder 1971). Lower respiratory rate, which is about one-third of that in mammals (Calder 1968), and greater tidal volume, which is about four times more than for a mammal of equivalent size (Lasiewski and Calder 1971; Pandey et al. 2015), compensate for the large tracheal dead space in birds: the lower respiratory frequency of birds ensures that the limitation caused by the large tracheal dead space is reduced to a level where the minute tracheal ventilation rate is ~1.5 to 1.9 times greater than that of a mammal (Lierz and Rüdiger 2012).

The Trumpeter Swan (*Cygnus buccinators*), the Whooping Crane (*Grus americana*), and the birds of paradise (Family: Paradisaeidae) have tracheal loops and coils that make the trachea as much as four times longer than those of similar-sized birds without such modifications (Clench 1978; Welty 1979; Welty and Baptista 1988). Clench (1978) observed that the trachea of the Trumpet Manucode (*Phonygammus keraudrenii*) elongated with age by broadening of existing tracheal cartilages and formation of new cartilages but it (trachea) did not increase in diameter. Hinds and Calder (1971) determined that for the Trumpeter Swan, the length of the trachea is 3.3 times longer than that of a bird of equivalent body mass and that tracheal volume was 3.8 times greater: they deduced that if the tidal volume of a swan was equivalent to that of a goose of similar body mass (a bird that lacks tracheal elongations), the ventilation of the swan would be reduced to that of the dead air space of the coiled trachea. The adaptive advantages, if any, of the particularly long trachea of certain species of birds is still debatable (Yarrell 1833; Wetmore 1918; Clench 1978; Gaunt 1986; Gaunt et al. 1987): it is, however, generally believed that the property is an acoustic adaptation that lowers the pitch of the sound and/or even amplifies it (Gaunt et al. 1987; Frith 1994; Fitch 1999). Schmidt-Nielsen et al. (1969), Bech et al. (1979), and Bech and Johansen (1980) suggested that tracheal elongation (with or without coiling) may increase the tracheal epithelial surface area for more efficient evaporative cooling: panting may occur without causing respiratory alkalosis. In cranes (Family: Gruidae), Prange et al. (1985), however, showed that although a long trachea may be advantageous, the state is not critical to the prevention of respiratory alkalosis in birds. The trachea of species of birds such as the Emu (*Dromaius novaehollandiae*) and the Ruddy Duck (*Oxyura jamaicensis*) has an inflatable sac-like diverticulum while others such as the males of the anseriform species have a rigid bulbous tracheal expansion (Murie 1867; Coughtrey 1873; Beddard 1886; Clench 1978). The booming call of the Emu has been attributed to the existence of a tracheal diverticulum (Murie 1867). In some penguin species (Family: Spheniscidae) and petrel ones (Family: Procellariidae), a double trachea, where the airway is divided by a median septum for some part of their length, has been reported (Zeek 1951).

In birds, the internal surface of the trachea (the lumen) is lined by combination of simple and pseudostratified ciliated columnar epithelia with numerous simple alveolar mucous glands and goblet cells that preponderate in the distal epithelial part of the airway (Chandra and Bharadwaj 1971; Hodges 1974). Resembling the epithelia which line the airways of the mammalian lungs that proximodistally decrease in complexity as the airways decrease in size (diameter) and proximate the respiratory surface (Weibel 1984), those of birds are lined by complex ciliated pseudostratified columnar epithelium with goblet (mucous secreting) cells at the level of the intrapulmonary primary bronchus, a simple columnar epithelium on the secondary bronchi, generally a simple cuboidal one at the parabronchial level and a squamous one over the air capillaries (Plate 3.4). Basal cells, nonciliated columnar cells as well as scattered granular endocrine cells exist in the epithelia of the upper airways (Walsh and McLelland 1974a, b; McLelland 1989a, b). The conducting airways of the mammalian lung are covered by a thin layer of fluid that contributes to the protection of the lung (Hartmann and Verkman 1990; Govindaraju et al. 1997). The epithelium of the trachea of a developing chick embryo is first enervated at about day 15 of incubation (Walsh and McLelland 1974c, 1978) and the epithelial cells rest on a lamina propria that comprises dense irregular connective tissue components, many blood vessels, lymphocytes and dense nerve plexuses of unmyelinated axons and ganglionic cells (Cover 1953; Hodges 1974; Walsh and McLelland 1974c). Although birds are known to exhibit cough-like action (Birdcage 2021), it is doubtful whether the behavior comprises an effective mechanism of mucous and/or debris (particulate) clearance from their longer trachea (Brown et al. 1997). Recently, Burn et al. (2022) determined that with transport speeds of 130–260 $\mu m\ s^{-1}$, birds have faster respiratory mucociliary clearance compared with mammals (20–80 $\mu m\ s^{-1}$).

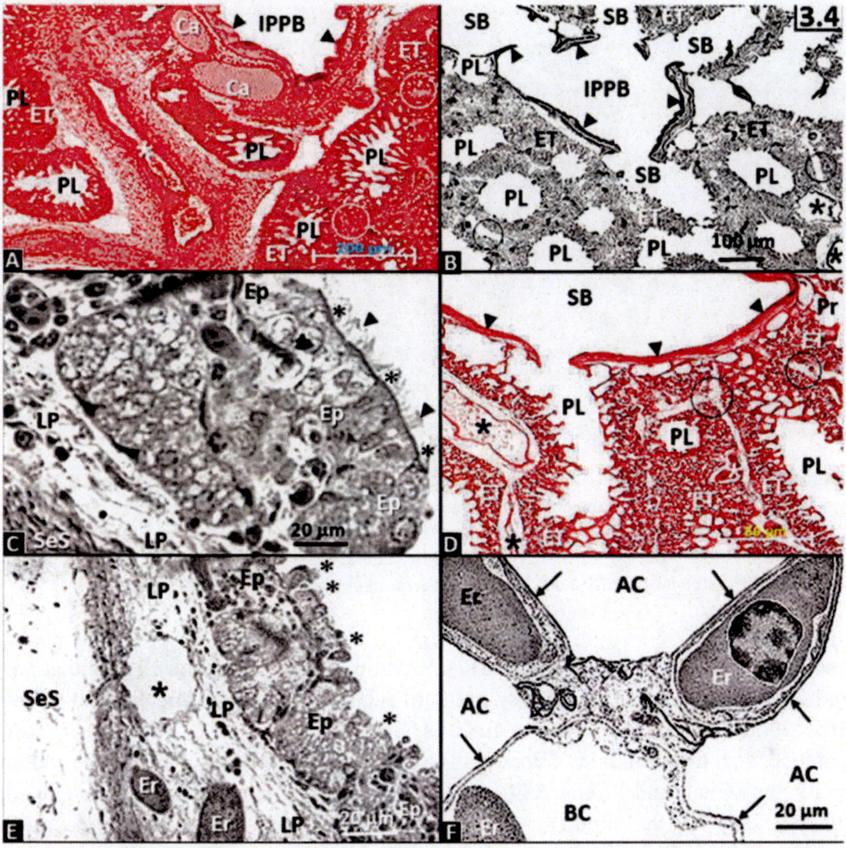

Plate 3.4 (**a–c**) Structure and epithelial lining (arrowheads, ▶) of the intrapulmonary primary bronchus (IPPB) of the lung of the Domestic Fowl (*Gallus gallus* variant *domesticus*): proximodistally, the airway is lined by a pseudostratified columnar ciliated epithelium (Ep) (**c**). (**d**) For a short distance after branching from the intrapulmonary primary bronchus, the secondary bronchi (SB) are lined by a pseudostratified columnar ciliated epithelium that changes to a simple columnar one (**e**). (**f**) The air capillaries (AC), where gas exchange occurs across the blood-gas barrier (arrows, ↓), are lined by a simple squamous epithelium. PL, parabronchial lumen; ET, exchange tissue; stars (⁂), blood vessels; LP, lamina propria; SeS, subepithelial space; asterisks (*), cilia; Er, erythrocytes; BC, blood capillary

3.2.3 *Syrinx*

In vertebrates, vocalization is an important means of communication and interaction. It considerably influences the ecological diversification and social organizations of birds (Xu et al. 2014; Riede et al. 2019). The capacity originated from complex coevolution between anatomical structures that produced sound and centrally and peripherally coordinated neuromuscular regulatory circuits (Nottebohm et al. 1976, 1982; McCasland 1987; Vu et al. 1994; Galis 1996; Dickinson et al. 2000; Fuxjager

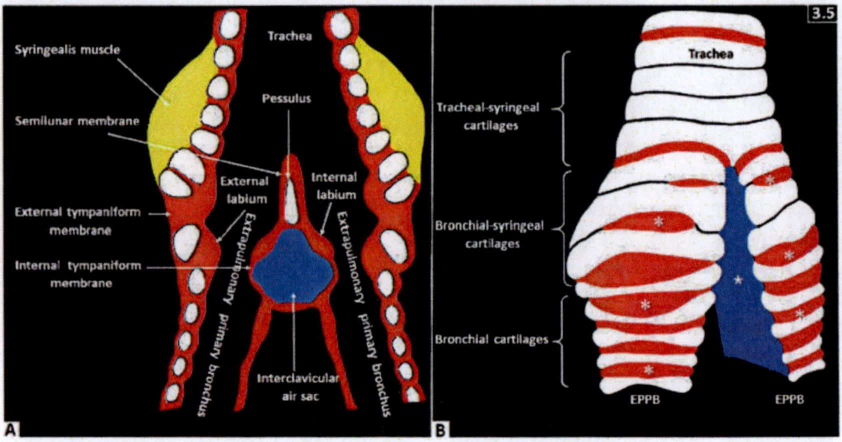

Plate 3.5 **(a, b)** Illustrations a tracheal-bronchial syrinx of the Domestic Fowl (*Gallus gallus* variant *domesticus*). Soft tympaniform (vibrating) membranes generate the primary sound. The pessulus is a wedge-shaped cartilage that divides the trachea into left and right extrapulmonary primary bronchi. B: star (⚲), medial tympaniform membrane; an asterisk (*), lateral tympaniform membrane. Figures modified from King and McLelland (1984) and McLelland (1989a)

and Schlinger 2015; Garcia et al. 2017). Rhythm-generating neural circuits that are found in the hindbrain and the spinal cord regulate vertebrate vocalization (Bass et al. 2008; Bass 2014; Barkan et al. 2018; Barkan and Zornik 2020). Among most tetrapod taxa, the evolution of vocalization possibly happened more than 100 million years ago (Chen and Wiens 2020). It is supposed that vocal structures developed independently of the circuits that regulate them (Senter 2008; Albersheim-Carter et al. 2016). With the exception of animals such as rabbits, giraffes, whales, and dolphins that lack vocal cords in the larynx, other mammals vocalize to various extents. Practically all extant species of birds that number ~11,000 (Brusatte et al. 2015) produce some kind of sound from the syrinx (King 1989; Goller and Larsen 1997; Riede et al. 2019; Elemans et al. 2015) and notable sexual dimorphism exists in the syringeal morphologies (Appel 2005). The morphologies of the syrinxes have been used for sex determination, classification, and evaluation of phylogenetic statuses of birds (Gaban-Lima and Höfling 2006).

In most species of birds, a syrinx, which is termed "a tracheal-syringeal syrinx", is located at the trachea bifurcation (King and McLelland 1984; Brackenbury 1989; McLelland 1989b) (Plates 1.1a–d and 3.5a, b). In birds, vocalization is essential in life's activities such as attracting sexual partners, asserting territory, escaping from predators and social and parent-offspring identification and communication (Seddon 2005). With an analogous structure lacking in the same anatomical location in reptiles, especially the crocodilians (Russell et al. 2000; Sacchi et al. 2004; Riede et al. 2015), the taxon that is evolutionary the most proximate vertebrate group to birds (Shine 2005; Crawford et al. 2012), the syrinx is a unique structure/organ in the avian clade. While vertebrates organ(s) of vocalization are typically located at the cranial end of the trachea, in birds, the syrinx is situated at its distal end (King 1989; Brackenbury 1989; Habib 2019) (Plates 1.1 and 3.5). The oldest fossil find of the avian syrinx is that of *Vegavis iaai*, a specimen dated back to the Late Cretaceous of Antarctica, i.e., 66 to 69 million years ago (Clarke et al. 2016). While a very interesting discovery that importantly informed on the evolution of the syrinx in birds, the investigators determined that the structure (syrinx of *Vegavis*) that was highly derived was not the actual original organ of the avian clade. Clarke et al. (2016) suggested that the syrinx appeared after the realization of powered flight and the development of the highly specialized avian respiratory system. In birds, originally, regarding sound production, the now defunct avian larynx appears to have been a fully functional structure, i.e., a sound-producing organ, before the role shifted to the syrinx (Kingsley et al. 2018). Long coevolution of two sound producing organs, i.e., the larynx and the syrinx, may have coexisted over a long period of time (Kingsley et al. 2018).

Syringeal morphologies enable many species of birds to generate a repertoire of sounds (King and McLelland 1984; King 1989; McLelland 1989b; Brackenbury 1989; Goller and Larsen 1997; Doĝan and Takici 2018; Riede et al. 2019) which correlate with ecological settings (Seddon 2005). While laryngeal vocalization is mostly regulated by the control of a pair of vocal folds, the syrinx utilizes a variety of paired membranes and soft tissue structures (labia) that to various extents tense and relax under various intrinsic and extrinsic muscles (King and McLelland 1984; McLelland 1989b; King 1989; Brackenbury 1989; Larsen and Goller 2002). The tracheobronchial syrinx is the most common form in birds: it is located at the bifurcation of the trachea into left and right extrapulmonary bronchi, i.e., the point where the extrapulmonary primary bronchi converge (Plate 3.5a, b). The tracheobronchial syrinx allows some species of birds to synchronously generate various sounds from the left and the right sides of the syrinx (Suthers 1990). In songbirds, the left and right parts of the syrinx can be used independently and in some cases, each side is specialized for different sound frequency ranges, thus increasing the overall frequency range attainable (Suthers 1990; Schmidt and Martin 2014). According to Suthers and Zollinger (2004), songbirds have a relatively uniform duplex vocal organ from which sound is generated by the oscillation of a pair of thickened labia on either side of the syrinx (Plate 3.5a, b). Paired syringeal muscles serve as flexible and autonomous regulators of sound frequency and amplitude and each side of the syrinx displays a degree of acoustic specialization (King and McLelland 1984). Songbirds differ from many non-songbirds as well as vocal learners such as parrots that have relatively fewer syringeal muscles and apply syringeal membranes to generate sound (Suthers and Zollinger 2004). Notwithstanding the fact that songbirds are considered to have the most extensive range of acoustic qualities of their large vocal inventories, interestingly, the syrinxes of the oscines are remarkably morphologically similar (Raikow and Bledsoe 2000): in the taxon, the broad range of sounds has mainly been ascribed to vocal learning and complex neural control of the structures of vocalization (Jarvis et al. 2014; Whitney et al. 2014; Pfenning et al. 2014). At variance with the oscines, the syrinxes of their sister group, suboscines, show remarkable morphological diversity (Ames 1971). Using downstream and upstream acoustic models, Riede et al. (2019) suggested that since a sound-producing organ converts aerodynamic energy into acoustic energy and that sound originates from its interaction with the primary air passageways, e.g., the trachea of the air-breathing vertebrates, location of a sound-producing structure (e.g., the syrinx) at the base of the air passage, i.e., in an upstream position, confers superior sound producing efficiency. Compared with the larynx that receives only one air stream from the trachea, for the syrinx, the vocal tract comprises two air streams that issue from two extrapulmonary primary bronchi which converge at the organ (Plate 3.5a, b). In contrast to other vertebrates, the longer trachea of birds (Hinds and Calder 1971; Clench 1978; Gaunt et al. 1987; McLelland 1989b; Frith 1994; Fitch 1999) (Plate 1.1a) may have compelled birds and probably the dinosaurian proto-birds to evolve an upstream vocal organ (Riede et al. 2019).

The structure and function of syrinxes differ between species and sexes of birds (Miskimen 1951; George and Berger 1966; Chamberlain et al. 1968; Ames 1971; Warner 1972, 2009; King and McLelland 1984; King 1989; Brackenbury 1989; Riede et al. 2010; Riede and Goller 2010; Prince et al. 2011; Erdoğan et al. 2015; Abdel-Maksoud et al. 2020; Ibrahim et al. 2020). Structurally, the syrinx comprises a number of irregularly ossified cartilages that comprise its body, i.e., the skeletal framework, and the flexible (vibrating) membranes and soft tissues (labia) that interconnect the skeletal support (Plate 3.5a, b). The syrinx is complexly associated with intrinsic and extrinsic syringeal muscles which change the tension of the flexible structures as well altering the physical location, shape, and size of the organ (Gaunt et al. 1973, 1982; Gaunt and Gaunt 1985). The tympanum is the space that is surrounded by the syringeal cartilages and the membranous/soft tissue elements (King and McLelland 1984; King 1989; McLelland 1989b): the space is a direct continuation of the tracheal passageway and occurs in the tracheobronchial, and in some cases, in the tracheal types of syrinxes, the space is largely lacking in the bronchial type of syrinx.

In birds, the classification of the types of syrinxes is based on the origins of the constructing cartilages, i.e., tracheal or bronchial cartilages for the respective syrinxes or a combination of the two for the tracheobronchial syrinx. Most birds, e.g., the Domestic Fowl (*Gallus gallus* variant *domesticus*), have a tracheobronchial syrinx (Nickel et al. 1977; King and McLelland 1984; Baumel et al. 1993; Yildiz et al. 2003; Kabak et al. 2007) (Plate 3.5a, b); tracheal syrinxes occur in the suboscine superfamily Furnarioidea (e.g., woodcreepers, ovenbirds, antpipits, gnateaters, and tapaculos) and in some Ciconiidae (Stocks) while; bronchial syrinxes occur in some cuculiform species and owls (Beddard 1898; Greenewalt 1968; Warner 1971, 1972; King and McLelland 1984). The pessulus is a wedge-shaped cartilage that projects from the caudal border of the tympanum (Plate 3.5a), dividing the airway (tracheal lumen) into two separate passageways, namely, the right and left extrapulmonary primary bronchi.

The primary vibrating structures of the syrinx (Plate 3.5a, b) are the following: (a) generally well-developed paired medial typhaniform membrane that stretches between the free ends of the bronchial-syringeal cartilages: it is considered to be the most important vibrating membrane (Warner 1972; Nickel et al. 1977; King and

McLelland 1984; Baumel et al. 1993); (b) paired lateral tympaniform membranes that extend between the cartilages on the lateral aspect of the syrinx; (c) lateral labium which is a pad of elastic tissue that protrudes into the lumen of the syrinx (tympanum); and (d) the medial labium that projects into the lumen from the pessulus. Compared with the medial one, the lateral labium is less well-developed and is often lacking. The intrinsic and extrinsic syringeal muscles modulate sound production by tensing and/or relaxing the vibrating membranes and the labia while extrinsic muscles move the organ (syrinx) from one location to another (Fee et al. 1998). Hersch (1966) observed that when the tracheolateralis muscle contracts and the sternotrachealis one is relaxed, the medial tympaniform membranes tighten, and airflow through the syrinx creates either no sound or only low-frequency ones: reversal of the process, however, causes the membrane to bulge into the bronchial lumen and high pitched sound is produced by airflow. For the Common Crow (*Corvus brachyrhynchos*), Chamberlain et al. (1968) observed that syringeal function was mainly based on the activity of the syringreal muscles. In birds, vocalization occurs only during expiration (Miskimen 1951; Brackenbury 1989). Experiments on anesthetized House Sparrows (*Passer domesticus*) and freshly euthanized Stirlings (Family: **Sturnidae**) showed that the production of sound originates from the vibration of the medial tympaniform membranes (Miskimen 1951). In the Northern Cardinal (*Cardinalis cardinalis*) and the House Sparrow (*Passer domesticus*), Cozic and Homberger (2015) observed that pressure changes in the cutaneous cervicocephalic air sacs that are not connected to the respiratory system (i.e., lung-air sac system) alter the resonance of sounds generated by the syrinx. Clavicular and cervical air sacs that physically relate to the syrinx (Stein 1968; Lockner and Murrish 1975; Gaunt and Gaunt 1985; Kabak et al. 2007; Cozic and Homberger 2016) (Plate 1.1d) and the suprasyringeal structures and organs of the neck modulate the raw sound produced by the syrinx (Cozic et al. 2014). The pitch of the sound created by the syrinx correlates with abdominal muscle activity and variations in air sac driving pressures (Gaunt et al. 1976; Gaunt and Gaunt 1977; Brackenbury 1977, 1978; Gillooly and Ophir 2010; Erdoğan et al. 2015). In the Domestic Fowl (*Gallus gallus* variant *domesticus*) (Youngren et al. 1974) and the female Mallard Duck (*Anas platyrhynchos*) (Lockner and Youngren 1976), synchronization between air sac pressures and airflow rate is critical to sound production.

3.2.4 Lung

The left and right extrapulmonary primary bronchi run from the tracheal bifurcation, pass through the horizontal septum, and approach the lungs at an angle of ~45° to enter them at the hilum which is located on the septal surface between the cranial and the middle third of the organ (McLelland 1989a) (Plates 1.1a, b, d, 3.6a–e, 3.7a, b, 3.8a–e and 3.9a–e). Structures such as airways (the left and right extrapulmonary primary bronchi), blood vessels (pulmonary artery and vein), lymphatic vessels, and nerves enter or leave the lung at the hilum (Plates 1.1a, b, d and 3.6a–e). The avian

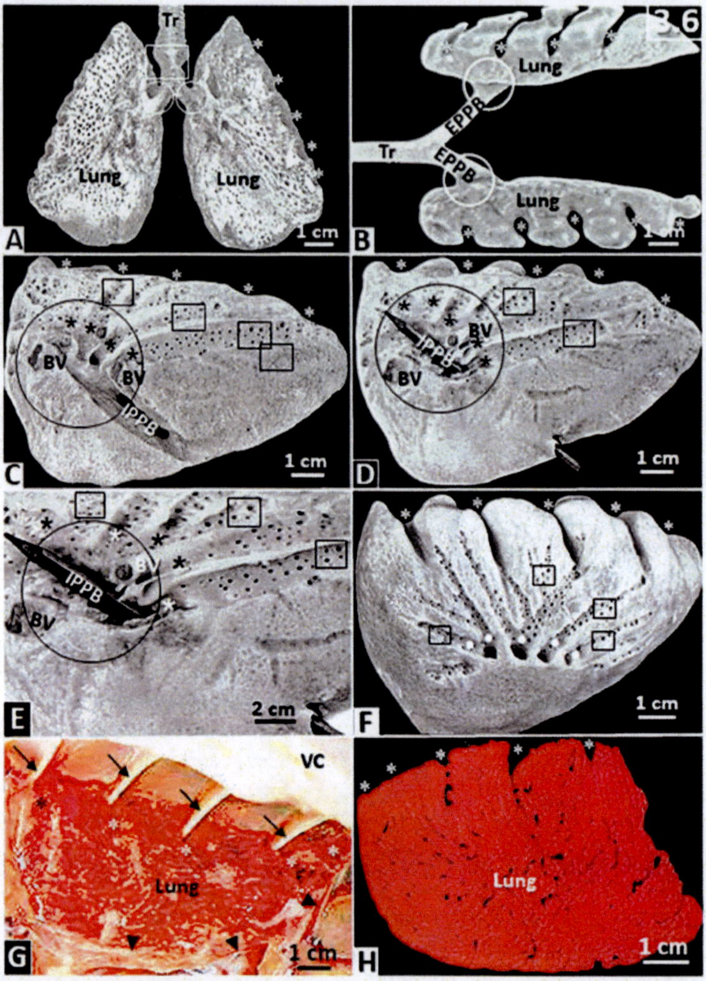

Plate 3.6 (**a**) Ventral view of lungs of the Domestic Fowl (*Gallus gallus* variant *domesticus*). Tr, trachea; circles (O), hilus; boxed area, location of a tracheobronchial syrinx. (**b**) Dorsal view of the lungs of the Ostrich (*Struthio camelus*) showing costal impressions (sulci) (asterisks, *). Tr, trachea; EPPB, extrapulmonary primary bronchi; encircled area (O) (**a**, **b**), hilus. (**c**–**e**) Medial view of the Ostrich lung showing various airways. Stars (⦶), medioventral secondary bronchi; IPPB, intrapulmonary primary bronchus marked by a pencil (**d**, **e**); BV, pulmonary blood vessel; asterisks (*), costal impressions (sulci) (**c**, **d**); boxed areas (□) (**c**–**e**), parabronchi; encircled areas (O), hilus. (**f**) Lateral view of the Ostrich lung showing mediodorsal secondary bronchi (dots, •). Boxed areas (□), parabronchi; asterisks (*), costal impressions (sulci). (**g**) Insertion of the lungs of the Domestic Fowl to the ribs (arrows, ←). Asterisks (*), costal impressions (sulci); arrowheads (▶), horizontal septum. (**h**) Lateral view of a latex rubber cast preparation of the lung of the Domestic Fowl showing the parabronchial complexity. Asterisks (*), costal impressions (sulci)

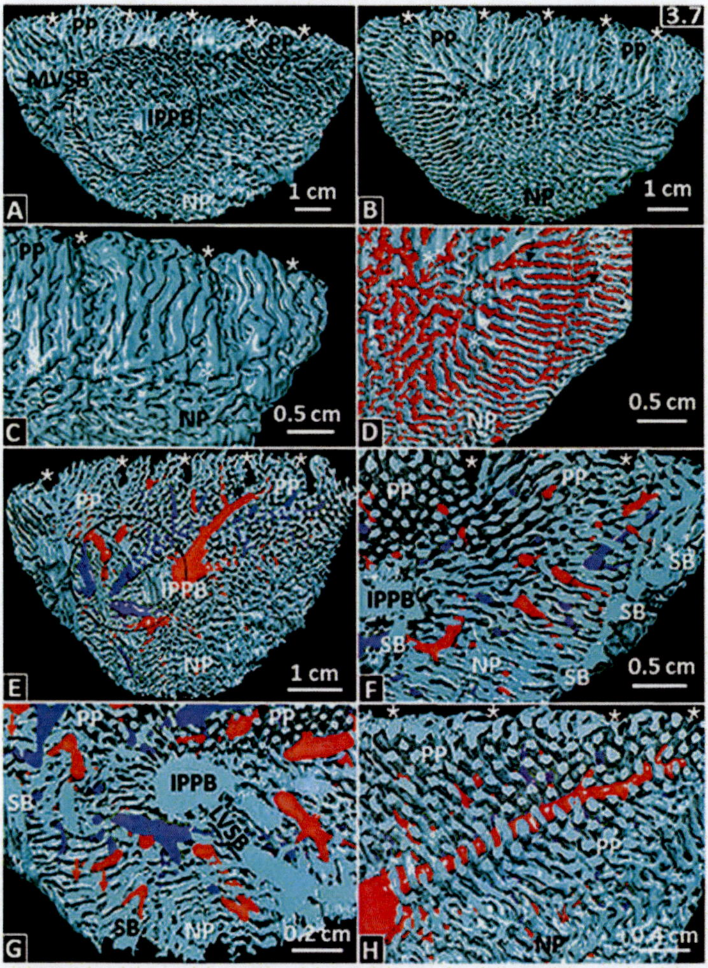

Plate 3.7 Three-dimensional serial section computer reconstruction of the lung of the Domestic Fowl (*Gallus gallus* variant d*omesticus*) showing the complexity of the airway system (**a**–**h**) and its topographical relationship with blood vessels (**d**–**h**). (**a**) Medial (**a**) and lateral (**b**) views showing airways. PP, paleopulmonic parabronchi; NP, neopulmonic parabronchi; stars (), costal impressions; IPPB, intrapulmonary primary bronchus; MVSB, medioventral secondary bronchi; asterisks (*), mediodorsal secondary bronchi; arteries are shown in red and veins in blue; SB, secondary bronchi and LVSB, lateroventral secondary bronchi (**g**)

lungs are small, compact (solid = unlobed), wedge-shaped paired organs (Plates 1.1i and 3.6a, b–d, f–h) that are located on the dorsal part of the coelomic cavity (King 1966; Duncker 1971; McLelland 1989a; Maina 2005) (Plates 2.5h and 3.6g) where they are deeply attached to the ribs and the thoracic vertebrae respectively on the lateral and the dorsal aspects. The impressions (costal sulci) are conspicuous from early in the development of the lung (Plates 2.5i, j, 2.6a–c and 3.6a–d, f–h).

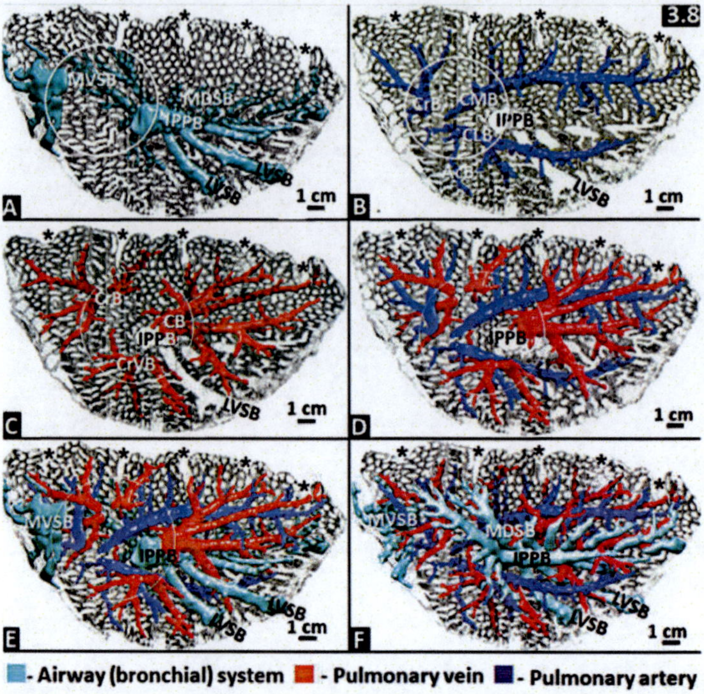

Plate 3.8 Three-dimensional serial section computer reconstruction of the lung of the Domestic Fowl (*Gallus gallus* variant *domesticus*) showing the topographical relationships between the airway (bronchial) and the vascular systems. (**a**) The airway system (shown in cyan). MVSB, medioventral secondary bronchi; MDSB, mediodorsal secondary bronchi; LVSB, lateroventral secondary bronchi; IPPB, intrapulmonary primary bronchus; circle (O), hilum; asterisks (*), costal sulci (**a–f**). (**b**) Branches of the pulmonary vein (shown in blue). CrB, cranial branch; CMB, caudal medial branch; CLB, caudal lateral branch; AcB, accessory branch. LVSB, lateroventral secondary bronchus; IPPB, intrapulmonary primary bronchus; asterisks (*), costal sulci. (**c**) Branches of the pulmonary artery (shown in red). CrB, cranial branch; CB, caudal branch; CrVB, cranial ventral branch; IPPB, intrapulmonary primary bronchus; LVSB, lateroventral secondary bronchus. (**d**) Topographic relationship between branches of the pulmonary artery and vein. (**e, f**) Medial (**e**) and lateral (**f**) views of the lung show topographic relationship between branches of the pulmonary artery, pulmonary vein, and the airway (bronchial) system

Depending on the shapes of the lungs, the number of costal sulci varies between avian species (Plates 1.1a, 3.7a–f, h, 3.8 and 3.9). Species such as the Domestic Fowl (*Gallus gallus* variant *domesticus*) (Plates 3.6a, g, h, 3.7a–f, h, 3.8 and 3.9), the Turkey (*Melleagris gallopavo*) and the Ostrich (*Struthio camelus*) (Plate 3.6b–d, f–h) have an average of five costal sulci while those of the Common Gull (*Larus ridibundus*) and the Mallard Duck (*Anas platyrhynchos*), of which the lungs are long and narrow, have as many as six costal sulci (McLelland 1989a). The avian lung has the following surfaces: the costal surface that contacts the ribs, the vertebral surface that relates to the vertebrae, the septal surface that borders the horizontal septum, and

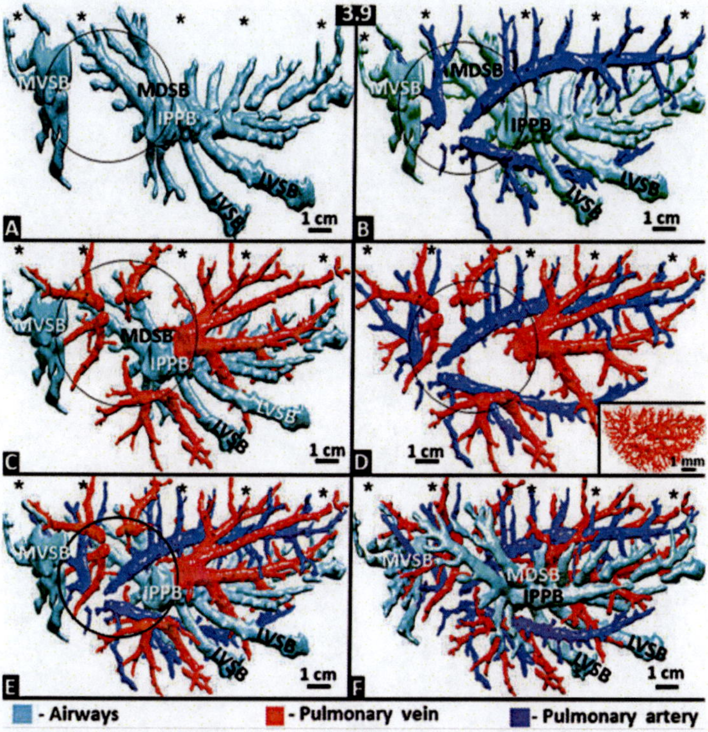

Plate 3.9 Three-dimensional serial section computer reconstruction of the airway and vascular systems of the lung of the Domestic Fowl (*Gallus gallus* variant d*omesticus*) of which the terminal branches and the rest of the structures have been cropped to better well-show the structure, topographical relationships and morphological features of the main branches of the bronchial (shown in cyan), the pulmonary artery (shown in red), and the pulmonary vein (shown in blue) systems. MVSB, medioventral secondary bronchi; MDSB, mediodorsal secondary bronchi; LVSB, lateroventral secondary bronchi; IPPB, intrapulmonary primary bronchus; circle (O), hilum; stars (*), costal sulci; inset (**d**), "unpruned" branches of the reconstructed pulmonary artery

the lateral surface that relates to the ribs. In birds, the parietal and the visceral pleurae lie close to each other: in some areas, the pleural cavity is traversed by collagen fibers while in others the space (cavity) is totally obliterated after fusion of the pleural coverings, a process that occurs from very early in the embryonic development of the chick lung (Plate 3.10a–f). The pleural "cover" comprises a rich network of connective tissue elements such as collagen, smooth muscle, and elastic tissue fibers (Plate 3.10f–h). In some areas, the pleura connects directly to the air and the blood capillaries of the lung (Plate 3.10c–f) while in other places secondary bronchi and parabronchi connect to pleura (Plate 3.10a, b).

With certain variations between species, about one-quarter of the volume of the lung is contained between the vertebrocostal attachments (King and McLelland 1984), and for the Domestic Fowl (*Gallus gallus* variant *domesticus*), between

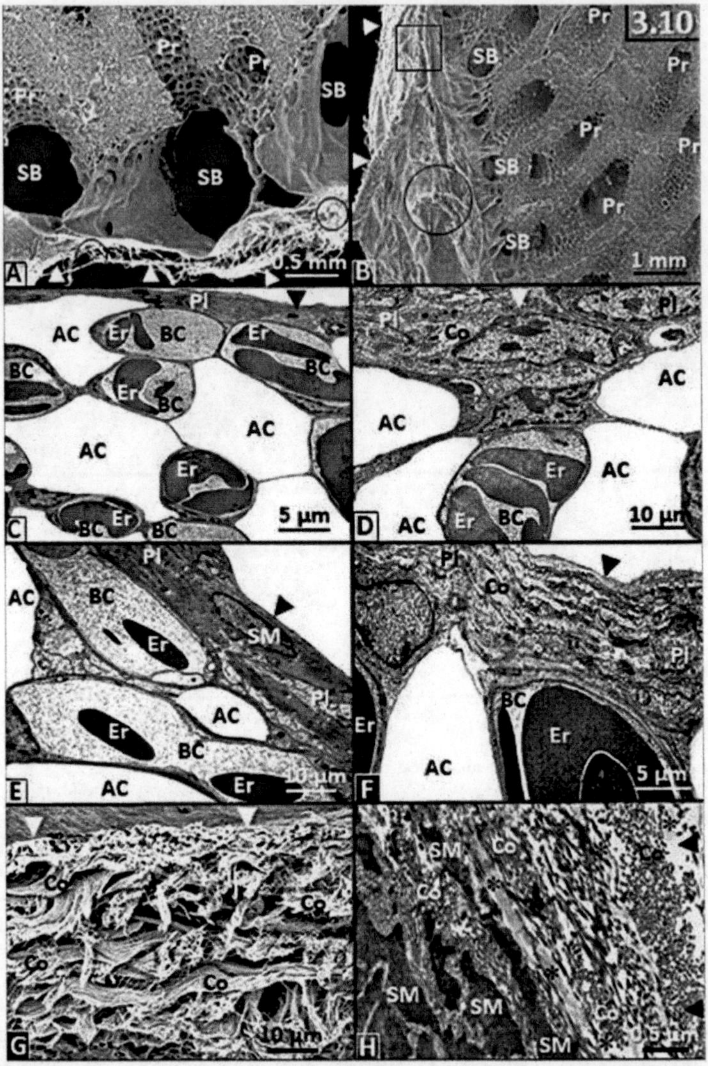

Plate 3.10 (**a**, **b**) Pleural coverings (arrowheads, ►) of the lungs of the Andean Goose (*Chloephaga melanoptera*) and the Domestic Fowl (*Gallus gallus* variant *domesticus*) (Pl) (**c**–**h**). For the Andean Goose, the parietal and the pleural coverings (parietal and visceral ones) are interconnected by collagen fibers but in some areas they connect directly, obliterating the pleural cavity (square, □) (**a**, **b**). (**c**–**h**) In the Domestic Fowl, the pleural cavity (space) is obliterated, i.e., the parietal and visceral layers are fused into a pleural covering (Pl). In the domestic fowl, the pleura covering contains collagen fibers (co), smooth muscle fibers (SM), and elastic tissue fibers (asterisks, *) (**d**, **f**, **g**, **h**). Blood and air capillaries of the lung of the Domestic Fowl connect directly to the pleural covering (**c**–**f**). SB, secondary bronchi; Pr, parabronchi; Pl, pleural covering; AC, air capillaries; BC, blood capillaries; Er, erythrocytes

one-fifth to one-third of the volume of the lung is held between the ribs (King and Molony 1971): the value increases during expiration. Although deep vertebral and costal impressions exist on the avian lung (Plates 1.1a, 1.6i, 2.2a, 2.5i, j, 2.6, 2.17a, h and 3.6a–d, f–g), unlike the mammalian lung (Plates 1.6j, 2.2b and 2.12), the avian one is not lobulated. In most species of birds, the lungs extend from about the first mobile rib (that marks the last cervical vertebrae) and caudally reach the cranial border of the ilium (King and McLelland 1984; McLelland 1989a; Maina 2005). In some species of birds such as geese, hoatzin, and stocks, however, the lungs reach close to the level of the hip joint (King and McLelland 1984; McLelland 1989a). Because the diaphragm or a structure homologous to it is lacking in birds, the liver and not the lungs surround the heart. With the exception of the intrapulmonary primary bronchus that runs deep through the substance of the lung in a craniocaudal direction, airways such as secondary bronchi and parabronchi (tertiary bronchi) are located superficially (Plates 2.6 and 3.6c–f, h). For the Ostrich (*Struthio camelus*) lung, the medioventral secondary bronchi (Plate 3.6c–e) and the mediodorsal secondary bronchi lie close to the surface of the lung for most of their lengths (Plate 3.6f) and for species of birds in groups such as penguins (Family: Spheniscidae), the Emu (Family: **Dromaiidae**), commorants (Family: Phalacrocoracidae) and to some extent in the Domestic Fowl (*Gallus gallus* variant *domesticus*), the intrapulmonary primary bronchus, the lateroventral secondary bronchi and the parabronchi are visible from the surface of the lung (McLelland 1989a). Latex rubber casting, three-dimensional reconstruction, and scanning electron microscopic methods well-show the location and the arrangement of the airway and the vascular systems of the avian lung (Plates 3.6h, 3.7, 3.8 and 3.9).

3.2.4.1 Airway (Bronchial) System

The airway (bronchial) system of the avian lung comprises a three-tiered assemblage of air conduits (Plate 2.2a) that consist of the intrapulmonary primary bronchus, secondary bronchi, and parabronchi (tertiary bronchi). As the intrapulmonary primary bronchus passes through the lung, from various parts/sites/aspects of its luminal surface, it gives rise to four sets of secondary bronchi, namely, the medioventral, the mediodorsal, the lateral ventral, and the laterodorsal secondary bronchi (Plates 2.2a, 3.8a, e, f and 3.9a–c, e, f): topographically, the airways relate closely with the pulmonary blood vessels (Plates 3.8a, d–f and 3.9b, d–f). The secondary bronchi are named according to the part of the lung they supply air (King 1966, 1975, 1979; King and McLelland 1984; McLelland 1989a; Baumel et al. 1993).

The arrangement of the bronchial (airway) systems of the mammalian and the avian lungs differ remarkably (Plate 2.2a, b). In contrast to the mammalian lung where the airways form by dichotomous branching (bifurcation) (Plates 2.2b, 2.4c and 2.12a), for the avian lung, the bronchial system develops by monopodial sprouting of the secondary bronchi from the intrapulmonary primary bronchus (Plates 2.4f and 2.7g) and the parabronchi develop from the secondary bronchi in the same manner (Plate 2.8a–g) to form densely packed passageways (Plate 2.2a). While in the so-called "respiratory tree" or the "bronchioalveolar tree" of the mammalian lung the airway system ends blindly in alveoli (Plate 2.2b—insets), the analogous structures of the parabronchial (avian) lung, i.e., the air capillaries, are rather globular (rotund) microstructures that interconnect by narrow channels (Woodward and Maina 2005, 2008; Maina and Woodward 2009): strictly, the airway system of the avian lung does not terminate blindly.

The evolved morphological differences between the avian and the mammalian respiratory systems have determined the ventilatory mechanisms that have evolved in the designs of the two gas exchangers. Because the mammalian and the avian lungs develop from the primitive pharynx by invagination of reserved epithelial cells into the surrounding mesenchyme (Maina 2003a, b; Schittny and Burri 2004), the organs are ventilated tidally, i.e., in-and-out. However, because of the exceptional morphological complexity of the avian respiratory system, where the gas-exchanging part (lung) is physically separated from the ventilating parts (air sacs) (King 1966; Duncker 1971; Maina 2005; Pandey et al. 2015), the lungs of birds are continuously and unidirectionally ventilated (in a caudocranial direction) by synchronized activities of the cranial and caudal groups of air sacs (King 1966; Scheid 1979; Fedde 1980).

3.2.4.2 Intrapulmonary Primary Bronchus

On entering the lung at the hilum, the extrapulmonary primary bronchus becomes the intrapulmonary primary bronchus that was in the past termed mesobronchus (King 1979; Baumel et al. 1993; McLelland 1989a). During its passage through the lung, the intrapulmonary primary bronchus markedly changes in its course, shape, and size (Plate 2.2a, 3.8a, e, f, 3.9a–c, e, f and 3.11a–g): it gives rise to four sets of secondary bronchi and terminates on the caudal margin of the lung where it opens into the caudothoracic and the abdominal air sacs (King 1966; King and Molony 1971; McLelland 1989a; Maina 2005). To maintain its patency, the airway is supported by cartilaginous plates (Plate 3.4a), and its luminal surface is lined by a pseudostratified columnar epithelium (Plate 3.4c) that is organized in epithelial longitudinal folds (Plate 3.11h–j): the epithelium largely comprises intraepithelial mucous secreting goblet glands and ciliated cells (McLelland 1989a). In groups of birds such as penguins (Family: Spheniscidae), the Emu (Family: Dromaiidae), stocks (Family: Ciconiidae), and cormorants (Family: Phalacrocoracidae), the terminal part of the intrapulmonary primary bronchus lies close to the costal surface and is visible and easily accessible. In species of birds where the neopulmonic parabronchi are well-developed, the secondary bronchi and the parabronchi displace the intrapulmonary primary bronchus medially and deeper into the substance of the lung (McLelland 1989a): the airway (intrapulmonary primary bronchus) is not therefore easily accessible from the surface of the organ. In birds, for experimental studies such as those that entail measurement of airflow dynamics and its

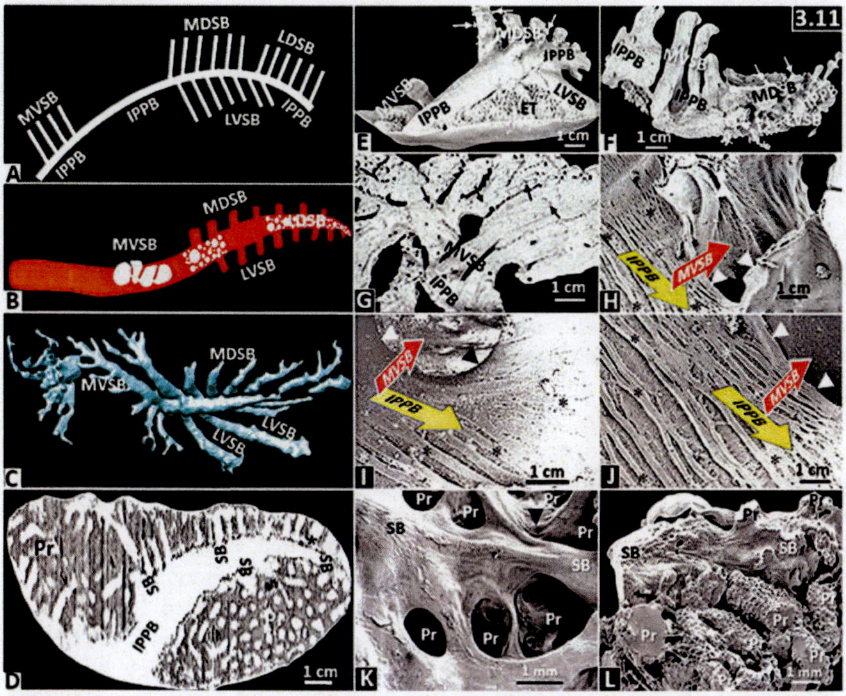

Plate 3.11 The airways of the avian lung. Four sets of secondary bronchi, namely, medioventral (MVSB), mediodorsal (MDSB), lateroventral (LVSB), and laterodorsal secondary bronchi (LDSB) originate from various luminal aspects of the intrapulmonary primary bronchus (IPPB) and the secondary bronchi (SB) give rise to parabronchi (arrows, ↓). SB, secondary bronchi; asterisks (*), epithelial mucosal folds (ridges) (**h–j**); arrowheads (▶) (**h–j**), narrowing and angulation of the orifices of the secondary bronchi. (**h–j**) The large yellow arrows (marked IPPB) show the direction of airflow in the intrapulmonary primary bronchi and the red one (marked MVSB) into the medioventral secondary bronchi

composition, the species of birds used should be carefully chosen for instructive morphological arrangement and placement of the airways, especially the secondary bronchi.

3.2.4.3 Secondary Bronchi

The secondary bronchi comprise the second-order airways of the avian lung (Plates 2.2a, 2.7c, h, 2.9b and 3.11a–l). For a short distance from their origin from the intrapulmonary primary bronchus, the airways are lined by pseudostratified columnar and ciliated and mucous goblet cells which are scattered among the epithelial cells (McLelland 1989a; Maina and Africa 2000). Further into the lumen of the secondary bronchus, the epithelium changes into a cuboidal one, especially at the parabronchial level (Duncker 1971; McLelland 1989a; Kandyel et al. 2021).

In the lungs of most species of birds, there are largely four medioventral secondary bronchi (King 1966; Duncker 1971; King and McLelland 1984; McLelland 1989a; Maina 2005): they are the largest group of secondary bronchi. They originate along a medially directed line from the dorsomedial luminal wall of the cranial third of the intrapulmonary primary bronchus near the hilum (King and McLelland 1984; McLelland 1989a; Maina and Africa 2000; Maina 2005) (Plates 3.8a, e, 3.9a–c, e, f and 3.11a–c, e–g). In the cassowaries (Family: Casuariidae) (Dotterweich 1930) and the pelicans (Family: Pelecanidae) (Schulze 1910), respectively, between five and six medioventral secondary bronchi exist. For the Ostrich (*Struthio camelus*) lung, Maina and Nathaniel (2001) and Maina et al. (2009) determined three medioventral secondary bronchi. In the Domestic Fow (*Gallus gallus* variant *domesticus*), the openings of the medioventral secondary bronchi have rather oval openings and are positioned at an angle relative to the course of the intrapulmonary primary bronchus (Plate 3.11h–j). The medioventral secondary bronchi spread out on the septal and the vertebral surfaces of the lung, with the first and the second airways arching forwards to deliver air to the ventral region of the lung. Depending on the species, the medioventral secondary bronchi open into the clavicular, cervical, and/or craniothoracic air sacs at the costoseptal border. Compared with the other secondary bronchi, for the Domestic Fowl (*Gallus gallus* variant *domesticus*), the fourth medioventral secondary bronchus has fewer branches (McLelland 1989a) and is lacking in penguins (Family: Sphenisdae) (Fischer 1905). In the lung of the Ostrich (*Struthio camelus*) (Maina and Africa 2000; Maina and Nathaniel 2001), the medioventral secondary bronchi lie close to the surface of the medial aspect of the lung (Plate 3.6c–e): they are easily reachable for experiments that involve aspects such as sampling and analysis of inspired air.

The mediodorsal secondary bronchi, the lateroventral secondary bronchi, and the laterodorsal secondary bronchi originate from the caudal two-thirds of the intrapulmonary primary bronchus (Plates 2.2a, 3.8a, e, f, 3.9a–c, e, f and 3.11a–f). Originating in a straight line from the dorsal luminal surface of the caudal part of the intrapulmonary primary bronchus, six to ten mediodorsal secondary bronchi exist in the lungs of birds (Vos 1934; Payne and King 1960) (Plate 3.11a–f). With the increasing development of the neopulmonic part of the lung, the mediodorsal secondary bronchi are displaced deeper in the material of the lung, making the airways less conspicuous and accessible from the costal surface. In most species of birds, with the first airway the largest, the sizes of the mediodorsal secondary bronchi decrease craniocaudally (Plates 3.8f, 3.9a, b, f and 3.11b–f) and the last airway is unbranched (Payne and King 1960). Most of the mediodorsal secondary bronchi are narrow at their origin and with the exception of the last two or three of them; their origins are inclined cranially at an angle of ~45° relative to the direction of the intrapulmonary primary bronchus (Plate 3.11e, f). The mediodorsal secondary bronchi supply air to the medial and the dorsal parts of the lung and do not open into any of the air sacs (McLelland 1989a). In the Ostrich (*Struthio camelus*) and the Domestic Fowl (*Gallus gallus* variant *domesticus*) lungs, the mediodorsal secondary bronchi lie close to the costolateral surface of the lung (Plates 3.6f and 3.8f): they are easily accessible from the surface of the lung for physiological experiments.

The lateroventral secondary bronchi of the avian lung are highly variable in number and size. They originate in a straight line from the caudoventral luminal surface of the intrapulmonary primary bronchus, directly opposite the openings of the mediodorsal secondary bronchi (Plates 2.2a, 3.9a–c, e, f and 3.11a–c, e, f). In the Domestic Fowl (*Gallus gallus* variant *domesticus*), where the first lateroventral secondary bronchus is particularly large (Plate 3.11e), a total of eight such airways (lateroventral secondary bronchi) exist (Payne and King 1960). The Mallard Duck (*Anas platryhynchos*) lung has about nine lateroventral secondary bronchi (Vos 1934) and from their very small size, the airways resemble parabronchi (tertiary bronchi) (Duncker 1971, 1972, 1974, 1978a, b, 1979). The lateroventral secondary bronchi supply air to the ventral and lateral regions of the lung. With certain variations between species, the lateroventral secondary bronchi indirectly or directly open into the caudothoracic air sacs. Together with the parabronchi that derive from them (Plate 3.11k, l), the lateroventral secondary bronchi, are located on the caudoventral part of the lung, i.e., between the intrapulmonary primary bronchus and the costoseptal border (Plate 3.7).

Compared with the other secondary bronchi, the laterodorsal secondary bronchi are the smallest and most variable secondary bronchi: they originate from the lateral aspect of the lumen of the intrapulmonary primary bronchus between the mediodorsal secondary bronchi and the lateroventral secondary ones. Both from the size and the presence of an exchange tissue mantle, in many important aspects, the laterodorsal secondary bronchi bear a resemblance to the parabronchi. In species of birds such as the Domestic Fowl (*Gallus gallus* variant *domesticus*), the first laterodorsal secondary bronchus is more prominent and lies superficially while the rest of them are covered by a network of neopulmonic parabronchi (Plate 2.2a). In contrast to the other secondary bronchi that originate in an orderly, i.e., successive manner, the origins of the laterodorsal secondary bronchi are scattered on the luminal surface of the intrapulmonary primary bronchus. The parabronchi that interconnect the medioventral secondary bronchi with the mediodorsal ones constitute the paleopulmonic part of the lung while those that interconnect the mediodorsal, the lateroventral, and the laterodorsal secondary bronchi form the neopulmonic one (Plates 2.2a and 3.7). In the lungs of penguins (Family: Spheniscidae), where the neopulmonic part is lacking, laterodorsal secondary bronchi are also absent (McLelland 1989a). In the Domestic Fowl (*Gallus gallus* variant *domesticus*), where the neopulmo is well-developed (Plates 3.6h and 3.7), 25–30 laterodorsal secondary bronchi exist (Payne and King 1960) while for the Mallard Duck (*Anas platryhynchos*) there are ~40 laterodorsal secondary bronchi (Vos 1934). The laterodorsal secondary bronchi supply air to the laterodorsal parts of the lung and do not directly connect to the air sacs (McLelland 1989a).

The origins of the secondary bronchi from the intrapulmonary primary bronchus greatly determine the shape and thickness of the bird lung (Duncker 1971). Together with their interconnecting parabronchi (Plates 2.2a, 3.6 and 3.7), the medioventral secondary bronchi and the mediodorsal secondary bronchi form the thick medial part of the avian lung and the entire cranial part of the organ while the lateroventral secondary bronchi and the laterodorsal secondary bronchi, together with their interconnecting parabronchi, comprise the relatively thinner lateral region of the lung (Plates 2.2a, 3.6c, d, h, 3.7a, b, e, 3.8e, f and 3.9e, f). At their openings from the intrapulmonary primary bronchus, most of the larger secondary bronchi display a narrowing (Plate 3.11h–j): in the Domestic Fowl (*Gallus gallus* variant *domesticus*) similar narrowings occur (Plate 3.11k, l). While for the mammalian lung diametric and histological changes occur progressively from the larger airways (bronchi) to the smaller ones (Weibel 1984; Khan and Lynch 2021), in the avian lung, the histological changes that occur from the secondary bronchi to the parabronchi are rather immediate (del Corral 1995). In some cases, however, with most of it comprising a cuboidal epithelium that is occasionally ciliated and has squamous cells, the epithelium lining some parts of the secondary bronchi histologically resembles some of that of the parabronchus (McLelland 1989a; del Corral 1995).

In species of birds such as the Emu (*Dromius novaehollandiae*), the auks (Family: Alcidae), the Ostrich (*Struthio camelus*), and the cormorants (Family: Phalacrocoracidae) where the laterally situated neopulmonic part of the lung lies close to the costal surface, the mediodorsal secondary bronchi lie superficially but in those of avian species such as the Domestic Fowl (*Gallus gallus* variant *domesticus*), ducks (Family: Anatidae), grebes (Family: Podicipedidae), loons (Family: Gaviidae) and passerine species (Order: Passeriformes), where the neopulmo and paleopulmo are well-developed (Maina 1988; Heard 1997; King and McLelland 1984; McLelland 1989a; Powell 2000), the airways (secondary bronchi) are located deeper into the "body" of the lung and are therefore not easily accessible from the surface of the lung. In the lung of the Mallard Duck (*Anas platyrhynchos*), where the neopulmo is poorly developed, the mediodorsal secondary bronchi lie close to the surface of the lung and are therefore easily reachable. (Duncker 1971). Compared with that of the Domestic Fowl (*Gallus gallus* variant *domesticus*), the lung of the Mallard Duck is more suitable for physiological experiments.

3.2.4.4 Parabronchi (Tertiary Bronchi)

The parabronchi originate from the secondary bronchi close to their (secondary bronchi) origin from the intrapulmonary primary bronchus (Plates 2.2a and 3.11k, l): they range in diameter from ~0.5 mm in lungs of hummingbirds (Family: Trochilidae), ~2 mm in those of the Coot (*Fulica atra*), the Mute Swan (*Cygnus olor*) and the Domestic Fowl (*Gallus gallus* variant *domesticus*) (Duncker 1971; Maina et al. 1982a, b; McLelland 1989a) to ~2.5 mm in the Ostrich (*Struthio camelus*) (McLelland 1989a). In the lungs of the 19 species of birds examined by Maina et al. (1982a, b), the parabronchial diameters ranged from 0.34 mm in the passerine species to 0.63 mm in those of the Greylag Goose (*Anser anser*) and the parabronchi constituted much of the volume of the avian lung. In the lungs of species of birds that have been investigated, the thickness of the exchange tissue mantle ranges from 0.2 to 1.0 μm. Because they contain much of the gas exchange tissue of the lung, at some level of organization, the parabronchi serve as respiratory units. In the lung of a mature Domestic Fowl (*Gallus gallus* variant *domesticus*), from latex

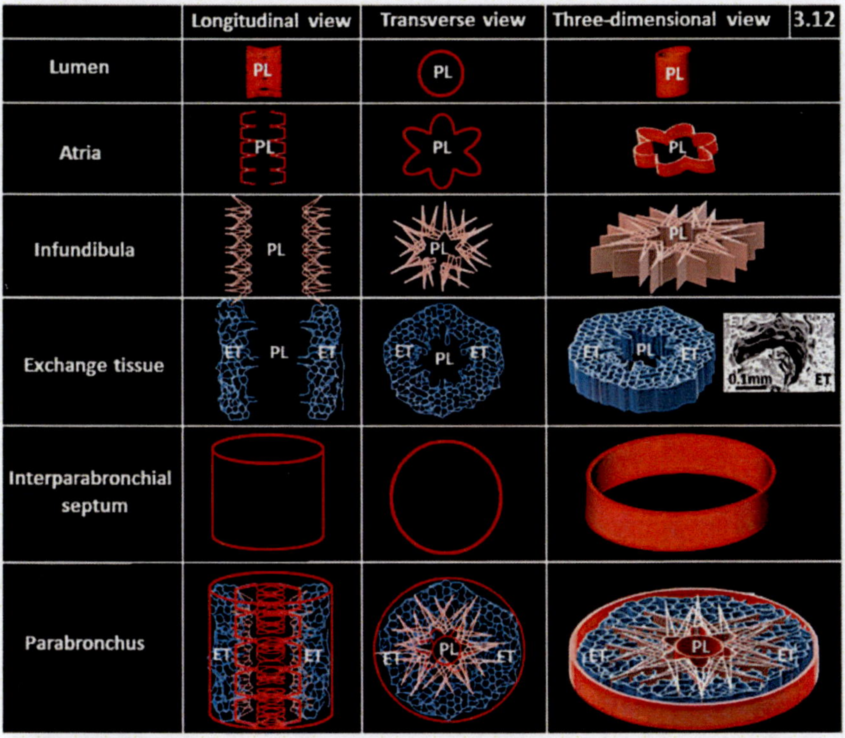

Plate 3.12 Schematic illustration of the arrangement of the main structural parts of a parabronchus. ET, exchange tissue; PL, parabronchial lumen. Inset: Fourth row (left end)—Scanning electron micrograph of a parabronchus of the lung of the Domestic Fowl (*Gallus gallus* variant *domesticus*)

rubber cast preparations (Plates 1.1d and 3.6h), measurements of individual parabronchi gave a total length of ~32 m (Maina, unpublished). Structurally, the parabronchi comprise five main parts, namely, a lumen, atria, infundibula, gas exchange tissue, and in the lungs of some species of birds an interparabronchial septa (King 1966; Duncker 1974, 1979; Maina 2005) (Plate 3.12). In the lungs of precocial and altricial species of birds (Duncker 1978a), differences exist in the development of the parabonchi: in precocial species such as the Domestic Fowl (*Gallus gallus* variant *domesticus*) and the Mallard Duck (*Anas platyrhynchos*), at hatching, the parabronchi are well-developed with thick gas exchange tissue mantle and the atria are conspicuous (Plate 2.11f–k) while for the altricial species such as the Australian pelican (*Pelecanus conspicillatus*), the parabonchi are small, the atria are shallow and the exchange tissue mantle is thin (Seymour et al. 2004; Runciman et al. 2005).

To various extents, parabronchi interconnect the secondary bronchi and form the bulk of the lung volume (Plates 2.2a and 3.13). Those parabronchi that connect the medioventral and the mediodorsal secondary bronchi comprise the paleopulmonic

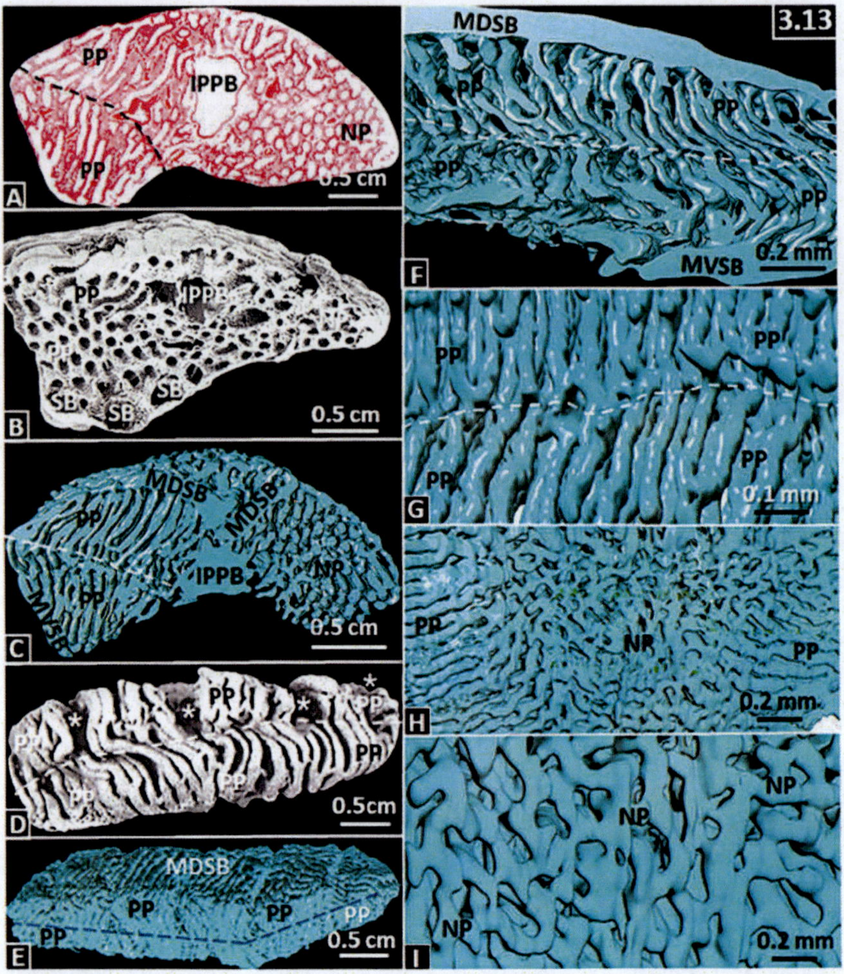

Plate 3.13 The paleopulmonic (PP) and neopulmonic parabronchi (NP) of the lung of the Domestic Fowl (*Gallus gallus* variant *domesticus*). (**a**) A histological preparation; (**b**) Scanning electron micrograph (**b**); (**c**, **e**, **f**–**i**) Three-dimensional serial section computer reconstructions and; (**d**) Latex rubber cast preparation scanning electron micrograph. The paleopulmonic parabronchi (PP) are arranged as stacks which are located on the craniodorsal part of the lung: they mainly comprise the parabronchial interconnections between the medioventral (MVSB) to mediodorsal secondary bronchi (MDSB) while the neopulmonic parabronchi (NP), which anastomore profusely, are largely located on the caudoventral aspect of the lung and consist of the parabronchial connections between the MDSB, the lateroventral secondary bronchi, and the laterodorsal secondary bronchi. SB, secondary bronchi; stars (⃞), costal sulci (impressions); dashed lines (**a**–**g**), line of anastomoses

part of the lung that is mostly found on the cranial half of the lung while those that connect the mediodorsal, the lateroventral, and the laterodorsal secondary bronchi comprise the neopulmonic part of the lung which is mostly located on the lateroventral region of the lung (Duncker 1971; King and McLelland 1984; McLelland 1989a; Maina 2005) (Plate 2.2a). For the Mute Swan (*Cygnus olor*), Duncker (1971) observed that the parabronchi of the first and the second medioventral secondary bronchi connect the parabronchi of the first two mediodorsal secondary bronchi while the parabronchi of the third and fourth medioventral secondary bronchi join those of the rest of the mediodorsal secondary bronchi as well as those of the lateroventral secondary bronchi. The part on the vertebral surface of the lung where the parabronchi from the medioventral and the mediodorsal secondary bronchi connect is termed the plane of anastomosis (*planum anastomotica*) and the actual point where they meet is marked by a perceptible stripe-like extent called the line of anastomosis (*linea anastomoticum*) (Plate 3.13a–g). While in the lungs of species of birds such as the Domestic Fowl (*Gallus gallus* variant *domesticus*) (Plates 3.7a, e and 3.13f, g) and those of swans (Family: Anatidae), the parabronchi anastomose profusely, and the sites of anastomoses are noticeable (McLelland 1989a) for the cranes (Family: Gruidae), the points of connection are not clear (Duncker 1971; McLelland 1989a). In the paleopulmonic part of the avian lung, parabronchial anastomoses are regular (Plate 3.13f–h) while those of the neopulmo occur profusely (Duncker 1971) (Plates 3.13i and 3.14a–d). While all the parabronchi have a parenchymal, i.e., gas exchange tissue, mantle, for the secondary bronchi, only the very small laterodorsal secondary bronchi (that are of equivalent size to the parabronchi) have a gas exchange tissue mantle (Plates 3.14, 3.15 and 3.16).

Atria evaginate from the parabronchial lumen and give rise to a variable number (as many as 8 in number) of infundibula (Plates 3.14, 3.15a–j, 3.16 and 3.17) which in turn give rise to air capillaries that together with the blood capillaries comprise most of the gas exchange tissue of the avian lung (Plate 3.15k, l). In the lungs of some species of birds, interparabronchial septa mark the external limits of parabronchi (Plates 3.4a, b, d, 3.15a–c and 3.16) and where they are lacking, interparabronchial arteries and veins mark the extents of the individual parabronchi (Plate 3.15d, g, h, i). Hexagonal shaped parabronchi (Plate 3.18a–d) noticeably exist in lungs of especially the galliform group of birds such as the Domestic Fowl (*Gallus gallus* variant *domesticus*) (Plates 3.4a, b, d, 3.15a, b, 3.16 and 3.18e–g): they are relatively less well developed in the lung of the Andean Goose (*Chloephaga melanoptera*) (Maina et al. 2017) (Plates 3.15c–f and 3.18h–k). In the lungs of the passerine species (Plate 3.15g, h) (Maina et al. 1982a, b; Maina 1984), the Ostrich (*Struthio camelus*) (Maina and Nathaniel 2001) (Plate 3.15i) and the Emu (*Dromaius novaehollandiae*) (Maina and King 1987) the interparabronchial septa are poorly developed and even lacking (Plate 3.15j). Hexagonal structures occur in nature and biology: the bees' wax honeycomb cells are a well-known example (Plate 3.18l). For the avian lung, the gas exchange tissue is located between the interparabronchial septa (in lungs of species where they exist) or the interparabronchial blood vessels (where interparabronchial septa are lacking) and the parabronchial lumen, precisely

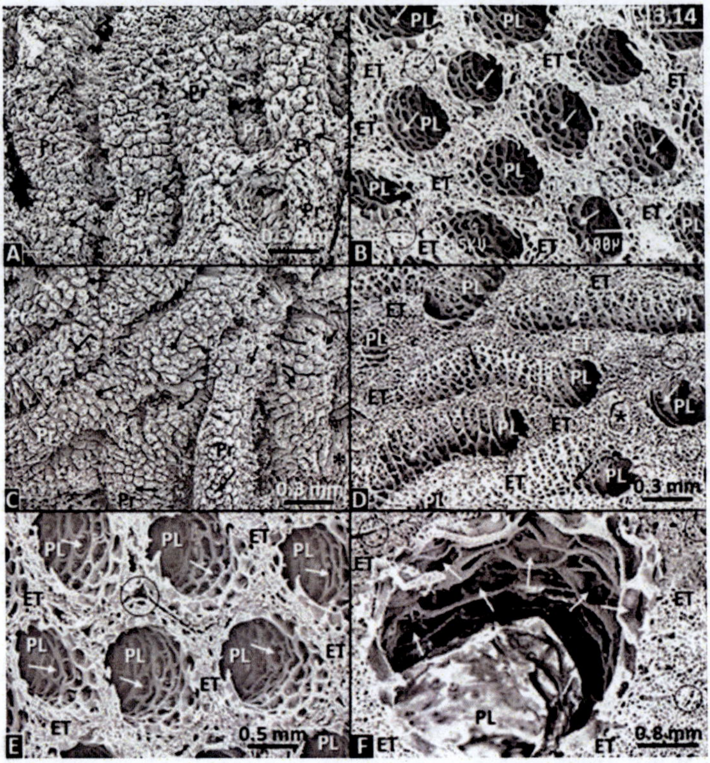

Plate 3.14 Morphology of the parabronchi of the lung of the Domestic Fowl (*Gallus gallus* variant *domesticus*). The paleopulmonic parabronchi are arranged as stacks of parallel air conduits (**a**, **b**) that anastome sporadically while the neopulmonic ones (**c**, **d**) anastomose profusely. The parabronchi (Pr) comprise parabronchial lumina (PL) that are surrounded by gas exchange tissue (ET) (**e**, **f**). The airways are separated by a connective tissue band, the interparabronchial septum (circles, O) (**b**, **e**, **f**). From the parabronchial lumen, atria (arrows, ←) (**a**–**f**) invaginate into the exchange tissue

from the terminal part of the infundibula, where air capillaries start (Plate 3.16 and 3.17).

Variations of the morphologies of the parabronchi exist between species of birds: interparabronchial septa are lacking in lungs of some species (Duncker 1971, 1979; Maina et al. 1982a, b; Mclelland 1989a; Maina 2005). In seven orders of birds examined by Maina et al. (1982a, b), the passeriform, columbiform, cuculiform, and psittaciform species lacked interparabronchial septa; lungs of the anseriform group lacked interparabronchial septa or they were poorly developed and; the partitionings were well developed in the charadriiform and the galliform species. For the avian lung, the hexagonal parabronchial shape may allow stronger airways (parabronchi) and structural components such as the air capillaries (Maina 2007a, b) to form and more intense packing of the respiratory units to occur in a small fixed (rigid) lung

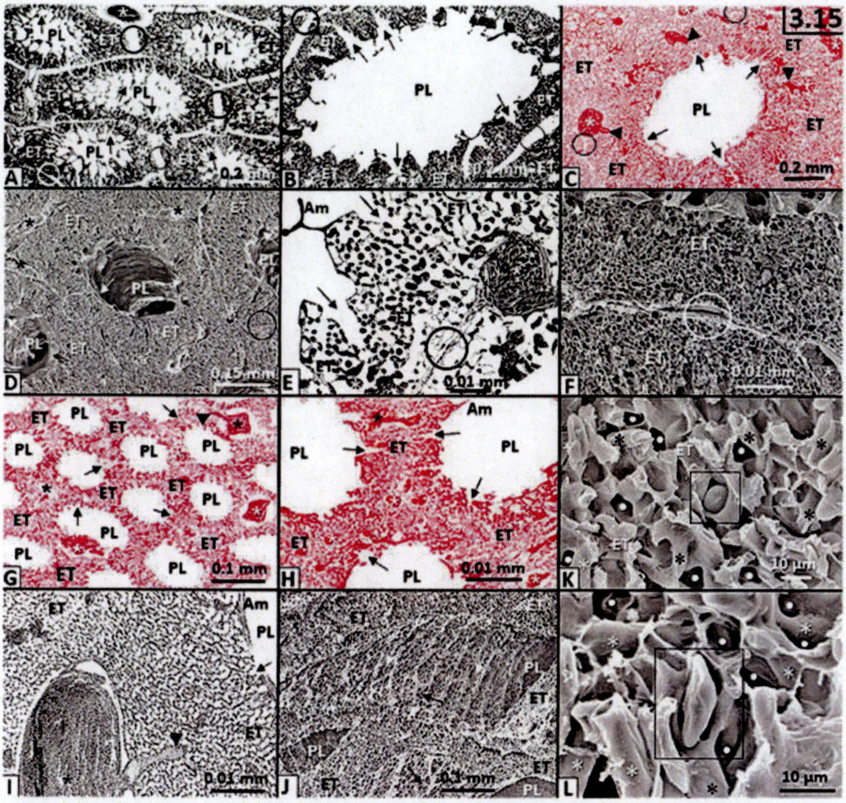

Plate 3.15 The structure of the parabronchi of the avian lung. They comprise a parabronchial lumen (PL) that is surrounded by an exchange tissue mantle (ET). In certain species of birds such as the Domestic Fowl (*Gallus gallus* variant *domesticus*) (**a, b**) and the Andean Goose (*Chloephaga melanoptera*) (**c–f**) the extents of the parabronchi are marked by interparabronchial septa (circles, O) in which interparabronchial blood vessels (stars, ▯) are located. Lungs of birds such as the passerine species (**g, h**), the Ostrich (*Struthio camelus*) (**i**) and the the Emu (*Dromaius novaehollandiae*) (**j**) lack interparabronchial septa. Arrows (←), atria; arrowheads (▶), intraparabronchial blood vessels. The air capillaries (dots, •) and the blood capillaries (asterisks, *) of the lung of the Andean goose comprise the gas exchange tissue (**e, f, k, l**). The red blood cells (squares, □) that occupy air capillaries (**k, l**) indicate how small the terminal respiratory units of the avian lung (the air capillaries) are

(Plates 1.1a, 2.6, 3.6 and 3.7): it provides a larger volume of gas exchange tissue and consequently vast respiratory surface area and large pulmonary capillary blood volume, structural features that provide large pulmonary diffusing capacity for oxygen (Maina 1989c, 2005; Maina et al. 1989b). The lungs of small, energetic birds such as the passerine species generally possess smaller parabronchi with a narrower lumen and greater volume density of exchange tissue while the interparabronchial septa are poorly developed or lacking (Duncker 1971; Maina et al. 1982a, b). Lungs of weak flyers and flightless species such as the galliform and

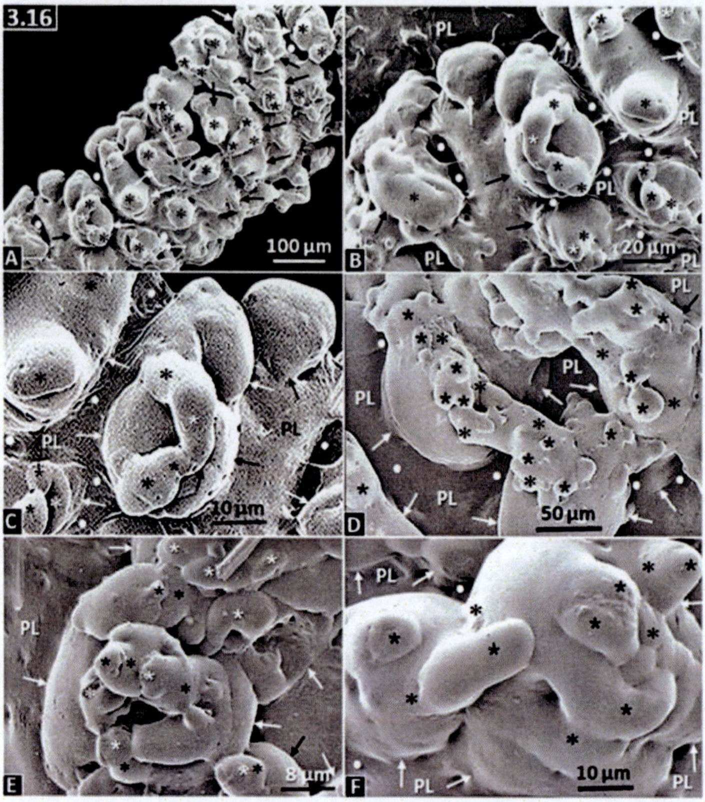

Plate 3.16 Latex rubber cast preparations of the parabronchi of the lung of the Domestic Fowl (*Gallus gallus* variant *domesticus*). The atria (arrows, ←) are separated by interatrial septa (dots, •). They evaginate outwards from the parabronchail lumen (PL) and give rise to infundibula (asterisks, *) that in turn generate air capillaries (stars, ⚹)

the penguin species have relatively wider parabronchi that have large atria, thinner gas exchange mantle, and thin to well-developed interparabronchial septa (Abdalla and Maina 1981; Abdalla et al. 1982; Maina et al. 1982a, b; Maina and King 1987). Among the bird species, there does not appear to be a correlation between development of interparabronchial septa in the lung with flight capacity: the lungs of the small highly energetic passerine species lack interparabronchial septa (Maina 1984) (Plate 3.15g, h), those of the Andean Goose (*Chloephaga melanoptera*), a champion high altitude flyer, have poorly developed interparabronchial septa (Maina et al. 2017) (Plates 3.15c–f and 3.18b, k) that correspond to those of the lungs of the flightless Ostrich (*Struthio camelus*) (Maina and Nathaniel 2001) (Plate 3.15i) while the lungs of the flightless Emu (*Dromaius novaehollandiae*) lack interparabronchial septa (Maina and King 1989) (Plate 3.15j).

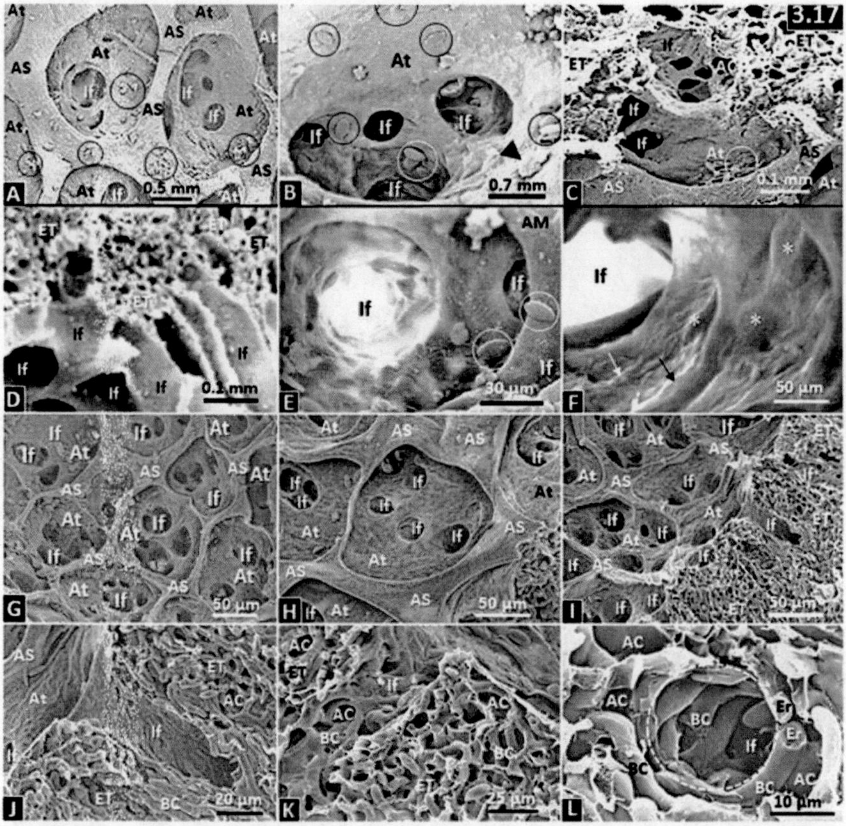

Plate 3.17 Atria and infundibula of the lungs of the Domestic Fowl (*Gallus gallus* variant *domesticus*) (**a–f**) and the Andean Goose (*Chloephaga melanoptera*) (**g–l**). For the lungs of the two species of birds, the atria (At) and the interatrial septa (AS) are well-developed. Unique to other birds that have so far been investigated and compared with those of the lungs of the Domestic Fowl (**a–f**), the infundibula of the lung of the Andean Goose are well-vascularized (**j–l**). If, infundibula; circles (O), extravasated erythrocytes lying on the atrial and infundibular epithelia (**a–c, e**); asterisks (*), putative type-II (granular) pneumocytes (**h**); arrowhead (▶), macrophage (**c**); arrows (←), surface macrophages in an infundibulum (**h**); AC, air capillaries; BC, blood capillaries; ET, exchange tissue; dashed encircled area (**l**), the opening of an infundibulum

3.2.4.5 The Paleo and Neopulmo

Morphologically, the avian lung displays heterogeneity (Duncker 1971; McLelland 1989a; Maina 2005). It consists of two developmentally and structurally distinctive parts which are designated paleopulmo (ancient or old lung) and neopulmo (modern or new lung). The paleopulmo occurs in the lungs of all species of birds while the neopulmo is lacking in the less-derived ones. Uncertainties exist on the validity of categorizing the avian lung into paleo and neopulmonic parts as well the evolution of the two parts of the lung (López 1995; Maina 2003a, b; Moura and Correia-Pinto

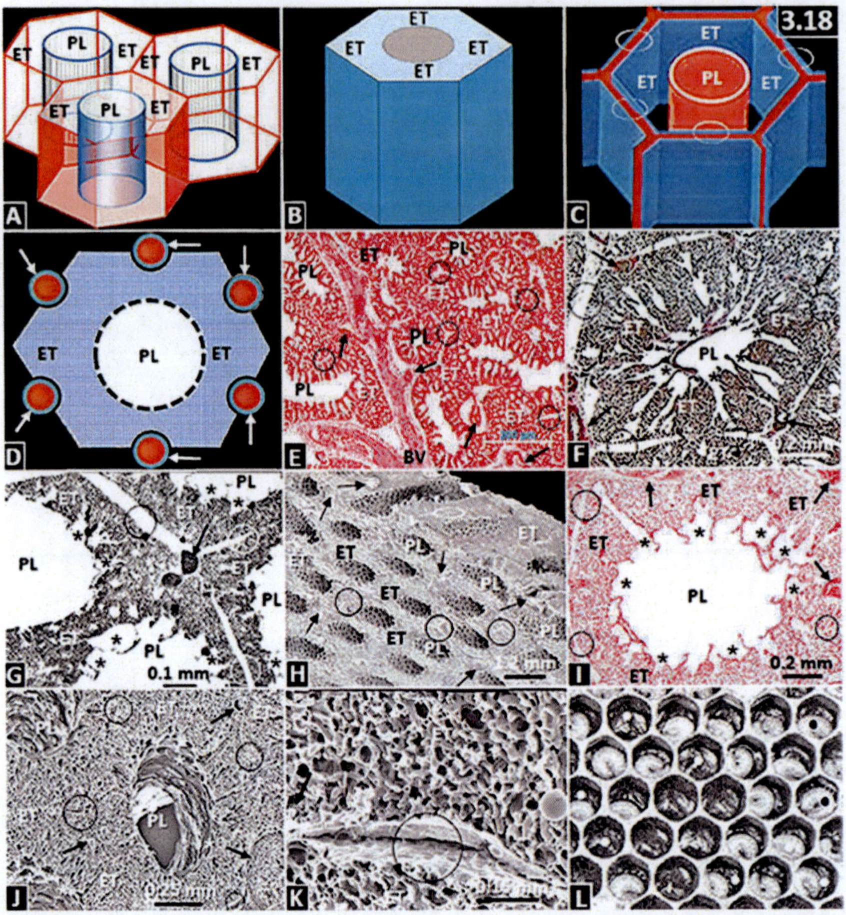

Plate 3.18 Hexagonal parabronchial shape (**a–d**) is displayed in the lungs of species of birds such as the Domestic Fowl (*Gallus gallus* variant ***domesticus***) (**e–g**), and the Andean Goose (*Chloephaga melanoptera*) (**h–k**), where interparabronchial septa (circles, O) are well-developed. Where interparabronchial septa are lacking, the interparabronchial blood vessels (arteries and veins) (arrows, ←) mark the outer limits of parabronchi (**d–k**). The wax cells of the honeycomb of bees are hexagonal (**l**) (dots, •). Pl, parabronchial lumen; ET, exchange tissue; arrows (↓), interparabronchial blood vessels; stars (⦀), atria

2017). Notwithstanding the equivocality, a distinction of the avian lung into paleo and neopulmonic parts has contributed to a certain understanding of some aspects of the function of the avian respiratory system. For example, for a long time, the difference between the concentration of CO_2 in the inspired air (i.e., the atmospheric air) and that reaching the caudal air sacs could not be explained (Scheid 1979; Fedde 1980, 1998; Scheid and Piiper 1989; Jaensch 2015). It is now known that the inspired air collects CO_2 from the exchange tissue of the neopulmonic parabronchi, as it flows towards the caudal air sacs, especially the abdominal air sacs (Piiper 1978). In penguins (Family: Spheniscidae), a group of birds whose lungs lack neopulmonic parabronchi, interestingly, gas tension measurements made by Powell and Hempleman (1985) showed higher concentrations of CO_2 in air in the caudal air sacs compared with the inspired one: the investigators proposed that in the penguin lung, airflow in all the paleopulmonic parabronchi may not occur in caudocranial direction. In the lungs of those species of birds where neopulmo is lacking or is poorly developed, the medioventral secondary bronchi are located more superficially and are therefore easily accessible from the costal surface of the lung for physiological studies (McLelland 1989a).

According to Duncker (1971), King and McLelland (1984), McLelland (1989a) and Maina (2005), the paleopulmo largely comprises the parabronchi that mainly join the medioventral and the mediodorsal secondary bronchi while the parabronchial interconnections between the mediodorsal, lateroventral, and the laterodorsal secondary bronchi constitute the neopulmo (Plates 2.2a, 3.6h, 3.7b, c, f–h, 3.13a–g and 3.14a, b). The most important structural, functional and developmental differences between the paleopulmonic parabronchi and the neopulmonic ones are the following: (a) the paleopulmonic parabronchi are largely found on the craniodorsal part of the lung where from their sizes and numbers make that part of the lung thicker while the neopulmonic parabronchi are located at the thin caudoventral part of the organ (Plates 3.6h, 3.7 and 3.13a–f); (b) in species of birds where it is best developed, the paleopulmonic parabronchi comprise as much as one-third of the volume of the lung (Duncker 1971; King and McLelland 1984; McLelland 1989a); (c) the paleopulmonic parabronchi are arranged as arch-like parallel stacks of airways that anastomose regularly (Plates 3.6h, 3.7f, h, 3.13a–h and 3.14a) while the relatively smaller neopulmonic parabronchi anastomose very profusely (Plates 3.6h, 3.7, 3.13h, i and 3.14c, d); (d) in the paleopulmonic parabronchi, the air flows continuously and unidirectionally in a caudocranial direction, i.e., from the mediodorsal to the medioventral secondary bronchi (Plate 3.13f) while that in the neopulmonic parabronchi the flow is tidal or bidirectional: the air flows towards the caudal air sacs during inspiration and towards the lung during expiration; and (e) the paleopulmo develop earlier than the neopulmo (Romanoff 1960; Maina 2003a, b). From the elaborate anastomoses of the neopulmonic parabronchi, the part of the avian lung that forms the neopulmo is designated "*pulmo reteformis*" (Duncker 1971; King 1979; López 1995) (Plates 3.7e–h and 3.13h, i) while because of the curved manner of the paleopulmonic parabronchi (Plates 3.6h, 3.7b, c, h and 3.13a–g), the part of the lung where they airways are found is termed "*pulmo arcuiformis*" (McLelland 1989a).

The degrees of development of the paleopulmonic and the neopulmonic parabronchi vary between the lungs of various species of birds (Duncker 1971; McLelland 1989a). The neopulmonic parabronchi are poorly developed in the lungs of birds such as stocks (Family: Ciconiidae) and the Emu (*Dromaius novaehollandiae*), where the airways comprise only a few laterodorsal secondary bronchi that originate from the most caudal part of the intrapulmonary primary bronchus. The size, the number, and the dorsocranial expansion of the neopulmonic parabronchi make the intrapulmonary primary bronchus curve as it passes through the lung. In cormorants (Family: Phalacrocoracidae) and auks (Family: Alcidae), the neopulmo includes the lateroventral secondary bronchi and their laterally directed parabronchi. In the cranes (Family: Gruidae), the laterodorsal secondary bronchi, that originate from the caudal part of the intrapulmonary primary bronchus, are covered by the neopulmonic parabronchi, displacing the other airways deeper into the substance of the lung. In plovers (Family: Charadriiformes), ducks (Family: Anatidae), gulls (Family: Laridae), owls (Family: Strigidae), buzzards (Family: Accipitridae) and parrots (Family: Psittacidae), the neopulmonic parabronchi include laterally directed parabronchi that originate from the proximal parts of the mediodorsal secondary bronchi. In pigeons (Family: Columbidae), galliform species, and passerine birds, the neopulmonic and the paleopulmonic parabronchi interconnect (McLelland 1989a; Maina 2005). In the galliform and the passerine species, where the neopulmonic parabronchi are well-developed, the diameter of the terminal part of the intrapulmonary primary bronchus is presumably from compression by such airways smaller than that of a single parabronchus. In the lungs of penguins (Family: Spheniscidae), the paleopulmo constitutes the entire lung, i.e., the neopulmo is lacking (Duncker 1971; Heard 1997; O'Connor 2004). In most species of birds, on average, the neopulmo comprises ~10% of the volume of the lung, and where it is most developed it may constitute as much as 20–25% of the volume of the lung.

Except for the topographical locations, sizes, and geometrical arrangement, there are no other significant differences between the morphologies of the paleopulmonic and the neopulmonic parabronchi: the two types of airways are involved in gas exchange. In the exchange tissue of the two parts of the avian lung, namely, the neo and the paleopulmo, the thicknesses of the blood-gas barrier and the length densities of the blood capillaries are comparable (Maina 1982b; Maina et al. 1983). Jones (1982) suggested that the tidally (bidirectionally) ventilated neopulmonic parabronchi may serve as reservoirs of CO_2 and by preventing excessive wash out of the gas from the exchange tissue (especially during such as panting under thermal stress and exercise), respiratory alkalosis is prevented. The Ostrich (*Struthio camelus*) can pant for as long as 8 h without suffering respiratory alkalosis Schmidt-Nielsen et al. (1969). In unanesthetized spontaneously breathing ducks, Holle et al. (1977) observed that compared with the whole lung, the neopulmo was better-perfused and together with the paleopulmonic parabronchi, blood flow decreased in the direction of the flow of ventilatory gas, reducing the PO_2 and increasing PCO_2 in the gas in the lung. Without providing physiological evidence, Duncker (1972) suggested that where well-developed in the avian lung, the neopulmo may serve as the most important or even as the exclusive site of gas exchange during rest while the paleopulmo may only become significantly involved during exercise. Since the evolution of the neopulmonic parabronchi afforded an alternative route for airflow to the caudal air sacs, (Duncker 1971) speculated that the reduced volume of air flowing through the intrapulmonary primary bronchus may explain the narrowness of the terminal part of the airway.

3.2.4.6 Atria and Infundibula

Duncker (1971, 1974, 1979) observed that for the avian lung, the conventional parabronchus is one that is characterized by a large diameter, prominent atria, thin gas exchange tissue mantle, and interparabronchial septa. The atria are rather polygonal air spaces that originate from the parabronchial lumen and project outwards into the periparabronchial gas exchange tissue mantle (Duncker 1971, 1974, 1979; Maina et al. 1982a, b; King and McLelland 1984; Mclelland 1989a; Maina 2005) (Plates 3.14, 3.15a–e, g–j, 3.16, 3.17a–c, g–k and 3.18a–c, j, k). From the floor of each atrium, 3 to 8 funnel-shaped passages called infundibula start (Plates 3.16 and 3.17d–f): in turn, the infundibula give rise to air capillaries which together with the blood capillaries comprise the gas exchange tissue (Plates 3.15, 3.17c, d, i–l and 3.19). In the lungs of the 19 avian species of birds that were investigated by Maina et al. (1982a, b), the atrial diameters ranged from 0.1 to 0.2 mm, and for the Rock Dove (*Columba livia*) and the Mallard Duck (*Anas platryhynchos*), respectively, the atria measured from 60 to 100 μm by 120 to 130 μm in diameter (West et al. 1977). Interatrial septa separate the atria (Plates 3.6, 3.14 and 3.17a, c, g–j) which together with the infundibula are lined by a cuboidal or rather squamous epithelia that contains scattered type-II cells (granular pneumocytes) (Plates 2.11j, k and 3.17h). Type-II cells exist in the epithelium that lines the parabronchial lumen, the atria, and the infundibula (Plate 2.11d). In lungs of the galliform species of birds such as the Domestic Fowl (*Gallus gallus* variant *domesticus*) (Plates 3.14, 3.15a, b, 3.16 and 3.17a–c), poor flyers such as coots (Family: Raliidae) (McLelland 1989a) and the Andean Goose (*Chloephaga melanoptera*) (Maina et al. (2017) (Plates 3.15d, e and 3.17g–j), the atria are surrounded by conspicuous interatrial septa (King and Cowie 1969; Gerisch and Schwartz 1972; West et al. 1977). Maina et al. (1982a, b) observed that atria, interatrial septa, and atrial muscles were well-developed in galliform, anseriform, and charadrifom orders of birds while the structures were poorly developed in the passeriform (Plate 3.15g, h), the columbiform, the cuculiform, and the psitaciform orders.

In birds, there does not seem to be a direct correlation between the degrees of the development of atria and flight capability. The lung of the high-altitude champion flyer, the Andean Goose (*Chloephaga melanoptera*) (Maina et al. 2017), has conspicuous atria (Plate 3.17g–i); the lung of the flightless Ostrich (*Struthio camelus*) (Plate 3.15i) and the Emu (*Dromaius novaehollandiae*) (Plate 3.15j) have very shallow atria (Maina and Africa 2000; Maina and Nathaniel 2001); lungs of the poorly or weakly flying galliform species have well-developed atria (Plates 3.16 and 3.17a–c) while; in lungs of the small passerine (songbird) species of birds the atria are poorly developed (Maina et al. 1982a, b) (Plate 3.15g, h). Amongst the lungs of species of birds that have been investigated and compared with those of the Domestic Fowl (*Gallus gallus* variant *domesticus*) (Plate 3.17d–f), the infundibulum of the lung of the Andean Goose (*Chloephaga melanoptera*) are unique in that they are very well-vascularized (Maina et al. 2017) (Plate 3.17j–l): the morphological feature may be an adaptive respiratory trait that reduces the atrial-infundibular respiratory

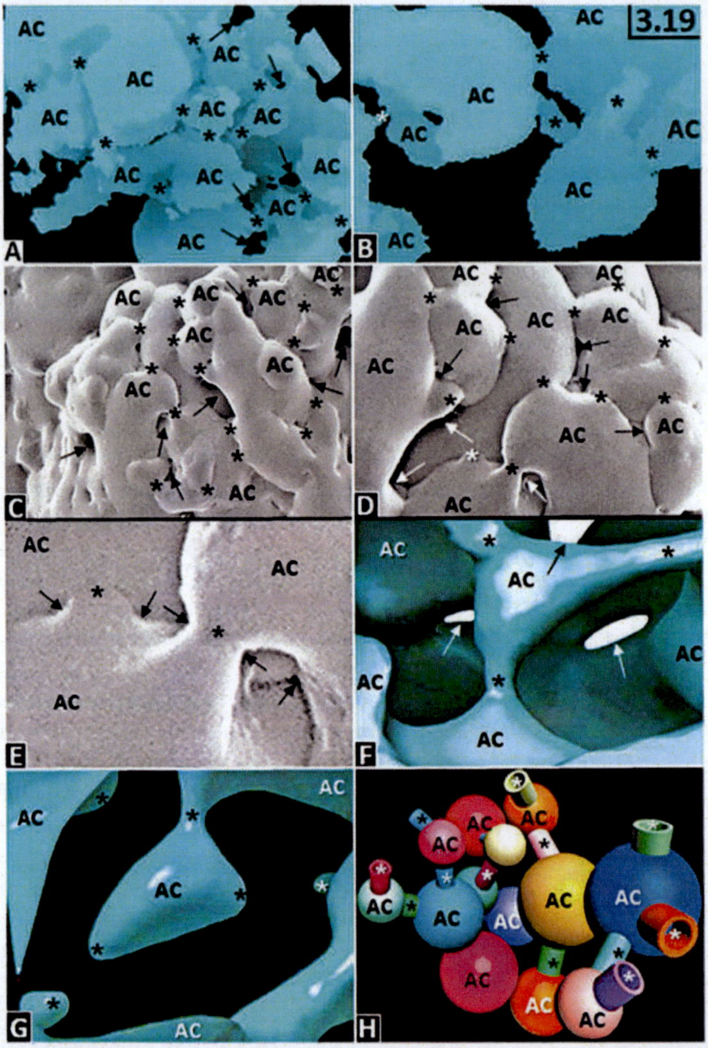

Plate 3.19 Morphology of the air capillaries of the avian lung shown by three-dimensional serial section computer reconstruction (**a**, **b**, **f**, **g**), latex rubber casting (**c**–**e**), and a schematic illustration (**h**). The air capillaries (AC) are rather spherical (globular) structures that are interconnected by narrow passageways (star, *). Arrows (⟵) are locations of blood capillaries that interdigitate closely with air capillaries

dead space while increasing the respiratory surface area and the pulmonary capillary blood volume in the lung. Among the species of birds that have been investigated, the Andean Goose (*Chloephaga melanoptera*), a high-altitude resident and flyer, has the highest mass-specific morphometric pulmonary diffusing capacity for oxygen (DLO_2) (Maina et al. 2017).

3.2.4.7 Air and Blood Capillaries

The air and the blood capillaries are the terminal respiratory units of the avian lung and comprise the exchange tissue of the avian lung (Plates 3.15, 3.17c, d, i–l, 3.18a–c, h, j, k, 3.19, 3.20 and 3.21): respectively, they originate from the infundibula and the intraparabronchial arterioles (Abdalla and King 1975, 1976a, b; Abdalla 1989). The diameters of the air capillaries range from ~3 μm in lungs of the small songbirds (passerine species) (Maina 1982a, b, 1989c; Maina 2005), 10 μm in penguins (Spheniscidae), swans (Anatidae), and coot (*Fulica atra*) (Duncker 1972) to ~20 μm in the Ostrich (*Struthio camelus*) (Maina and Nathaniel 2001). For the mammalian lung, the smallest alveoli (50 μm) that have been reported are those of an unnamed species of bat (Tenney and Remmers 1963). On account of their very small size, for a long time, the morphology of the air capillaries was misinterpreted: the structures were presumed to be "capillary-like" and hence the term "air capillary" was given to them (Rainey 1849; Eberth 1863; Huxley 1882; Schulze 1908, 1910;

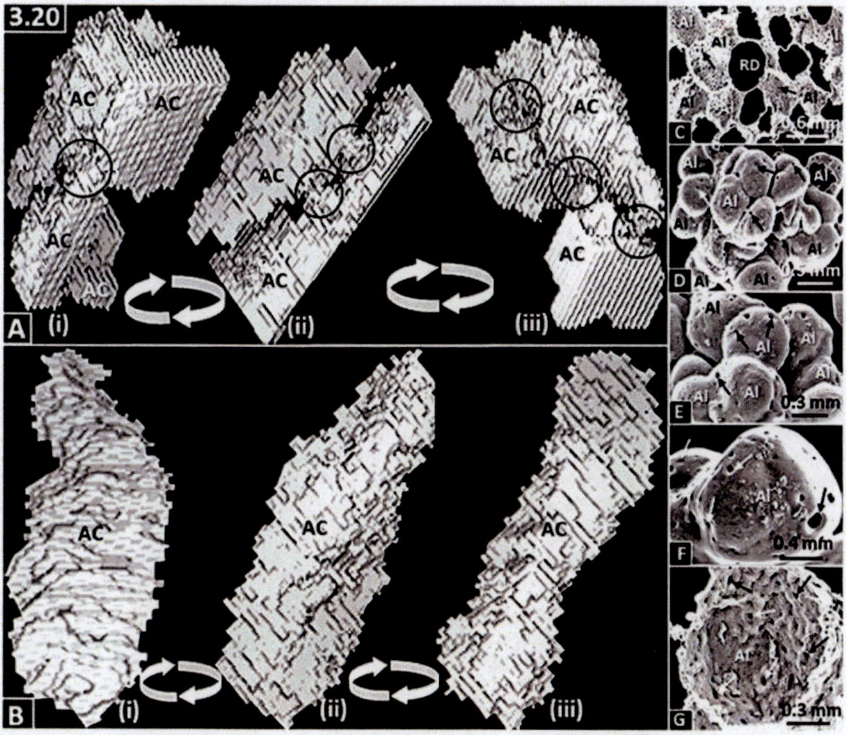

Plate 3.20 While most of the air capillaries (AC) of the lung of the Domestic Fowl (*Gallus gallus* variant *domesticus*) comprise rather globular structures that are interconnected across narrow passageways (circles, O) (**a**, i–iii), in rare cases, isolated air capillaries exist (**b**, i–iii). For the mammalian lung (**c–g**), the terminal respiratory units (alveoli, Al) are rather spherical structures that are interconnected by interalveolar pores (arrows (↓). RD, respiratory duct

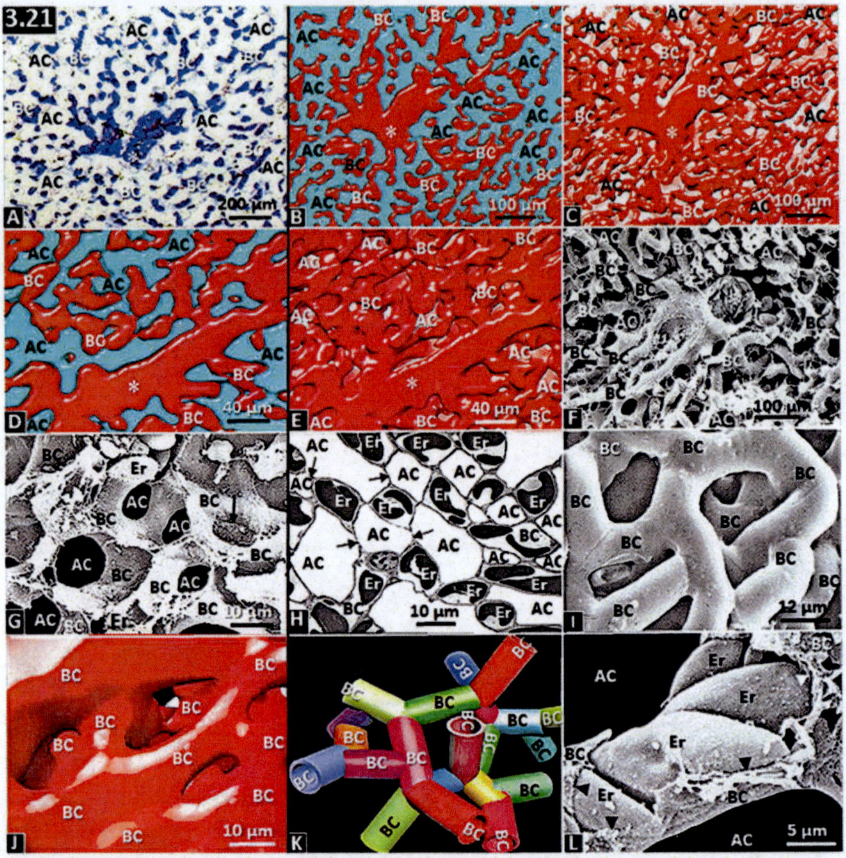

Plate 3.21 Morphology of the blood capillaries (BC) of the lung of the Domestic Fowl (*Gallus gallus* variant *domesticus*). (**a**) Toluidine stained preparation; (**b–e**) Three-dimensional (3D) serial section computer reconstructions—(**b**) and (**d**) show air capillaries (AC) (shown in cyan) and blood capillaries (shown in red) while for (**c**) and (**e**) the air capillaries have been extracted from the 3D preparation. (**f**) Scanning electron micrographs of a small blood vessel (arteriole) (asterisk, *) giving to blood capillaries (BC) that interdigitate with air capillaries (AC) (**f**) and close-up of blood capillaries in the exchange tissue (**g**); (**h**) Transmission electron micrograph of the exchange tissue showing air capillaries AC) and blood capillaries (BC) forming a dense network; (**i**) Scanning electron micrograph of a latex rubber cast showing that various parts of blood capillaries comprise parts that interconnect in 3D; (**j**) 3D serial section computer reconstruction of the blood capillaries that closely resembles the latex cast preparation (**i**); (**k**) Schematic illustration of the arrangement of the blood capillaries in the exchange tissue, where the capillary system comprises parts that interconnect in three-dimensions and; (**l**) Close-up of an opened part of a blood capillary showing erythrocytes (Er) tightly squeezing through the blood vessel. Arrowheads (▶), blood-gas barrier; arrows (↓) (**g**, **h**), epithelial-epithelial connections or epithelial laminae; asterisks (*), (**a–f**), arterioles

Krause 1922; Cover 1953). An account of the misidentification and misnaming of the air capillaries was discussed in Maina (2015a). By three-dimensional serial section computer reconstruction, it has now been shown that the air capillaries are not "capillary-like" or "tube-like": they are rather globular structures that are interconnected by very narrow passageways (Woodward and Maina 2005, 2008; Maina and Woodward 2009; Maina 2015b) (Plates 3.19 and 3.20a, b). Interestingly, morphologically, the air capillaries resemble the corresponding structures of the bronchioalveolar (mammalian) lung, i.e., alveoli, that are rather spherical and are interconnected by interalveolar pores, formerly pores of Kohn (Plates 2.2b and 3.20c–g), passageways that reportedly allow collateral ventilation (Terry et al. 1978; Menkes et al. 1979; Voshaar 2008): structurally and functionally, the interalveolar pores appear to be analogous to the narrow passageways that interconnect the globular parts of the air capillaries (Plates 3.19 and 3.20a). While most air capillaries are interconnected and are therefore directly supplied with inhaled air (Woodward and Maina 2005, 2008; Maina and Woodward 2009) (Plate 3.19), in rare cases, totally isolated air capillaries, that are not connected to the rest of the respiratory system and are therefore not ventilated with air, exist at least in the lung of the Muscovy Duck (*Cairina moschata*) that was studied by Woodward and Maina (2005) (Plate 3.20b). Interestingly, in avian anatomy, this state corresponds with the topographical anatomy of the air sacs were the so-called "cutaneous air sacs" do not connect to the much larger respiratory ones (Akester et al. 1973; Cozic et al. 2014; Cozic and Homberger 2015). The process by which the isolated air capillaries develop and their functional significance, if any, is currently unknown.

Until recently, the morphology of the blood capillaries of the avian was not well-known. By three-dimensional serial section computer reconstruction, Woodward and Maina (2005), Woodward and Maina (2008), and Maina and Woodward (2009) showed that in contrast to the typical structure of the blood capillaries such as those of the vertebrate skeletal muscles that are much longer than they are wide and largely run parallel to each other while anastomosing regularly (Pietschmann et al. 1982; Groom et al. 1995; Mathieu-Costello et al. 1992; Poole and Mathieu-Costello 1992), for the exchange tissue of the avian lung, the blood capillaries comprise individual parts that are about as long as they are wide and are interconnected three-dimensionally (Plate 3.21): the blood capillaries interdigitate very closely with the air capillaries (Plates 3.19a–g and 3.21a–k) and contain tightly packed erythrocytes which squeeze through them (Plate 3.21l). In the exchange tissue, the blood capillaries are literally "suspended" by epithelial-epithelial cells connections (epithelial retinaculae) and in rarer cases connect directly (Plate 3.22a–c). Folds of the epithelial laminae and the blood-gas barrier that are occasionally observed (Maina 2005) (Plate 3.22a–c) suggest that the air and the blood capillaries, may not be absolutely rigid, as generally suggested (Macklem et al. 1979; Powell et al. 1985; Maina 2007a, b: West et al. 2006, 2007a, b; West 2009, 2015, 2017). Connective tissue fibers that comprise mostly collagen, elastic, and smooth muscle fibers support the atria, the infundibula, and the air and blood capillaries (Plate 3.22d–o). At the parabronchial level, the exchange tissue is located between two conspicuous structural columns, namely, the atrial septa on the inside (Plates 3.22d–f and 3.23a) and the interparabronchial septa

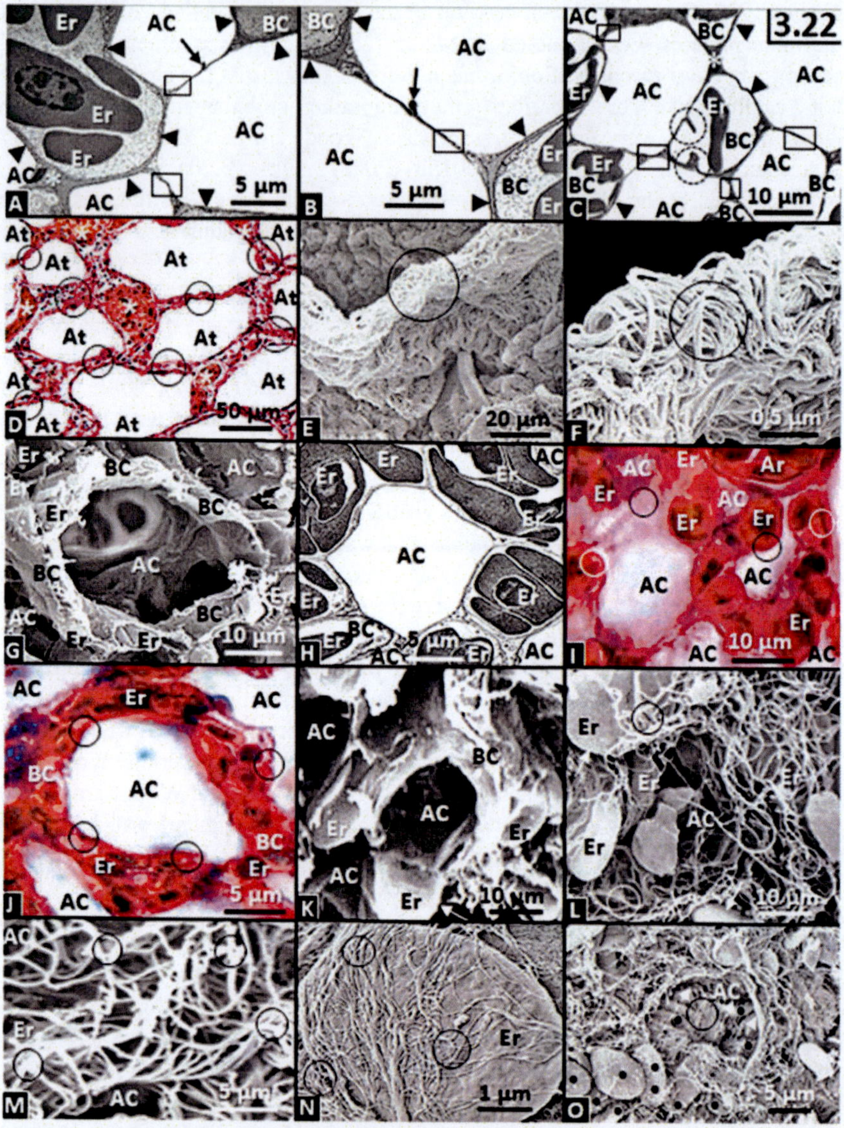

Plate 3.22 Organization of the exchange tissue of the lungs of the Graylag Goose (*Anser anser*) (**a–c**) and the Domestic Fowl (*Gallus gallus* variant *domesticus*) (**d–o**). The epithelial-epithelial connections (epithelial laminae) (areas marked with rectangles, □) and the blood-gas barrier (areas marked with arrowheads, ▶) of the exchange tissue of the lung of the Graylag Goose (*Anser anser*) (**a–c**). Folds that occur on the epithelial laminae (arrows, ←) (**a**, **b**) and the blood-gas barrier (dashed encircled areas) (**c**) suggest that the air capillaries (AC) and the blood capillaries (BC) may not be absolutely rigid as suggested but may to an extent vary in size. (**d–f**) Interatrial septa that separate atria (At) largely comprise collagen fibers (circles, O) and smooth muscle fibers (stars, ▯). (**g–l**) Air and blood capillaries (**g**, **h**) that comprise the gas exchange tissue of the avian lung. The epithelial cell laminae and the blood gas barrier contain collagen fibers that strengthen them (**i–o**). Collagen fibers preponderate in the walls of the blood vessels (**g–o**) and the blood-gas barrier

and/or interparabronchial blood vessels on the outside (Plate 3.23b–f). Although they entwine very closely (Plates 3.19, 3.21a–h and 3.24), the air and the blood capillaries of the exchange tissue of the avian lung are not morphological mirror images (Maina 1982a, 1988; Woodward and Maina 2005, 2008; Maina and Woodward 2009): in general, the air capillaries are larger than the blood capillaries (Fujii et al. 1981; Maina 1982a).

The respiratory surfaces of the gas exchangers (lungs) of air-breathing vertebrates are lined by a surface-active substance, a surfactant (Pattle 1978; Daniels et al. 1990, 1994, 1995; Sullivan et al. 1998; Wood et al. 2000; Bernhard et al. 2001, 2004; Orgeig et al. 2007, 2016; Bernhard 2016). A phospholipid-protein complex, the material is secreted by the type-II (granular or cuboidal) pneumocytes. While for the mammalian lung, the cells are located on the alveolar surface, for the avian lung, the cells are found on the surfaces of the atria and infundibula (Plates 2.11j, k and 3.17f): the cells are not found on the surface of the air capillaries (Maina 2005) (Plate 2.11l). It is likely that type-II cells are too large to "fit" in the minuscular air capillaries and that if they were it would greatly increase the thickness of the blood-gas barrier. Pulmonary surfactant enhances the flow of O_2 across the thin aqueous film that covers the respiratory surface (Olmeda et al. 2010). The occurrence of surfactant in the rigid avian lung, where the air capillaries are very has been deemed enigmatic and strong (Maina 2007a, b, 2008, 2015a; West et al. 2007a, b). Bernhard et al. (2001) observed that a surfactant composed primarily of dipalmitoylphosphatidylcholine and SP-B was sufficient to preserve the patency of the air capillaries of the avian lung. In contrast to the compliant mammalian lung where surfactant lowers surface tension and therefore stabilizes the alveoli (Daniels and Orgeig 2003; Orgeig et al. 2007, 2016), for the rigid avian lung where in the gas exchange tissue the blood capillaries are literally "suspended" by the epithelial-epithelial cells connections (epithelial retinacula) (Maina et al. 2010; West et al. 2010; Jimoh and Maina 2013) (Plates 3.21h and 3.22a–c) and are particularly strong (Powell et al. 1985; Maina 2007a, b; West et al. 2006, 2007a, b; West 2015), a surfactant should not be needed, at any rate to serve a surface tension reducing role. The surfactant is a multipurpose substance which in the avian lung may play other important functions such as prevention of exudation of blood plasma onto the respiratory surface (Keough et al. 1988) and protection of the blood-gas barrier from pathogenic microorganisms and injurious particulates (Pison et al. 1994). The blood-gas barriers of lungs and the water-blood barrier of fish gills comprise an epithelial cell, a basement membrane, and an endothelial cell (Plates 2.1 and 3.25a, b). In contrast to the mammalian lung, where thick and thin parts of the interalveolar septum exist (Plate 3.25b), for the avian lung, the blood-gas barrier is rather uniform in thickness and the pulmonary capillary blood is exposed to air across most of the blood-gas barrier (Maina 1989c, 2005, 2015c; Maina et al. 1989b) (Plate 3.25c, d).

Plate 3.22 (continued) contains collagen fibers (**i**–**o**). Er, erythrocytes; dots (•) (**o**), erythrocytes seen flowing onto and across the respiratory surface of an air capillary (dashed arrows)

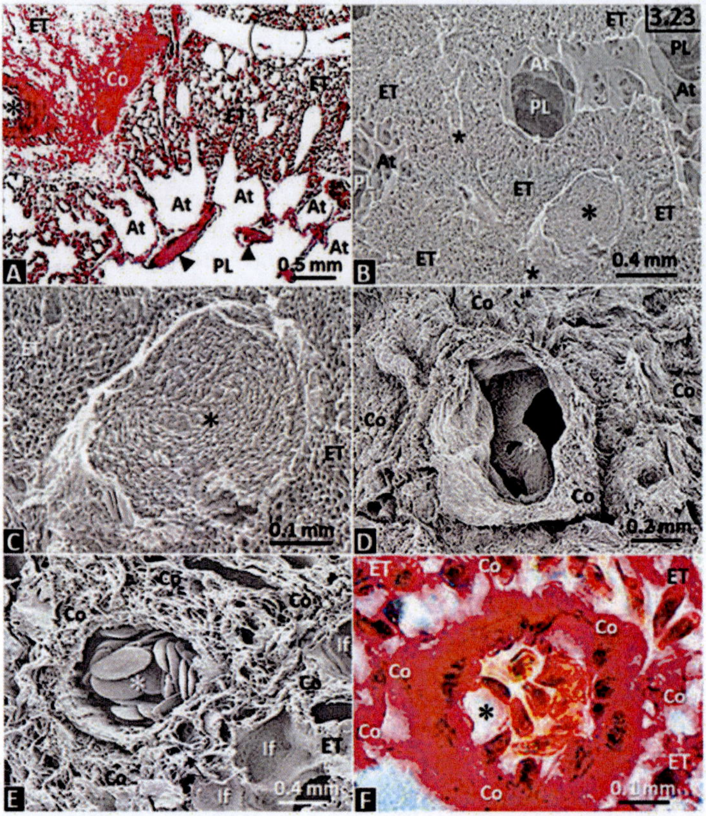

Plate 3.23 (**a–f**) The parabronchial gas exchange tissue (ET) of the avian lung is suspended between two supporting tissue columns: the external one comprises the interparabronchial blood vessels (asterisks, *) that are located in the interparabronchial septum (circle, O) (where it exist) and the internal one consists of the connective tissue elements such as smooth muscle (arrowheads, ►) (**a**) and collagen (Co) and elastic tissue fibers that form the atrial septa. At, atria; PL, parabronchial lumen; stars (✩), intraparabronchial blood vessesl (**a–f**); If, infundibula (**e**). Preparations (**a**) and (**f**) have been immunocytochemically stained to show collagen type-IV fibers

3.2.5 Pulmonary Circulation

Like the mammalian lung (Deffebach et al. 1987; Baile 1996; Rizzo et al. 2015; Jain et al. 2021), the avian pulmonary circulation comprises the following parts: (a) the pulmonary artery that delivers blood from the heart to the lung for oxygenation, (b) the pulmonary vein that returns oxygenated blood from the lungs to the heart and, (c) bronchial artery that supplies O_2 and nutriments to the main parts of branches of the airways and the blood vessels of the lung (Abdalla and King 1976b, 1977; Abdalla 1989; Wagner and Weir 1994; Shimoda 2006; Morrell 2014; Huetsch and Shimoda 2022). In vertebrates, in certain important aspects, the pulmonary

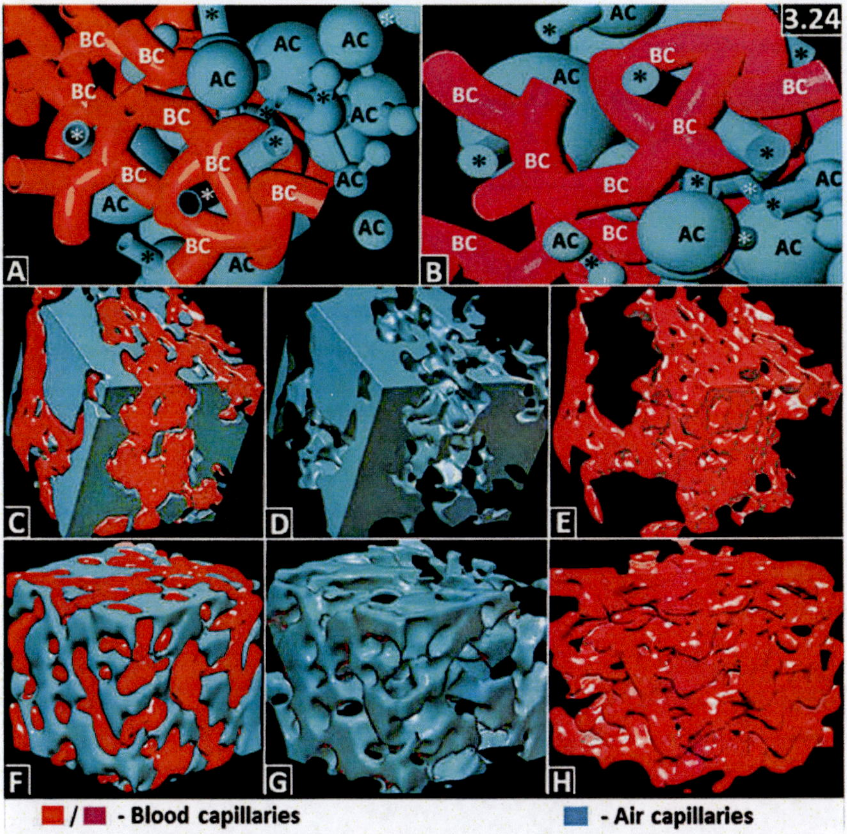

Plate 3.24 The air capillaries (AC) and the blood capillaries (BC) are the terminal respiratory units of the avian lung. The air capillaries are rather rotund structures that are interconnected by very narrow passageways (asterisks, *) (**a**, **b**) while the blood capillaries comprise parts that are connected in three dimensions (**a**, **b**). Although they interdigitate very closely (**a**–**h**), the air and the blood capillaries are not mirror images. (**a**, **b**) are computer-generated schematic illustrations while (**c**–**h**) are three-dimensional serial section computer reconstructions

circulation differs from the systemic one (Kardong 2009; Powell 2015) in the following respects: (a) with the volume of the capillary bed about equal to the stroke volume, the whole capillary blood volume is replaced with each heartbeat; (b) serving as a reservoir of blood, the pulmonary blood capillary system for a moment supports the left ventricular output during incidents of reduced right ventricular output; and (c) despite receiving large blood volume, the pulmonary circulatory system is both a low pressure and a low resistance vascular system, properties that minimize physical injury to the thin and delicate blood-gas barrier while decreasing or eliminating exudation of pulmonary capillary blood plasma onto the respiratory surface. (Weibel 1973; Meban 1980; Maina and King 1982a, b; Maina and West 2005). In the dog lung, Brody et al. (1968) determined that the

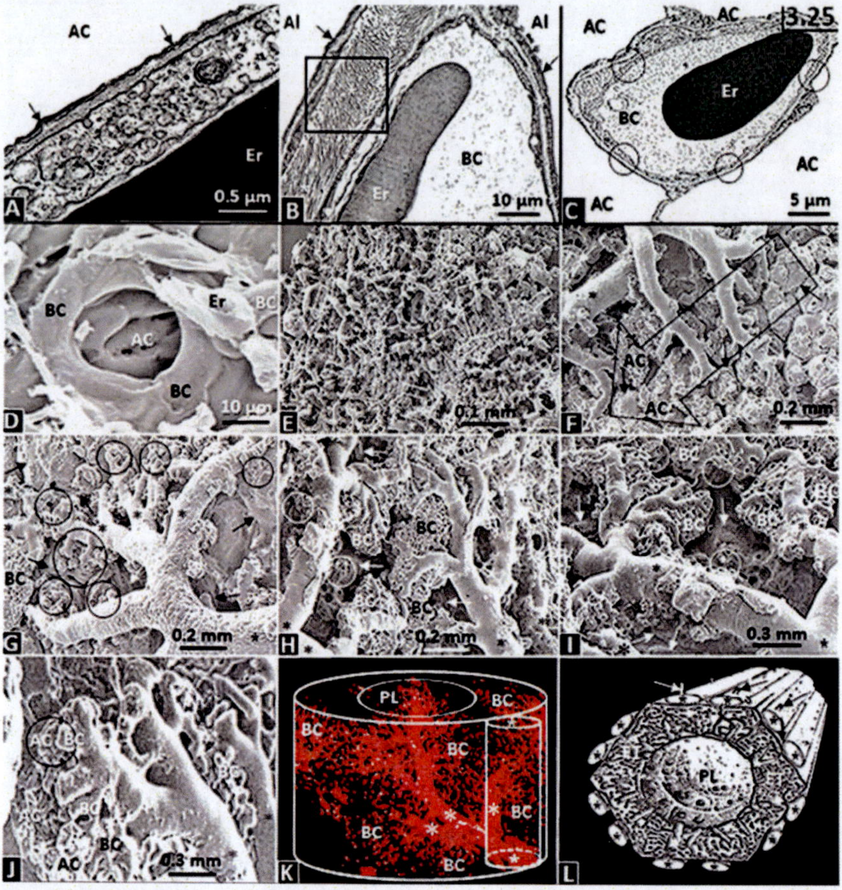

Plate 3.25 (**a**) The blood gas barrier of the avian lung is a tripartite structure that comprises a very thin epithelial cell (arrows, ↓), a basement membrane (star, ▯), and an endothelial cell (dots, •). AC, air capillary; Er, erythrocyte. For the mammalian lung, thicker (boxed area, □) and thinner (arrow, ←) sides of the interalveolar septum exist (**b**). Al, alveoli; BC, blood capillary; Fb, fibrocyte. (**c, d**) For the avian lung, blood in the blood capillaries (BC) is exposed to air about all around (**c, d**). Circles (O) (**c**), blood-gas barrier. (**e–j**) Double latex rubber cast preparations of the airway (bronchial) and the pulmonary artery of the lung of the Domestic Fowl (*Gallus gallus* variant *domesticus*) showing the topographic relationships between the structures. The directions of the flow of venous blood into the exchange tissue of the parabronchi from the interparabronchial arteries (stars, ▯) through the intraparabronchial ones (asterisks, *) relative to that of the flow of air in the parabronchial lumen (large open arrows, ⇨) (**e, f**) is perpendicular (orthogonal): the arrangement constitutes the crosscurrent gas exchange system. Solid thin black arrows (←) (**f–i**), atria; circles (O) (**g–j**), sites where air capillaries (AC) and blood capillaries (BC) connect to constitute a countercurrent gas exchange system. (**k**) Schematic illustration of a parabronchus with an interparabronchial artery (star, ▯) running alongside it and a three-dimensional reconstruction of an interparabronchial artery giving rise to intraparabronchial arteries (asterisks, *) that were included to show the topographic relationship between the flow of air in the parabronchial lumen (PL) and that of blood in the blood vessel. BC, blood capillaries. (**l**) Schematic illustration of a parabronchus showing the parabronchial lumen (PL) that is surrounded by exchange tissue (ET). Stars (▯), interparabronchial blood vessels; arrows, intraparabronchial arteries

midcapillary pressure averaged 13.3 mbar while the pulmonary artery and vein pressures were 20.4 and 9.2 mbar, respectively. For the Domestic Fowl (*Gallus gallus* variant *domesticus*), doubling the volume of blood flowing through one lung, an experimental procedure that about doubled the mean pulmonary artery pressure did not alter resistance and the dimensions of the blood capillaries remained unchanged (Powell et al. 1985): the study showed the remarkable strength of the blood capillaries of the avian lung (Maina 2007a, b; Watson et al. 2008; West 2017).

The structure of the avian functional (gas exchange) pulmonary circulatory system has been studied by among others Radu and Radu (1971), Abdalla and King (1975, 1976a, b), West et al. (1977), Maina (1988) and Abdalla (1989) while the bronchial one was investigated by Abdalla and King (1976b) and Abdalla and King (1977). Unlike the mammalian lung where particularly the pulmonary artery and its branches track and pattern the airways (bronchial system) closely (Weibel 1984; Maina and van Gils 2001) (Plate 2.12), for the avian lung, the patterning is not as close (Abdalla and King 1975; Abdalla 1989; Woodward et al. 2019; Maina et al. 2021) (Plates 3.8, 3.9 and 3.26). On entering the lung (at the hilum), the pulmonary artery accompanies the extrapulmonary part of the intrapulmonary primary bronchus and lies ventral to the main part of the first medioventral secondary bronchus while the pulmonary vein lies dorsal to the airway (Maina et al. 2021) (Plates 3.6c–e, 3.8c–e and 3.9c, e). The pulmonary artery gives rise to four main branches (rami), namely, the accessory, the cranial, the caudomedial, and the caudolateral branches that supply blood to various territories of the lung (Abdalla 1989; Woodward et al. 2019; Maina et al. 2021). The accessory branch, that is the first and smallest branch of the pulmonary artery, supplies deoxygenated blood to a very small cranial part of the lung ventral to the hilum: the branch is lacking in lungs of some species such as the Turkey (*Meleagris gallopavo*), the Guinea Fowl (*Numida meleagris*), the Mallard Duck (*Anas platyrhynchos*), the Graylag Goose (*Anser anser*), and the Muscovy Duck (*Cairina moschata*) (Abdalla and King 1977). In a plane that is about parallel to and lateral to the origins of the medioventral secondary bronchi, the cranial branch of the pulmonary artery passes both dorsally and caudally to deliver blood to the craniodorsal part of the lung, in front of the third costal sulcus (Abdalla and King 1975; Abdalla 1989). The caudomedial branch conveys blood to most of the lung and from its location and orientation the branch is the most direct continuation of the pulmonary artery, and the caudolateral branch of the pulmonary artery supplies blood to the caudolateral, the ventral, and the caudoventral parts of the lung (Plates 3.8b, d, e, 3.9b, d, e and 3.26). According to Abdalla and King (1975) and Abdalla (1989), for the lung of the Domestic Fowl, a transverse plane that roughly passed along the third costal sulcus separated the organ into cranial and caudal vascular territories. By three-dimensional computer serial section reconstruction study of the lung of the same species, however, Woodward et al. (2019) and Maina et al. (2021) determined that the transverse plane passed along the second and not the third costal sulcus (Plates 3.8 and 3.9). For the Domestic Fowl, Bennett (1971) observed adrenergic innervation of the pulmonary artery and its branches. At parabronchial level, the pulmonary artery gives rise to interparabronchial arteries which in turn give rise to intraparabronchial arteries and arterioles which terminate in blood

3.26

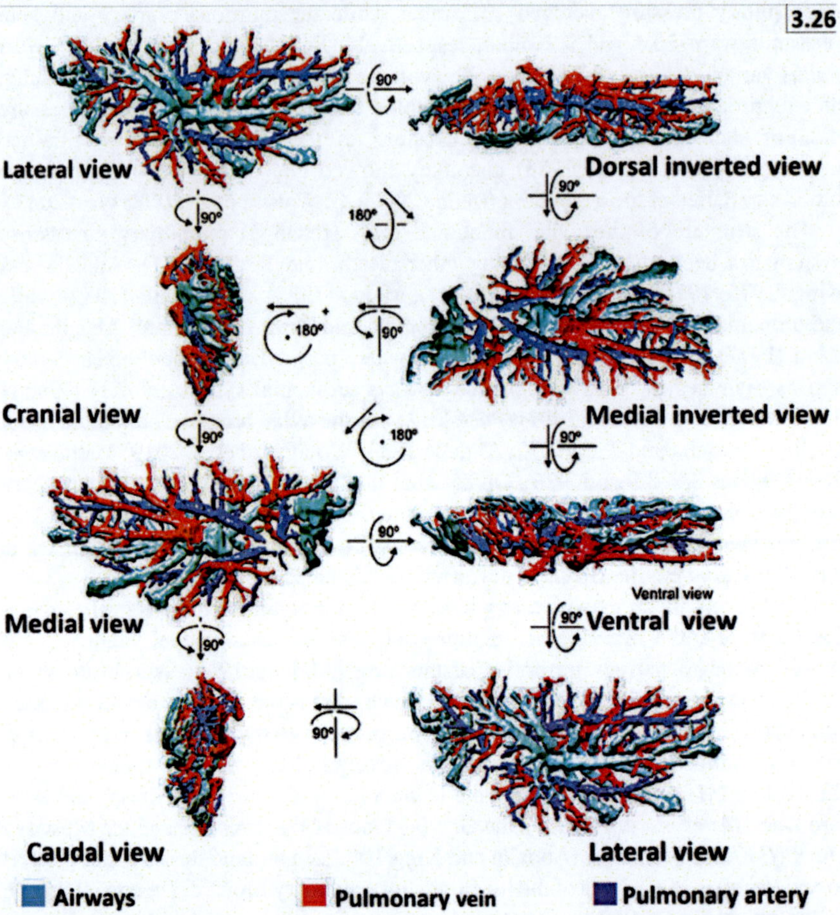

Plate 3.26 Three-dimensional serial section computer reconstructions of the lung of the Domestic Fowl (*Gallus gallus* variant *domesticus*) rotated to show various views of the topographical relationships between the airway (bronchial) and vascular systems. The three parts do not track and pattern each as closely as those of the mammalian lung do

capillaries (Plates 3.15c, e, g, i, 3.23b and 3.25e–l). The walls of the interparabronchial blood vessels are well-endowed with collagen fibers (Plate 3.23a–d).

Abdalla and King (1975) and Abdalla (1989) reported that the pulmonary vein starts from the blood capillaries that drain directly or via intraparabronchial venules into large atrial veins which are located in the interatrial septa. The blood vessels pass through the exchange tissue to drain into interparabronchial veins which are located at the periphery of a parabronchus. For the Domestic Fowl (*Gallus gallus* variant *domesticus*), according to Abdalla and King (1975) and Abdalla (1989), two branches (rami = radices) of the pulmonary vein drain the lung, returning oxygenated blood back to the left atrium of the heart. However, in the same species of bird, by serial section three-dimensional computer reconstruction, Woodward et al. (2019) and Maina et al. (2021) determined that three conspicuous branches of the pulmonary vein converged at the hilum. The existence of smooth muscle in the atrial septa (Duncker 1974; Abdalla and King 1975; King and McLelland 1984; Abdalla 1989; Maina et al. 2010) suggests that the parabronchial luminal diameters and hence the flow of air through the passageway may be regulated by the state of contractility of the smooth muscle and elastic tissue elements. The exchange tissue is well provided with collagen fibers that should strengthen the air and the blood capillaries (Plate 3.22g–o). King et al. (1978) observed that the parabronchial blood vessels are well-enervated (by adrenergic nerves) and therefore blood flow through the exchange tissue may be well-regulated by matching parabronchial ventilation with perfusion. Intriguingly, King et al. (1978) and Abdalla (1989) reported a muscular venous valve with smooth muscle tissue elements that was located between the opening of a small interparabronchial vein and a large interparabronchial vein which they suggested possibly controlled blood flow in parts of the lung which may comprise as many as three parabronchi. By radioactive intravenous microsphere injection followed by histological study (Abdalla and King 1975, 1976a, b; James et al. 1976; Holle et al. 1978; Parry and Yates 1979; Abdalla 1989), scanning electron microscopy (West et al. 1977) and three-dimensional serial section computer reconstruction (Woodward et al. 2019; Maina et al. 2021), no arteriovenous anastomoses between the pulmonary artery and vein have been found in lungs of avian species.

3.2.6 *Air Sacs*

The air sacs of the avian respiratory system are large, thin-walled, transparent membranous structures that connect to lungs at different sites called Ostia (Hamlet and Fischer 1967; Kloek and Casler 1972; McLelland 1989a; Demirkan et al. 2006a, b; Orhan et al. 2009; El-Bably et al. 2014; Ragab and Reem 2016; El-Sayed and Hassan 2020; Bejdić et al. 2021a, b) (Plates 2.17, 2.23a, b—inset, c–f and 3.27). Developing early during embryogenesis (Plates 2.17 and 2.23), at maturity, the air sacs occupy much of the coelomic cavity and constitute the greatest part of that of the avian respiratory system. Compared with mammals, where the volume of the respiratory system per unit body mass ranges from 35 to 70 ml kg^{-1}, for birds, the equivalent values (volumes of the lung and air sacs) range from 100 to 200 ml kg^{-1} (Jaensch 2015). The air sacs are on the whole avascular (Fletcher 1980) and are therefore not meaningfully involved in gas exchange (Magnussen et al. 1976; Powell and Hempleman 1985). This was determined by Soum (1896) who performed an experiment in which the large abdominal air sacs were isolated from the lung (by ligating the ostial connections) and carbon monoxide (CO) was introduced into it (air sac): the experimental birds showed no distress from the lethal gas. Dating back to ~1911 until quite recently (1986), when it was outlawed and

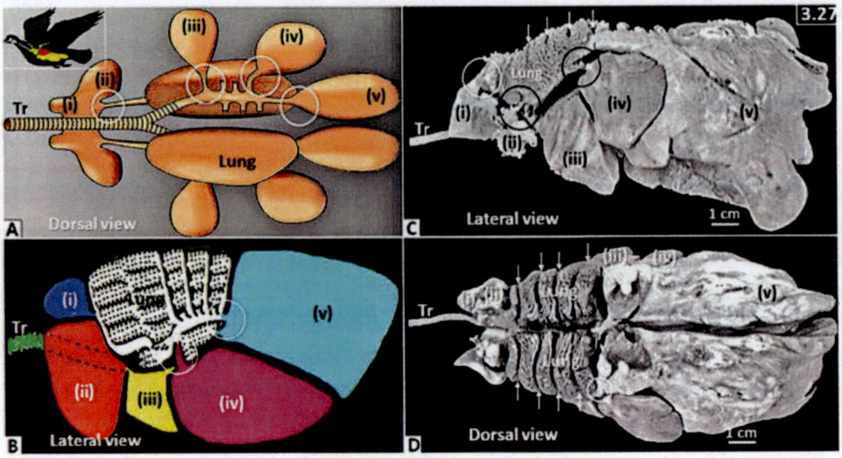

Plate 3.27 The lung-air sac system of the Domestic Fowl (*Gallus gallus* variant *domesticus*) showing lungs intercalated between the cranial (i–iii) and the caudal (iv, v) groups of air sacs: (**a**) and (**b**) are schematic illustrations of the lung-air sac system and (**c** and **d**) are latex rubber cast preparations. The inset (**a**) shows the location of the lung-air sac system of a bird. The cranial air sacs are the cervical air sacs (i), the interclavicular air sacs (ii), and the cranial thoracic air sacs (iii) while the caudal ones are the caudothoracic air sacs (iv) and abdominal air sacs (iv). Tr, trachea; circles (O), Ostia; arrows, costal sulci (**c**, **d**)

replaced with electronic CO detectors, canaries in particular were used in coal mines to detect harmful levels of poisonous gases such as CO and methane (Reiblein 2017).

For the Domestic Fowl (*Gallus gallus* variant *domesticus*), Rawal (1976) and Cook et al. (1987) observed adrenergic and cholinergic nerve plexuses in the walls of the air sacs and Cook et al. (1987) demonstrated vasoactive intestinal polypeptide (VIP, a neuropeptide), somatostatin, and enkephalin-immunoreactive fibers in the walls of the air sacs. In the chick embryo, six primordial pairs of air sacs namely, cervical, lateral clavicular, medial clavicular, cranial thoracic, caudal thoracic, and abdominal air sacs exist (King and McLelland 1984; McLelland 1989a). With the development of the embryo, some of the air sacs connect and change in sizes, numbers, and locations. The definitive number of air sacs decreases from the primordial six pairs as some of them, especially the cranial ones, fuse to form single air sacs (Gilbert 1939; Gier 1952; Crespo et al. 1998; King and McLelland 1984). In most species, the craniothoracic, the caudothoracic, and the abdominal air sacs generally remain paired (McLelland 1989a). For such species, a total of eight air sacs, namely, one cervical, one clavicular, two craniothoracic, caudothoracic, and abdominal air sacs exist. For the Ostrich (*Struthio camelus*), from the presence of a paired clavicular air sac, a total of ten air sacs exist (Bezuidenhout 1998). During development, the air sacs, except for the abdominal one, invade the postpulmonary septum of the embryo and separate it into the horizontal (postpulmonary) and the oblique septa (King and McLelland 1984): the abdominal air sacs pass through the horizontal septum into the peritoneal cavity (Locy and Larsell 1916a, b; Duncker 1978a; Kitazawa et al. 1977) (Plate 2.17i). Cook et al. (1986, 1987) observed that the ostium of the abdominal air sac of the Domestic Fowl (*Gallus gallus* variant *domesticus*) has a sphincter-like ring of well-innervated smooth muscle fibers.

In clinical veterinary medicine, during endoscopy, the air sacs allow radiographic discrimination of internal organs. Also, they provide alternative means of administering gaseous anesthesia where and when tracheal administration is problematic (Gunkel and Marin 2005; Lierz and Korbel 2012; Jaensch 2015). In birds, rupture of air sacs may occur in cases where a bird in fast flight strikes a transparent window pane at high speed or is attached by a predator such as a cat (Petevinos 2006; Kamani et al. 2009; Hollinger 2017). Depending on the severity of the injury, the pathology occurs as abnormal swelling of a part of the body. Where the injury is not large or complicated, the problem can be relieved by needle aspiration of air under sterile conditions or by surgical intervention (Clippinger 1997; Jung 2001; (Hollinger 2017; Browning et al. 2019). Interestingly, the so-called cutaneous air sacs, such as the cervicocephalic air sac, that have no direct connection with the lung and its associated air sacs exist in some species of birds (Akester et al. 1973; Cozic et al. 2014; Cozic and Homberger 2015). It has been reported that such air sacs are better developed in powerful/efficient flyers (Jaensch 2015). The subcutaneous air sacs of the pelicans (Pelecaniformes) have been supposed to serve as shock absorbers when a bird strikes the surface of the water at speed in a dive and afterward help the bird remain afloat on the surface of the water (https://www.britannica.com/animal/pelecaniform/Form-and-function; Redrobe 2015). The cervical air sac largely comprises a median chamber that is located between the lungs and occupies the craniodorsal part of the colelomic cavity dorsal to the oesophagus and ventral to the base of the neck (McLelland 1989a). Such air sacs are lacking in the loons (Family: Gaviidae) and in grebes (Family: Podicipedidae) (Duncker 1971). In the Turkey (*Meleagris gallopavo*), the connection between the left and right cervical air sacs occurs on day 17 of incubation (King and Atherton 1970). In some species of birds such as the Gannet (*Morus bassanus*) and the Ostrich (*Struthio camelus*), the cervical air sacs form extensive subcutaneous extensions (Bezuidenhout 2005) that pneumatized the vertebral column as far up the neck as the axis cervical vertebra. In their caudal part, dorsolaterally, the cervical air sacs contact the lung and the horizontal septum (McLelland 1989a). The clavicular air sac occupies the cranioventral area of the thorax (the thoracic inlet), the base of the neck, and most of the right and left axillary space, extending to surround the heart, the sternum and muscles of the pectoral girdle: to various extents, the air sacs pneumatize certain proximate bones. In postcranial skeletal pneumaticity (PCP) study of the Ostrich (*Struthio camelus*), Apostolaki et al. (2015) concluded that the basis of the invaginations and their correlation with respiratory function was indeterminate.

The trachea, the syrinx, and the extrapulmonary primary bronchi are held by folds of the clavicular air sac (Plate 1.1d). In most birds, the air sac is relatively large and unpaired and is formed by the fusion of the right and left primordia (McLelland 1989a). In groups of birds such as gulls (Family: Laridae), however, the primordia remain separate (Locy and Larsell 1916a, b) while in species such as the Rock Dove

(*Columba livia*) (Müller 1908) and in the House Sparrow *(Passer domesticus)* (Wetherbee 1951), the clavicular and the cervical air sacs connect to form a single, large cervicoclavicular air sac. In passerine species and hummingbirds (Family: Trochilidae), the cervical air sacs connect to the craniothoracic air sacs (Wetherbee 1951; Duncker 1971). The clavicular air sacs, that are compliant and change in size with ventilatory movements, modulate sound production in birds (King 1989; McLelland 1989a; Brackenbury 1989). Several diverticula extend outwards from the clavicular air sacs to the heart, to the surface of the body, and into adjacent bones (McLelland 1989a). In most birds, without diverticula, the craniothoracic air sacs lie between the horizontal and the oblique septum and are generally located on the dorsolateral position of the thoracic part of the coelomic cavity: they lie caudal to the craniothoracic air sacs and cranial to the abdominal air sacs. In most species of birds, the craniothoracic air sacs and the caudothoracic ones are paired and are located in the subpulmonary cavity, i.e., in the space ventral to the lung and the horizontal septum (McLelland 1989a): the craniothoracic air sacs are generally smaller than the caudothoracic air ones. In some species of birds, e.g., the Common Coot (*Fulica atra*), however, the craniothoracic air sac is much larger (Duncker 1971) while in others, e.g., the Common Moorhen (*Gallinula chloropus*), the two sets of air sacs are about equal in size (Groebbels 1932). In the penguin (Spheniscidae) and the passerine species of birds, the craniothoracic air sacs are particularly small while in hummingbirds (Trochlids), the caudothoracic air sac is the largest (Stanislaus 1937). In the Turkey *(Meleagris gallopavo)*, making the total number of air sacs seven, the caudothoracic air sac is lacking (King and Atherton 1970). Normally, the craniothoracic air sacs and the caudothoracic ones have few, if any, diverticula (King 1966, 1979). The abdominal air sacs, that in most species of birds are the largest set of air sacs, are paired and occupy the intestinal peritoneal cavity and normally lie dorsal to the viscera as they spread between loops of intestines and organs such as kidneys: they pneumatize bones such as the femur and the synsacrum (King 1957; El-Sayed and Hassan 2020). In groups of birds such as the Domestic Fowl (*Gallus gallus* variant *domesticus*), the Gannet (*Morus bassanus*), the Rock Dove (*Columba livia*) (Müller 1908), and the Ostrich (*Struthio camelus*), the air sacs extend out of the coelomic cavity, coming to lie subcutaneously (King and McLelland 1984; Bezuidenhout et al. 1999). In hummingbirds (Family: Trochilidae) (Stanislaus 1937), the caudothoracic air sac is the largest. In the kiwi (Family: Apterygidae), the abdominal air sac is located in the subpulmonary cavity (Duncker 1979): from the asymmetry of the viscera, the left abdominal air sac is commonly smaller than the right one. In some species of birds such as the Common Loon (*Fulica atra*) and the Herring Gull (*Larus argentatus*), however, the left abdominal air sac is larger (Gier 1952). In certain species of birds, to various extents, the left and right abdominal air sacs connect directly (Duncker 1971). While they are remarkably large in some species, e.g., the Greater Flamingo (*Phoenicopterus roseus*) (Groebbels 1932), the abdominal air sacs are poorly developed in birds such as penguins (Family: Spheniscidae) and the Greater Rhea (*Rhea americana*) (Duncker 1971) and are remarkably small in the hummingbirds (Family: Trochilidae) (Groebbels 1932; Stanislaus 1937). In many species of birds, the abdominal air sacs pneumatize the synsacrum, the pelvic girdle, the femur, and the caudal vertebrae (King 1975, 1979; Bezuidenhout et al. 1999; Cevik-Demirkan et al. 2006; Casteleyn et al. 2018; El-Sayed and Hassan 2020) and in some species, e.g., the Rock Dove (*Columba* livia) they extend out of the coelomic cavity to lie under the skin above the pelvis (Müller 1908).

Most of the internal surface of air sacs is covered by a simple and thin (squamous) epithelium that rests on a layer of connective tissue, a basement membrane (Carlson and Beggs 1973; Walsh and McLelland 1974a; Bezuidenhout 2005; Jaensch 2015). Close to the Ostia, patches of ciliated cuboidal and columnar epithelia exist (Fletcher 1980). Compared with nondiving species of birds, the walls of the air sacs of diving species are relatively thicker and a columnar type of epithelial cells exist (Jaensch 2015). In the Domestic Fowl (*Gallus gallus* variant *domesticus*), a broad band of pseudostratified ciliated columnar epithelium with goblet cells extends from the intrapulmonary primary bronchus into the surface of the abdominal air sacs and onto those of the caudothoracic air sac (Cook and King 1970). In penguins (Family: Spheniscidae), the epithelium that lines the abdominal air sacs is generally cuboidal and the epithelial cells are interconnected by junctional complexes that are located at the luminal aspect. Microvilli project into the luminal space and electron-dense lysosome-like granules exist in the cytoplasm (Carlson and Beggs 1973; Walsh and McLelland 1974b). Fletcher (1980) observed that scarce smooth muscle cells and clusters of fat cells existed in the walls of the abdominal air sacs of some species of birds.

3.2.6.1 Ostia

Ostia are the sites on the cranial, ventral, and caudal borders of the avian lung where air sacs connect to it (King 1966; Duncker 1971; King and McLelland 1984; McLelland 1989a; Maina 2005) (Plates 2.2a, 2.17 and 2.23): two types of ostial connections, namely, direct and indirect, ones exist. Direct connections occur where air sacs join the lung via the intrapulmonary primary bronchus and/or by secondary bronchi while indirect ones exist where parabronchi are found (McLelland 1989a): most of the air sacs have one or two direct connections and many indirect ones. In penguins (Spheniscidae), indirect connections are lacking (Vos 1937). With a single ostium that is located on the septal surface of the lung, the cervical air sac of the Domestic Fowl (*Gallus gallus* variant *domesticus*) has a direct connection that derives from the first mediodorsal secondary bronchus and indirect connections are lacking (Biggs and King 1957). The clavicular air sac has two direct ostial connections, namely, the medial ostium that connects directly to the third mediodorsal secondary bronchus and a lateral one that connects to the first mediodorsal secondary bronchus; an indirect connection occurs between the parabronchi of the medioventral secondary bronchi and the clavicular air sac (McLelland 1989a). In the lungs of hummingbirds (Family: Trochlidae), a lateral ostium is lacking (Stanislaus 1937). Interestingly, Juillet (1912) reported that in columbiform and psittaciform species of birds, the clavicular air sacs do not connect to the lung. This was, however, shown to be wrong by El-Sayed and Hassan (2020) who noted that the clavicular air sac connects with the lung through the medioventral secondary bronchi and the subscapular, the axillary, the humeral, the subpectoral, and the sternal diverticula. The cranial and the caudal thoracic air sacs connect to the lung through the lateroventral secondary bronchi and the two air sacs lack diverticula. (McLelland 1989a): the craniothoracic air sacs are typically associated with two Ostia, namely, the medial ostium that connects directly to the third mediodorsal secondary bronchus while the lateral one is indirectly connected via the parabronchi of one or more mediodorsal secondary bronchi. In hummingbirds (Family: Trochlidae), the craniothoracic air sacs are indirectly connected to the lung (Stanislaus 1937). In many species of birds, the craniothoracic air sacs have a single ostium that is found at or near the caudal part of the costal septal border of the lung: direct and indirect connections exist. In groups of birds such as penguins (Spheniscidae), where the neopulmo is lacking, indirect connections to the caudothoracic air sacs do not exist. The abdominal air sacs have one ostium that is located on the lateral aspect of the caudal border of the lung, where both direct and indirect connections exist: the direct connection occurs through the intrapulmonary primary bronchus while the indirect ones occur via the parabronchi of the laterodorsal secondary bronchi, the lateroventral secondary bronchi and the last mediodorsal secondary bronchus. In hummingbirds (Trochlidae), only one indirect connection to the abdominal air sacs exists (Stanislaus 1937).

References

Abdalla MA (1989) The blood supply to the lung. In: King AS, McLelland J (eds) Form and function in birds, vol 4. Academic Press, London, pp 281–306

Abdalla MA, King AS (1975) The functional anatomy of the pulmonary circulation of the domestic fowl. Respir Physiol 23:267–290

Abdalla MA, King AS (1976a) Pulmonary arteriovenous anastomoses in the avian lung: do they exist? Respir Physiol 27:187–191

Abdalla MA, King AS (1976b) The functional anatomy of the bronchial circulation of the domestic fowl. J Anat 121:537–550

Abdalla MA, King AS (1977) The avian bronchial arteries: species variations. J Anat 123:697–704

Abdalla MA, Maina JN (1981) Quantitative analysis of the exchange tissue of the avian lung (Galliformes). J Anat 134:677–680

Abdalla MA, Maina JN, King AS, King DZ, Henry J (1982) Morphometrics of the avian lung. 1. The domestic fowl, Gallus domesticus. Respir Physiol 47:267–278

Abdel-Maksoud FM, Hussein MM, Hamdy A, Ibrahim IA (2020) Anatomical, histological, and electron microscopic structures of syrinx in male budgerigars (*Melopsittacus undulatus*). Microsc Microanal 26:1226–1235

Aboelkassem Y, Staples AE (2013a) Stokeslets-meshfree computations and theory for flow in a collapsible microchannel. Theor Comput Fluid Dyn 27:681–700

Aboelkassem Y, Staples AE (2013b) Selective pumping in a network: insect-style microscale flow transport. Bioinspiration Biomimetics 8:026004. https://doi.org/10.1088/1748-3182/8/2/026004

Abourachid A, Hackert R, Herbin M, Libourel PA, Lambert F, Gioanni H, Provini P, Blazevic P, Hugel V (2011) Bird terrestrial locomotion as revealed by 3D kinematics. Zoology 114:360–368

Akester AR, Pomeroy DE, Purton MD (1973) Subcutaneous air pouches in marabu stock (*Leptoptios curmeniferus*). J Zool 170:493–499

Albersheim-Carter J, Blubaum A, Ballagh IH, Missaghi K, Siuda ER, McMurray G, Bass AH, Dubuc R, Kelley DB, Schmidt MF et al (2016) Testing the evolutionary conservation of vocal motoneurons in vertebrates. Respir Physiol Neurobiol 224:2–10

Ames PL (1971) The morphology of the syrinx in passerine birds. Bull Peabody Mus Nat Hist 37: 1094

Angilletta MJ (2009) Thermal adaptation: a theoretical and empirical synthesis. Oxford University Press, Oxford (UK)

Angilletta MJ, Niewiarowski PH, Navas CA (2002) The evolution of thermal physiology in ectotherms. J Therm Biol 27:249–268

Apostolaki NE, Rayfield EJ, Barrett PM (2015) Osteological and soft-tissue evidence for pneumatization in the cervical column of the ostrich (*Struthio camelus*) and observations on the vertebral columns of non-volant, semi-volant and semiaquatic birds. PLoS One 10(12): e0143834. https://doi.org/10.1371/journal.pone.0143834

Appel FW (2005) Sex dimorphism in the syrinx of the fowl. J Morphol 47:497–517

Araújo R, David R, Benoit J, Lungmus JK, Stoessel A, Barrett PM, Maisano JA, Ekdale E, Orliac M, Luo ZX et al (2022) Inner ear biomechanics reveals a Late Triassic origin for mammalian endothermy. Nature 409. https://doi.org/10.1038/s41586-022-04963-z

Badlangana NL, Adams JW, Manger PR (2009) The giraffe (*Giraffa camelopardalis*) cervical vertebral column: a heuristic example in understanding evolutionary processes? Zool J Linn Soc 155:736–757

Baile EM (1996) The anatomy and physiology of the bronchial circulation. J Aerosol Med 9:1–6

Bailey L (1954) The respiratory currents in the tracheal system of the adult bee. J Exp Biol 31:589–595

Bairlein F, Fritz J, Scope A, Schwendenwein I, Stanclova G, van Dijk G, Meijer HAJ, Verhulst S, Dittami J (2015) Energy expenditure and metabolic changes of free-flying migrating northern bald ibis. PLoS One 10:e0134433. https://doi.org/10.1371/journal.pone.0134433

Banzett RB, Nations CS, Barnas JL, Lehr JL, Jones JH (1987) Inspiratory aerodynamic valving in goose lungs depends on gas density and velocity. Respir Physiol 70:287–300

Banzett RB, Nations CS, Wang N, Fredberg JJ, Butler PJ (1991) Pressure profiles show features essential to aerodynamic valving in geese. Respir Physiol 84:295–309

Barkan CL, Zornik E (2020) Inspiring song: the role of respiratory circuitry in the evolution of vertebrate vocal behavior. Dev Neurobiol 80:31–41

Barkan CL, Kelley DB, Zornik E (2018) Premotor neuron divergence reflects vocal evolution. J Neurosci 38:5325–5337

Bass AH (2014) Central pattern generator for vocalization: is there a vertebrate morphotype? Curr Opinion Neurobiol 28:94–100

Bass AH, Gilland EH, Baker R (2008) Evolutionary origins for social vocalization in a vertebrate hindbrain-spinal compartment. Science 321:417–421

Baumel JJ, King AS, Breazile JE, Evans HE, Vanden Berge JC (1993) Handbook of avian anatomy: Nomina anatomica avium, 2nd edn. Publications of the Nuttall Ornithological Club, Cambridge (MA)

Bech C, Johansen K (1980) Ventiltatory and circulatory responses to hyperthermia in the mute swan (*Cygnos olor*). J Exp Biol 88:195–204

Bech C, Johansen K, Maloiy GMO (1979) Ventilation and expired gas composition in the flamingo, *Phoenicopterus ruber*, during normal respiration and panting. Physiol Zool 52:313–328

Beddard FE (1886) Notes on the convoluted trachea of a curassow (*Northocrax urumutum*), and on the syrinx in some stocks. Proc Zool Soc (Lond) 1886:321–325

Beddard FE (1898) The structure and classification of birds. Longmans, London

Bejdić P, Hadzimusic N, Seric-Haracic S, Maksimovic A, Lutvikadic I, Hrkovic-Porobija A (2021a) Morphology of the air sacs in crimson rosella (*platycercus elegans*) parrots. Adv Anim Vet Sci 9:1959–1963

Bejdić P, Katica A, Mlaćo N, Velić L, Ćutuk A, Čengić B (2021b) Gross morphological studies on the air sacs in rosy-faced parrots (*Agapornis roseicollis*). Adv Anim Vet Sci 9:989–993

Bennett T (1971) The adrenergic enervation of the pulmonary vasculature, the lung and the thoracic aorta, and on the presence of aortic bodies in the domestic fowl (*Gallus gallus domesticus* L.). Z Zellforsch mikrosk Anat 114:117–134

Bennett A, Ruben J (1979) Endothermy and activity in vertebrates. Science 206:649–654

Benson RBJ, Butler RJ, Carrano MT, O'Connor PM (2012) Air-filled postcranial bones in theropod dinosaurs: physiological implications and the 'reptile' - bird transition. Biol Rev 87:168–193

Benton MJ (2021) The origin of endothermy in synapsids and archosaurs and arms races in the Triassic. Gondwana Res 100:261–289

Bernhard W (2016) Lung surfactant: function and composition in the context of development and respiratory physiology. Ann Anat 208:146–150

Bernhard W, Gebert A, Vieten G, Rau GA, Hohlfeld JM, Postle AD, Freihorst J (2001) Pulmonary surfactant in birds: ccoping with surface tension in a tubular lung. Am J Physiol Regul Integr Comp Physiol 281:R327–R337

Bernhard W, Haslam PI, Floros J (2004) From birds to humans new concepts on airways relative to alveolar surfactant. Am J Respir Cell Mol Biol 30:6–11

Bewig M, Mitchell MA (2009) Wildlife. In: Mitchell MA, Tully TN (eds) Manual of exotic pet practice. WB Saunders, Saint Louis, pp 493–529

Bezuidenhout AJ (1998) Anatomy. In: Deeming DC (ed) The ostrich: biology, production and health. CABI, Washington, DC, pp 13–50

Bezuidenhout AJ (2005) Light and electron microscopic study of the thoracic respiratory air sacs of the fowl. Anat Histol Embryol 34:185–191

Bezuidenhout AJ, Groenewald HB, Soley JT (1999) An anatomical study of the respiratory air sacs in ostriches. Onderst J Vet Res 66:317–325

Biggs PM, King AS (1957) A new experimental approach to the problem of the air pathway within the avian lung. J Physiol (Lond) 138:282–289

Birdcage (2021) Why is my parrot sneezing and coughing? https://www.birdcagesnoe.com/blogs/bird-blog/parrot-sneezing-and-coughing. Retrieved 21-06-2021

Biro GP (2013) From the atmosphere to the mitochondrion: the oxygen cascade. In: Kim H, Greenburg A (eds) Hemoglobin-based oxygen carriers as red cell substitutes and oxygen therapeutics. Springer, Berlin, pp 27–53

Bishop CM, Butler PJ (2015) Flight. In: Scanes CG (ed) Sturkie's avian physiology, 6th edn. Academic Press, New York, pp 919–974

Bock WJ (1978) Morphology of the larynx of *Corvus brachyrhynchos* (Passeriformes: Corvidae). Wilson Bull 90:553–565

Bódi I, Kocsis K, Benyeda Z, Fejsz N, Molnár D, Nagy N, Olah I (2016) Dual secretion locations on type-II cells in the avian lung suggest local as well as general roles of surfactant. J Morphol 277:1062–1071

Böhmer C, Plateau O, Cornette R, Abourachid A (2019) Correlated evolution of neck length and leg length in birds. R Soc Open Sci 6:181588. https://doi.org/10.1098/rsos.181588

Bonebrake TC, Rezende EL, Bozinovic F (2020) Climate change and thermoregulatory consequences of activity time in mammals. Am Nat 196:45–56

Bourbon JR, Chailley-Heu B (2001) Surfactant proteins in the digestive tract, mesentery, and other organs: evolutionary significance. Comp Biochem Physiol A Mol Integr Physiol 129:151–161

Bout RG (1997) Postures of the avian craniocervical column. J Morphol 231:287–295

Brackenbury JH (1977) Physiological energetics of cock crow. Nature 270:433–435

Brackenbury JH (1978) Respiratory mechanics of sound production in chickens and geese. J Exp Biol 72:229–250

Brackenbury JH (1989) Functions of the syrinx and the control and sound production. In: King AS, McLelland J (eds) Form and function in birds, vol 4. Academic Press, London, pp 193–220

Brackenbury J, Amaku J (1990) Effects of combined abdominal and thoracic air-sac occlusion on respiration in domestic fowl. J Exp Biol 152:93–100

Brackenbury JH, Darby C, El-Sayed MS (1989) Respiratory function in exercising fowl following occlusion of the thoracic air sacs. J Exp Biol 145:227–237

Bradley TJ (2007) Control of the respiratory patterns in insects. In: Roach RC, Wagner PD, Hackett PH (eds) Hypoxia and the circulation. Springer, New York (NY), pp 211–220

Bretz WL, Schmidt-Nielsen K (1971) Bird respiration: flow patterns in the duck lung. J Exp Biol 54:103–118

Brocklehurst RJ, Schachner ER, Sellers WI (2018) Vertebral morphometrics and lung structure in non-avian dinosaurs. Open Sci 5:180983. https://doi.org/10.1098/rsos.180983

Brocklehurst RJ, Schachner ER, Codd JR, Sellers WI (2020) Respiratory evolution in archosaurs. Philos Trans R Soc (Lond) B 375:20190140. https://doi.org/10.1098/rstb.2019.0140

Brody JS, Stemmler EJ, DuBois AB (1968) Longitudinal distribution of vascular resistance in the pulmonary arteries, capillaries, and veins. J Clin Invest 47:783–799

Brown RE, Kovacs CE, Butler JP, Wang N, Lehr J, Banzett RB (1995) The avian lung: is there an aerodynamic expiratory valve? J Exp Biol 198:2349–2357

Brown RE, Brain JD, Wang N (1997) The avian respiratory system: a unique model for studies of respiratory toxicosis and for monitoring air quality. Environm Hlth Perspect 105:188–200

Browning GR, Eshar D, Tucker-Mohl K, Berke K (2019) Diagnosis and surgical repair of a chronic ruptured cervical air sac in a double yellow-headed amazon parrot (*Amazona Ochrocephala Oratrix*). J Exotic Pet Med 29:45–50

Brusatte SL, O'Connor JK, Jarvis ED (2015) The origin and diversification of birds. Curr Biol 25: 888–898

Buck J (1962) Some physical aspects of insect respiration. Ann Rev Entomol 7:27–56

Burn A, Schneiter M, Ryser M, Gehr P, Rička J, Frenz M (2022) A quantitative interspecies comparison of the respiratory mucociliary clearance mechanism. Europ Biophys J 51:51–65

Butler PJ (1991) Exercise in birds. J Exp Biol 160:233–262

Butler PJ (2016) The physiological basis of bird flight. Philos Trans R Soc (Lond) B 371:20150384. https://doi.org/10.1098/rstb.2015.0384

Buttemer WA, Abele D, Costantini D (2010) The ecology of antioxidants and oxidative stress in animals: from bivalves to birds: oxidative stress and longevity. Funct Ecol 24:971–983

Calder WA (1968) Respiratory and heart rates of birds at rest. Condor 70:358–365

Cameron JN (1989) The respiratory physiology of animals. Oxford University Press, New York

Carlson HC, Beggs EC (1973) Ultrastructure of the abdominal air sac of the fowl. Res Vet Sci 14: 148–150

Casteleyn C, Cornillie P, Van Cruchten S, Van den Broeck W, Van Ginneken C, Simoens P (2018) Anatomy of the lower respiratory tract in domestic birds, with emphasis on respiration. Anat Histol Embryol 47:89–99

Castranova V, Rabovsky J, Tucker JH, Miles PR (1988) The alveolar type-II epithelial cell: a multifunctional pneumocyte. Toxicol Appl Pharmacol 93:472–483

Cevik-Demirkan A, Kürtül I, Haziroglu RM (2006) Gross morphological features of the lung and air sac in the Japanese quail. J Vet Med Sci 68:909–913

Chamberlain DR, Gross WR, Cornwell GW, Mosby HS (1968) Syringeal anatomy, in the common crow. Auk 89:244–252

Chandra P, Bharadwaj MB (1971) Histology and certain histochemical studies on the respiratoty system of chicken. II. Trachea, syrinx bronchi and lungs. Indian J Anim Sci 41:37–45

Chapman RF (1998) The insects: structure and function. Cambridge University Press, Cambridge

Chen Z, Wiens JJ (2020) The origins of acoustic communication in vertebrates. Nature Commun 11:369. https://doi.org/10.1038/s41467-020-14356-3

Cieri RL, Farmer CG (2016) Unidirectional pulmonary airflow in vertebrates: a review of structure, function, and evolution. J Comp Physiol B 186:541–552

Cieri RL, Farmer CG (2020) Computational fluid dynamics reveals a unique net unidirectional pattern of pulmonary airflow in the savannah monitor lizard (*Varanus exanthematicus*). Anat Rec 303:1768–1791

Cieri R, Craven B, Schachner E, Farmer CG (2014) New insight into the evolution of the vertebrate respiratory system and the discovery of unidirectional airflow in iguana lungs. Proc Natl Acad Sci (USA) 111:17218–17223

Cieri RL, Moritz S, Capano JG, Brainerd EL (2018) Breathing with floating ribs: XROMM analysis of lung ventilation in savannah monitor lizards. J Exp Biol 221:jeb.189449. https://doi.org/10.1242/jeb.189449

Clarke A, Rothery P (2008) Scaling of body temperature in mammals and birds. Funct Ecol 22:58–67

Clarke JA, Chatterjee S, Zhiheng L, Riede T, Agnolin F, Goller F, Isasi MP, Martinioni DR, Mussel FJ, Novas FE (2016) Fossil evidence of the avian vocal organ from the Mesozoic. *Nature*538:502–505

Clench M (1978) Tracheal elongation in birds of paradise. Condor 80:423–430

Clippinger TL (1997) Diseases of the lower respiratory tract of companion birds. Semin Avian Exotic Pet Med 6:201–208

Coitier V (1573) (Cited by Campana 1875) Anatomia avium. In: Externum et internarum praecipalium humani corporis partium tabulae atque anatomicae exercitationes observationesque varieae. Norimbergae, pp 130–133

Comroe JH (1974) Physiology of respiration: an introductory text. YearBook Medical Publishers, Chicago (IL)

Cook RD, King AS (1970) Observations on the ultrastructure of the smooth muscle and its innervation in the avian lung. J Anat 106:273–283

Cook RD, Vaillant CR, King AS (1986) The abdominal air sac ostium of the domestic fowl: a sphincter regulated by neuro-epithelial cells? J Anat 149:101–111

Cook RD, Vaillant CR, King AS (1987) The structure and innervation of the saccopleural membrane of the domestic fowl, *Gallus gallus*: an ultrastructural and immunohistochemical study. J Anat 150:1–9

Coughtrey M (1873) Note respecting the tracheal pouch of the emu. Ann Mag Nat Hist 12:217–218

Cover MS (1953) Gross and microscopic anatomy of the respiratory system of the turkey. II. The larynx, trachea, syrinx bronchi and lungs. Am J Vet Res 14:239–238

Cozic AM, Homberger DG (2015) The paired cervico-cephalic air sacs and their role in the vocalizations of songbirds. FASEB J 29(S1):867.9. https://doi.org/10.1096/faseb.29.1_supplement.867.9

Cozic AM, Homberger DG (2016) The role of cervical air sacs in the vocalization of songbirds. Anat Rec 299 (Special Feature: ICVM11-2016 Program and Abstracts) 41:258–259

Cozic AM, Suthers RA, Homberger DG (2014) The cervical cutaneous and subcutaneous structures in songbirds and their possible roles in vocalization. FASEB J 918:10. https://doi.org/10.1096/fasebj.28.1_supplement.918.10

Crawford NG, Faircloth BC, McCormack JE, Brumfield RT, Winker K, Glenn TC (2012) More than 1000 ultraconserved elements provide evidence that turtles are the sister group archosuars. Biol Letts 8:783–786

Crespo R, Yamashiro S, Hunter DB (1998) Development of the thoracic air sacs of turkeys with age and rearing conditions. Avian Dis 42:35–44

Daniels CB, Orgeig S (2001) The comparative biology of pulmonary surfactant: past, present and future. Comp Biochem Physiol A Mol Integr Physiol 129:9–36

Daniels CB, Orgeig S (2003) Pulmonary surfactant: the key to the evolution of air breathing. News Physiol Sci 18:151–157

Daniels CB, Barr HA, Power JHT, Nicholas TE (1990) Body temperature alters the lipid composition of pulmonary surfactant in the lizard *Ctenophorus nuchalis*. Exp Lung Res 16:435–449

Daniels CB, Orgeig S, Wilsen J, Nicholas TE (1994) Pulmonary-type surfactants in the lungs of terrestrial and aquatic amphibians. Respir Physiol 95:249–258

Daniels CB, Orgeig S, Smits AW (1995) The composition and function of reptilian pulmonary surfactant. Respir Physiol 102:121–135

De Kok-Mercado F, Habib M, Phelps T, Gailloud P (2013) Adaptations of the owl's cervical and cephalic arteries in relation to extreme neck rotation. Science 339:514–514

De Souza RBB, Bonfim VMG, Rios VP, Klein W (2021) Allometric relations of respiratory variables in Amniota: effects of phylogeny, form, and function. Comp Biochem Physiol A: Mol Integr Physiol 252:110845. https://doi.org/10.1016/j.cbpa.2020.110845

Deffebach ME, Charan NB, Lakshminarayan S, Butler J (1987) The bronchial circulation. Small, but a vital attribute of the lung. Am Rev Respir Dis 135:463–481

del Corral JPD (1995) Anatomy and histology of the lung and air sacs of birds. In: Pastor LM (ed) Histology, ultrastructure and immunohistochemistry of the respiratory organs in non-mammalian vertebrtaes. Publicaciones de la Universitatd de University of Murcia, Murcia, pp 179–233

Demirkan AC, Kurtul I, Hazroglu RM (2006a) Gross morphological features of the lung and air sac in the Japanese quail. J Vet Med Sci 68:909–913

Demirkan AC, Haziroglu RM, Kurtul I (2006b) Air sacs (*sacci pneumatici*) in mallard ducks (*Anas platyrhynchos*). Ankara Univ Vet Fak Derg 53:75–78

Dickinson MH, Farley CT, Full RJ, Koehl MAR, Kram R, Lehman S (2000) How animals move: an integrative view. Science 288:100–106

Doĝan GK, Takici I (2018) Anatomy of respıratory system in poultry. MAE Vet Fak Derg 3:141–147

Dominelli P, Wiggins C, Roy T, Secomb T, Curry T, Joyner M (2021) The oxygen cascade during exercise in health and disease. Mayo Clinic Proc 96:1017–1032

Dotterweich H (1930) Die Bahnhofstauben und die Frage nach dem Weg der Atemlauft. Zool Anz 90:253–262

Dudley R (1998) Atmospheric oxygen, giant Paleozoic insects, and the evolution of aerial locomotor performance. J Exp Biol 201:1043–1050

Dudley R (2000) The Biomechanics of insect flight: form, function, evolution. Princeton University Press, Princeton (NJ)

Duncan FD, Byrne MJ (2002) Respiratory airflow in a wingless dung beetle. J Exp Biol 205:2489–2497

Duncker HR (1971) The lung-air sac system of birds. A contribution to the functional anatomy of the respiratory apparatus. Ergebn Anat EntwGesch 45:1–171

Duncker HR (1972) Structure of the avian lung. Respir Physiol 14:4–63

Duncker HR (1974) Structure of the avian respiratory tract. Respir Physiol 22:1–34

Duncker H-R (1978a) Development of the avian respiratory and circulatory systems. In: Piiper J (ed) Respiratory function in birds, adult and embryonic. Springer, Berlin, pp 260–273

Duncker H-R (1978b) General morphological principles of amniotic lungs. In: Piiper J (ed) Respiratory function in birds, adult and embryonic. Springer, Heidelberg, pp 1–18

Duncker HR (1979) Die funktionelle Anatomie des Lungen-Luftsack-Systems der Vögel - mit besonderer Berücksichtigung der Greivögel. Der Prakt Tierarzt 60:209–218

Eberth CJ (1863) Uber den feineren bau der lunge. Z Wiss Zool 12:427–454

El-Bably SH, Rezk HM, Tolba AR (2014) Gross morphological studies on the air sacs (saccipneumatici) of Golden Pekin duck (*Anas platyrhynchus*). Haryana Vet 53:13–17

Elemans C, Rasmussen J, Herbst C, Düring DN, Zollinger SA, Brumm H, Srivastava K, Svane N, Ding M, Larsen ON, Sober SJ, Švec JG (2015) Universal mechanisms of sound production and control in birds and mammals. Nat Commun 6:8978. https://doi.org/10.1038/ncomms9978

Ellington CP (1984) The aerodynamics of hovering insect flight. VI. Lift and power requirements. Philos Trans R Soc (Lond) B 305:145–181

Ellington CP, Machin KE, Casey TM (1990) Oxygen consumption of bumblebees in forward flight. Nature 347:472–473

El-Sayed AK, Hassan S (2020) Gross morphological features of the air sacs of the hooded crow (*Corvus cornix*). Anat Histol Embryol 49:159–166

Erdoğan S, Sağsöz H, Paulsen F (2015) Functional anatomy of the syrinx of the chukar partridge (galliformes: *Alectoris chukar*) as a model for phonation research. Anat Rec 298:602–617

Erwin TL (1991) How many species are there? Revisited. Conserv Biol 5:330–333

Farmer CG (2006) On the origin of avian air sacs. Respir Physiol Neurobiol 154:89–106

Farmer CG (2010) The provenance of alveolar and parabronchial lungs: Insights from paleoecology and the discovery of cardiogenic, unidirectional airflow in the American alligator (*Alligator mississippiensis*). Physiol Biochem Zool 83:561–575

Farmer CG (2015a) The evolution of unidirectional pulmonary airflow. Physiology 30:260–272

Farmer CG (2015b) Similarity of crocodilian and avian lungs indicates unidirectional flow is ancestral for archosaurs. Integr Comp Biol 55:962–971

Farmer CG (2017) Pulmonary transformations of vertebrates. In: Maina JN (ed) The biology of the avian respiratory system: evolution, development, structure and function. Springer, Heidelberg, pp 99–112

Farmer CG, Sanders K (2010) Unidirectional airflow in the lungs of alligators. Science 327:338–340

Fedde MR (1980) The structure and gas flow pattern in the avian lung. Poult Sci 59:2642–2653

Fedde MR (1998) Relationship of structure and function of the avian respiratory system to disease susceptibility. Poult Sci 77:1130–1138

Fee MS, Shraiman B, Pesaran B, Mitra PP (1998) The role of nonlinear dynamics of the syrinx in the vocalizations of a songbird. Nature 395:67–71

Fischer G (1905) Vergleichende anatomische Untersuchungen über den Bronchial baum der Vögel. Zoologica Stuttg 19:1–46

Fitch WT (1999) Acoustic exaggeration of size in birds via tracheal elongation: comparative and theoretical analyses. *J Zool (Lond)*248:31–48

Fletcher OJ (1980) Pathology of the avian respiratory system. Poult Sci 59:2666–2679

Fristoe TS, Burger JR, Balk MA, Khaliq I, Hof C, Brown JH (2015) Metabolic heat production and thermal conductance are mass-independent adaptations to thermal environment in birds and mammals. Proc Natl Acad Sci (USA) 112:15934–15939

Frith CB (1994) Adaptive significance of tracheal elongation in manucodes (Paradisaeidae). *Condor*96:552–555

Fujii S, Tamura T, Okamoto T (1981) Microarchitecture of air capillaries and blood capillaries in the respiratory area of the hen's lung examined by scanning electron microscopy (*Gallus gallus*). Jap J Vet Sci 43:83–88

Fuxjager MJ, Schlinger BA (2015) Perspectives on the evolution of animal dancing: a case study of manakins. Curr Opin Behav Sci 6:7–12

Gaban-Lima R, Höfling E (2006) Comparative anatomy of the syrinx in the Tribe Arini (Aves: Psittacidae). Braz J Morphol Sci 23:501–512

Galis F (1996) The application of functional morphology to evolutionary studies. Tree 11:124–129

Garcia SM, Kopuchain C, Mindlin GB, Fuxjager MJ, Tubaro PL, Goller F (2017) Evolution of vocal diversity through morphological adaptation without learning of complex neural control. Curr Biol 27:2677–2683

Gatesy SM, Dial KP (1996) Locomotor modules and the evolution of avian flight. Evolution 50: 331–340

Gaunt AS (1986) Phonation. In: Seller TJ (ed) Bird respiration, vol 1. CRC Press, Boca-Raton (FL), pp 71–96

Gaunt AS, Gaunt SLL (1977) Mechanics of the syrinx in *Gallus gallus*. II. Electromyographic studies of *ad libitum* vocalization. J Morphol 152:1–20

Gaunt AS, Gaunt SLL (1985) Syringeal structure and avian phonation. In: Johnston RF (ed) Current ornithology, vol 2. Plenum Press, New York, pp 213–245

Gaunt AS, Stein RC, Gaunt SLL (1973) Pressure and airflow during stress calls of the starling, *Sturnus vulgaris*. J Exp Zool 183:241–262

Gaunt AS, Gaunt SLL, Hector DH (1976) Mechanics of the syrinx in *Gallus gallus*. I. A comparison of pressure events in chickens to those of Oscines. Condor 78:208–223

Gaunt AS, Gaunt SLL, Casey RM (1982) Syringeal mechanics reassessed: evidence from *Streptopelia*. Auk 99:474–494

Gaunt AS, Gaunt SLL, Prange HD, Wasser JS (1987) The effect of tracheal coiling on the vocalization of cranes (Aves: Gruidae). J Comp Physiol A 161:43–58

Gehr P, Mwangi DK, Amman A, Maloiy GMO, Taylor CR, Weibel ER (1981) Design of the mammalian respiratory system. V. Scaling morphometric diffusing capacity to body mass: wild and domestic animals. Respir Physiol 44:41–86

Geipel I, Kalko EKV, Wallmeyer K, Knörnschild M (2013) Postweaning maternal food provisioning in a bat with a complex hunting strategy. Anim Behav 85:1435–1441

George JC, Berger AJ (1966) Avian myology. Academic Press, New York

Gerisch D, Schwartz R (1972) Morphologische Befunde an den Bronchi arteriales des Hayshuhnes (*Gallus gallus* dom.). Dt tierärztl Wschr 79:573–612

Gier HT (1952) The air sacs of the loon. Auk 69:40–49

Gilbert PW (1939) The avian lung and air-sac system. Auk 56:57–63

Gillooly JF, Ophir AG (2010) The energetic basis of acoustic communication. Proc Biol Sci 277: 1325–1331

Goller F, Larsen ON (1997) A new mechanism of sound generation in songbirds. Proc Natl Acad Sci (USA) 94:14787–14791

Govindaraju K, Cowley EA, Eidelman DH, Lloyd DK (1997) Microanalysis of lung airw ay surface fluid by capillary electrophoresis w ith conductivity detection. Anal Chem 69:2793–2797

Greenewalt CH (1968) Bird song: acoustics and physiology. Smithosonian Institution Press, Washington DC

Grigg GC, Beard LA, Augee ML (2004) The evolution of endothermy and its diversity in mammals and birds. Physiol Biochem Zool 77:982–997

Groebbels F (1932) De Vögel. Bau, Funktion, Lebenserscheinung, Einpassung, vol 1. Borntraeger

Groenewald B, Hetz SK, Chown SL, Terblanche JS (2012) Respiratory dynamics of discontinuous gas exchange in the tracheal system of the desert locust, *Schistocerca gregaria*. J Exp Biol 215: 2301–2307

Groom AC, Ellis CG, Wrigley SJ, Potter RF (1995) Capillary network morphology and capillary flow. Int J Microcirc Clin Exp 5:223–230

Gunkel C, Marin ML (2005) Current techniques in avian anesthesia. Semi Avian Exotic Pet Med 14:263–276

Habib MB (2019) New perspectives on the origins of the unique vocal tract of birds. PLoS Biol 17: e3000184. https://doi.org/10.1371/journal.pbio.3000184

Hamlet MP, Fischer HI (1967) Air sacs of respiratory origin in some Procellariiform birds. Condor 69:586–595

Hansell M (2005) Animal architecture. Oxford University Press, New York (NY)

Harris WA (2022) Zero to birth. Princeton University Press, Princeton (NJ)

Harrison JF (1997) Ventilatory mechanism and control in grasshopers. Am Zool 37:73–81

Harrison JF, Woods HA, Roberts SP (2012) Ecological and environmental physiology of insects. Oxford Univ Press, Oxford

Harrison JF, Waters JS, Cease AJ, Brooks JMV, Callier V, Klok CJ, Shaffer K, Socha JJ (2013) How locusts breathe. Physiology 28:18–27

Hartmann T, Verkman AS (1990) Model of ion transport regulation in chloride-secreting airway epithelial cells: integrated description of electrical, chemical, and fluorescence measurements. Biophys J 58:391–401

Heard DJ (1997) Avian respiratory anatomy and physiology. Respir Dis 6:172–179

Henderson DR, Prange HD (1995) Contributions of spiracle movement and airway resistance to unidirectional flow in grasshoppers, *Melanoplus differentialis*. FASEB J 9:A642. Corpus ID: 32295181

Hersch GL (1966) Bird voices and resonant tuning in helium-air mixtures. PhD Thesis, University of California

Hinds DS, Calder WA (1971) Tracheal dead space in the respiration of birds. Evolution 25:429–440

Hochgraf JS, Waters JS, Socha JJ (2018) Patterns of tracheal compression in the thorax of the ground beetle, *Platynus decentis*. Yale J Biol Med 91:409–430

Hodges RD (1974) The histology of the fowl. Academic Press, London

Hoffstetter R, Gasc JP (1969) Vertebrae and ribs of modern reptiles. Biol Reptil 1:201–310

Hofman MA (2014) Evolution of the human brain: when bigger is better. Front Neuroanat 8:15. https://doi.org/10.3389/fnana.2014.00015

Hogg DA (1982) Ossification of the laryngeal, tracheal and syringeal cartilages in the domestic fowl. J Anat 134:57–71

Holle JP, Meyer M, Scheid P (1977) Oxygen affinity of duck blood determined by *in vivo* and *in vitro* techniques. Respir Physiol 29:355–361

Holle JP, Heisler N, Scheid P (1978) Blood flow distribution in the bird lung and its controls by respiratory gases. Am J Physiol 234:R146–R154

Hollinger H (2017) Air sac rupture in birds. https://wagwalking.com/bird/condition/air-sac-rupture#symptoms. Veterinary aspects. Reviewed by Michele K on 06-03-2021. Retrieved 03-07-2021

Homberger DG (1979) Functional morphology of the larynx in the parrot, *Psittacus erithacus*. Am Zool 19:988. https://doi.org/10.1007/BF00312112

Homberger DG (2017a) Functional Functional-morphological diversity and complexity of the larynx as a basis for vocal complexity in birds: analogies to human speech. FASEB J 31: 577.8. https://doi.org/10.1096/fasebj.31.1_supplement.577.8

Homberger DG (2017b) The avian lingual and laryngeal apparatus within the context of the head and jaw apparatus, with comparisons to the mammalian condition: functional morphology and biomechanics of evaporative cooling, feeding, drinking, and vocalization. In: Maina J (ed) The biology of the avian respiratory system. Springer, Cham, pp 27–97

Hoppeler H, Weibel ER (2000) Structural and functional limits or oxygen supply to muscle. Acta Physiol 168:445–456

Hsia CC, Schmitz A, Lambertz M, Perry SF, Maina JN (2013) Evolution of air breathing: Oxygen homeostasis and the transitions from water to land and sky. Compr Physiol 3:849–915

Huetsch JC, Shimoda LA (2022) Pulmonary circulation. In: Janes SM (ed) Encyclopedia of respiratory medicine, 2nd edn. Academic Press, Oxford, pp 81–89

Huey RB, Stevenson RD (1979) Integrating thermal physiology and ecology of ectotherms: a discussion of approaches. Am Zool 19:357–366

Huxley TH (1882) On the respiratory organs of *Apteryx*. Proc Zool Soc (Lond) 1882:560–569

Ibrahim IA, Hussein MM, Hamdy A, Abdel-Maksoud FM (2020) Comparative morphological features of syrinx in male domestic fowl *Gallus gallus domesticus* and male domestic pigeon *Columba livia domestica*: a histochemical, ultrastructural, scanning electron microscopic and morphometrical study. Microsc Microanal 26:326–347

Jaensch S (2015) Inspirational evolution: the avian lower respiratory tract. https://www.aavac.com.au/files/2015-01.pdf. Accessed 27-06-2021

Jain V, Bordes S, Bhardwaj A (2021) Physiology, pulmonary circulatory system. [Updated 2021 May 9]. StatPearls [Internet]. StatPearls Publishing, Treasure Island (FL). https://www.ncbi.nlm.nih.gov/books/NBK525948/. Accessed 28-06-2021

James AE, Hutchins G, Bush M, Natarajan TK, Burns B (1976) How birds breathe: correlation of radiographic and anatomical and pathological studies. J Vet Radiol Soc 17:77–86

Jarvis ED, Mirarab S, Aberer AJ, Li B, Houde P, Li C, Ho SYW, Faircloth BC, Nabholz B, Howard JT et al (2014) Whole-genome analyses resolve early branches in the tree of life of modern birds. Science 346:1320–1331

Jenni-Eiermann S (2017) Energy metabolism during endurance flight and the post-flight recovery phase. J Comp Physiol A Neuroethol Sens Neural Behav Physiol 203:431–438

Jimoh SA, Maina JN (2013) Immuno-localization of type-IV collagen in the blood-gas barrier and the epithelial - epithelial cell connections of the avian lung. Biol Lett 9:20120951. https://doi.org/10.1098/rsbl.2012.0951

Jones JH (1982) Pulmonary blood flow distribution in panting ostriches. J Appl Physiol 53:1411–1417

Juillet A (1912) Recherches anatomiques, embryologiques, histologiques et comparatives sur le poumon des oiseaux. Arch Zool Exp Gen IX:207–371

Jung B (2001) Surgical repair of air sac rupture in a parrot. Tierärztliche Umschau 56:35–37

Jürgens JD, Gros G (2002) Phylogeny of gas exchange systems. Anasthesiol Intensivmed Notfallmed Schmerzther 37:185–198

Kabak M, Orhan IO, Haziroglu RM (2007) The gross anatomy of larynx, trachea and syrinx in the long-legged buzzard (*Buteo rufinus*). Anat Histol Embryol 36:27–32

Käfer H, Kovac H, Stabentheiner A (2013) Respiration patterns of resting wasps (Vespula sp.). J Insect Physiol 59:475–486

Kamani J, Tijjani A, Yidawi JP, Gana AL, Egwu OK, Gusi AM (2009) Subcutaneous emphysema (windpuff) in a 13 weeks old pullet: case report. Int J Poult Sci 8:1121–1122

Kambic RE, Biewener AA, Pierce SE (2017) Experimental determination of three-dimensional cervical joint mobility in the avian neck. Front Zool 14:1–15. https://doi.org/10.1186/s12983-017-0223-z

Kandyel RM, El Basyouny HA, El Nahas EE, Madkour F, Haddad S, Massoud D, Morsy K, Madkour N, Abumandour M (2021) A histological and immunohistochemical study on the parabronchial epithelium of the domestic fowl's (*Gallus gallus domesticus*) lung with special reference to its scanning and transmission electron microscopic characteristics. Microsc Res Tech 2021:1–12

Kardong KV (2009) Vertebrates: comparative anatomy, function, evolution, 5th edn. McGraw Hill, New York

Keough KM, Parsons CS, Phang PT, Tweeddale MG (1988) Interactions between plasma proteins and pulmonary surfactant: surface balance studies. Can J Physiol Pharmacol 66:1166–1173

Khan YS, Lynch DT (2021) Histology, lung. [Updated 2021 May 10]. In: StatPearls [Internet]. StatPearls Publishing, Treasure Island (FL). https://www.ncbi.nlm.nih.gov/books/NBK534789/. Accessed 03-07-2021

King AS (1957) The aerated bones of *Gallus domesticus*. Acta Anat 31:220–230

King AS (1966) Structural and functional aspects of the avian lung and its air sacs. Intern Rev Gen Exp Zool 2:171–267

King AS (1975) Aves respiratory system. In: Getty R (ed) Sisson and Grossman's The anatomy of the domestic animals, vol 2, 5th edn. Saunders, Philadelphia, pp 1011–1075

King AS (1979) Systema respiratorium. In: Baumel JJ, King AS, Lucas AM, Breazile JE, Evans HE (eds) Nomina anatomica avium. Academic Press, London, pp 227–265

King AS (1989) Functional anatomy of the syrinx. In: King AS, McLelland J (eds) Form and function in birds, vol 4. Academic Press, London, pp 105–192

King AS, Atherton JD (1970) The identity of the air sacs of the turkey (*Melleagris gallopavo*). Acta Anat 77:78–91

King AS, Cowie AF (1969) The functional anatomy of the bronchial muscle of the bird. J Anat 105: 323–336

King AS, King DZ (1979) Avian morphology: general principles. In: King AS, McLelland J (eds) Form and function in birds, vol I. Academic Press, London, pp 1–38

King AS, McLelland J (1984) Birds: their structure and function. Bailliére Tindall, Eastbourne (Sussex - England)

King AS, McLelland J (eds) (1989) Form and function in birds, vol 4. Academic Press, London

King AS, Molony V (1971) The anatomy of respiration. In: Bell DF, Freeman BM (eds) Physiology and biochemistry of the domestic fowl, vol 1. Academic Press, London, pp 347–384

King AS, King DZ, Abdalla MA (1978) The structure of the intrapulmonary vaculature of the domestic fowl. In: Piiper J (ed) Respiratory function in birds, adult and emberyonic. Springer, New York, pp 112–124

Kingsley EP, Eliason CM, Riede T, Li Z, Hiscock TW, Farnsworth M, Thomson SL, Goller F, Tabin CJ, Clarke JA (2018) Identity and novelty in the avian syrinx. Proc Natl Acad Sci (USA) 115:10209–10217

Kirkton SD (2007) Effects of insect body size on tracheal structure and function. In: Roach RC, Wagner PD, Hackett PH (eds) Hypoxia and the circulation. Springer, New York (NY), pp 221–228

Kitazawa S, Fuchina S, Susaki N (1977) Studies on air sacs in domestic fowls. I. Development of air sacs in embryos. Res Bull Obihiro Univ Series I 10:407–427

Kloek GP, Casler CL (1972) The lung and air-sac system of the common gracle. Auk 89:817–825

Klowden MJ (2002) Physiological systems in insects. Academic Press, New York

Krause R (1922) Mikroskopische Anatomie der Wirbeltiere in Einzeldarstellungen. 11. Vögel und Reptilien. Berlin-Leipzig: De Gruyter and Co

Krings M, Nyakatura JA, Fischer MS, Wagner H (2014) The cervical spine of the American barn owl (*Tyto furcata pratincola*): I. Anatomy of the vertebrae and regionalization in their S-shaped arrangement. PLoS One 9:e91653. https://doi.org/10.1371/journal.pone.0091653

Krings M, Nyakatura JA, Boumans MLLM, Fischer MS, Wagner H (2017) Barn owls maximize head rotations by a combination of yawing and rolling in functionally diverse regions of the neck. J Anat 231:12–22

Kuethe DO (1988) Fluid mechanical valving of airflow in bird lungs. J Exp Biol 136:1–12

Labocha MK, Hayes JP (2019) Endotherm. In: Fath B (ed) Encyclopedia of ecology, 2nd edn. Elsevier BV, Amsterdam, pp 368–374

Larsen ON, Goller F (2002) Direct observation of syringeal muscle function in songbirds and a parrot. J Exp Biol 205:25–35

Lasiewski RC, Calder WA (1971) A preliminary allometric analysis of respiratory variables in resting birds. Respir Physiol 11:152–166

Lierz M, Korbel R (2012) Anesthesia and analgesia in birds. J Exotic Pet Med 21:44–58

Lierz M, Rüdiger K (2012) Anesthesia and analgesia in birds. J Exotic Pet Med 21:44–58

Lockner FR, Murrish DE (1975) Interclavicular air sac pressures and vocalization in mallard ducks, *Anas platyrhynchos*. Comp Biocem Physiol 52A:183–187

Lockner FR, Youngren OM (1976) Functional syringeal anatomy of the mallard. I. *In situ* electromyograms during ESB elicited calling. Auk 93:324–342

Locy WA, Larsell O (1916a) The embryology of the birds lungs. Based on observations of the domestic fowl. Part I. 1. The external aspects of lung development. 2. The development of the bronchial tree. Am J Anat 19:447–504

Locy WA, Larsell O (1916b) The embryology of the bird's lung based on observations of the domestic fowl. Part II. Am J Anat 20:1–44

López J (1995) Anatomy and histology of the lung and air sacs of birds. In: Pastor LM (ed) Histology, ultrastructure, and immunohistochemistry of the respiratory organs in non-mammalian vertebrates. Publicaciones de la Universitatd de University of Murcia, Murcia, pp 179–233

Lovegrove BG (2012) The evolution of endothermay in Cenozoic mammals: a plesiomorphic-apomorphic continuum. Biol Rev 87:128–162

Lovegrove BG (2017) A phenology of the evolution of endothermy in birds and mammals. Biol Rev 92:1213–1240

Lovegrove BG (2019) Fires of life: endothermy in birds and mammals. Yale University Press, New Haven

Luis da Cruz A, Pedretti ACE, Fernandes MN (2009) Stereological estimation of the surface area and oxygen diffusing capacity of the respiratory stomach of the air-breathing armored catfish, *Pterygoplichthys anisitsi* (Teleostei: Loricariidae). J Morphol 270:601–614

Macklem P, Bouverot P, Scheid P (1979) Measurement of the distensibility of the parabronchi in duck lungs. Respir Physiol 33:23–35

Magnussen H, Willmer H, Scheid P (1976) Gas exchange in the air sacs: contribution to respiratory gas exchange in ducks. Respir Physiol 26:129–146

Maina JN (1982a) A scanning electron microscopic study of the air- and blood capillaries of the lung of the domestic fowl (*Gallus domesticus*). Experientia 35:614–616

Maina JN (1982b) Stereological analysis of the paleopulmo and neopulmo respiratory regions of the avian lung (*Streptopelia decaocto*). IRCS Med Sci 10:328. http://erepository.uonbi.ac.ke:8080/xmlui/handle/123456789/50100

Maina JN (1984) Morphometrics of the avian lung. 3. The structural design of the passerine lung. Respir Physiol 55:291–309

Maina JN (1985) A scanning and transmission electron microscopic study of the bat lung. J Zool (Lond) 205B:19–27

Maina JN (1988) Scanning electron microscopic study of the spatial organization of the air- and blood conducting components of the avian lung *(Gallus gallus domesticus)*. Anat Rec 222:145–153

Maina JN (1989a) The morphology of the lung of the black mamba *Dendroaspis polylepis* (Reptilia: Ophidia: Elapidae): a scanning and transmission electron microscopic study. J Anat 167:31–46

Maina JN (1989b) A scanning and transmission electron microscopic study of the tracheal air-sac system in a grasshopper (*Chrotogonus senegalensis*, Kraus)- (Orthoptera: Acrididae: Pygomorphinae). Anat Rec 223:393–405

Maina JN (1989c) The morphometry of the avian lung. In: King AS, McLelland J (eds) Form and function in birds, vol 4. Academic Press, London, pp 307–368

Maina JN (1998) The gas exchangers: structure, function, and evolution of the respiratory processes. Springer, Berlin

Maina JN (2000a) Comparative respiratory morphology: themes and principles in the design and construction of the gas exchangers. Anat Rec 261:25–44

Maina JN (2000b) What it takes to fly: the novel respiratory structural and functional adaptations in birds and bats. J Exp Biol 203:3045–3064

Maina JN (2000c) Is the sheet-flow design a 'frozen core' (a Bauplan) of the gas exchangers? Comparative functional morphology of the respiratory microvascular systems: Illustration of the geometry and rationalization of the fractal properties. Comp Biochem Physiol 126A:491–515

Maina JN (2002a) Structure, function and evolution of the gas exchangers: comparative perspectives. J Anat (Lond) 201:281–304

Maina JN (2002b) Fundamental structural aspects in the bioengineering of the gas exchangers: comparative perspectives. Adv Anat Embryol Cell Biol 163:1–112

Maina JN (2002c) Functional morphology of the vertebrate respiratory organs. Oxford and IBH Publishing Company, Lebanon

Maina JN (2003a) A systematic study of the development of the airway (bronchial) system of the avian lung from days 3 to 26 of embryogenesis: a transmission electron microscopic study on the domestic fowl, *Gallus gallus* variant *domesticus*. Tissue Cell 35:375–391

Maina JN (2003b) Developmental dynamics of the bronchial (airway)- and air sac systems of the avian respiratory system from days 3 to 26 of life: a scanning electron microscopic study of the domestic fowl, *Gallus gallus* variant *domesticus*. Anat Embryol 207:119–134

Maina JN (2005) The lung-air sac system of birds: development, structure and function. Springer, Berlin

Maina JN (2006) Development, structure, and function of a novel respiratory organ, the lung-air sac system of birds: to go where no other vertebrate has gone. Biol Rev 81:545–579

Maina JN (2007a) Minutialization at its extreme best! The underpinnings of the remarkable strengths of the air- and the blood capillaries of the avian lung: a conundrum. Respir Physiol Neurobiol 159:141–145

Maina JN (2007b) Spectacularly robust! Tensegrity principle explains the mechanical strengths of the avian lung. Respir Physiol Neurobiol 155:1–10

Maina JN (2008) Structure of the air- and blood capillaries of the avian lung and the debate regarding the basis of their astounding strengths. In: Morris S, Vosloo A (eds) Proceedings of the 4th comparative physiology and biochemistry conference in Africa, Maasai Mara Game Reserve, Kenya. Medimond S.r.l, via Maserati, Bologna, pp 304–313

Maina JN (2009) The morphology of the lung of a tropical terrestrial slug, *Trichotoxon copleyi* (Mollusca: Gastropoda: Pulmonata): a scanning and transmission electron microscopic study. J Zool (Lond) 217:355–366

Maina JN (2011) Bioengineering aspects in the design of gas exchangers: comparative evolutionary, morphological, functional, and molecular perspectives. Springer, Heidelberg

Maina JN (2012) Comparative molecular developmental aspects of the mammalian- and the avian lungs, and the insectan tracheal system by branching morphogenesis: recent advances and future directions. Front Zool 9:16. https://doi.org/10.1186/1742-9994-9-16

Maina JN (2015a) Structural and biomechanical properties of the exchange tissue of the avian lung. Anat Rec 298:1673–1688

Maina JN (2015b) The design of the avian respiratory system: development, morphology and function. J Ornithol 156:41–63

Maina J (2015c) Morphological and morphometric properties of the blood-gas barrier: comparative perspectives. In: Makanya AN (ed) The vertebrate blood-gas barrier in health and disease. Springer, Cham, pp 15–39

Maina JN (2017a) Pivotal debates and controversies on the structure and function of the avian respiratory system: setting the record straight. Biol Rev 92:1475–1504

Maina JN (2017b) Functional design of the mature avian respiratory system. In: Maina JN (ed) The biology of the avian respiratory system: evolution, development, structure and function. Springer, Cham, pp 191–218

Maina JN (2017c) Critical appraisal of some factors pertinent to the functional designs of the gas exchangers. Cell Tissue Res 367:747–767

Maina JN (2022) Perspectives on the structure and function of the avian respiratory system: functional efficiency built on structural complexity. Front Anim Sci 3:851574. https://doi.org/10.3389/fanim.2022.851574

Maina JN, Africa M (2000) Inspiratory aerodynamic valving in the avian lung: functional morphological study of the extrapulmonary primary bronchus. J Exp Biol 203:2865–2876

Maina JN, King AS (1982a) The thickness of the avian blood-gas barrier: qualitative and quantitative observations. J Anat 134:553–562

Maina JN, King AS (1982b) Morphometrics of the avian lung. 2. The wild mallard *(Anas platyrhynchos)* and greylag goose *(Anser anser)*. Respir Physiol 50:299–313

Maina JN, King AS (1984) The structural functional correlation in the design of the bat lung. A morphometric study. J Exp Biol 111:43–63

Maina JN, King AS (1987) A morphometric study of the lung of the Humboldt penguin (*Spheniscus humboldti*). Zentralb Vet Med C Anat Histo Embryol 16:293–297

Maina JN, King AS (1989) The lung of the emu, *Dromaius novaehollandiae*: A microscopic and morphometric study. J Anat 163:67–74

Maina JN, Nathaniel C (2001) A qualitative and quantitative study of the lung of an ostrich, *Struthio camelus*. J Exp Biol 204:2313–2330

Maina JN, van Gils P (2001) Morphometric characterization of the airway and vascular systems of the lung of the domestic pig, *Sus scrofa*: comparison of the airway, arterial, and venous systems. Comp Biochem Physiol 130A:781–798

Maina JN, West JB (2005) Thin and strong! The bioengineering dilemma in the structural and functional design of the blood-gas barrier: comparative and evolutionary perspectives. Physiol Rev 85:811–844

Maina JN, Woodward JD (2009) Three-dimensional serial section computer reconstruction of the arrangement of the structural componenets of the parabronchus of the ostrich, *Struthio camelus*. Anat Rec 292:1685–1698

Maina JN, Abdalla MA, King AS (1982a) Light microscopic morphometry of the lungs of 19 avian species. Acta Anat 112:264–270

Maina JN, King AS, King DZ (1982b) A morphometric analysis of the lung of a species of bat. Respir Physiol 50:1–11

Maina JN, Howard CV, Scales L (1983) Length densities and maximum diameter distribution of the air capillaries of the paleopulmo and neopulmo region of the avian lung. Acta Stereol 2:101–107

Maina JN, Maloiy GMO, Warui CN, Njogu EK, Kokwaro ED (1989a) A scanning electron microscopic study of the morphology of the reptilian lung: the savanna monitor lizard *(Varanus exanthematicus)* and the pancake tortoise *(Malacochersus tornieri)*. Anat Rec 224:514–522

Maina JN, King AS, Settle G (1989b) An allometric study of the pulmonary morphometric parameters in birds, with mammalian comparison. Philos Trans R Soc (Lond) 326B:1–57

Maina JN, Thomas SP, Dallas DM (1991) A morphometric study of bats of different size: correlations between structure and function of the chiropteran lung. Philos Trans R Soc (Lond) B 333:31–50

Maina JN, Veltcamp CJ, Henry J (1999) A study of the spatial organization of the gas-exchange components of a snake lung - the sandboa *Eryx colubrinus* (Reptilia: Ophidia; Corubridae) by latex casting. J Zool (Lond) 247:81–90

Maina JN, Singh P, Moss EA (2009) Inspiratory aerodynamic valving occurs in the ostrich, *Struthio camelus* lung: a computational fluid dynamics study under resting unsteady state inhalation. Respir Physiol Neurobiol 169:262–270

Maina JN, Jimoh SA, Hosie M (2010) Implicit mechanistic role of the collagen, smooth muscle, and elastic tissue components in strengthening the air- and blood capillaries of the avian lung. J Anat 217:597–608

Maina JN, McCracken KG, Chua B, York JM, Milsom WK (2017) Morphological and morphometric specializations of the lung of the Andean goose, *Chloephaga melanoptera*: a lifelong high-altitude resident. PLoS One 12:e0174395. https://doi.org/10.1371/journal.pone.0174395

Maina JN, Ramonisi R, Mashiteng R, Mokae L, Woodward JD (2021) 3D Computer reconstruction of the airway and the vascular systems of the lung of the domestic fowl, *Gallus gallus* variant *domesticus*. J Appl Math Comput 5:89–104

Marek RD, Falkingham PL, Benson R, Gardiner JD, Maddox TW, Bates KT (2021) Evolutionary versatility of the avian neck. Proc Biol Sci 288:20203150. https://doi.org/10.1098/rspb.2020.3150

Mason RJ (2006) Epithelial cells: Type II cells. In: Laurent GJ, Shapiro SD (eds) Encyclopedia of respiratory medicine. Academic Press, Oxford, pp 138–142

Mathey WJ (1965) Avian tracheal rings. Poult Sci 44:1465–1467

Mathieu-Costello O, Szewczak JM, Logermann RB, Agey PJ (1992) Geometry of blood-tissue exchange in bat flight muscle compared with bat hindlimb and rat soleus muscle. Am J Physiol 262:R955–R965

May RM (1988) How many species are there on earth? Science 241:1441–1449

May RM (1992) How many species inhabit the earth? Sci Am 10:18–24

McCasland JS (1987) Neuronal control of bird song production. J Neurosci 7:23–39

McLelland J (1989a) Anatomy of the lungs and air sacs. In: King AS, McLelland J (eds) Form and function in birds, vol IV. Academic Press, London, pp 221–279

McLelland J (1989b) Larynx and trachea. In: King AS, McLelland J (eds) Form and function in birds, vol IV. Academic Press, London, pp 69–103

Meban C (1980) Thicknesses of the air-blood barriers in vertebrate lungs. J Anat 131:299–307

Menkes H, Traystman R, Terry P (1979) Collateral ventilation. Fed Proc 38:22–26

Miller PL (1974) Respiration - aerial gas transport. In: Rockstein M (ed) The physiology of insects, 2nd edn. Academic, New York, pp 346–402

Milsom WK, Scott GR, Frappell PB, McCracken KG (2021) Different strategies for convective O2 transport in high altitude birds: a graphical analysis. Comp Biochem Physiol A: Mol Integr Physiol 253:110871. https://doi.org/10.1016/j.cbpa.2020.110871

Miskimen M (1951) Sound production in passerine birds. Auk 68:493–504

Misof B, Liu S, Meusemann K, Peters RS, Donath A, Mayer C, Frandsen PB, Ware J, Flouri T, Beutel RG et al (2014) Phylogenomics resolves the timing and pattern of insect evolution. Science 346:763–767

Mitchell C (2016) The evolution of brains and cognitive abilities. In: Pontarotti P (ed) Evolutionary biology. Springer, Cham, pp 73–87

Mora C, Tittensor DP, Adl S, Simpson AGB, Worm B (2011) How many species are there on earth and in the ocean? PLoS Biol 9:e1001127. https://doi.org/10.1371/journal.pbio.1001127

Moreira MO, Qu YF, Wiens JJ (2021) Large-scale evolution of body temperatures in land vertebrates. Evol Lets 5:484–494

Morrell NW (2014) Pulmonary circulation: a reference module in biomedical sciences. Elsevier, New York

Morris CR, Nelson FE, Askew GN (2010) The metabolic power requirements of flight and estimations of flight muscle efficiency in the cockatiel (*Nymphicus hollandicus*). J Exp Biol 213:2788–2796

Mortola JP (2004) Pulmonary adaptation to sustained changes in metabolic rate. In: Massaro DJ, Massaro GD, Chambon P (eds) Lung development and regeneration. Marcel Dekker, New York, pp 525–571

Moura RS, Correia-Pinto J (2017) Molecular aspects of avian lung development. In: Maina JN (ed) The biology of the avian respiratory system: evolution, development, structure and function. Springer, Cham, pp 129–146

Müller B (1908) The air sacs of the pigeon. Smithson misc Colls 50:365–414

Murie J (1867) On the tracheal pouch of the emu (*Dromaius novaehollandiae* Vieill). Proc zool Soc (Lond) 1867:405–415

Nation JL (2002) Insect physiology and biochemistry. CRC Press, Boca Raton (FL)

Nation JL (2008) Tracheal system and respiratory gas exchange. In: Capinera JL (ed) Encyclopedia of entomology. Springer, Dordrecht, pp 245–262

Nespolo RF, Bacigalupe LD, Figueroa CC, Koteja P, Opazo JC (2011) Using new tools to solve an old problem: the evolution of endothermy in vertebrates. Trends Ecol Evol 26:414–423

Nickel R, Schummer A, Seiferle E (1977) Anatomy of the domestic birds. Verlag Paul Parey, Berlin

Niven JE, Scharlemann PW (2005) Do insect metabolic rates at rest and during flight scale with body mass? Biol Lett 1:346–349

Nottebohm F, Stokes TM, Leonard CM (1976) Central control of song in the canary, *Serinus canarius*. J Comp Neurol 165:457–486

Nottebohm F, Kelley DB, Paton JA (1982) Connections of vocal control nuclei in the canary telencephalon. J Comp Neurol 207:344–357

Nudds RL, Bryant DM (2000) The energy cost of short flights in birds. J Exp Biol 203:1561–1582

Nyakatura JA, Andrada E (2014) On vision in birds: coordination of head-bobbing and gait stabilises vertical head position in quail. Front Zool 11:27. https://doi.org/10.1186/1742-9994-11-27

O'Connor PM (2004) Pulmonary pneumaticity in the postcranial skeleton of extant aves: a case study examining Anseriformes. J Morphol 261:141–161

O'Connor PM (2009) Evolution of archosaurian body plans: skeletal adaptations of an air-sac-based breathing apparatus in birds and other archosaurs. J Exp Zool 311A:629–646

O'Connor PM, Claessens LPAM (2005) Basic avian pulmonary design and flow-through ventilation in non-avian theropod dinosaurs. Nature 436:253–256

Olmeda B, Villén L, Cruz A, Orellana G, Perez-Gil J (2010) Pulmonary surfactant layers accelerate O_2 diffusion through the air-water interface. Biochim Biophys Acta Biomemb 1798:1281–1284

Orgeig S, Bernhard W, Biswas SC, Daniels CB, Hall SB, Hertz SK, Lang CJ, Maina JN, Panda AK, Perez-Gil J et al (2007) The anatomy, physics, and physiology of gas exchange surfaces: is there a universal function for pulmonary surfactant in animal respiratory structures? Integr Comp Biol 47:610–627

Orgeig S, Morrison JL, Danield CB (2016) Evolution, development, and function of the pulmonary surfactant system in normal and perturbed environments. Compr Physiol 6:363–422

Orhan İÖ, Kabak M, Oto Ç, Haziroğlu RM (2009) Air sacs (sacci pneumatici) in the long-legged buzzard (*Buteo Rufinus*). Ankara Üniv Vet FakDerg 56:7–11

Osilla EV, Marsidi JL, Sharma S (2022) Physiology, temperature regulation. [Updated 2022 May 8]. In: StatPearls [Internet]. Treasure Island (FL): StatPearls Publishing; 2022 Jan. https://www.ncbi.nlm.nih.gov/books/NBK507838/

Padian K, de Ricqlès AJ (2020) Inferring the physiological regimes of extinct vertebrates: methods, limits and framework. Philos Trans Royal Soc (Lond) B 375:20190147. https://doi.org/10.1098/rstb.2019.0147

Pandey AK, Praveen PK, Ganguly S, Para PA, Saro RW, Mahajan T (2015) Avian respiratory anatomy and physiology with its interspecies variations: a review. World J Pharmaceut Life Sci 3:137–148

Parry K, Yates MS (1979) Observations on the avian pulmonary and bronchial circulation using labeled microspheres. Respir Physiol 38:131–140

Parry LA, Smithwick F, Nordén KK, Saitta ET, Lozano-Fernandez J, Tanner AR, Caron JB, Edgecombe GD, Briggs DGE, Vinther J (2018) Soft-bodied fossils are not simply rotten carcasses - toward a holistic understanding of exceptional fossil preservation. BioEssays 40: 1700167. https://doi.org/10.1002/bies.201700167

Pass G (2018) Beyond aerodynamics: the critical roles of the circulatory and tracheal systems in maintaining insect wing functionality. Arthropod Struct Dev 47:391–407

Patt DI, Patt GR (1969) Comparative vertebrate histology. Harper and Row, New York

Pattle RE (1978) Lung surfactant and lung lining in birds. In: Piiper J (ed) Respiratory function in birds, adult and embryonic. Springer, Berlin, pp 23–32

Payne DC, King AS (1960) The lung of *G. domesticus* secondary bronchi. J Anat 94:292

Perry SF (1988) Functional morphology of the lungs of the Nile crocodile *Crocodylus niloticus*: non-respiratory parameters. J Exp Biol 143:99–117

Perry SF (1990) Recent advances and trends in the comparative morphometry of vertebrate gas exchange organs. In: Boutilier RG (ed) Vertebrate gas exchange. Adv Comp Environml Physiol, vol 6. Springer, Berlin, pp 45–71. https://doi.org/10.1007/978-3-642-75380-0_3

Perry SF (1992) Gas exchange strategies in reptiles and the origin of the avian lung. In: Wood SC, Weber RE, Hargens AR, Millard RW (eds) Physiological adaptations in vertebrates: respiration, circulation, and metabolism. Marcel Dekker Inc., New York, pp 149–167

Perry SF (1998) Lungs: comparative anatomy, functional morphology, and evolution. In: Gans C, Gaunt AS (eds) Biology of the reptilia, vol 19: Morphology G visceral organs. Society for the Study of Amphibians and Reptiles, St. Louis (MO), pp 1–92

Perry SF (2001) Functional morphology of the reptilian and avian respiratory systems and its implications for theropod dinosaurs. In: Gauthier J, Gall LF (eds) New perspectives on the origin and early evolution of birds: Proceedings of the International Symposium in honor of John H. Ostrom. Peabody Museum of Natural History, Yale University, New Haven (CT), pp 429–441

Perry SF, Duncker HR (1978) Lung architecture, volume and static mechanics in five species of lizards. Respir Physiol 34:61–81

Perry SF, Sander M (2004) Reconstructing the evolution of the respiratory apparatus in tetrapods. Respir Physiol Neurobiol 144:125–139

Perry SF, Darian-Smith C, Alston D, Limpus CJ, Maloney JE (1989) Histological structure of the lungs of the loggerhead turtle, *Caretta caretta*, before and after hatching. Copeia 1989:1000–1010

Perry SF, Lambertz M, Schmitz A (2019) Respiratory biology of animals: evolutionary and functional morphology. Oxford University Press, Oxford (UK)

Pete AE, Kress D, Dimitrov MA, Lentink D (2015) The role of passive avian head stabilization in flapping flight. J R Soc Interface 12:0508. https://doi.org/10.1098/rsif.2015.0508)

Petevinos H (2006) A method for resolving subcutaneous emphysema in a griffon vulture chick (*Gyps fulvus*). J Exotic Pet Med 15:132–137

Pfenning AR, Hara E, Whitney O, Rivas MV, Wang R, Roulhac PL, Howard JT, Wirthlin M, Lovell PV, Ganapathy G et al (2014) Convergent transcriptional specializations in the brains of humans and song-learning birds. Science 346:1256846. https://doi.org/10.1126/science.1251385

Pickard WF (1974) Transition regime diffusion and the structure of the insect tracheolar system. J Insect Physiol 20:947–956

Pietschmann M, Bartels H, Fons R (1982) Capillary supply of heart and skeletal muscle of small bats and non-flying mammals. Respir Physiol 50:267–282

Piiper J (1978) Origin of carbon dioxide in caudal air sacs of birds. In: Piiper J (ed) Respiratory function in birds, adult and embryonic. Springer, Berlin, pp 221–248

Pison U, Max M, Neuendank A, Weissbach S, Pietschmann S (1994) Host defence capacities of pulmonary surfactant: evidence for 'non-surfactant' functions of the surfactant system. Eur J Clin Invest 24:586–599

Poelmann RE, Gittenberger-de Groot AC, Vicente-Steijn R, Wisse LJ, Bartelings MM, Everts S, Hoppenbrouwers T, Kruithof BP, Jensen B, de Bruin PW et al (2014) Evolution and development of ventricular septation in the amniote heart. PLoS One 9:e106569. https://doi.org/10.1371/journal.pone.0106569

Poole DC, Mathieu-Costello O (1992) Capillary and fiber geometry in rat diaphragm perfusion-fixed *in situ* at different sarcomere lengths. J Appl Physiol 73:151–159

Powell FL (1983) Respiration. In: Abs M (ed) Physiology and behaviours of the pigeon. Academic, New York, pp 73–95

Powell FL (2000) Respiration. In: Whittow CG (ed) Sturkie's avian physiology, 5th edn. Academic, San Diego, pp 233–264

Powell FL (2015) Respiration. In: Scanes CG (ed) Sturkie's avian physiology, 6th edn. Academic, San Diego, pp 301–336

Powell FL, Hempleman SC (1985) Sources of carbon dioxide in penguin air sacs. Am J Physiol 248:R748–R752

Powell FL, Hopkins SR (2004) Comparative physiology of lung complexity: implications for gas exchange. News Physiol Sci 19:55–60

Powell FL, Hastings RH, Mazzone RW (1985) Pulmonary vascular resistance during unilateral arterial pulmonary occlusion in ducks. Am J Physiol 249:R39–R43

Prange HD, Wasser JS, Gaunt AS, Gaunt SLL (1985) The respiratory responses to acute heat stress in cranes (Gruidae): the effects of tracheal coiling. Respir Physiol 62:95–103

Prince B, Riede T, Goller F (2011) Sexual dimorphism and bilateral asymmetry of syrinx and vocal tract in the European starling (*Sturnus vulgaris*). J Morphol 272:1527–1536

Radu C, Radu L (1971) Le dispositif vasculaire du poumon chez les oiseaux domestiques (coq, dindon, oie, canard). Revue Med Vet 122:1219–1226

Ragab SA, Reem RT (2016) Macroscopical anatomy of the air sacs of the turkey. Int J Adv Res Biol Sci 3:149–159

Raikow RJ, Bledsoe AH (2000) Phylogeny and evolution of the passerine birds: independent methods of phylo-genetic analysis have produced a well-supported hypothesis of passerine phylogeny, one that has proved particularly useful in ecological and evolutionary studies. BioScience 50:487–499

Rainey G (1849) On the minute anatomy of the lung of the bird, considered chiefly in relation to the structure with which the air is in contact whilst traversing the ultimate subdivisions of the air passages. Med Chir Trans 82:47–58

Rawal UM (1976) Nerves in the avian air sacs. Pavo 14:57–60

Rayner JMV (1982) Avian flight energetics. Annu Rev Physiol 44:109–119

Redrobe S (2015) Pelecaniformes (Pelicans, Tropicbirds, Cormorants, Frogatebirds, Anhingas, Gannets). In: Miller R, Fowler E, Murray E (eds) Fowler's zoo and wild animal medicine, vol 8. WB Saunders, St Louis, pp 96–99

Reiblein E (2017) Canary in a coal mine. Marine Log 122:9. Gale OneFile: Business. https://link.gale.com/apps/doc/A512184846/GPS?u=webdemo&sid=GPS&xid=086f9b8b

Riede T, Goller F (2010) Functional morphology of the soundgenerating labia in the syrinx of two songbird species. J Anat 216:23–36

Riede T, Fisher JH, Goller F (2010) Sexual dimorphism of the zebra finch syrinx indicates adaptation for high fundamental frequencies in males. PloS One 5:e11368. https://doi.org/10.1371/journal.pone.0011368

Riede T, Li Z, Tokuda IT, Farmer CG (2015) Functional morphology of the *Alligator mississippiensis* larynx with implications for vocal production. J Exp Biol 218:991–998

Riede T, Thomson SL, Titze IR, Goller F (2019) The evolution of the syrinx: an acoustic theory. PLoS Biol 17:e2006507. https://doi.org/10.1371/journal.pbio.2006507

Rizzo AN, Fraidenburg DR, Yuan JXJ (2015) Pulmonary vascular anatomy. In: Lanzer P (ed) Panvascular medicine. Springer, Berlin. https://doi.org/10.1007/978-3-642-37078-6_201

Romanoff AL (1960) The avian embryo. Macmillan, New York

Rothe HJ, Biesel W, Nachtigall W (1987) Pigeon flight in a wind tunnel. II. Gas exchange and power requirements. J Comp Physiol B 157:99–109

Roux E (2002) Origin and evolution of the respiratory tract in vertebrates. Rev Mal Respir 19:601–615

Ruben J (1995) The evolution of endothermy in mammals and birds: From physiology to fossils. Annu Rev Physiol 57:69–95

Runciman S, Seymour RS, Baudinette RV, Pearson JT (2005) An allometric study of lung morphology during development in the Australian pelican, *Pelicanus conspicillatus*, from embryo to adult. J Anat 207:365–380

Russell AP, Rittenhouse DR, Bauer AM (2000) Laryngotracheal morphology of Afro-Madagascan geckos: a comparative survey. J Morphol 245:241–268

Sacchi R, Galeotti P, Fasola M, Gerzeli G (2004) Larynx morphology and sound production in three species of Testudinidae. J Morphol 261:175–183

Sackton TB, Grayson P, Cloutier A, Hu Z, Liu JS, Wheeler NE, Gardner PP, Clarke JA, Baker AJ, Clamp M, Edwards SV (2019) Convergent regulatory evolution and loss of flight in paleognathous birds. Science 364:74–78

Sanders K, Farmer CG (2012) The pulmonary anatomy of *Alligator mississippiensis* and its similarity to the avian respiratory system. Anat Rec 295:699–714

Sappey PC (1847) Recherches sur L'appareil Respiratoire des Oiseaux. Bailliere, Paris

Schachner ER, Lyson TR, Dodson P (2009) Evolution of the respiratory system in nonavian theropods: evidence from rib and vertebral morphology. Anat Rec 292:1501–1513

Schachner ER, Farmer CG, McDonald AT, Dodson P (2011) Evolution of the dinosauriform respiratory apparatus: new evidence from the postcranial axial skeleton. Anat Rec 294:1532–1547

Schachner ER, Hutchinson JR, Farmer C (2013) Pulmonary anatomy in the Nile crocodile and the evolution of unidirectional airflow in Archosauria. PeerJ 1:e60. https://doi.org/10.7717/peerj.60

Schachner ER, Cieri RL, Butler JP, Farmer C (2014) Unidirectional pulmonary airflow patterns in the savannah monitor lizard. Nature 506:367–370

Schachner ER, Hedrick BP, Richbourg HA, Hutchinson JR, Farmer C (2021) Anatomy, ontogeny, and evolution of the archosaurian respiratory system: a case study on *Alligator mississippiensis* and *Struthio camelus*. J Anat 238:845–873

Scheid P (1979) Mechanisms of gas exchange in bird lungs. Rev Physiol Biochem Pharmacol 86: 137–186

Scheid P, Piiper J (1989) Respiratory mechnics and airflow in birds. In: King AS, McLelland J (eds) Form and function in birds, vol 4. Academic, London, pp 364–391

Scheid P, Hook C, Bridges CR (1981) Diffusion in gas exchange of insects. Fed Proc 41:2143–2145

Schittny J, Burri PH (2004) Morphogenesis of the mammalian lung: aspects of structure and extracellular matrix. In: Massaro DJ, Massaro GC, Chambon P (eds) Lung development and regeneration. Marcel Dekker, New York, pp 275–316

Schmidt MF, Martin WJ (2014) The respiratory-vocal system of songbirds: anatomy, physiology, and neural control. Prog Brain Res 212:297–335

Schmidt-Nielsen K (1971) How birds breathe. Sci Am 225:72–79

Schmidt-Nielsen K (1972) Locomotion: energy cost of swimming, flying, and running. Science 177:222–228

Schmidt-Nielsen K (1983) Animal physiology: adaptation and environment, 3rd edn. Cambridge University Press, Cambridge

Schmidt-Nielsen K (1990) Animal physiology: adaptation and environment, 4th edn. Cambridge University Press, Cambridge

Schmidt-Nielsen K, Kanwisher J, Lasiewski RC, Cohn JE, Bretz WL (1969) Temperature regulation and respiration in the ostrich. Condor 71:341–352

Schmitz A, Perry SF (1999) Stereological determination of tracheal volume and diffusing capacity of the tracheal walls in the stick insect, *Carausius morosus* (Phasmatodea, Lonchodidae). Physiol Biochem Zool 72:205–218

Schulze FE (1908) Die lungen des afrikanisches straußes. S-B Preus Akad Wiss 1:416–431

Schulze FE (1910) Über die Bronchi Saccales und den Mechanismus der Atmung bei den Vögeln. S.-B preuss Akad Wiss. Berlin 1910:537–538

Seddon N (2005) Ecological adaptation and species recognition drives vocal evolution in neotropical suboscine birds. Evolution 59:200–215

Seller TJ (ed) (1987) Bird respiration, vol I and II. CRC Press, Boca Raton (FL)

Senter P (2008) Voices of the past: a review of paleozoic and mesozoic animal sounds. Historical Biol 20:255–287

Seymour RS, Runciman S, Baudinette RV, Pearson JT (2004) Developmental allometry of pulmonary structure and function in the altricial Australian pelican, *Pelecanus conspicillatus*. J Exp Biol 207:2663–2669

Sherwood CC (2018) Are we wired differently? Sci Am 2018:52–55

Shimoda LA (2006) Pulmonary circulation. In: Laurent GJ, Shapiro SD (eds) Encyclopedia of respiratory medicine. Academic Press, London, pp 537–544

Shine R (2005) Life-history evolution in reptiles. Annu Rev Ecol Evol Syst 36:23–46

Simelane SM, Abelman S, Duncan FD (2016) Gas exchange models for a flexible insect tracheal system. Acta Biotheor 64:161–196

Snelling EP, Seymour RS, Runciman S, Matthews PGD, White CR (2011) Symmorphosis and the insect respiratory system: allometric variation. J Exp Biol 214:3225–3237

Snelling EP, Seymour RS, Runciman S, Matthews PGD, White CR (2012) Symmorphosis and the insect respiratory system: a comparison between flight and hopping muscle. J Exp Biol 215: 3324–3333

Snyder GK, Sheafor B, Scholnick D, Farrelly C (1995) Gas exchange in the insect tracheal system. J Theor Biol 172:199–207

Soum JM (1896) Recherches physioloqiques sur l'appareil respiratoire des oiseaux. Ann Univ Lyon 28:1–126

Stanislaus M (1937) Untersuchungen an der Kolibrilunge. Z Morph Ökol Tiere 33:261–289

Stein RC (1968) Modulation in bird sounds. Auk 85:229–243

Stork NE (1994) How many species are there? Biodivers Conserv 3:204. https://doi.org/10.1007/BF02291892

Stork NE (2018) How many species of insects and other terrestrial arthropods are there on earth? Annu Rev Entomol 63:31–45

Sullivan L, Daniels CB, Phillips I, Orgeig S, Whitsett J (1998) Conservation of surfactant protein A: evidence for a single origin for vertebrate pulmonary surfactant. J Mol Evol 46:131–138

Suthers RA (1990) Contributions to birdsong from the left and right sides of the intact syrinx. Nature 347:473–477

Suthers RA, Zollinger SA (2004) Producing song: the vocal apparatus. Ann N Y Acad Sci (USA) 1016:109–129

Tambussi CP, de Mendoza R, Degrange FJ, Picasso MB (2012) Flexibility along the neck of the neogene terror bird *Andalgalornis steulleti* (Aves Phorusrhacidae). PLoS One 7:e37701. https://doi.org/10.1371/journal.pone.0037701

Taylor CR, Weibel ER (1981) Design of the mammalian respiratory system. I. Problem and strategy. Respir Physiol 44:1–10

Tenney SM, Remmers JE (1963) Comparative quantitative morphology of the mammalian lung: diffusing area. Nature 197:54–56

Terblanche JS, Woods HA (2018) Why do models of insect respiratory patterns fail? J Exp Biol 221:jeb130039. https://doi.org/10.1242/jeb.130039

Terray L, Plateau O, Abourachid A, Böhmer C, Delapré A, de la Bernardie X, Cornette R (2020) Modularity of the neck in birds (Aves). Evol Biol 47:97–110

Terry PB, Traystman RJ, Newball HH, Batra G, Menkes HA (1978) Collateral ventilation in man. N Engl J Med 298:10–15

Ton R, Boyce AJ, Mitchell AE, Mouton JC, Gobbo NR, Blake W, Tobalske B (2021) Galliformes exhibit reduced cardiorespiratory morphology yet similar skeletal mass compared with other gamebirds. Wilson J Ornithol 133. https://doi.org/10.1676/20-00077

Tucker V (1971) Flight energetics in birds. Am Zool 11:115–124

Van der Leeuw AHJ, Bout RG, Zweers GA (2001a) Control of the cranio-cervical system during feeding in birds. Am Zool 41:1352–1636

Van der Leeuw A, Bout RG, Zweers GA (2001b) Evolutionary morphology of the neck system in ratites, fowl and waterfowl. Netherlands J Zool (LOnd) 51:243–262

Varela-Lasheras I, Bakker AJ, van der Mije SD, Metz JAJ, van Alphen J, Gallis F (2011) Breaking evolutionary and pleiotropic constraints in mammals: on sloths, manatees and homeotic mutations. EvoDevo 2:11. https://doi.org/10.1186/2041-9139-2-11

Vaughan TA, Bateman GC (1970) Functional morphology of the forelimb of mormoopid bats. J Mammal 51:217–235

Vos HJ (1934) Über die Vege der Atemluft in der Entenlunge. Z vergl Physiol 21:552–578

Vos HJ (1937) Über das Fehlen der rekurrenten Bronchien beim Pinguin und bei den Reptilien. Zool Anz 117:176–181

Voshaar TH (2008) Kollaterale Ventilation [Collateral ventilation]. Pneumologie 62:355–360

Vu ET, Mazurek ME, Kuo YC (1994) Identification of a forebrain motor programming network for the learned song of zebra finches. J Neurosci 14:6924–6934

Wagner WW, Weir EK (eds) (1994) The pulmonary circulation and gas exchange. Futura Publishing, Armonk (NY)

Walsh C, McLelland J (1974a) The ultrastructure of the avian extrapulmonary respiratory epithelium. Acta Anat 89:412–422

Walsh C, McLelland J (1974b) Granular 'endocrine' cells in avian respiratory epithelia. Cell Tiss Res 153:269–276

Walsh C, McLelland J (1974c) Intraepithelial axons in the avian trachea. Z Zellforsch mikrosk Anat 147:109–117

Walsh C, McLelland J (1978) The development of the epithelium and its enervation in the avian extra-pulmonary respiratory tract. J Anat 125:171–182

Wang N, Banzett RB, Butler JP, Fredberg JJ (1988) Bird lung models show that convective inertia effects inspiratory aerodynamic valving. Respir Physiol 73:111–124

Wang N, Banzett RB, Nations CS, Jenkins EA (1992) An aerodynamic valve in the avian primary bronchus. J Exp Biol 262:441–445

Wang X, O'Conner JK, Maina JN, Pan Y, Wang M, Wang Y, Zheng X, Zhou Z (2018) *Archaeorhynchus* preserving significant soft tissue including probable fossilized lungs. Proc Natl Acad Sci (USA) 115:11555–11560

Warner RW (1971) The structural basis of the organ of voice in the genera *Anas* and *Aythya* (Aves). J Zool (Lond) 164:197–207

Warner RW (1972) The anatomy of the syrinx in passerine birds. J Zool (Lond) 168:381–393

Warner RW (2009) The anatomy of the sytinx in passerine birds. J Zool, Lond 168:381–393

Warrick DR, Bundle MW, Dial KP (2002) Bird maneuverin flight: blurred bodies, clear heads. Integr Comp Biol 42:141–148

Waters PD, Patel HR, Ruiz-Herrera A, Alvarez-González L, Listera NC, Simakov O, Ezaz T, Kaur P, Frere C, Grutzner F et al (2021) Microchromosomes are building blocks of bird, reptile, and mammal chromosomes. Proc Natl Acad (USA) 118:e2112494118. https://doi.org/10.1073/pnas.2112494118

Watson RR, Fu Z, West JB (2008) Minimal distensibility of pulmonary capillaries in avian lungs compared with mammalian lungs. Respir Physiol Neurobiol 160:208–214

Webster MR, Socha JJ, Teresi L, Nardinocchi P, De Vita R (2015) Structure of tracheae and the functional implications for collapse in the American cockroach. Bioinspir Biomim 10:066011. https://doi.org/10.1088/1748-3190/10/6/066011

Wedel MJ (2009) Evidence for bird-like air sacs in saurischian dinosaurs. J Exp Zool 311A:611–628

Weibel ER (1973) Morphological basis of the alveolar-capillary gas exchange. Physiol Rev 53: 419–495

Weibel ER (1984) The pathways for oxygen. Harvard University Press, Harvard (MA)

Weibel ER (2000) Symmorphosis: on form and function in shaping life. Harvard University Press, Cambridge (MA)

Weis-Fogh T (1964a) Diffusion in insect flight muscle, the most active tissue known. J Exp Biol 41: 229–256

Weis-Fogh T (1964b) Functional design of the tracheal system of flying insects as compared with the avian lung. J Exp Biol 41:207–228

Weis-Fogh T (1967) Respiration and tracheal ventilation in locusts and other flying insects. J Exp Biol 47:561–587

Welty JC (1979) The life of birds, 2nd edn. Saunders, Philadelphia (PA)

Welty JC, Baptista L (1988) The life of birds, 4th edn. Saunders College Publishing, New York (NY)

West JB (2000) Respiratory physiology: the essential, 6th edn. Lippincott Williams and Wilkins, Philadelphia (PA)

West JB (2009) Comparative physiology of the pulmonary blood-gas barrier: the unique avian solution. Am J Physiol Regul Integr Comp Physiol 297:R1625–R1634

West JB (2015) Stress failure of the pulmonary blood-gas barrier. In: Makanya A (ed) The vertebrate blood-gas barrier in health and disease. Springer, Cham (Switzerland), pp 135–158

West JB (2017) Structure and function of avian pulmonary capillaries: comparison with mammals. In: Maina JN (ed) The biology of the avian respiratory system: evolution, development, structure and function. Springer, Heidelberg, pp 179–190

West NH, Bamford OS, Jones DR (1977) A scanning electron microscope study of the microvasculature of the avian lung. Cell Tiss Res 176:553–564

West JB, Watson RR, Fu Z (2006) The honeycomb-like structure of the bird lung allows a uniquely thin blood-gas barrier. Respir Physiol Neurobiol 152:115–118

West JB, Watson RR, Fu Z (2007a) Support of pulmonary capillaries in avian lung. Respir Physiol Neurobiol 159:146

West JB, Watson RR, Fu Z (2007b) Major differences in the pulmonary circulation between birds and mammals. Respir Physiol Neurobiol 157:382–390

West JB, Fu Z, Deerinck TJ, Mackey MR, Obayashi JT, Ellisman MH (2010) Structure-function studies of blood and air capillaries in chicken lung using 3d electron microscopy. Respir Physiol Neurobiol 170:202. https://doi.org/10.1016/j.resp.2009.12.010

Westneat MW, Betz O, Blob RW, Fezza K, Cooper WJ, Lee WK (2003) Tracheal respiration in insects visualized with synchrotron X-ray imaging. Science 299:558–560

Wetherbee DK (1951) Air sacs in the English sparrow. Auk 68:242–244

Wetmore A (1918) A note on the tracheal air sac in the ruddy duck. Condor 20:19–20

White SS (1975) The larynx. In: Getty R (ed) Sisson and Grossman's The anatomy of the domestic animals, vol 2. WB Saunders, Philadelphia, pp 1883–1918

White SS, Chubb (1967) The muscles and movements of the larynx of *Gallus domesticus*. J Anat 102:575

Whitney O, Pfenning AR, Howard JT, Blatti CA, Liu F, Ward JM, Wang R, Audet J, Kellis M, Mukherjee S et al (2014) Core and region-enriched networks of behaviorally regulated genes and the singing genome. Science 346:1256780. https://doi.org/10.1126/science.1256780

Wigglesworth VB (1972) The principles of insect physiology, 7th edn. Chapman and Hall, London

Wigglesworth VB (2008) The respiration of insects. Biol Rev 6:181–220

Wittenberg JB, Wittenberg BA (2007) Myoglobin-enhanced oxygen delivery to isolated cardiac mitochondria. J Exp Biol 210:2082–2090

Wood PG, Lopatko OV, Orgeig S, Joss JM, Smits AW, Daniels CB (2000) Control of pulmonary surfactant secretion: an evolutionary perspective. Am J Physiol Regul Integr Comp Physiol 278: R611–R619

Woodward JD, Maina JN (2005) A 3D digital reconstruction of the components of the gas exchange tissue of the lung of the muscovy duck, *Cairina moschata*. J Anat 206:477–492

Woodward JD, Maina JN (2008) Study of the structure of the air and blood capillaries of the gas exchange tissue of the avian lung by serial section three-dimensional reconstruction. J Microsc 230:84–93

Woodward JD, Ramonisi Y, Mashiteng R, Mokae L, Maina JN (2019) Three-dimensional computer reconstruction of the airway- and the vascular systems of the lung of the domestic fowl, *Gallus gallus* variant *domesticus*. In: Proceedings of 10th Intern Conf Comput Meth, Vol. 6 (9th–13th July 2019), Singapore. Scientech Publisher LLC (USA), pp 36–48. ISSN 2374-3948

Xu X, Zhou Z, Dudley R, Mackem S, Chuong CM, Erickson GM, Varricchio DJ (2014) An integrative approach to understanding bird origins. Science 346:1253293. https://doi.org/10.1126/science.1253293

Yarrell W (1833) On the organs of voice in birds. Trans Linn Soc (Lond) 16:305–321

Yildiz H, Bahadir A, Akkoç A (2003) A study on the morphological structure of syrinx in ostriches (*Struthio camelus*). Anat Histol Embryol 32:187–191

Young MW, Dickinson E, Flaim ND, Granatosky MC (2022) Overcoming a 'forbidden phenotype': The parrot's head supports, propels and powers tripedal locomotion. Proc Royal Soc (Lond) B 289:20220245. https://doi.org/10.1098/rspb.2022.0245

Youngren OM, Peek FW, Phillips RE (1974) Repetitive vocalization evoked by local electrical stimulation of avian brains. III. Evoked activity in the tracheal muscles of the chicken (*Gallus gallus*). Brain Behav Evol 9:393–421

Zeek PM (1951) Double trachea in penguins and sea lion. Anat Rec 111:327–343

Zweers GA, van Pelt HC, Beckers A (1981) Morphology and mechanics of the larynx of the pigeon (*Columba livia* L.): a drill-chuck system (Aves). Zoomorphology 99:37–69

Zweers GA, Vanden BJC, Koppendraier R (1986) Avian cranio-cervical systems. Part I: anatomy of the cervical column in the chicken (*Gallus gallus* L.). Acta Morphol Neerl Scand 25:131–155

Zweers G, Bout R, Heidweiller J (1994) Motor organization of the avian head-neck system. In: Davies MNO, Green PR (eds) Perception and motor control in birds. Springer, Berlin, pp 201–221

Chapter 4
Function of the Avian Respiratory System

Abstract The gas exchange function of the avian respiratory system is explained. The following important structural—and functional aspects are presented: the mechanisms of airflow through the airway (bronchial) system and the air sacs; the four-phase respiratory modus operandi of the avian respiratory system and the inspiratory and the expiratory aerodynamic valving mechanisms by which air is shunted across the airways; the physiological adaptations of coping with hypoxia and hypocapnia, especially during high altitude residence and flight; the functional significance of the physical strength of the avian lung, the air and the blood capillaries and the blood-gas barrier; the gas exchange designs, namely, the countercurrent, the crosscurrent, and the multicapillary serial arterialization systems of the avian lung and; the morphometric specializations of phylogenetically different species of birds and those that lead various lifestyles. The exceptional functional efficiency of the avian respiratory system is highlighted.

4.1 Fundamental Aspects of the Functional Designs of Gas Exchangers

Life has existed on Earth for as long as ~4 billion years (Mojzsis et al. 1996; Nutman et al. 2015, 2016): it is envisaged to have formed in an anaerobic environment, a so-called "primeval soup" or "prebiotic broth" (Oro 1983; Plankensteiner et al. 2004; Peretó 2005). From its injurious nature, the accumulation of O_2 in the biosphere comprised a selective pressure on the emerging animal life. At a critical level, designated the Pasteur Point, i.e., a level of oxygen above which facultative aerobic microorganisms and facultative anaerobes adapt from fermentation to aerobic respiration, as life because structurally more complex. O_2 was commissioned to serve as a terminal electron acceptor in the important process of aerobic metabolism by the production of ATP (adenosine triphosphate) the high energy molecule (Schopf 1978; Castresana and Saraste 1995). In most of the modern animal life, the primary role of gas exchangers is acquisition O_2 from the external respiratory fluid medium (water and/or air) and discharge CO_2, which is a product of metabolism, back into them (Weibel 1984; Maina 1998, 2011; Mondrinos et al. 2014).

© The Author(s), under exclusive license to Springer Nature Switzerland AG 2023

J. N. Maina, *Current Perspectives on the Functional Design of the Avian Respiratory System*, Zoological Monographs 8, https://doi.org/10.1007/978-3-031-35180-8_4

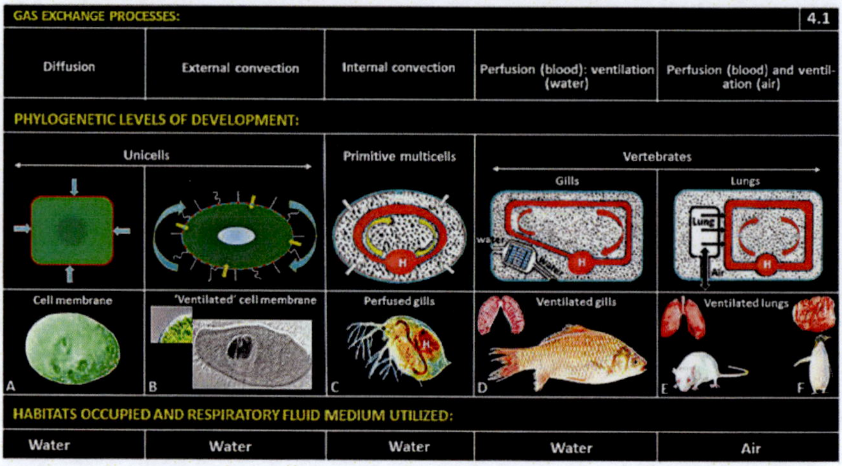

Plate 4.1 **(a–f)** Schematic illustration of the morphological diversity of gas exchangers, respiratory mechanisms and processes, phylogenetic levels of development, environments occupied, and respiratory fluid medium utilized in simple to complex animal life. Complexity and respiratory efficiency developed with evolutionary advancement and an increase in metabolic requirements for molecular oxygen. H, heart **(c–f)**

Adaptively, the functional design of gas exchangers came to relate with among other factors phylogeny, lifestyle, respiratory fluid medium utilized, and habitats occupied (Dejours 1981, 1988, 1989; Maina 1994, 1998, 2000a, 2002a–d, 2011; Figueroa et al. 2007) (Plate 4.1). Respiration comprises complex and highly integrated physiological, biochemical, mechanical, and behavioral processes by which O_2 is procured and transported to all parts of the body across various tissue spaces and barriers. Ultimately, O_2 crosses the cell membrane to contribute to energy production in the form of ATP in the mitochondria through the so-called citric acid or Kreb's cycle (Plate 4.2a–e). Ventilation, which is the product of breathing frequency (rate) and tidal volume, is the mechanism by which air is moved in-and-out of the gas exchanger. It is the first and critical step in the complex process of O_2 procurement and transport in animals' bodies (Weibel 1984, 2000; Weibel et al. 1998; Snelling et al. 2011, 2012; Biro 2013; Powell 2014) (Plate 4.2a–e) which ends at the mitochondrial level, the so-called "O_2 sink" (Wenger 2006): in complex animals, the partial pressure of O_2 (PO_2) at that point is nearly zero (Gayeski et al. 1987; Costa et al. 1997; Connett et al. 1990; Pittman 2011; Egbert 2013) (Plate 4.2e). On account of the mitochondria being the ultimate sites of O_2 exploitation, the organelles serve as the elemental drivers of O_2 flow from the external environment into the cells of the body (Taylor and Weibel 1981; Weibel 1984). According to among others Margulis (1981), Gray (2012), and Zachar and Boza (2020), mitochondria appeared in eukaryotic cells by an endosymbiotic process: in prokaryotes, mitochondrial endosymbiotic occurrences are rare (Garg and Martin 2016; Martin et al. 2017, 2021a; López-García et al. 2017). An increase in ventilatory activity is largely the first response/reaction to anoxic and/or hypoxic conditions (Powell et al.

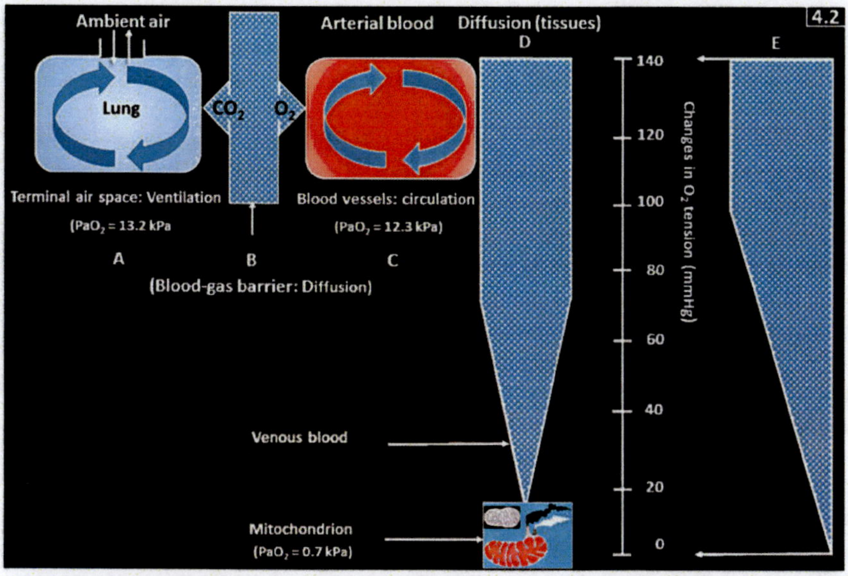

Plate 4.2 Schematic illustration of an oxygen transfer cascade where a gas exchanger is mechanically ventilated (**a**), oxygen diffuses across the blood-water/blood-gas (tissue) barrier (**b**), oxygen is transported by the circulatory system to different parts of the body (**c**) and ultimately delivered to the mitochondria where it is involved in the production of ATP (adenosine triphosphate) (**d**). Oxygen consumption at the mitochondrial level is the primary driver of the flow of oxygen along the cascade. The partial pressure of oxygen at the mitochondrial level (**e**)

1998), a state that can be ameliorated by either increasing the volume of breath (tidal volume) and/or respiratory frequency (rate) (York et al. 2017). Lasiewski and Calder (1971) and Jammes (1975) determined that the overall volume of the avian respiratory system is much greater than that of mammals, the tidal volumes are larger in birds compared with those of mammals, the respiratory rates are lower in birds and the minute volumes are slightly less in birds compared with those of similarly sized mammals.

Regarding respiratory mechanics, in birds, the visceral cavity, i.e., the thoracoabdominal or the coelomic space, functions as a single unit (Brackenbury 1973; King and McLelland 1984; McLelland 1989). Like for mammals (Comroe 1974; Weibel 1984; Hlastala and Berger 1996; West 2008), the flow of air in-and-out of the respiratory system fundamentally occurs from recurrent changes in pressure within and around the lung and the air sacs. During inspiration, by the so-called "pump-handle movement" that is similar to that which occurs in mammals, in birds, the ribs move craniolaterally pushing the sternum ventrally and cranially (King and McLelland 1984; McLelland 1989), a process that increases the dorsolateral, the transverse, and the craniocaudal diameters of the coelomic cavity: it causes the pressure in the coelomic cavity and consequently that in the air sacs to decrease, a change that makes air to flow into the respiratory system. Expiration entails a rather passive return of the thoracic cage to its resting position: the air sacs are compressed, driving air out of them. In the Domestic Fowl (*Gallus gallus* variant *domesticus*), electromyographic studies have identified the muscles that are directly involved in respiratory activities (Kadono and Okada 1962; Kadono et al. 1963; Fedde et al. 1964a–c). Among others, the inspiratory muscles include scalenus, external intercostal, costosternal major, levator costarium, and serratus superficialis muscles while the expiratory ones include part of the external intercostal, part of the intercostal, costosternalis minor, oblique external abdominis, oblique internal abdominis, transversus abdominis, rectus abdominis, serratus profundus, and costoseptalis muscles (King and McLelland 1984; McLelland 1989). Showing that in birds other peripheral receptors may be involved in the continuation of normal coordinated respiration, Fedde et al. (1963) observed that bilateral thoracic vagotomy (when innervation of the carotid bodies, common carotid arteries, and aortic bodies is conserved) caused changes of respiration similar to those caused by bilateral cervical vagotomy. When all the respiratory muscles are inactive, i.e., all the muscular forces are eliminated, unlike for the mammalian lungs where muscles rest at functional residual capacity (Powell 2014), in birds, the thoracic cage comes to rest about halfway between full inspiration and full expiration (Siefert 1896; Stübel 1910). It shows that in birds, passive elastic forces may significantly contribute to the onset of inspiratory and expiratory movements. Providing a smooth transition between the ventilatory phases, Fedde et al. (1964b, c) found that inspiratory muscles become active just before the onset of inspiration while for the expiratory ones, it occurs near the end of inspiration.

In birds and mammals, during eupneic breathing, the mechanics of ventilation of the lung are essentially similar (Scheid and Piiper 1989): air moves in-and-out of the gas exchangers from activities of the respiratory muscles that change the pressures prevailing around the respiratory organ, generating tidal ventilatory airflow (Zimmer 1935; King 1966; King and Molony 1971). Interestingly, on the whole, the mammalian and the avian respiratory systems are tidally ventilated (Weibel 1984; King and McLelland 1984; Powell 2014), because the avian lungs are practically rigid (Plates 1.1a, 2.6c, and 3.6b, g), synchronized bellows-like activity of the air sacs ventilates the lung (specifically the paleopulmonic part of the lung) continuously and unidirectionally back-to-front, i.e., in a caudocranial direction (Zimmer 1935; King and Molony 1971; Brackenbury 1972, 1973; Scheid 1979; Fedde 1980; Powell 2014). The main structural features that make the avian lung rigid are (a) its deep fixation (attachment) to the ribs and the vertebrae (Plates 2.6 and 3.6b, f, g); (b) fusion and/or connection of the parietal and the visceral pleurae by connective tissue fibers (collagen) (Plate 3.10a–f); (c), location of the lung more-or-less at the craniodorsal part of the body trunk which from the musculoskeletal assemblage is least movable (Duncker 1971); and (d) activities of the costoseptal muscles which prevent the lung from being physically compressed during expiration (Fedde et al. 1964a; Mackelprang and Goller 2013).

In birds, the process of breathing is distinctive for the following reasons: (a) each tidal volume delivers a volume of air equivalent to the complete throughput of the air spaces of the lungs and the trachea; (b) respiration provides more extensive CO_2

washout from the parabronchial exchange tissue compared with that which occurs in the mammalian lung parenchyma (Scheid and Piiper 1989); and (c) compared with those of mammals, displaying great functional efficiency, bird lungs extract more O_2 from the inspired air (Lasiewski and Calder 1971; Scheid 1979, 1987, 1990; Scheid and Piiper 1972; Scheid et al. 1972). The lung-air sac system of birds is contained in a continuous space (coelomic cavity) that is bordered by a skeletomuscular wall which is not divided by a muscular diaphragm as is the case in mammals (Brackenbury 1972; King and McLelland 1984; McLelland 1989). With such a structure (a diaphragm) lacking, in birds, the heart is bounded by the liver and not the lung (as is the case in mammals) and ventrally, the lungs contact the horizontal septum (Plate 3.6g): the tension existing in the horizontal membrane is regulated by the costoseptal muscles (McLelland 1989). However, unlike the muscular diaphragm which is the most important respiratory muscle in mammals, the horizontal septum does not change the volume of the lung (Fedde et al. 1964a). Developmentally, structurally and functionally, the horizontal and oblique septa of birds are not homologous or analogous to the mammalian diaphragm (King and McLelland 1984; McLelland 1989). As cautioned by King and McLelland (1984), the term "diaphragm," which occasionally appears, especially in popular literature, is wrong and misleading and should not be used to refer to any part of the avian anatomy. The parabronchi and the secondary bronchi of the duck lung have limited compliance and display variations in intrapulmonary pressures that result in compression of the lung (from inside) (Macklem et al. 1979): it does not cause a collapse of air capillaries, a condition that should compromise gas exchange efficiency. For the Domestic Fowl (*Gallus gallus* variant *domesticus*), Scheid and Piiper (1969) determined that the maximum compliance of the lung is 9.5 ml per mbar, a value equivalent to a relative compliance of 4% per mbar (cm H_2O). Although airways such as the primary bronchus, secondary bronchi and parabronchi (tertiary bronchi) may to some extent change in diameter and shape, the air and blood capillaries of the avian lung (Plates 3.19, 3.20 and 3.21) are very strong (West et al. 2006, 2007a, b; Maina 2007a, b; West et al. 2010; Watson et al. 2007, 2008; West 2009, 2011; Maina et al. 2010). However, the folds of the epithelial-epithelial cells connections (epithelial retinaculae = laminae = bridges) and the blood-gas barrier (Plate 3.22a–c) suggest that the terminal gas exchange components of the avian lung, the air- and the blood capillaries, may not be absolutely immovable. Smooth muscle and motor nerves exist in the internal and external parts of the parabronchial walls and the interatrial septa are well-endowed with smooth muscle, elastic tissue, and collagen fibers (Plate 3.22d–f).

By displacing the ribs and the sternum, in birds, the muscles associated with the uncinate processes of the rib cage (Plate 1.2c) provide a respiratory function (Welty 1979; Codd et al. 2005; Tickle et al. 2007, 2009; Tickle 2010; Tickle and Codd 2009; Codd 2010). As well, the uncinate processes physically strengthen the rib cage, making it possible for it (the skeletal framework) to tolerate the forces created by the beating of the wings during flight (Kardong 1988; Liem and Walker 1994). In the Turkey (*Meleagris gallopavo*), ossification of the uncinate processes starts before hatching and the sizes correlate with the commencement of air-breathing and an increase in sternal mass (Tickle and Codd 2009). The extents of the development of the uncinate processes vary between bird species. In general, the processes correlate with modes of life and are by-and-large completely ossified at maturity (Tickle et al. 2007; Codd 2010). The uncinate processes are small in nonvolant (flightless = walking) birds, long in diving species, and are of intermediate size in nonspecialist ones (Tickle et al. 2007). The lengths of the uncinate processes increase the mechanical advantage of the appendicocostales muscle during inspiration (Tickle et al. 2009; Tickle 2010). In screamers (Anhimidae), the uncinate processes are totally lacking (Fowler and Cubas 2001). Among the extant air-breathing vertebrates, boney uncinate processes are exclusive to birds but cartilaginous ones occur in crocodiles and reportedly existed in sphenodon and fossil vertebrates that include Caudipteryx, oviraptorids and dromaeosaurids, *Confuciusornis*, *Chaoyangia* and *Longipteryx* (Paul 1988; Codd et al. 2008; Codd 2010). Possibly, uncinate processes were lacking in *Archeopteryx lithographica* (Wieland 1978; Zhang et al. 2001).

The concept that in biological entities structure and function correlate is an old one: it has been appreciated since the time of the Greek philosopher and polymath Aristotle (384-322 BC). Much recently, with empirical evidence, awareness that structural designs of biological systems and their parts that are regulated by morphogenetic, growth, and maintenance/repair processes accord with functional requirements was termed "symmorphosis" (Weibel 1990; Weibel et al. 1991, 1998; Dudley and Gans 1991; Milsom 1991; Canals et al. 2010; Gifford et al. 2016; Weibel 2000): this state has been established in among others organs lungs and skeletal muscle (Gehr et al. 1980, 1981; Dubach 1981; Maina et al. 1982a, b, 1989a, 1991; Maina and King 1984; Suarez et al. 1986, 1991, 2009; Blake 1991; Suarez 1992; Weibel et al. 1998; Maina 2006; Figueroa et al. 2007; Canals et al. 2010; Gifford et al. 2016; Welch et al. 2018). For the acquisition of O_2 and its distribution to the various parts of the body, based on what gives the impression of a cost-benefit appraisal, animals have evolved the most efficient possible pulmonary and vascular systems that meet their metabolic requirements for O_2 (Murray 1926; Milsom 1991; Andrade et al. 1998; Maina 2000a–c; Canals et al. 2005, 2008; Mauroy et al. 2004; Lee et al. 2008).

In complex life forms, O_2 is a critical resource that has to be continuously acquired from outside in sufficient amounts at the best possible energy cost. At rest, breathing at a rate of ~16 times per minute, an adult human being breathes ~12,000 L of air over a period of 12 h (Burri 1985). While many groups of organisms/animals will live for weeks without food and days without water, most of them require a continuous source of O_2. Moreover, compared with activities such as feeding, thermoregulation, locomotion, and reproduction that can be fine-tuned or abandoned altogether without permanent ill effects, continuous acquisition of O_2 is vital to life. Depending on environmental conditions and age, irreversible cellular damages generally occur after about 3 min of termination of breathing (apnea) and death is definite after about 6 min (Brunelle and Chandel 2002; Miller and Zachary 2017). One of the life history strategies that animals such as those that undertake arduous journeys like migration or face extremely cold winters have adopted

(Eikenaar et al. 2020) is that of strategically storing energy in the form of adipose (fat) tissue in the body (Young 1976; Fuller et al. 2016; Araújo et al. 2019). Birds are unique in that they largely fuel their highly energetically costly activities, such as migratory flights, by metabolizing fat (McWilliams et al. 2004; Jenni-Eiermann 2017; Jenni-Eiermann et al. 2014; Guglielmo et al. 2001; Guglielmo 2018). The "weightlessness" and the large quantity of energy generated through oxidizing fat compared with carbohydrates or proteins makes adipose tissue the choice form of energy storage. Furthermore, fat adds much less to body mass per unit amount of energy contained in it and therefore reduces the energetic costs of tasks such as flight. In the migratory Garden Warblers (*Sylvia borin*), during spring migration, subcutaneous fat stores determine the lengths of the stopover periods (Goymann et al. 2010): birds with large quantities of fat stores embark on their migratory flights earlier than the relatively lean ones. Among vertebrates, birds are unique in that while they have relatively high metabolic rates that are accompanied by increased production of reactive species, they live relatively longer than mammals (Barja 1998; Buttemer et al. 2010; Munshi-South and Wilkinson 2010; Jimenez 2018; Jimenez et al. 2019). A common mutation in the main antioxidant control pathway that lowers the risk of macromolecular oxidative damage exists in avian taxa with the highest metabolic rates (Castiglione et al. 2020).

While O_2 comprises ~64% of the human body by weight (Wang 1997), much of it is locked up in molecular complexes, making it inaccessible for metabolic roles and therefore unavailable for physiological functions. It was conceivably from its toxic effects (Davies 1995; Maltepe and Saugstad 2009; Zhu et al. 2016) and the large amounts of it that are needed for oxidative processes that during the evolution of complex life forms storage of O_2 in the body in large amounts was not a possibility. During the apneneic state, animals rely on O_2 in the air that is inhaled into the lungs and that which becomes bound to hemoglobin in the blood, in myoglobin, and the smaller quantity which is physically dissolved in blood plasma. For an individual weighing 70 kg, only ~1.6 L of available O_2 exists in the body (Farhi and Rahn 1955; Rahn 1964; Farhi 1964; Cherniack and Longobardo 1970): at functional residual capacity, ~370 cm^3 of O_2 in the body is located in the gas cavities (mainly in the alveoli of the lung); ~280 cm^3 is found in the arterial blood (largely dissolved and bound to hemoglobin), ~600 cm^3 exists in the capillary and venous blood and ~60 cm^3 is dissolved in the body tissue water and; ~240 cm^3 is chemically bound to myoglobin. As most of the O_2 which is stored in the body is bound to hemoglobin, only a small quantity has to be released from it for a large decrease in the arterial PO_2 to occur. For example, when hemoglobin is 50% saturated with O_2, the arterial partial pressure of O_2 (PaO_2) is only 26 mmHg (3.5 kPa). Because one of the vital roles of the respiratory system is to remove CO_2, the lung plays an important role in the regulation of the concentration of hydrogen ions (H^+) in blood and other body fluids. Stores of CO_2 in solution and the form of bicarbonate ions (HCO_3^{-1}) in the body considerably exceed those of O_2. For a 70 kg person, with stores changing constantly, the stored amount of CO_2 is 120 L (Lin et al. 2009): CO_2 is the most important acid end-product of metabolism and is greatly concerned with the regulation of body fluids pH.

While the brain does not function physically, e.g., in the manner skeletal muscles do, it consumes ~20% of the total O_2consumption by the body (Javed et al. 2010; Rink and Khanna 2011; Liu et al. 2022): the brain is both the most metabolically active organ of the body and the most sensitive one to lack O_2 (Sharp and Bernaudin 2004; Casas et al. 2017). At rest, the brain consumes 3.5 ml of O_2 per 100 g of tissue per minute (Klabunde 2005; Butterworth 2006). Respiratory efficiency is a measure of the ratio of gas (O_2) transfer into the body versus the energetic cost of obtaining it (Otis 1954; Zakynthinos and Roussos 1991; Vella et al. 2006). For human beings, depending on factors such as environmental conditions and metabolic conditions, the energetic cost of air-breathing has been reported to range from ~0.5 ml O_2 to as high as 10 ml O_2 per liter of ventilation (Otis 1954; Bartlett and Specht 1957). Indicating that the metabolic cost of ventilation is relatively low in birds and that it falls within the range of the values for the other amniotes, for the Guinea Fowl (*Numida meleagris*), the measure was determined to be 0.48 ml O_2 per liter of the ventilated volume of air (Markley and Carrier 2010). York et al. (2017) estimated that the metabolic cost of breathing ranges between 1% and 3% of the basal metabolic rate, values that fall in the lower end of those that have been determined for the other tetrapod species. The cost of breathing water per unit volume of O_2 uptake ranges from 0.5% to 70% of the total O_2consumption (VO_2) (Alexander 1967; Dejours 1988; Schumann and Piiper 1966; Wood and Lenfant 1979).

Among the extant animal life, an assortment of gas exchangers that support diverse metabolic needs for O_2 exists (Plates 1.6, 1.7 and 4.1a–f). Factors such as body size, phylogenetic level of development, habitat occupied, lifestyle pursued, and respiratory fluid medium utilized have together determined the development, structure, and function of respiratory structures/organs (Graham 1997; Maina 1998, Maina 2000a, c, 2006, 2011; Jürgens and Gros 2002; Hsia et al. 2013; Figueroa et al. 2007): they range from the cell membranes in microorganisms, surfaces such as skin and buccal cavity (e.g., in amphibians), gills (e.g., in fish), lungs (e.g., in air-breathing vertebrates such as mammals and birds) and the tracheal system (e.g., in insects) (Maina 1989b–e, 1994, 1998, 2002a–d; Graham 1997; Tavolzhanova et al. 2004) (Plates 2.2, 2.4, 2.12, 3.1a, 3.2 and 3.6). The so-called intermediate, transitional, bimodal, and trimodal breathers (e.g., the amphibian species that use the skin, the lungs, and the gills) are a group of broadly specialized animals that utilize different respiratory structures and methods to acquire O_2 from both water and air (Phillips et al. 2022). In catfishes (Family: Clariidae), the respiratory system comprises gills that are water-breathing organs, and accessory respiratory organs that are air-breathing structures (Maina and Maloiy 1986; Mbanga et al. 2018; Munshi and Hughes 1992). A profoundly fascinating respiratory stratagem termed “rebreathing” was recently reported in the semi-aquatic *Anolis* lizards that inhabit neotropical streams (Boccia et al. 2021): to escape from predators or to forage for a longer time underwater, the animals that are model air-breathers, prolong their diving times by rebreathing exhaled air that is trapped by the nasal hairs and under their rough skin. Interestingly, the respiratory strategy is similar to the plastron/physical gill breathing one that occurs, e.g., in insects that among the animal life evolved much earlier (Ege 1915; Thorpe and Crisp 1941; Thorpe 1950; Flynn and Bush 2008; Pedersen and Colmer 2012; Seymour and Matthews 2013).

Development of efficiency gas exchangers was critical to existence of complex animal life, occurrence of important events such as the transition from water-breathing to air-breathing and subsistence under hypoxic conditions that have occurred at certain points across the evolutionary time (Randall et al. 1981; Berner and Canfield 1989; Graham et al. 1995; Dejours 1988; Little 1990; Kasting 1993; Maina 1998; Planavsky et al. 2014; Sosa Torres et al. 2015). The "Great Oxidation Event," "O_2 Catastrophe," or the "O_2 Crisis," which occurred ~2.45 billion years ago in the Paleoproterozoic, is probably the earliest large mass extinction event (Lyons et al. 2014). Although at macroscopic (gross) level the morphologies of gas exchangers vary greatly (Plates 1.6), interestingly, at microscopic level, certain shared structural and functional features exist (Maina 1998, 2005, 2014) (Plate 2.1). Among these are the following: (a) the flux (transfer) of O_2 and CO_2 occurs across a thin tripartite (three-ply = laminated) water-blood or blood-gas (air) tissue barriers by passive diffusion along existing partial pressure gradients; (b) the diffusing capacity of a tissue barrier for O_2 (DO_2) correlates inversely with the thickness of the tissue barrier and directly with that of the respiratory surface area; and (c) in the complex animal life, convective, and diffusive processes are employed to transport and deliver O_2 to the cells and eliminate CO_2 that is produced during cellular metabolism. Over long distances, convective transport of O_2 requires energy while across very short distances, such as the thickness of the water-blood and the blood-gas tissue barriers, the transfer occurs by diffusion, a passive process.

Fundamentally, determined by the environment occupied and consequently the respiratory fluid medium utilized, the evolved gas exchangers have formed by two processes, namely, evagination or invagination (Maina 1994, 19982002a–c, 2011). While in microorganisms gas exchange occurs across a simple plain cell membrane/wall, in complex animals (Plate 4.3a), the water-breathing organs that are generally termed gills form by evagination (Plate 4.3b, c) while the air-breathing ones (lungs) develop by invagination (Plate 4.3d): the insectan tracheal system is a good example of an invaginated respiratory structure (Plate 4.3d—inset). Invagination of gas exchangers allowed conservation of water, an adaptation that supported transition from water onto land which is a more desiccating environment (Randall et al. 1981; Little 1990; Maina 1998). With the airways entering the organ and the system terminating blindly (Plates 2.2 and 2.4a, d), in general, invaginated gas exchangers are ventilated tidally. However, interestingly, albeit being an invaginated gas exchanger which is overall tidally ventilated, on account of the morphological complexity of the lung-air sac system of birds (Plates 2.17, 2.23, 3.27, 4.4a—inset, b—inset and 4.5), the avian lung (specifically the paleopulmonic part) is ventilated continuously and unidirectionally in a caudocranial direction while the neopulmonic one is ventilated tidally (Plate 4.5).

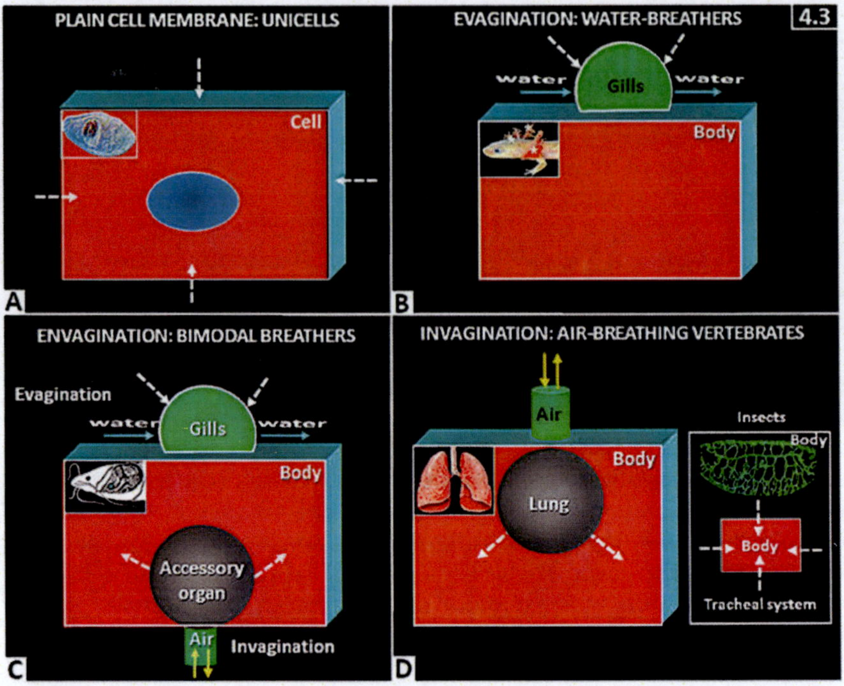

Plate 4.3 Forms of evolved gas exchangers. (**a**) The simplest gas exchanger is the plain cell membrane of unicellular organisms. (**b**) Gills, that are used for water-breathing, develop by evagination. (**c**) In bimodal or transitional breathers gills (dot, •) are evaginated gas exchangers that serve a water-breathing role and the accessory respiratory organs (asterisk, *) that are invaginated organs that are utilized for air-breathing. (**d**) In air-breathing vertebrates, lungs, that are invaginated organs, are the gas exchangers. (**d**)—inset: The insectan tracheal system delivers oxygen directly to the cells of the body: it is an invaginated structure. Dashed arrows, the direction of diffusion of oxygen in a gas exchanger

4.2 Structural and Functional Correlations in Properties of Gas Exchangers: General Considerations

In air-breathing vertebrates, the primary role of lungs is gas exchange, i.e., maintenance of proper concentrations of O_2 and CO_2 and hydrogen ions (H^+) in the cells and tissues of the body (Burggren and West 1982; Liem 1988; Brainerd 1994, 1999; Maina 1998; Graham 1997; Graham and Wegner 2010; Hsia et al. 2013). Excess hydrogen ions and CO_2 stimulate the respiratory center, triggering corrective changes by ventilatory activities. Interestingly, in regulating respiration, O_2 does not have a meaningful direct effect on the respiratory center of the brain: it acts on peripheral chemoreceptors that are located in the carotid and aortic bodies that convey relevant nervous signals to the respiratory center for processing and control of respiration. Among the air-breathing vertebrates, morphologically, gas

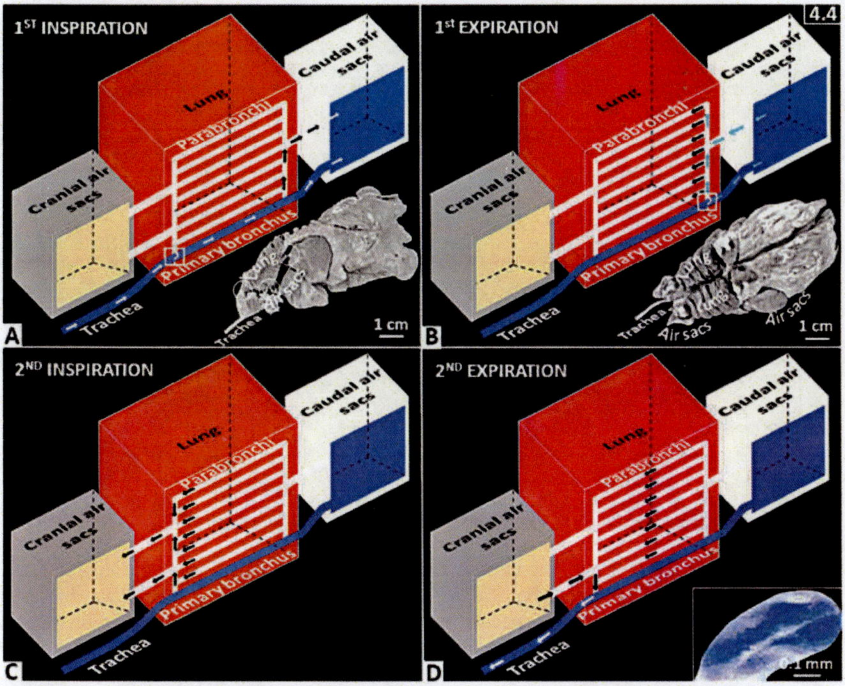

Plate 4.4 Airflow through the avian respiratory system (the lung-air sac system): it takes two inspiratory and two expiratory cycles for inspired air to enter and leave the respiratory system. (**a**) During the first inspiratory cycle, the air (shown in blue) flows from outside down the trachea and the primary bronchus to enter the caudal air sacs. (**b**) In the first expiratory cycle, the air moves from the abdominal air sacs into the parabronchial system of the lung via the mediodorsal secondary bronchi. (**c**) During the second inspiratory cycle, the air (in the lung) relocates to the cranial air sacs. (**d**) In the second expiratory cycle, the air in the cranial air sacs moves to the outside via the primary bronchus and the trachea. (**a, b**) Boxed areas (□) with question marks (?), respectively, show sites where inspiratory aerodynamic valving and expiratory aerodynamic valving occur. (**a, b**) Respectively, lateral and medial views of a latex rubber cast preparation of the lung-air sac system of the Domestic Fowl (*Gallus gallus* variant *domesticus*) showing the physical separation of the gas exchanger (lung) and ventilators (air sacs). Circles (O) (**a**—Inset), ostia, i.e., sites where air sacs connect to the lung. (**d**)—Inset: Delicate and transparent embryonic abdominal air sac of the Domestic Fowl

exchangers range in structural complexity from simple rather smooth-surfaced saccular organs of some amphibian and lungfish (Dipnoan) species (Maina and Maloiy 1988) (Plate 1.6a–d) as well as some reptilian ones such as the Tokay Gecko (*Gekko gecko*) (Perry et al. 1994) to heterogenous ones in animal species such as the Nile crocodile (*Crocodylus niloticus*) (Perry 1988, 1989a), the Pancake Tortoise (*Malacochersus tornieri*) (Maina et al. 1989b) (Plates 1.6e and 1.7c), the Black Mamba (*Dendroaspis polylepis*) (Stinner 1982; Maina 1989d; Wallach 1998; Champini et al. 2022) (Plates 1.6g and 1.7e), the Sandboa (*Eryx colubrinus*) (Maina et al. 1999) and the Monitor Lizard (*Varanus exanthematicus*) (Maina et al. 1989b)

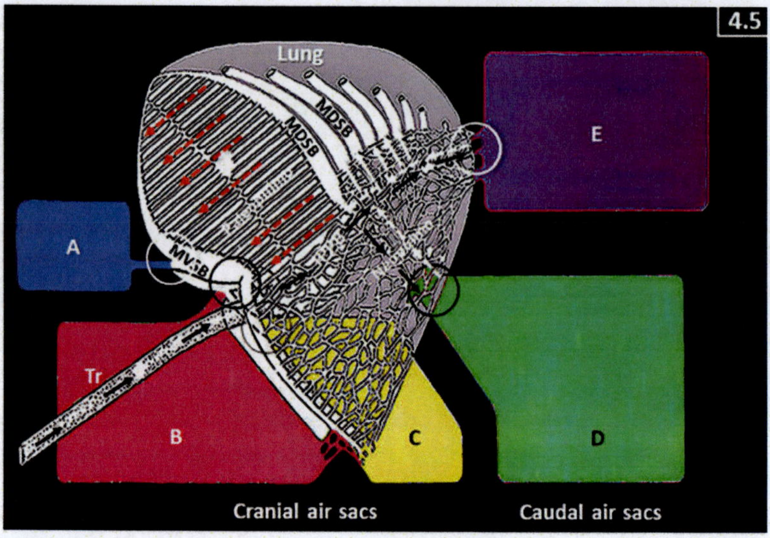

Plate 4.5 Airflow through the lung-air sac system of birds. Functionally and structurally, the air sacs comprise a cranial group, i.e., cervical (**a**), clavicular (**b**), and craniothoracic (**c**) air sacs and a caudal group that consists of caudothoracic (**d**) and abdominal (**e**) air sacs: the lungs are intercalated between the cranial and the caudal groups of air sacs. The air sacs connect to the lung at sites called ostia (circles, O). During the first inspiratory cycle, the air flows down the trachea (Tr) and the intrapulmonary primary bronchus (IPPB) to enter the caudal air sacs (solid black arrows, →). In the following respiratory cycle, the air flows from the caudal air sacs across the neopulmonic parabronchi (dashed white arrows) and then into the mediodorsal secondary bronchi (MDSB) (solid white arrows). Subsequently, the air flows through the paleopulmonic parabronchi (dashed red arrows) to enter the medioventral secondary bronchi (MVSB)

(Plates 1.6h and 1.7d) to the extremely complicated ones of mammals (alveolar lungs) (Plate 1.7b) and birds (parabronchial lungs) (Weibel 1984; King and McLelland 1984; McLelland 1989; Maina et al. 1989a; Maina 2005) (Plates 1.7a, 3.7, 3.8 and 3.9). Albeit the morphological differences that exist between gills and lungs, O_2 and CO_2 diffuse between the external respiratory fluid media (water or air) and the internal respiratory fluid medium (blood) across thin tissue barriers (Meban 1980; Maina and King 1982a; Weibel 1973; Maina and West 2005) (Plate 2.1): to maintain a partial pressure gradient across the tissue barriers, the gas exchangers are ventilated with fluid medium and perfused with blood (Maina 2000a, 2005; Maina and West 2005). The tripartite respiratory tissue barriers of the vertebrate gas exchangers (Plate 2.1) have morphologically been considerably conserved. Metabolically active animals with high O_2 demand have evolved structurally more complex and functionally efficient gas exchangers (Plates 1.6, 1.7 and 4.1a–e) with thin and extensive water-blood/blood-gas (air) barriers (Hughes and Morgan 1973; Maina 1998; Maina and King 1984; Maina 1989a; Maina et al. 1991). Mammals and birds are among the amniotes the only endothermic-homeotherms and among the air-breathing vertebrates have the highest metabolic rates (McNab

2002) and have the most efficient respiratory systems (Maina 1998, 2005, 2011). In biology, few organs have attracted and held as much scientific interest as the avian respiratory apparatus (King 1957; Akester 1960; Farner 1970; Duncker 1974; McLelland 1989; Scheid 1990). Studies span nearly five centuries from Coitier (1573) and subsequently by among others Campana (1875), Huxley (1882), Selenka (1866), Newton (1896), Beddard (1898), Ross (1899), Müller (1908), Juillet (1911), Gilbert (1939), McLelland (1989) to the present day Maina (2022): although investigated over a long period of time, certain important questions on its development, structure, and function still remain unresolved (McLelland 1989; Maina 2017a–c).

Among the extant air-breathing vertebrates, the avian respiratory system is structurally the most complicated and functionally most efficient (King 1966; Duncker 1971, 1972, 1974; Piiper and Scheid 1972; Scheid 1979, 1990; Fedde 1980; Kardong 1988; McLelland 1989; Powell and Hopkins 2004; Maina 2005, 2006, 2022; Powell 2014; Pandey et al. 2015). One of the early experiments that illustrated the complexity of the avian respiratory system was done by Biggs and King (1957) who determined that a chicken continued breathing through a severed (cut) humerus, a bone which in the chicken is well-pneumatised by the proximate air sac that in turn connects to the lung (Bremer 1940; Hogg 1984; McLelland 1989). Considering the elementary tools and the crude methods that were used by the early investigators, the findings made on the structure and function of the avian respiratory system are impressive for their meticulous attention to detail and interpretative skills (Berger 1960). During the investigation of the avian respiratory system, research questions have been formulated and concepts advanced, challenged, and tested (Maina 2017a): suggestions and theories have been debunked after empirical evidence has been adduced. For example, for a long time, an intuitively appealing suggestion was that anatomical valves or sphincters governed airflow across the avian lung. The supposition which was championed by among others Dotterweich (1930) and Vos (1934) was much later found to be flawed (King 1966; Duncker 1971; Scheid 1979; King and McLelland 1984; McLelland 1989; Powell et al. 1981; Powell 2014). In the recent past, experimental, morphological, mathematical, and computational modeling techniques (Banzett et al. 1987, 1991; Wang et al. 1988, 1992; Maina and Africa 2000; Maina and Nathaniel 2001; Maina et al. 2009; Harvey and Ben-Tal 2016) have shown that convective momentum of inspired air shunts it across the orifices (openings) of the medioventral secondary bronchi into the intrapulmonary primary bronchus (Plate 4.6a–c, e).

4.3 Air Flow Dynamics in the Avian Respiratory Systems

The pattern of airflow in the avian respiratory system is a greatly complicated process (Plate 4.5). On account of the morphological complexity of the airway (bronchial) system in particular (Plates 2.2a, 3.7, 3.8 and 3.9) and the respiratory system in general, infinitely many likely paths across which inspired air can pass through the lung to reach the air sacs and exit them exist. The actual route taken by

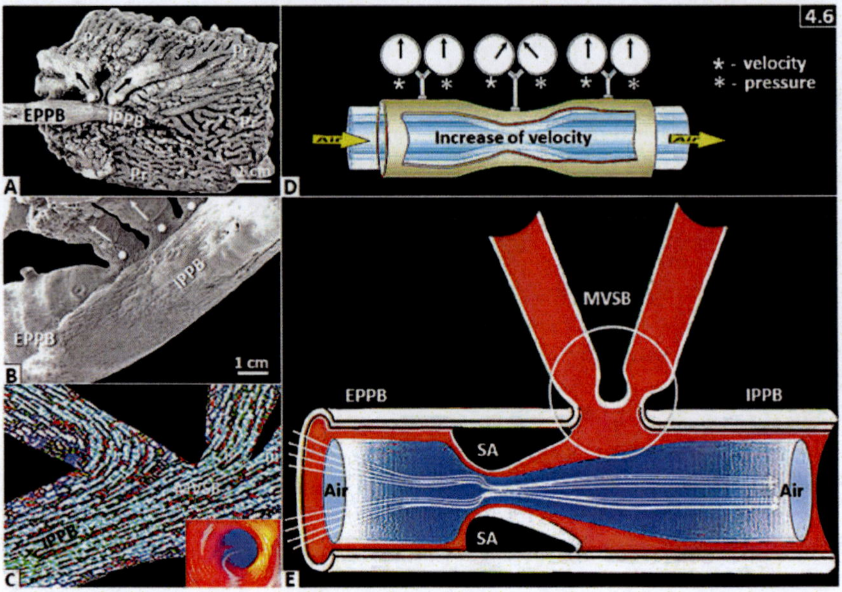

Plate 4.6 **(a, b)** Latex rubber cast preparations showing orientations of the medioventral secondary bronchi (arrows, ←) of the lung of the Domestic Fowl (*Gallus gallus* variant *domesticus*) relative to the position of the intrapulmonary primary bronchus (IPPB). Dots (•), constrictions at the orifices of the medioventral secondary bronchi; EPPB, extrapulmonary primary bronchus; Pr, parabronchi. **(c)** Computational fluid dynamics (CFD) simulation of air in the lung of the Ostrich (*Struthio camelus*) showing angulation and the flow of air from the intrapulmonary primary bronchus into the medioventral secondary bronchi (MVSB—i–iii). **(c)**—inset: Illustration of nature of airflow through an angulated airway. **(d)** Diagrammatic illustration of the Bernoulli's effect, where an increase in the flow velocity of fluid results in the decrease in pressure. **(e)** A segmentum accerelans (SA) that exists at the terminal part of the extrapulmonary primary bronchus (EPPB) of the lung of the Domestic Fowl (*Gallus gallus* variant *domesticus*) convectively accelerates the flow of inhaled air (grouped arrows), convectively thrusting it (air) past the orifices of the medioventral secondary bronchi (MVSB) encircled area, O). IPPB, intrapulmonary primary bronchus

the inhaled air cannot be deduced by physical examination of the very complicated airway and anatomical systems. In contrast to the mammalian lung where air flows in and out of the lung, for the avian respiratory system, it takes two inspiratory and two expiratory cycles for inspired air enter and leave (Plate 4.4): for a specified volume of air, the phases are as follows: (a) during the first inspiratory cycle, the air flows from outside down the trachea, the extrapulmonary primary bronchus and the intrapulmonary primary bronchus to enter the caudal air sacs (Plate 4.4a and 4.5); (b) in the first expiratory cycle, the air flows from the caudal air sacs into lung via the mediodorsal secondary bronchi (Plates 4.4b and 4.5); (c) the second inspiratory cycle drives the air from the lung into the cranial air sacs (Plate 4.4c) and; (d) in the second expiratory cycle, the air flows from the cranial air sacs to the outside, via the trachea (Plate 4.4d). Because the avian lung is fixed (rigid) and noncompliant, the flow of air across the lung is driven by synchronized actions of the cranial and caudal groups of air sacs which respectively comprise the cervical, the clavicular, and the craniothoracic air sacs and the later the caudothoracic and the abdominal air sacs (Plate 3.27).

In phylogenetically more derived species of birds, the paleopulmonic part of the lung is continuously and unidirectionally ventilated in a caudocranial direction (back-to-front) by the air sacs while the flow in the neopulmonic part is tidal, i.e., is oscillatory or bidirectional (Plate 4.5) (King 1966; Fedde 1980; Mclelland 1989). Transparent, delicate, and largely avascular (Fletcher 1980) (Plate 4.4d—inset), the air sacs do not significantly participate in gas exchange (Magnussen et al. 1976). An experiment performed nearly one-and-half centuries ago by Soum (1896) is commonly cited in scientific literature to support this deduction (e.g. Schmidt-Nielsen 1983, pp. 48–49): albeit birds being highly sensitive to carbon monoxide (CO) poisoning (https://kdvr.com/news/1-dead-10-hospitalized-from-carbon-monoxide-leak-after-birds-alert-firefighters/—accessed 25-10-2022), when the connection of the large abdominal air sacs to the lung was blocked and CO introduced into the air space (abdominal air sac), no distress was observed or harm detected to the experimental animals. The air sacs of the Domestic Fowl (*Gallus gallus* variant *domesticus*) and Mallard Duck (*Anas platyrhychos*) have a rich nerve supply (Rawal 1976). The complex mode of ventilation of the avian respiratory system correlates with its elaborate morphology (King and McLelland 1984; Fedde 1980; McLelland 1989; Maina 2005, 2006) (Plates 3.7, 3.8, 3.9, 3.13, 3.14, 3.15, 3.16, 3.17, 3.18, 3.19, 3.20, 3.21, 3.22, 3.23, 3.24, 3.25, 3.26 and 3.27). The continuous unidirectional mode of ventilation that occurs in the avian lung is possible because of the following main reasons: (a) for the avian respiratory system, the gas exchanger (the lung) is physically separated from the compliant, i.e., ventilatory part (the air sacs) (Plates 3.27, 4.4 and 4.5) and (b) the paleopulmonic parabronchi (Plates 3.7 and 3.13a–c) are arranged in parallel relative to the placement of the intrapulmonary primary bronchus: in contrast, the neopulmonic parabronchi, that are arranged in series relative to the intrapulmonary primary bronchus (Plate 4.5), are ventilated tidally or bi-directionally. Unlike the blind-ended airway (bronchial) system of the mammalian lung (Maina and van Gils 2001) (Plates 2.2b, 2.4a, b and 2.12a, d), for the avian lung, the parabronchi form a continuous hoop-like arrangement (Plate 2.2a). Radiographic imaging of the orifices of the medioventral secondary bronchi of lungs of resting ducks showed that the openings did not change in size or shape during respiratory cycles (Jones et al. 1981): indicated that in directing airflow through the lung, the airways behaved as aerodynamic rather than as valvular structures. Without clear evidence of their existence, over the many years that airflow in the avian lung was investigated, largely because of its theoretical appeal and from the fact that valves were well-known to exist in biological structures such as the heart, the blood vessels (especially veins) and intestines, it was supposed that valvular or sphincteric anatomical structures controlled the flow of air across the respiratory organ. Later on, a discovery that unidirectional airflow occurs in pump-ventilated paralyzed and also fixed (i.e., dead) avian lungs (Scheid et al. 1972) showed that structure-specific aerodynamic properties/features (and not valves or sphincters) regulated the airflow pattern in the airways of the avian lung.

The mechanism by which inspired air is shunted across the openings of the medioventral secondary bronchi has been investigated (Wang et al. 1988, 1992; Kuethe 1988; Butler et al. 1988; Banzett et al. 1987, 1991; Maina and Africa 2000; Maina and Nathaniel 2001; Maina et al. 2009; Harvey and Ben-Tal 2016; Nguyen et al. 2021). While some investigators have suggested that some inspired air enters the medioventral secondary bronchi (Portier 1928; Vos 1934, 1935; Dotterweich 1930, 1933, 1936; Zeuthen 1942; Cohn and Shannon 1968; Schmidt-Nielsen et al. 1969; Scheid et al. 1972), others have reported that the air completely bypasses the orifices of the airways (Hazelhoff 1943). The so-called "inspired loop hypothesis of Hazelhoff" (Hazelhoff 1943), in which it was supposed that during inspiration air flowed twice through the parabronchi, has been debunked (Scheid et al. 1972; Powell et al. 1981; Scheid and Piiper 1971, 1972). After no anatomical valves were located in the airways of the lung, Dotterweich (1936) proposed the "fluid valve theory" that suggested that aerodynamic forces and not anatomical valves or sphincters and sizes and geometries of the airways created inspiratory aerodynamic valving in the avian lung, i.e., the process by which inspired air was shunted past the orifices of the medioventral secondary bronchi (Banzett et al. 1987, 1991; Butler et al. 1988; Kuethe 1988; Wang et al. 1988, 1992; Maina and Africa 2000; Maina and Nathaniel 2001; Maina et al. 2009) (Plates 3.11h–j and 4.6a, b). Interestingly, at the terminal part of the intrapulmonary primary bronchus, the process by which expired air from the caudal air sacs is directed into the mediodorsal secondary bronchi, instead of it leaving the way it comes in, i.e., through the intrapulmonary primary bronchi, the extrapulmonary primary bronchi and the trachea (Plate 4.4b), is termed "expiratory aerodynamic valving" (IAV) (Brown et al. 1995). Scanty experimental data and morphological details exist on the process of expiratory aerodynamic valving (Scheid et al. 1972; Scheid 1979; Kuethe 1988; Brown et al. 1995; Maina et al. 2009): sizes and geometries of the airways of the avian lung (Plates 3.8a, e, f, 3.9a–c, e, f, 3.11a–j and 4.7) vitally determine the direction and the pattern of airflow in the organ. Molony et al. 1976) and Powell et al. (1981) determined that CO_2 is a significant factor in the regulation of airflow in the avian lung. The openings and angulation of the medioventral secondary bronchi relative to the orientation of the intrapulmonary primary bronchus (Plates 3.11a–j and 4.6a–c) may meaningfully generate inspiratory valving. Constriction of the orifices of the medioventral secondary bronchi (Plates 3.11g, h, i, j and 4.6a, b) may participate in regulating the flow of inspired air into and across the airways. The avian lung is well-endowed with connective tissue elements such as collagen, smooth muscle, and elastic tissue fibers (King and Cowie 1969; King and Molony 1971; Maina et al. 2010) (Plates 3.10 and 3.22d–h) that may cause small dimensional changes. At the parabronchial level, the contractile elements may change parabronchial luminal diameters and modify airflow through them. For the Duck lung, Macklem et al. (1979) observed that the parabronchi and the secondary bronchi have finite compliance. The occasional folds of the epithelial-epithelial cells connections (epithelial retinacula) (Plate 3.22a, b) and the blood-gas barrier (Plate 3.22c), that has been observed in the gas exchange tissue of the avian lung, suggest that the air and the blood capillaries, which have

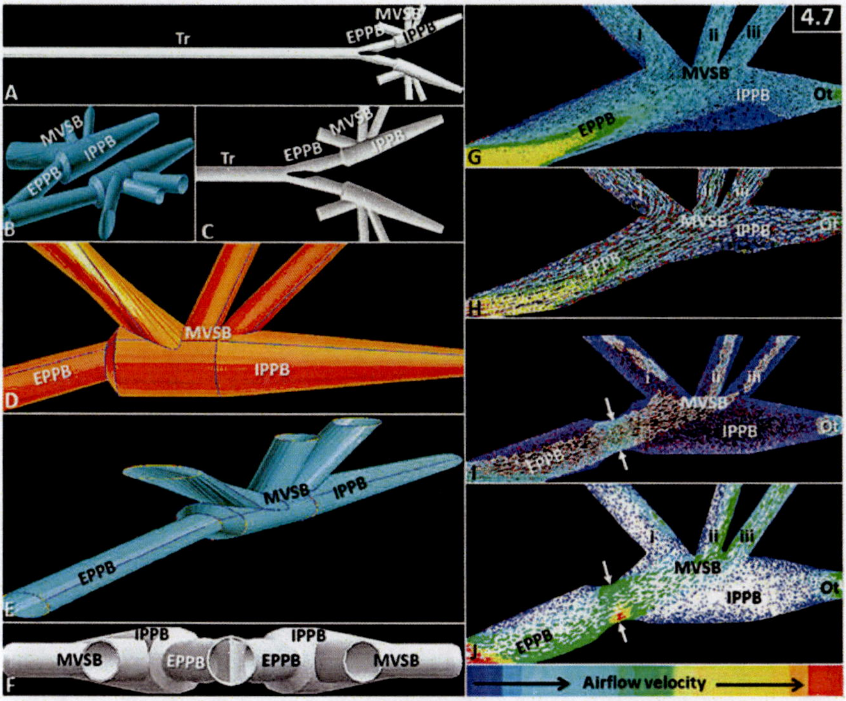

Plate 4.7 **(a–f)** Computer-generated structural models (based on actual measurements of the airways) showing sizes, orientations, and shapes of some of the airways of the lung of the Ostrich (*Struthio camelus*) prepared from observations and actual measurements of the anatomical preparations. Tr, trachea; EPPB, extrapulmonary primary bronchus; IPPB, intrapulmonary primary bronchus; MVSB, medioventral secondary bronchi. **(g–j)** Computational flow dynamics (CFD) illustrations of inspiratory aerodynamic valving analysis of the lung of the Ostrich showing airflow dynamics. **(i)** and **(j)** show airflow after a constriction, a segmentum accerelans (arrows, ↓) was created at the distal part of the extrapulmonary primary bronchus (EPPB). Of the three medioventral secondary bronchi (MVSB, i–iii), airflow largely occurred through the third positively inclined airway (MVSB, iii) (**g**, **h**) while after creating a constriction (**i**, **j**), a similar airflow pattern was replicated. IPPB, intrapulmonary primary bronchus; Ot, ostium

been reported to be very strong (Macklem et al. 1979; Maina 2007a, b; West et al. 2006; Watson et al. 2008), may not be absolutely rigid.

In the lungs of the Domestic Fowl (*Gallus gallus* variant *domesticus*), Maina and Africa (2000) observed a swelling that protruded (bulged) into the lumen at the terminal part of the extrapulmonary primary bronchus close to the origin of the first medioventral secondary bronchus (Plate 4.8a–d): the structure was intensely vascularized (Plate 4.8d, e) and with a "blunt" leading part and a tapering end (tail), it showed a characteristically aerodynamic cross-sectional profile (Plate 4.8a, c). Connective tissue elements such as collagen and possibly elastic tissue fibers appear to exist in the subepithelial of the protuberance, especially around the blood vessels (Plate 4.8e). By investigating the flow of inspired radio-opaque gas in the

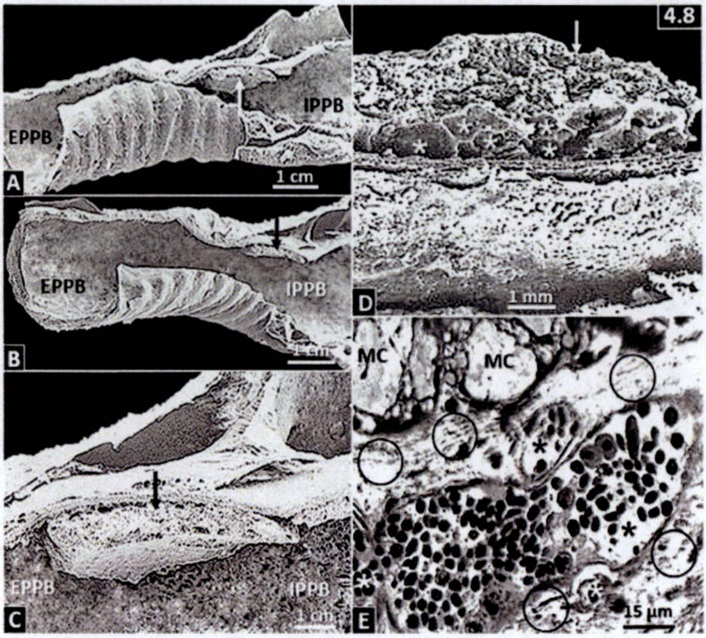

Plate 4.8 **(a–c)** A segmentum accerelans of the lung of the Domestic Fowl (*Gallus gallus* variant d*omesticus*) (arrow, ↓) that is located at the distal end of the extrapulmonary primary bronchus (EPPB), i.e., at the entrance of the intrapulmonary primary bronchus (IPPB) and close to the point the first medioventral secondary bronchus originates: the structure narrows the lumen of the airway. With a blunt front and tapering end (**a, c**), the segmentum accerelans has an aerodynamic shape: it is highly vascularized (**d–e**). Stars (*) (**d, e**), blood vessels; MC (**e**), mucus secretory (goblet) cells; circles (O) (**e**), collagen fibers

lung of the Graylag Goose (*Anser anser*), Wang et al. (1992) determined that a constriction existed at the distal part of extrapulmonary primary bronchus. The narrowing was observed to change in diameter with respiratory rate: the lumen widened during ventilatory hyperpnea and constricted during eupneic breathing while it widened under high CO_2 levels. From its envisaged function of accelerating the flow of inspired air and convectively propelling it across the openings of the medioventral secondary bronchi, Wang et al. (1992) termed the constriction "segmentum accelerans." More recently, applying mathematical and computational fluid dynamics methods (Maina et al. 2009; Harvey and Ben-Tal 2016; Nguyen et al. 2021) illustrated the mechanism of inspiratory aerodynamic valving (Harvey and Ben-Tal 2016; Nguyen et al. 2021).

While the actual mechanism by which a segmentum accerelans contributes to shunting of air across the openings of the medioventral secondary bronchi is largely still largely unclear, Maina and Africa (2000) suggested that the structure may work as an erectile (carvernous) tissue, much like the human penile one functions, where the degree of swelling (tumescence = turgidity = thickness = enlargement) and collapse (folding up = flopping) are determined by the influx and the efflux of blood into and out of the structure. Swelling of the segmentum accerelans should narrow the lumen at the distal part of the extrapulmonary primary bronchus while its collapse should dilate (widen) it. In fluid flow dynamics, according to Bernoulli's Principle (Batchelor 2000; Anderson 2016), an increase in the speed of airflow is caused by a reduction in static pressure or a decrease in the fluid's potential energy (Plate 4.6d). When the flow velocity or driving pressure is constant, constriction of an airway or a tube (conduit) accelerates the speed of flow of a fluid (Plate 4.6d, e), a process that consequences in pressure reduction (Batchelor 2000; Anderson 2016). At least for the Domestic Fowl (*Gallus gallus* variant *domesticus*) that has been better investigated, inspiratory aerodynamic valving occurs at the point where the medioventral secondary bronchi originate from the intrapulmonary primary bronchus (Plates 4.6e and 4.7i, j): through the Bernouli's effect the narrowing caused by a segmentum accelerans increases the velocity of the inspired air, convectively thrusting it (air) past the orifices of the angulated medioventral secondary bronchi (Plates 4.6e and 4.7). Banzett et al. (1991) and Wang et al. (1992) conjectured that during exercise and under conditions and states that cause hyperpnea, e.g., flight at high altitude under hypoxic and rarefied air, the segmentum accelerans may collapse to the level of the wall of the extrapulmonary primary bronchus, a state under which the velocity of airflow creates the requisite convective momentum that propels the air past the orifices of the medioventral secondary bronchi. During eupneic (resting) breathing, when the velocity of the inspired air is slow, constriction of the distal part of the extrapulmonary primary bronchus may be a critical feature for the generation of inspiratory aerodynamic valving. Regarding expiratory aerodynamic valving (Kuethe 1988), Brown et al. (1995) suggested that the process may depend on dynamic compression of the intrapulmonary primary bronchus, a state that may largely stem from airflow resistance and flow velocity but not from gas density. Using dynamical systems techniques and numerical analysis, Harvey and Ben-Tal (2016) showed that unidirectional airflow can be created by inspiratory or effective expiratory valving and that the two processes are critical in the development of the complexities and the high efficiencies of the airflow patterns that occur in the avian lung. According to the observations made by Kuethe (1988), Harvey and Ben-Tal (2016) showed that the efficiencies of the inspiratory and the expiratory valving mechanisms depend on the compliances of the air sacs and the airflow resistances, events that occur far from the actual sites where aerodynamic valving mechanism happen.

A segmentum accelerans does not interestingly appear to exist in the lungs of some birds such as the Ostrich (*Struthio camelus*) (Maina and Nathaniel 2001). Since the pattern of airflow in the lungs of birds is reported to be similar (Scheid and Piiper 1972; Scheid 1979; Fedde 1980), a segmentum accelerans does not seem to be quintessential for generation of inspiratory aerodynamic valving. Structural features/properties such as sizes, shapes, and orientation of airways (Plates 3.11a–j, 4.6a, b and 4.7) as well as the presence of upstream structures such as the larynx (Plate 3.3a, b), the trachea (Plates 1.1a and 3.3c), the syrinx (Plates 1.1 and 3.5) and the cranial air sacs (Plates 3.27, 4.4 and 4.5) and downstream ones such as the caudal air sacs

(Plates 3.27, 4.4 and 4.5) may to various extents be involved creating and maintaining of the process of inspiratory aerodynamic valving. It has been reported that chemoreceptors that are sensitive to CO_2 (in the inspired air) and/or mechanoreceptors that may detect the airflow velocity and diametric changes of the extrapulmonary primary bronchus may also influence the flow of air in the avian lung. In the duck lung, Molony et al. (1976) determined that airflow resistance across the openings of the medioventral secondary bronchi was dependent on the partial pressures of CO_2 (PCO_2) in the inspired air, with resistance being high at low PCO_2 and vice versa; Barnas et al. (1978) observed that the smooth muscles that exist where the paleopulmonic parabronchi connect to the mediodorsal secondary bronchi respond to changes in the concentration of intrapulmonary CO_2 and; Ray and Fedde (1969) noted that airflow resistance across the avian respiratory system declined under hypoxia. In chickens and ducks, Ray and Fedde (1969) and Molony et al. (1976) observed that changes in the concentrations of intrapulmonary CO_2 altered the airflow resistance across the lungs. For the duck (Leitner and Roumy 1974) and the goose lungs (Barnas et al. 1978), it was determined that airways contracted forcefully under hypercapnia, and changes in CO_2 concentration in the intrapulmonary gas during eupneic breathing was regulated by the contractility of the intrapulmonary smooth muscles.

4.4 Gas Exchange: Coping with Hypoxia and Hypocapnia in Birds

For the air-breathing vertebrates, hypoxia importantly affects pulmonary vascular tone (Euler and Liljestrand 1946; Butler 1970; Ramirez et al. 2007; Laguë 1985; Scott and Milsom 2007; Scott 2011; Dzal et al. 2015; Sommer et al. 2008, 2016; Suresh and Shimoda 2016; Zheng and Zhou 2017; Scott and Dawson 2017; Dunham-Snary et al. 2017): it causes vasoconstriction that increases pulmonary blood resistance and pressure (hypertension) (Crossley et al. 1998; Skovgaard et al. 2005; Aaronson et al. 2006; Sylvester et al. 2012; Tarry and Powell 2017; Charolidi and Carroll 2017; Dunham-Snary et al. 2017). Hypoxic pulmonary vasoconstriction, that is also termed “the von Euler–Liljestrand mechanism” in recognition of the discoverers of the process (Euler and Liljestrand 1946), is a reflex contraction of vascular smooth muscle in the pulmonary circulation in response to the decreased regional partial pressure of O_2 (PO_2) (Aaronson et al. 2002; Sommer et al. 2008; Sylvester et al. 2012). During hypoxic pulmonary vasoconstriction, pulmonary blood flow is directed from the poorly ventilated and hypoxic parts of the lung to better-ventilated ones (Theissen and Meissner 1996; Brimioulle et al. 2002): the process is deemed to be important in the local matching of ventilation (with air) to perfusion with blood (Skovgaard et al. 2005; Tarry and Powell 2017; Dunham-Snary et al. 2017). By reducing pulmonary ventilation-perfusion inequalities, sufficient oxygenation of blood is provided. Skovgaard et al. (2005) and

Skovgaard and Wang (2006) suggested that hypoxic pulmonary vasoconstriction is more strongly expressed in species that have structurally complex lungs and functionally partitioned hearts.

For most air-breathing vertebrates, CO_2 is a powerful respiratory stimulant (Milsom 2002). Compared with mammals, birds are more tolerant to hypocapnia (Scheid 1990; Faraci 1986, 1991; Ramirez et al. 2007; Scott 2011; Parr et al. 2019). Notwithstanding their suffering respiratory alkalosis, birds can sustain elevated ventilatory rate that permits them to increase and maintain high O_2 uptake and transfer without suffering negative (downregulatory) chemoreceptor feedback (Laguë 1985; Scheid 1990; Scott 2011). During hypercapnia, in male ducks Pekin Ducks (*Anas platyrhynchos*), Dodd et al. (2007) determined that pH compensation was elicited by stable or rising arterial partial pressure of carbon dioxide ($PaCO_2$) working at central chemoreceptors. For the same species of birds, Grubb et al. (1977) observed that the lack of a hypocapnia response may underwrite the unique tolerance of birds to high altitudes by maintaining normal cerebral blood flow. Birds have the capacity to rapidly restore blood pH under challenges of CO_2 loading (hypercapnia) (Dodd et al. 2007). Birds hyperventilate more before CO_2 depletion in blood ($PaCO_2$) compromises the function(s) of critical organs such as the brain and the heart (Butler 2016): low PCO_2 in arterial blood (hypocapnia) that occurs with hyperventilation does not lower cerebral blood flow (Grubb et al. 1978; Faraci et al. 1984, 1985; Faraci and Fedde (1986) as occurs in mammals (Hansen et al. 1984; Rosenberg 1988). Compared with mammals, some investigators have reported that in hypoxic birds, O_2 delivery to the brain increases (Piiper and Scheid 1975; Faraci 1991; Butler 2016). For the Bar-Headed Goose (*Anser indicus*), Faraci and Fedde (1986) made the following important physiological observations: (a) the arterial blood pressure was not affected by hypocapnia but rose moderately during hypercapnia; (b) heart rate did not vary during changes in the $PaCO_2$; (c) extreme hypocapnia did not significantly change cerebral, coronary, or pectoral muscle blood supply; and d) hypercapnia increased cerebral and coronary blood flow, but pectoral muscle blood flow was not affected. For high-altitude flying birds, lack of decrease of blood flow to the brain during severe hypocapnia constitutes an important adaptation that allows them to increase O_2 uptake by hyperventing under very low PO_2 (Grubb et al. 1977, 1978; Faraci et al. 1984, 1985). Experimental comparison of the Pekin Duck (*Anas platyrhynchos domesticus*) and the Bar-Headed Goose (*Anser anser*) by Black and Tenney (1980) showed no altitude-induced behavioral consequences such as restlessness up to a simulated altitude of 7620 m for ducks while the geese tolerated hypoxia up to that of 10,668 m, with no evident behavioral changes. While neither the O_2-sensing mechanism nor the exact physiological and structural pathways that determine the process of hypoxic pulmonary vasoconstriction are well-known (Faraci 1986; Sommer et al. 2008), the adaptations may derive from the complex development (Chap. 2) and the elaborate structure (Chap. 3) of the avian respiratory system and the highly metabolically active lifestyles they pursue. With O_2-sensing and signal transduction machinery that is found in the pulmonary arterial smooth muscle cells of the pre-capillary vessels and possibly modulated by the endothelium, it has been suggested that reactive O_2 species, redox couples, and adenosine monophosphate-activated kinases and to an extent calcium sensitization and intracellular calcium stores may serve as mediators of hypoxic pulmonary vasoconstriction (Archer et al. 2000; Moudgil et al. 2006; Sommer et al. 2008, 2016). In eukaryotes, the hypoxia-induced factor (HIF)-based O_2-sensing mechanism developed ~800 million years ago, possibly in the ancestral metazoan ancestor (Martin et al. 2021a, b): happening at a time when the atmospheric O_2 level was about the level of the Pasteur Point or ~0.3% by volume of present atmospheric level (Rutten 1970), HIFs greatly influenced the evolution of animal life under shifting O_2 concentrations (Hammarlund et al. 2020; Hammarlund 2020).

4.5 Structural Complexity and Functional Efficiency of the Avian Respiratory System

In the avian lung, except for the very small amount that is found in the very small laterodorsal secondary bronchi, practically all the exchange tissue exists in the parabronchi (tertiary bronchi) (Plates 3.14b, d, e, f, 3.15, 3.17d, i–l and 3.18e–k). At the higher (gross = macroscopic) structural level of the avian lung, the parabronchi are the respiratory units while the air capillaries are the functional units at the lower (microscopic) one. In species of birds such as the Domestic Fowl (*Gallus gallus* variant *domesticus*) (Plates 3.15a, b and 3.18e–g) and the Andean Goose (*Chloephaga malanoptera*) (Plates 3.15c–f and 3.18h–k), where external boundaries of the parabronchi are marked by interparabronchial septa (Maina et al. 1982a, b; Kandyel et al. 2021), the parabronchi display a rather hexagonal shape (Plate 3.18a–d). Such shapes are a common feature of various natural units, objects, things, and creatures. Mathematically, it has been shown that only regular hexagons, squares, and equilateral triangles can be placed together, i.e., tessellated, to leave no unexploited space. Furthermore, when their areas are equal, of the three geometric figures, the hexagon has the smallest perimeter. In animal and plant life, the honeycomb cells of beeswax (Plate 3.18l), the patterns of turtle shells, and the conducting (vascular) components of plants, i.e., xylem and phloem (Sperry et al. 2006; Venturas et al. 2017; https://www.coursehero.com/file/121887836/91-Xylem-Transportpdf/—accessed on 04-03-22) are probably the best-known forms. Hexagonal shapes are also displayed in the famous interlocking columns of the volcanic rocks that form the Giant's Causeway in Ireland (https://rockdoctorcronkite.com/giants-causeway—accessed on 04-03-22) and in shapes of bonds of many organic molecules, e.g. benzene, and in many crystalline structures such as graphite, quartz, dolomite, and emerald. Hexagonal shapes appear to have evolved because they intermesh perfectly, therefore optimize the number of units and amount of substance that can be "packaged" (contained) in a given area or volume. Furthermore, the hexagonal shape confers the strongest form that can tolerate large external and internal forces and tensions (https://buffaloarchitecture.org/wp-content/uploads/2015/08/AEPresentation-ArchitecturalElements.pdf—

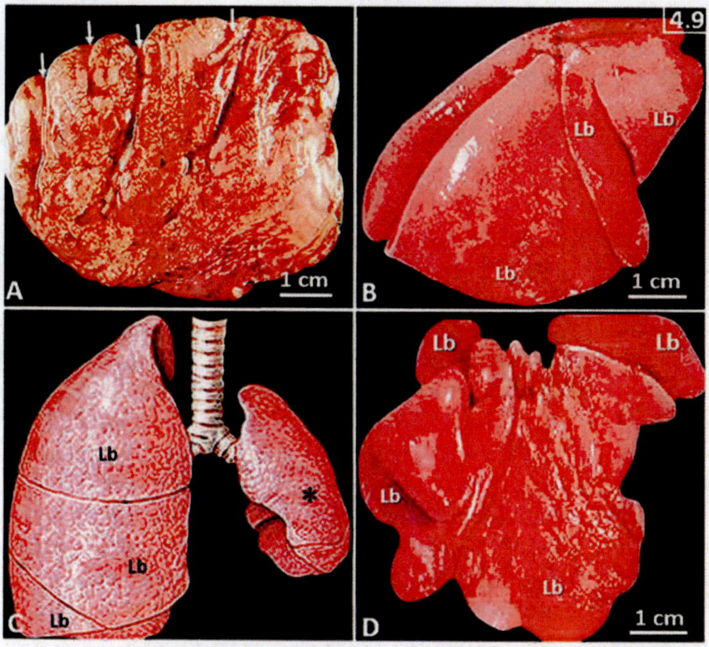

Plate 4.9 Strength of the avian lung from studies of the lung of the Domestic Fowl (*Gallus gallus* variant *domesticus*). (**a**) The avian lung well-retains its shape and size when it is removed from its costovertebral attachments while the mammalian one (shown inflated) (**b**) collapses substantially after pneumothorax (asterisk, *) (**c**) and when removed from the thorax (**d**). The mammalian lung is lobulated while the avian one is not. Lb, lobes (**b, d**)

accessed on 05-11-2021). Showing that structural components with hexagonal shapes require less material to make, compared with other forms that fit (mesh) in a similar manner, the lines that form a regular hexagon (or the boundaries that separate the individual units) are geometrically the shortest possible. It has been suggested that bees use less wax and do less work to construct cells in a unit wax space (volume) (https://mathinnature.blogspot.com/2009/02/hexagons-in-nature.html—accessed on 24-07-22). The hexagonal shape of the parabronchi of the lungs of some species of birds, where the units are marked by conspicuous interparabronchial septa (Maina et al. 1982a, b; Maina 2005; Tarnai et al. 1993) (Plates 3.15a–d and 3.18e–k) should allow more parabronchi and consequently more exchange tissue to be packed in the relatively small rigid avian lung, providing large pulmonary capillary blood volume and vast respiratory surface area (Dubach 1981; Duncker and Güntert 1985a, b; Maina 1989a, 2005; Maina et al. 1989a). As well, the hexagonal shape may explain the strength of the parabronchi in particular (Macklem et al. 1979) and the avian lung in general (Maina 2005) (Plate 4.9a). A non-compliant avian lung allowed extreme compartmentalization (subdivision) of the exchange tissue into minuscular terminal respiratory units, the air capillaries (Plates 3.19 and 3.20): it conferred extensive respiratory surface area which granted large diffusing capacity of the lung for O_2 (Maina 1989a, 2005; Maina et al. 1989a). In human engineering and architecture, the so-called geodesic shapes comprise geometric units that in many cases include hexagonal shapes (Craven 2021): such designs are light in weight and can be easily assembled (Miura 1990; Othilia 2013; Gythiel et al. 2020). Unlike the mammalian lung (Plate 4.9b) that noticeably collapses after pneumothorax (Plate 4.9c) or on its removal from the thoracic cavity (Plate 4.9d), the avian lung well-maintains its shape and size after it is "extracted" from its costovertebral attachments (Plate 4.9a).

In the vertebrate gas exchangers, the arrangement of the airways and the blood vessels greatly determine the delivery and the exposure of blood to the inspired external respiratory fluid medium (i.e., water or air) and hence the exchange between O_2 and CO_2 across the blood-gas barrier (Maina 1998, 2011). In the avian lung, at the parabronchial level, air convectively flows through the parabronchial lumen and as well outwards into the atria and infundibulae (Plates 3.14, 3.15a–j, 3.16, 3.17 and 3.18e–k): in the air capillaries of the exchange tissue, it moves by diffusion (Plates 3.19 and 3.20a, b). Deoxygenated (venous) blood flows centripetally (i.e., inwards) through the intraparabronchial arteries (Plates 3.15a–h, i, 3.18e, f, g, i, j and 3.23) into arterioles and blood capillaries (Plate 3.21). The arrangement between the flows of the deoxygenated (venous) blood (from the periphery of a parabronchus) to that of the flow of air in the parabronchial lumen are essentially orthogonal or perpendicular: they comprise a crosscurrent gas exchange system [Plates 4.10a (i, ii), 4.11a, b (i), c and 4.12a–e (i)]. During the early days of the study of the physiology of the avian respiratory system, from the well-known efficiency of human-engineered countercurrent exchange designs (Bosworth and Groden 1964; Dusenbery 1973; Aota et al. 2007a, b; Pushpavanam and Malengier 2012) and in the naturally evolved exchange systems such as fish gills (Hills and Hughes 1970; Scheid et al. 1986; Malte and Lomholt 1998) and the urine concentrating tubules of the vertebrate kidneys (Gottschalk and Myle 1958, 1959; Schmidt-Nielsen 1979, 1981), without empirical evidence, it was claimed that a countercurrent gas exchange system explained the incomparable gas exchange efficiency of the avian lung respiratory system (Schmidt-Nielsen et al. 1969; Bretz and Schmidt-Nielsen 1971). However, a brilliant experiment performed by Scheid and Piiper (1972) showed that that was not the case: after reversing the direction of airflow in the lung of a duck, during steady state breathing, pulmonary gas flow and CO_2 and O_2 concentrations and their partial pressures in air and arterial and mixed venous blood remained unchanged. Such a gas exchange profile cannot be explained by the existence of a countercurrent system, where the experiment manipulation would have created a concurrent system in which the respiratory fluid media (air and blood) flowed in the same direction and little if any, gas exchange would have occurred. In the avian lung, the existence of a crosscurrent system has now been well-morphologically corroborated (Abdalla and King 1975; West et al. 1977; Maina 1988; Abdalla 1989; Woodward and Maina 2005, 2008; Maina 2005; Maina and Woodward 2009).

Because in the avian lung at the parabronchial level deoxygenated blood flows "inwards," i.e., centripetally, into the gas exchange tissue while the air travels "outwards," i.e., centrifugally, convectively through the atria and infundibula and

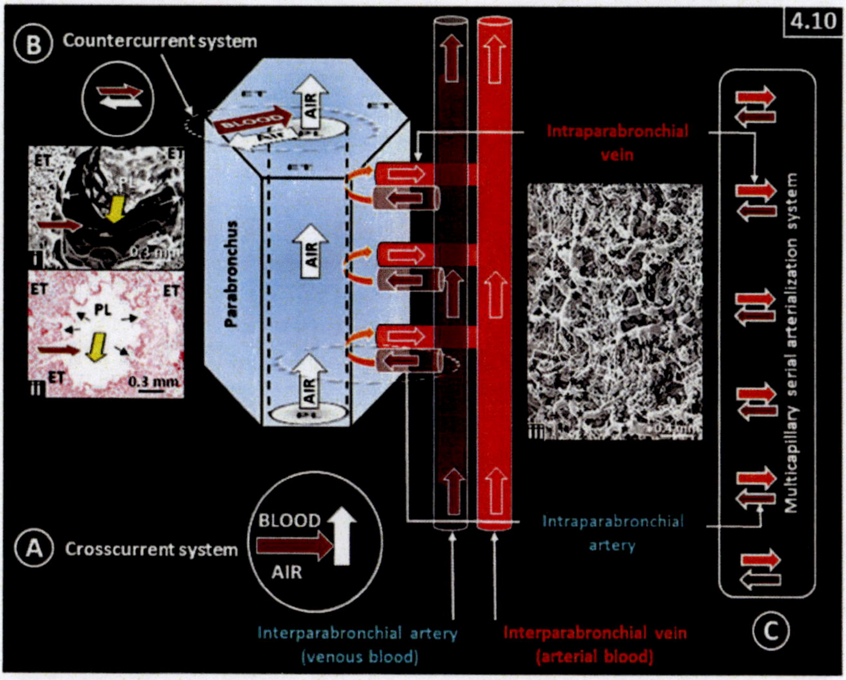

Plate 4.10 Schematic illustration showing the crosscurrent, the countercurrent, and the multicapillary serial arterialization systems (designs) in the avian lung. (**a**) The crosscurrent system comprises the disposition between the direction of the airflow in the parabronchial lumen (large white arrow) relative to that of venous blood from the periphery of a parabronchus (large brown arrow). (**b**) The countercurrent system consists of the arrangement between the inwards flow of venous blood in the exchange tissue (large brown arrow) relative to the outwards flow of air from the parabronchial lumen (large white arrow). (**c**) The multicapillary serial arterialization system consists of parabronchial blood supply to the exchange tissue by interparabronchial arteries and their branches and its interaction with air flowing outwards from the parabronchial lumen all along the lengths of the parabronchi. ET, exchange tissue; PL, pararonchial lumen (PL); arrows (←), atria; an asterisk (*), interparabronchial blood vessel; i, ii, iii, preparations showing arrangements of the parabronchi and blood vessels. Double latex cast of the lung of the Domestic Fowl (*Gallus gallus* variant *domesticus*) showing the locations of the interparabronchial arteries (asterisks,*) that give rise to intraparabronchial arteries (stars, ▯) which deliver blood to the exchange tissue of the parabronchi (open arrows). In succession, arterialization of blood (in the blood capillaries) occurs along the lengths of the parabronchi, a structural arrangement that comprises a multicapillary serial arterialization system of the avian lung (area enclosed by a dashed rectangle (**c**, iii)

diffusion in the air capillaries, the arrangement between the directions of the flows of the gas exchange media (air and blood) are considered to comprise a countercurrent gas exchange system (Plates 4.10b and 4.11a, b (ii), c). The role of such a countercurrent system in the general gas exchange efficiency of the avian lung is, however, uncertain. Piiper and Scheid (1973) dismissed the arrangement as a "counter-current-like mechanism" while Scheid (1979) set it aside as "an auxiliary mechanism superposed on, and independent of the basic cross-current arrangement between

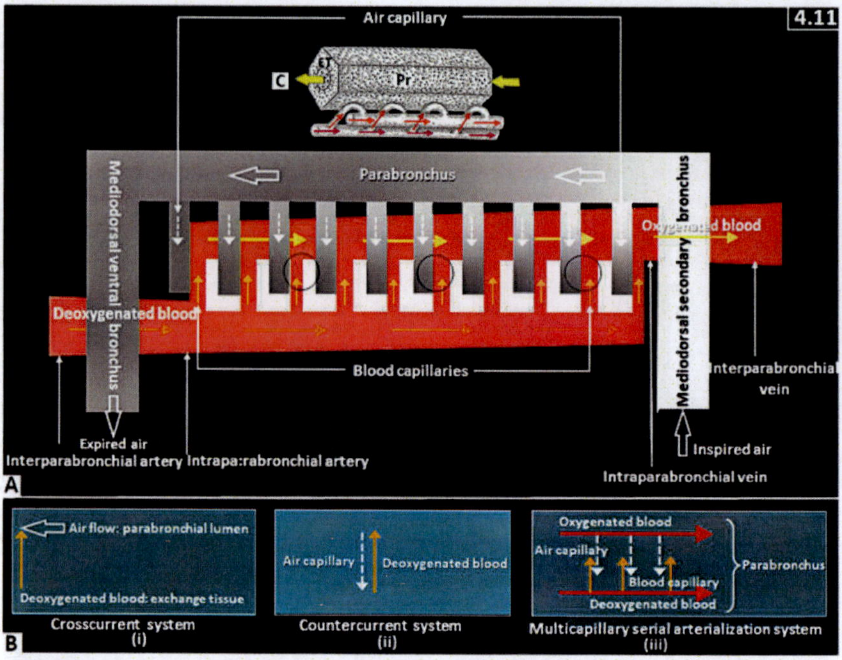

Plate 4.11 Schematic diagram showing the crosscurrent, countercurrent, and multicapillary serial arterialization gas exchange systems in the avian lung. (**a**, **b**) (i): the crosscurrent system comprises the arrangement between the direction of airflow along the parabronchial lumen and that of inwards (centrifugal) flow of venous (deoxygenated) blood from the periphery of a parabronchus across the exchange tissue. (**a**, **b**) (ii): The countercurrent system comprises the arrangement between the air and the capillaries, where in the exchange tissue oxygen diffuses outwards, i.e., centrifugally, and venous blood flows inwards, i.e., centripetally (parabrobronchial lumen). (**a**, **b**) (iii): The multicapillary serial arterialization system consists of a sequential supply of deoxygenated blood to the parabronchial gas exchange tissue by intraparabronchial arteries and exchange of oxygen from the air capillaries arising from the air spaces of the parabronchial lumen, i.e., atria and infundibula. (**c**) Schematic diagram showing the multicapillary serial arterialization of a parabronchus (Pr). The length of a parabronchus is supplied with deoxygenated blood (brown arrows) which is arterialized and pooled (red arrows). ET, exchange tissue; yellow arrows (**a**), air flowing through the parabronchial lumen. Red arrows, oxygenated blood in an interparabronchial vein

the capillary blood flow and the bulk parabronchial gas flow." The main reasons that should make the countercurrent-like design in the avian lung inconsequential in gas exchange are the following: (a) convective flow of blood occurs in the blood capillaries while diffusion of O_2 occurs in the air capillaries and; (b) because the blood and the air capillaries entwine very closely in three dimensions (Plates 3.19, 3.21a–j and 3.24), at any one point/moment, gas exchange units contact over very short distances that constantly changes in direction (Plate 3.24).

Blood and air are delivered at about the same time to all parts of the parabronchial exchange tissue (Plates 4.10c, iii, 4.11a, c and 4.12a–d): respiratory media with

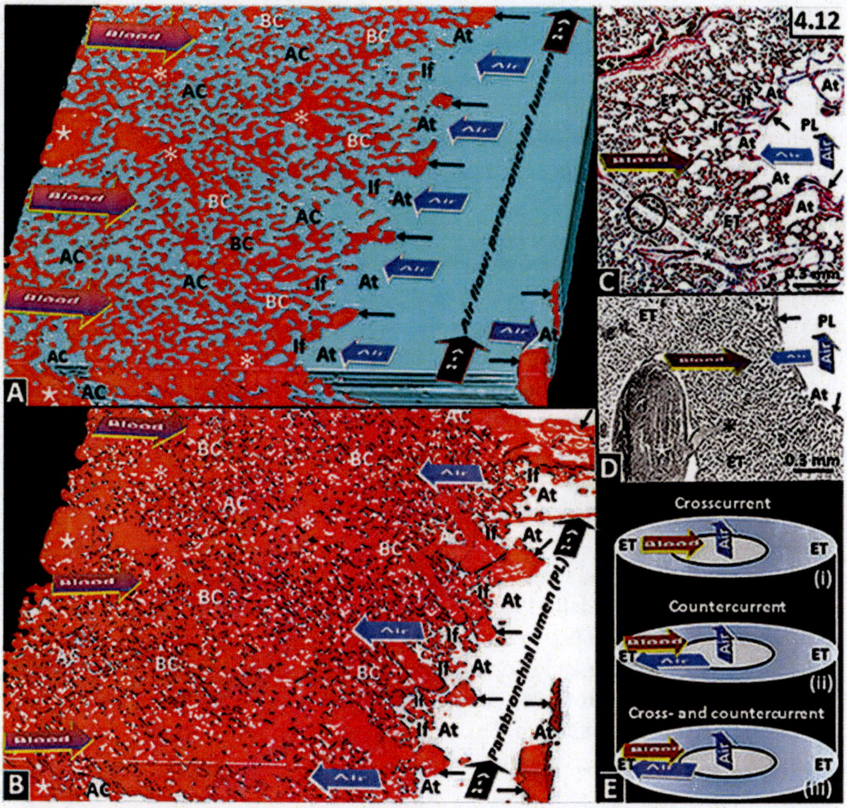

Plate 4.12 Explanations of crosscurrent, countercurrent, and multicapillary serial arterialization systems of the avian lung shown on three-dimensional computer reconstruction preparations (**a, b**), histological tissue preparations (**c, d**) and a schematic diagram of a parabronchus (**e**). The crosscurrent system comprises the perpendicular (orthogonal) disposition between the flow of venous (deoxygenated) blood) from the periphery of the parabronchus through the intraparabronchial arteries (brown arrows—denoted blood)] and that of air in the exchange tissue flowing from the parabronchial lumen (blue arrow—marked air). The crosscurrent system comprises the arrangement of air flowing outwards into the exchange tissue from the parabronchial lumen (PL) (blue arrows) and that of venous blood flowing from the periphery of the parabronchus into the exchange tissue towards the parabronchial lumen (brown arrows). The multicapillary serial arterialization system comprises the sequential supply of venous (deoxygenated) blood by the intraparanronchial arteries (brown arrows—marked blood) along the length of the parabronchus and the delivery of air into the exchange tissue from the parabronchial lumen (blue arrows—marked air). Stars (☆), interparabronchial arteries; asterisks (*), intraparabronchial arteries; PL, parabronchial lumen; BC, blood capillary; AC, air capillary; At, atria; If, infundibula; arrows ←, interatrial muscle

changing partial pressures of O_2 equilibrate along the lengths of the parabronchi. At the entrance (proximal end) of a paleopulmonic parabronchus, i.e., the part connecting to the mediodorsal secondary bronchus, blood is exposed to air with relatively high PO_2 and low CO_2 while at the opposite end, i.e., the distal end or the part connecting to the medioventral secondary bronchus, the reverse is the case (Plates 4.10, 4.11a and 4.12a, b). On the whole, for the avian lung, the quantity of O_2 in the oxygenated (arterial) blood returned to the heart (from the lung by the pulmonary vein) derives from an additive process of O_2 which is exchanged at the infinitely many points that air and blood capillaries contact [Plates 4.10c, 4.11a, b (iii), c and 4.12a, b]: the arrangement comprises the "multicapillary serial arterialization system" (Maina 2005) (Plate 4.10c, iii). The arrangement extends the time and distance across which deoxygenated blood is exposed to inspired air in the exchange tissue. Together with the present unidirectional and continuous ventilatory processes of the paleopulmonic parabronchi, the multicapillary serial arterialization system greatly enhances the oxygenation of blood in the avian lung (Scheid et al. 1972; Fedde 1980) and explain the exceptional functional efficiency of the gas exchanger. In birds, under hypoxic conditions and strenuous exercise, the $PaCO_2$ is lower than that in the end-expired air ($PeCO_2$) and the opposite gas exchange properties may occur for O_2, i.e., the PaO_2 surpasses the PEO_2 (Scheid 1979; Powell and Scheid 1989). Interestingly, arteriovenous anastomoses, that could allow shunting of blood once the partial pressures of respiratory gases in blood have equilibrated with those in the air flowing through the parabronchus do not appear to exist between the terminal branches of the pulmonary artery and vein (Abdalla and King 1975, 1976a, b; Abdalla 1989: West et al. 1977; James et al. 1976; Holle et al. 1978; Parry and Yates 1979). In the lung of the adult Domestic Fowl (*Gallus gallus* variant *domesticus*) King and Molony (1971) suggested that the total length of the parabronchi is ~300 m. However, measurements made on parabronchi isolated from a latex rubber cast preparation of the chicken lung (Plate 3.6h) gave a value of 32 m (Maina, unpublished observation).

4.6 Gas Exchange: The Inventory—Respiratory Mechanisms and Processes

In water-breathing vertebrates, gas exchange occurs across the water-blood (tissue) barrier of the gills (Plate 2.1a) while in the air-breathing ones, it happens through the blood-gas (air) tissue barrier of the lungs (Plates 2.1b–j and 3.25a–c). To generate and maintain a partial pressure gradient of O_2 (PO_2) across the tissue barrier, the gas exchangers are continuously ventilated with water or air and perfused with blood. Under changing internal and external pressures that respectively derive from the pumping activity of the heart and the ventilatory activities of respiratory muscles, although innately passive, the mammalian lung is a visibly "active" organ: in contrast, the avian lung is a rigid and fixed organ. For the human being, as much as 12,000 L of air and 6000 of blood pass through the lung per day (Burri 1985). Arranged in parallel relative to the systemic circulation, in mammals and birds, the pulmonary circulation receives the entire cardiac output from the right side of the heart and it is a low pressure-low resistance and high compliance system (Warrell et al. 1972; Weibel 1984; Townsley 2012; Powers and Dhamoon 2022). The lung is the only organ in the body that puts up with (holds) the entire cardiac output and all the systemic blood volume passes through it. Pulmonary arteries transport venous (deoxygenated) blood to arterioles that give rise to a dense blood capillary system which presents a vast respiratory surface area. With the dampening of the pressure wave from the contractions of the heart occurring in the pulmonary blood capillary system (Wienner et al. 1966), of the entire blood volume in the body, at any one instant, ~9% of blood is found in the lung (Dock et al. 1961). The pulmonary blood flow is pulsatile (Milnor 1982; Maarek and Chang 1991) and dampening of the pressure wave largely occurs in the vast blood capillary system (Wienner et al. 1966). The blood-gas barrier tolerates fluctuating hemodynamic pressures and surface tension forces (West 2011; Leiby et al. 2020). For the reason that the lung handles large amounts of generally polluted atmospheric air with injurious pathogenic microorganisms and particulate matter as well as toxic chemical and gaseous agents, the very thin and vast blood-gas barrier is protected by a highly specialized line of motile phagocytic cells called macrophages that are interspersed between surface pneumonocytes (Ginhoux and Jung 2014; Stark and Mueller 2012; Bissonnette et al. 2020; Woo et al. 2021; Martin et al. 2021b). In the avian lung, on the surface of the air capillaries, macrophages are lacking. they exist on the atria and infundibula (Nganpiep and Maina 2002) (Plate 3.17f). Lack of macrophages and type-II pneumocytes from the surface of the air capillaries may be explained by the necessity of development of extremely thin blood-gas barrier. Furthermore, the cells may be located upstream of the air capillaries to allow them (air capillaries) to be very small (Plate 3.19). In the lungs of air-breathing vertebrates, the type-II cells or granular pneumocytes secrete surfactant that covers the respiratory surface: it lowers surface tension, provides protection from injurious pathogens and particulate matter, and confers alveolar stability (Daniels and Orgeig 2001, 2003; Daniels et al. 1995, 1998a, b; Alcorn 2017; Mirastschijski et al. 2020).

Located in the coelomic cavity, the avian lung is firmly attached to the ribs and the vertebrae, respectively on the lateral and the dorsal aspects making them practically rigid (Jones et al. 1985; Maina 2005; Makanya et al. 2021) (Plates 2.5i, j, 2.6 and 3.6b, f–h). Morphologists, physiologists, biophysicists, pulmonologists, and evolutionary biologists are among others interested in areas of the development, structure, function, and pathology of lungs (Müller-Redetzky et al. 2015; Makanya 2015; Makanya et al. 2013; Harkema et al. 2018). In vertebrates, two main kinds of gas exchangers, namely, gills and lungs, have evolved. Their structural features and functional properties have largely stemmed from the various biophysical properties of the respiratory fluid media, respectively water and air that are exploited as a source of O_2. In gills, a water-blood (tissue) barrier separates water and blood, materials/substances of comparable specific densities while for lungs the blood-gas (tissue) barrier separates air and blood, materials/substances that are physically a great deal different. For lungs, on account of external support practically lacking or being very low, the integrity of the blood-gas barrier is continuously challenged by operational tensional and compressional forces that among others stem from gravity acting on the weight of the structural tissues, the blood in the lung and the blood pressures

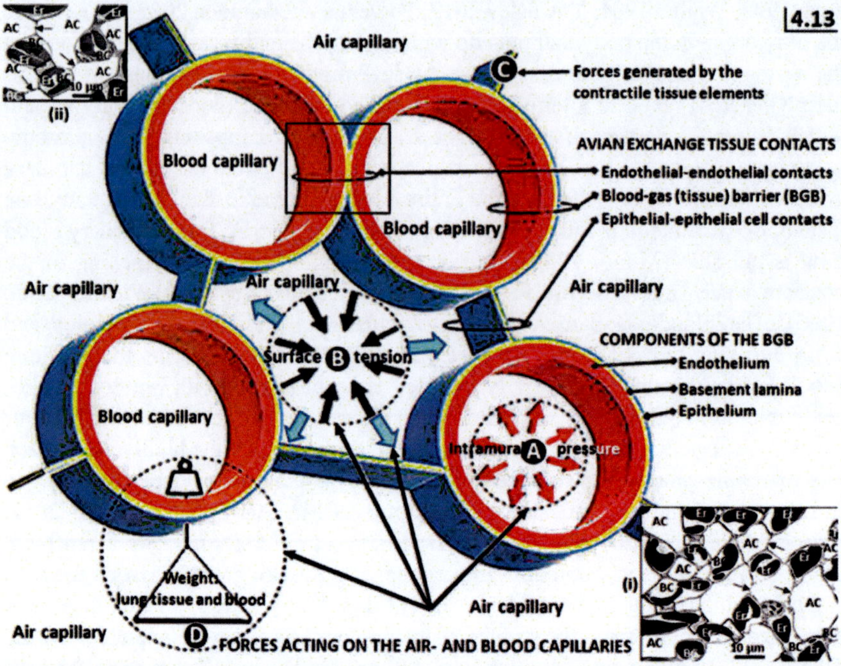

Plate 4.13 Forces acting on air and blood capillaries. (**a**) Transmural pressures in the blood capillaries. (**b**) Surface tension forces exist on the surface of the air capillaries. (**c**) Forces generated by contractile tissue elements such as smooth muscle and elastic tissue fibers. (**d**) Gravitational forces deriving from the weight of the lung tissue and that of the volume of blood contained in the lung which are dependent on body size and posture. Insets i and ii: Exchange tissue of the avian lung that comprises air capillaries (AC) and blood capillaries (BC). Boxed area (□) (main figure and insets i and ii), sites where blood capillaries connect directly; arrows (←), epithelial-epithelial cells connections (epithelial retinaculae), i.e., sites where blood capillaries are interconnected by epithelial retinaculae; Er, erythrocytes

stemming from the beating (pumping) heart (West and Mathews 1978) (Plate 4.13). Whereas the blood vessels of the systemic circulation are secured to proximate structures, the pulmonary blood vessels are practically suspended in the air by a diffuse fibroskeletal tissue network (Weibel 1984, 1989; Ohtani and Nakatani 1994; Toshima et al. 2004; Maina et al. 2010). Conditions such as degree of inflation, perfusion pressure, surface tension, and hydrostatic pressure determine the structure and the organization of the structural components of the pulmonary scaffold (Dunnill 1962; Klingele and Staub 1965; Kuno and Staub 1968; Untersee et al. 1972; Weibel 1984). In vertebrate lungs, the design of an extremely thin and strong blood-gas (tissue) barrier is a unique and highly conserved biological design (Weibel 1973; Meban 1980; Maina and King 1982a; Maina and West 2005; Maina 2011; Leiby et al. 2020). Morphologically, thin tripartite blood-gas barrier design (Weibel 1973; Meban 1980; Maina and King 1982a; Maina and West 2005) (Plates 2.1, 3.25a–c and 4.15) has evolved to optimize transfer of O_2 by diffusion under prevailing partial pressure gradient but must to be sufficiently strong to maintain structural integrity under forces and tensions it is constantly exposed to (West 1985, 2003, 2009; Leiby et al. 2020). Maina and West (2005) characterized the design of the blood-gas barrier as "a bioengineering dilemma" while Leiby et al. (2020) stated that "the pulmonary blood-gas barrier represents a remarkable feat of engineering." To provide a flow rate of 2.3×10^{-5} cm^2 s^{-1}, a process completed in 250 to 500 ms, a flow of 200 ml min^{-1} of O_2 across the blood-gas barrier requires a partial pressure gradient of only 0.057 kPa (Roughton and Forster 1957; Grote 1967). In biology, structures such as the tripartite water-blood and blood-gas barriers, that have remained unchanged for a long evolutionary time, i.e., have been conserved, are termed "Bauplans," "blueprints," or "frozen cores" (Wagner 1989, 1998; Maina 2000a, 2011; Marza et al. 2005; Fisher et al. 2006; Chan et al. 2009; Coronado-Zamora et al. 2019; Scanes et al. 2022). Such structures provide important solutions to particularly vital evolutionary requirements.

The adaptation of air-breathing is envisaged to have arisen in fishes during the Silurian (438–408 million years ago) (Graham 1997; Farmer and Jackson 1998; Icardo et al. 2015), several million years before the start of terrestrialization by the first tetrapods (Panchen 1980; Gardiner 1980; Liem 1988; Little 1990; Gordon and Olson 1994; Roux 2002). Lungs apparently originated from basal osteichthyans (Perry and Sander 2004; Roux 2002; Goujet 2011; Cupello et al. 2017). The organs may have first evolved for air-breathing in lungfishes (Dipnoi) ~400 million years ago (Hughes and Weibel 1976, 1978; Maina and Maloiy 1985; Bemis et al. 1987; Clement and Long 2010; Tatsumi et al. 2016; Cupello et al. 2022). In the evolved vertebrate gas exchangers, water-blood and blood-gas (tissue) barriers of various structural complexities exists in respiratory organs such as fish gills (Hughes and Wright 1970; Hughes and Morgan 1973) (Plates 2.1a and 4.14a), accessory respiratory organs of air-breathing fish (Maina and Maloiy 1986; Maina 2000a; Mbanga et al. 2018) (Plate 4.14b, c), lungs of pneumonate gastropods (Maina 2009a) (Plate 4.14d, e), respiratory (phyostomatous) swim (air) bladder of fish (Maina et al. 1996; Maina 2000c) (Plate 4.14f, g), gills of crabs (Plate 4.14h, i) and in vertebrate lungs (Weibel 1973; Hughes and Weibel 1978; Goniakowska-Witalinska 1978, 1986, 1995; Meban 1980; Maina and King 1982a, b; Maina and Maloiy 1988; Maina 1989a–d) (Plates 2.1b–j and 4.15). In the lungs and gills of invertebrates, generally, the tissue barriers are poorly developed (Taylor and Greenaway 1979; Bubel 1983; Maina 1989b, 1990a; Rodriguez et al. 2019; Lawson et al. 2019; Hou et al. 2021) (Plate 4.14h, i). The low hemodynamic pressures that exist in the open circulations of the low-metabolism invertebrates (Monahan-Earley et al. 2013) may explain the less well-organized morphology of the water-blood and the blood-gas (tissue) barriers of gas exchangers in the taxon. In insects, where open circulation does not generally play a meaningful respiratory function, O_2 is for the main transported by the trachea (Plates 3.1a and 3.2a–f) and directly delivered to the mitochondria (Maina 1989e; Hartung et al. 2004; Kirkton 2007; Greenlee et al. 2009) (Plate 3.2g, h): it explains how and why compared with other invertebrates that have an open circulation, insects have the highest metabolic (aerobic) capacities (Monahan-Earley et al. 2013). In mammals and birds, the evolution of endothermy occasioned

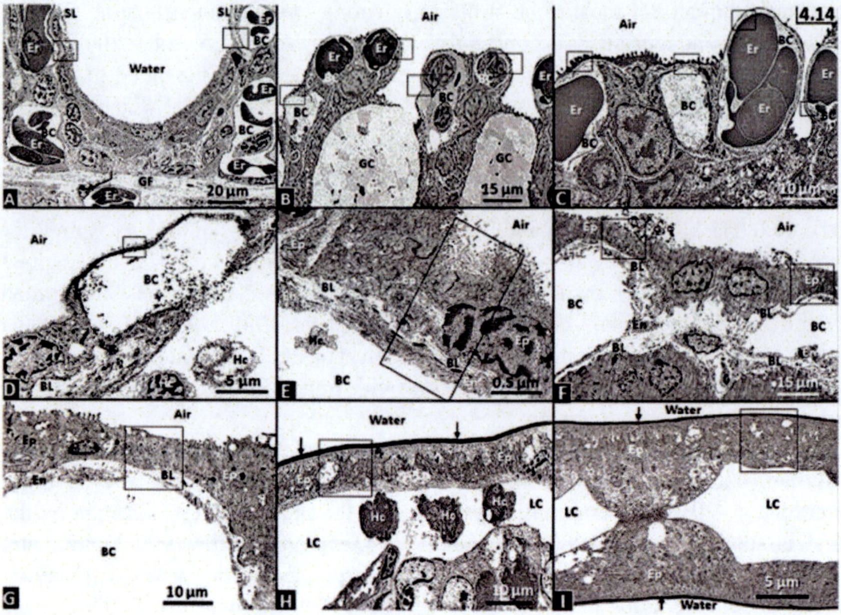

Plate 4.14 Presence and absence of tripartite design of water-blood and blood-gas (tissue) barriers in gas exchangers. (**a**) Fish, *Oreochromis alcalicus*, gills. (**b**, **c**) Accessory respiratory organs of a catfish, *Clarias aethiopicus*. (**d**, **e**) Lung of a pneumonate gastropod, *Trichotoxon copleyi*. (**f**, **g**) Pphysostomatous swimbladder of a fish, *Oreochromis alcalicus*. (**h**, **i**) Gills of a crab, *Potamon niloticus*. Although relatively less well-developed compared with the lungs of air-breathing vertebrates, a basement exists in the water blood barrier of fish gills (**a**), the accessory respiratory organs of fish (**b**, **c**), lungs of slugs (**d**, **e**) and gas (swim) bladders of teleosts fish (**f**, **g**). Basement membranes are, however, lacking in gas exchangers such as gills of crabs (**h**, **i**) that are covered by a cuticular lining (arrows, ↑). SL (**a**), secondary lamella; Er, erythrocyte; GF, gill filament (**a**); boxed area (□), water/air-blood barrier; GC, goblet (secretory) cells (**b**, **c**); BC, blood capillary; Ep, epithelial cell; LC, lymphatic channel (**h**, **i**); HC, hemocyte; En, endothelial cell

the development of exceptionally functionally efficient gas exchangers. For birds, ranging from 39 to 44 °C (McNab 1966; Prinzinger et al. 1991; Clarke and Rothery 2008; Zehnder et al. 2014), body temperatures are in general higher than those of mammals and hence their metabolic rates are largely greater (Butler 1991; Bishop 1997, 1999). Anatomical specializations such as the separation of pulmonary and systemic circulations allowed the pulmonary capillary blood pressure to be low so as to avert stress (structural) failure mainly of the blood-gas barrier (West 2011).

Structurally, the blood-gas (tissue) barrier of the lung basically comprises a thin outer epithelial cell that is covered by a thin aqueous layer that overlies a surfactant lining (Bastacky et al. 1995; Knudsen and Ochs 2018), an intermediate or extracellular matrix layer (the basement membrane) and the innermost layer, the endothelium that forms part of a blood capillary wall (Weibel 1973; Meban 1980; Maina and West 2005) (Plates 2.1 and 4.15). In the exchange tissue of the avian lung, the blood capillaries are literally suspended by the epithelial-epithelial cells connections

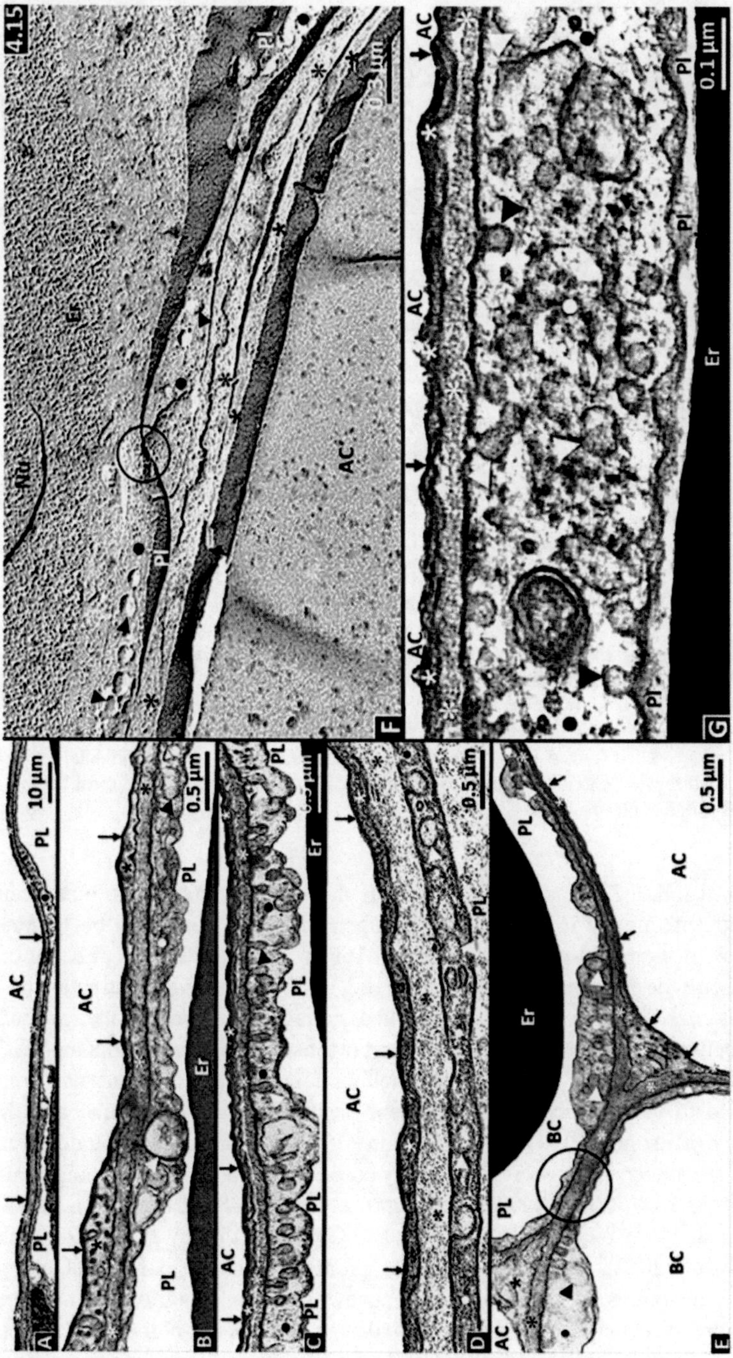

Plate 4.15 Structure of the blood-gas barrier of the lung of the Domestic Fowl (*Gallus gallus* variant *domesticus*) that comprises a thin surfactant film that covers the surface of the air capillaries (AC) (arrows, ↓), thin epithelial cell (star, ▯), a basement membrane (asterisk, *) and an endothelial cell (dot, •). PL, plasma layer; arrowheads (▶), micropinocytotic vesicles; BC, blood capillary. (**a–e, g**) are transmission electron micrographs, and (**f**) is a freeze-fracture preparation

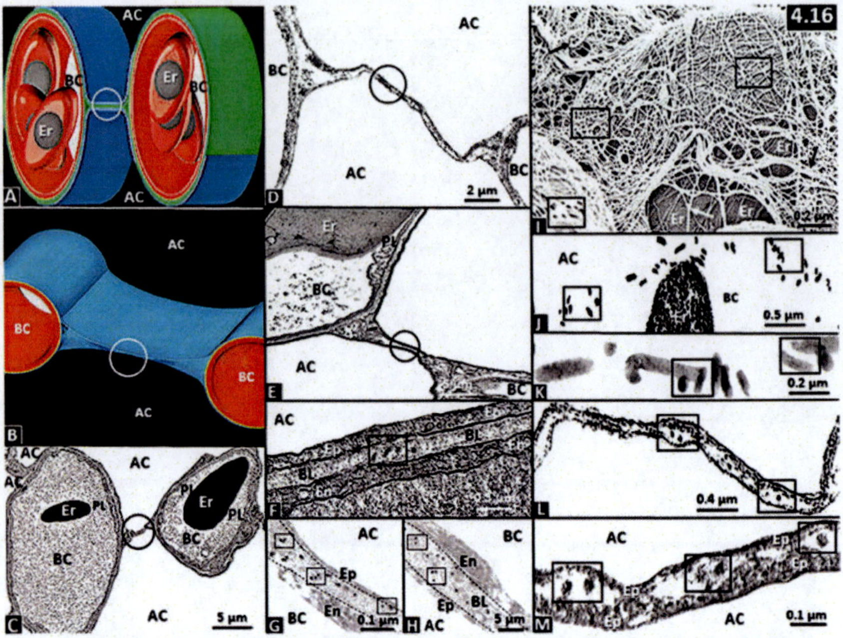

Plate 4.16 The arrangement of the blood capillaries (BC) and air capillaries (AC) in the exchange tissue of the lung of the Domestic Fowl (*Gallus gallus* variant *domesticus*) avian lung. **(a–e)** The blood capillaries are literally suspended in the periparabronchial exchange tissue by epithelial-epithelial connections (epithelial retinacula) (Circles, O) that separate air capillaries. **(f–m)** Distribution of collagen fibers (boxed areas, □) in the blood-gas barrier **(f–k)** and the epithelial retinacula **(l, m)**. Er, erythrocytes; Circle (O), epithelial retinacula; Ep, epithelium; BL, basement lamina; En endothelial cell; PL, plasma layer

(Plates 3.21g, h, 3.22a–c and 4.16a–e) while in relatively rarer cases, blood capillaries connect directly (Plate 4.15d). Collagen fibers are found in the basement membrane of the blood-gas barrier (Plate 4.16f–j) and between the epithelial cells that comprise the epithelial lamina (Plate 4.16d, k–m). Regarding the structure of the blood-gas barrier of the avian lung, marked sporadic attenuations, that are interspersed between relatively thicker parts, exist (Meban 1980; Maina and King 1982a) (Plates 2.1e, j, 4.15a–d, f and 4.16c, e). The functional designs of tissues and organs correspond to factors such as environment occupied, lifestyle pursued and their phylogeny (Milsom 1991; Gans 1991; Maina 1998). To optimize respiratory function, gas exchangers have evolved rather by cost-benefit trade-offs and compromises (West 1985, 2003, 2004, 2009; Maina 2002a, b, d, 1998, 2011, 2014, 2017a–c; Weibel et al. 1998; Weibel 2000; Jürgens and Gros 2002; Leiby et al. 2020; Husak and Lailvaux 2022). With some aspects promoting thinness (for efficient gas exchange) and others strength (for maintenance of structural integrity), counteracting selective pressures determined the tripartite design of the water-blood and the blood-gas barrier. Gas exchange passes through tissue barriers and compartments (spaces)

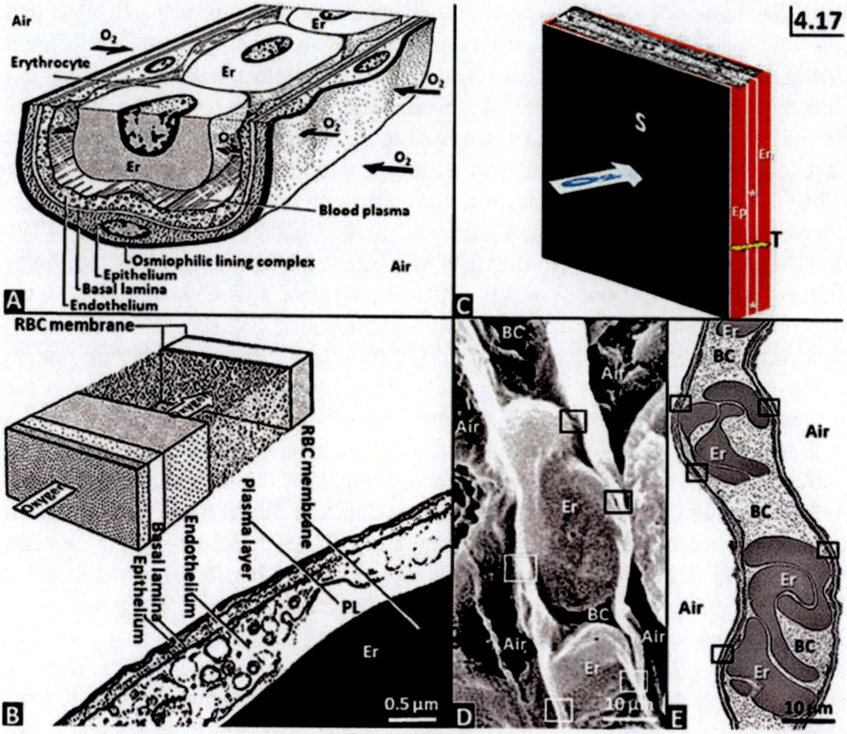

Plate 4.17 Schematic illustrations showing the diffusion pathway for oxygen molecules at the blood capillary level (**a**) and the blood-gas (tissue) barrier (**b**, **c**). Er, erythrocytes; PL, plasma layer: in (**c**), a blood-gas barrier is superimposed on the edge of the schematic diagram. S, respiratory surface area; Ep, epithelial cell; En, endothelial cell; star (⊕), basement membrane; T, thickness of the blood-gas (tissue) barrier. The diffusing capacity of the blood-gas (tissue) barrier correlates directly with surface area (S) and inversely with thickness. (**d**, **e**) Views of blood capillaries (BC) of the mammalian lung containing erythrocytes (Er) that are separated from the air by a blood-gas (tissue barrier (squares, □). (**d**) is a scanning electron micrograph while (**e**) is a transmission electron micrograph

(Plate 4.17) which occurs by physical diffusion (Krogh 1910): the diffusion capacity of a tissue barrier for O_2 (DtO_2) correlates inversely with its thickness and directly with surface area (Plate 4.17a–c).

4.7 Strength of the Avian Lung and the Blood-Gas Barrier

The vertebrate pulmonary blood-gas barriers are built adequately strong to bear changing tensions to which they are normally subjected (Maina 2007a, b; West and Mathieu-Costello 1994, 1995, 1998; West et al. 1997; West et al. 1991, 1993, 2006; West 2003). In the avian lung, lack of a distinctive basement membrane at the epithelial lamina, i.e., sites where the air capillaries lie next to each other and interconnect the blood capillaries (Jimoh and Maina 2013; Maina and Jimoh 2013a, b) (Plate 4.16d, e, j, k) may be explained by the fact that such parts are not directly subjected to intramural blood pressures. Compared with the blood-gas barrier, the epithelial lamina does not need to be particularly strong. The presence of a common basement membrane in areas where blood capillaries contact directly (Plate 4.15e) supports this conjecture. The epithelium that lines the air capillaries is covered by a surfactant lining (Maina and King 1982a) (Plates 3.25a and Plate 4.15b–e, g): a surfactant is envisaged to have existed before the evolution of lungs and swim-bladders (Daniels et al. 1998a, b; Daniels and Orgeig 2003) in the actinopterygian and sarcopterygian as well in the teleost fish (Daniels et al. 2004). In a comparative study, Kunchala et al. (2023) determined that surfactant protein (SP)-A2 originated in an early tetrapod before amphibians separated from the mainstream animal taxon: it [(SP)-A2] progressively lost the collagen domain.

Based on their structure, the epithelial and the endothelial cells of the blood-gas barrier cannot account for the considerable strength of the separation (West et al. 1991; West and Mathieu-Costello 1999; Maina and West 2005): the basement membrane, which contains collagen fibers (Plates 3.22i–o and 4.16f, g) is the main supporting (stress-bearing) part of the blood-gas (tissue) barrier (Jimoh and Maina 2013). The blood-gas (tissue) barriers of the vertebrate gas exchangers serve a dual role: they have to be sufficiently thin to allow the gas transfer by passive diffusion (Krogh 1908, 1910; Krogh and Krogh 1909a, b; Morrell 2015) and strong enough to tolerate stresses and strains it suffers during its function, especially from changes in intramural blood pressures (Manohar et al. 1993; Erickson and Lowe 1994; Hopkins et al. 1997; Hachenberg and Rettig 1998; Maina and Jimoh 2013a, b; Maina and West 2005). Healthy trained horses generally suffer hypoxemia and hypercapnia in the course of strenuous exercise (Franklin et al. 2012; Van Erck-Westergren et al. 2013). For the human lung, the hydrostatic pressures during exercise may exceed 30 mmHg (4 kPa) while in animals such as thoroughbred horses, the pulmonary capillary blood pressure rises to ~100 mmHg (13.3 kPa) (West et al. 1991): the pressures cause stress in the blood-gas barrier of $5–10 \times 10^4$ N m^{-2} (50–100 kPa), resulting in tearing (structural failure) of collagen fibers (West and Mathieu-Costello 1995). The cardiovascular and muscular systems of horses can consume more O_2 than the lungs are capable of acquiring, creating an O_2 deficit (Poole and Erickson 2011). In rabbit, dog, and horse lungs, the pressures that cause stress failure are ~50 cmH_2O (6.7 kPa), 90 cmH_2O (12 kPa), and 130 cmH_2O (17.3 kPa), respectively (Birks et al. 1994, 1997). Metadata analysis by Jones (1998) showed that over the past 70 to 150 years and especially in the last two decades, the winning times of the most elite British and American thoroughbred horses have about leveled out: it suggests possible structural and/or functional optimization of the equine respiratory system, especially at blood-gas barrier level. During extreme exercise, from structural failure of the blood gas barrier, more than 40 to 80% of thoroughbred horses suffer pulmonary hemorrhage (West et al. 1993; Epp et al. 2006). In anethetized rabbits, an experimental increase of transmural pulmonary capillary pressure to 40 mmHg (5.3 kPa) caused structural injuries of the blood-gas barrier (West et al. 1991; Tsukimoto et al. 1991).

The strength and the structural failure of the blood-gas barrier of the mammalian lung have been well-investigated (West 1985; West et al. 1991, 1993; West and Mathieu-Costello 1992a, 1994, 1998, 1999). Diseases and pathologies where the pulmonary capillary pressure is abnormally high, e.g., in neurogenic pulmonary edema, high altitude pulmonary edema, exercise-induced pulmonary hemorrhage (e.g., in racehorses), and in some cases of adult respiratory distress syndrome as well as conditions such as the Goodpasture Syndrome (GPS) or anti-glomerular basement membrane disease, which with a prevalence rate of 1.6 cases per million per year (Canney et al. 2016), is a rare autoimmune disease in which antibodies damage the basement membrane of lungs and kidneys resulting in hemorrhage from the lungs, glomerulonephritis and ultimately renal failure (Wilson and Dixon 1973; Goodpasture 2009; Greco et al. 2015; Thibaud et al. 2019). Through attack of the alpha-3 subunit of type-IV collagen by autoantibodies, the blood-gas barrier weakens and fails, causing bleeding. As well, hemorrhage from structural failure of the blood-gas barrier may occur from overinflation of the lung that may ensue from continued mechanical ventilation under high-inflation pressures during hospitalization (Pascoe et al. 1981; Sweeney et al. 1984; Fu et al. 1985; West and Mathieu-Costello 1992b–d, 1995; Dreyfuss and Saumon 1997; Hachenberg and Rettig 1998; West 2003, 2013; Agrawal et al. 2016; Bai et al. 2010; Swenson and Bärtsch 2012).

The strength of the blood-gas barrier may be attributed to the very high tensile strength of the type-IV collagen that exists in the basement membrane of the blood-gas barrier (Fisher and Wakely 1976; Kern et al. 1993; West and Mathieu-Costello 1999; West 2011; Maina and West 2005; Maina and Jimoh 2013a, b; Jimoh and Maina 2013; Loscertales et al. 2016). In engineering and materials science (Plate 4.18a, c) and in the natural/biological sciences (Plate 4.18b), structural (fatigue) failure occurs after cracks/breaks or tears appear and reaches a critical level (Timoshenko 1953; Cottrell 1981; Hull and Bacon 2001; Yong et al. 2016) (Plate 4.18c). For the avian lung, the structural failure of the blood-gas barrier (Plate 4.18d–f) and the epithelial retinacula (Plate 4.18g–i) exist in the form of cellular and tissue damages/injuries (West 1985; Jimoh 2012, Jimoh and Maina 2013, Maina and Jimoh 2013a, b). The pulmonary blood-gas barrier maintains its extreme thinness and profound strength through constant structural remodeling (West and Mathieu-Costello 1999; Makanya et al. 2013). In elite human athletes, physiological exertions such as those that occur during extreme exercise affect the structural properties of the blood-gas barrier (Hopkins et al. 1997; Hanel et al. 2003) but the integrity of the blood-gas barrier is altered only at extreme levels of exercise (Hopkins et al. 1998).

The firm attachment of the avian lungs to the vertebrae and the ribs on their dorsolateral aspects (Plates 1.1a, 1.6i, 3.6a–d, f–h, 3.7a–c, d, h and 3.8) which rendered the organ practically rigid (Jones et al. 1985; Maina 2005; Makanya et al. 2021) importantly determined the structure and function of of the avian lung: extremely small air capillaries could only form by subdivision (compartmentalization) of the exchange tissue in a rigid (fixed) i.e., noncompliant lung (Plates 3.19,

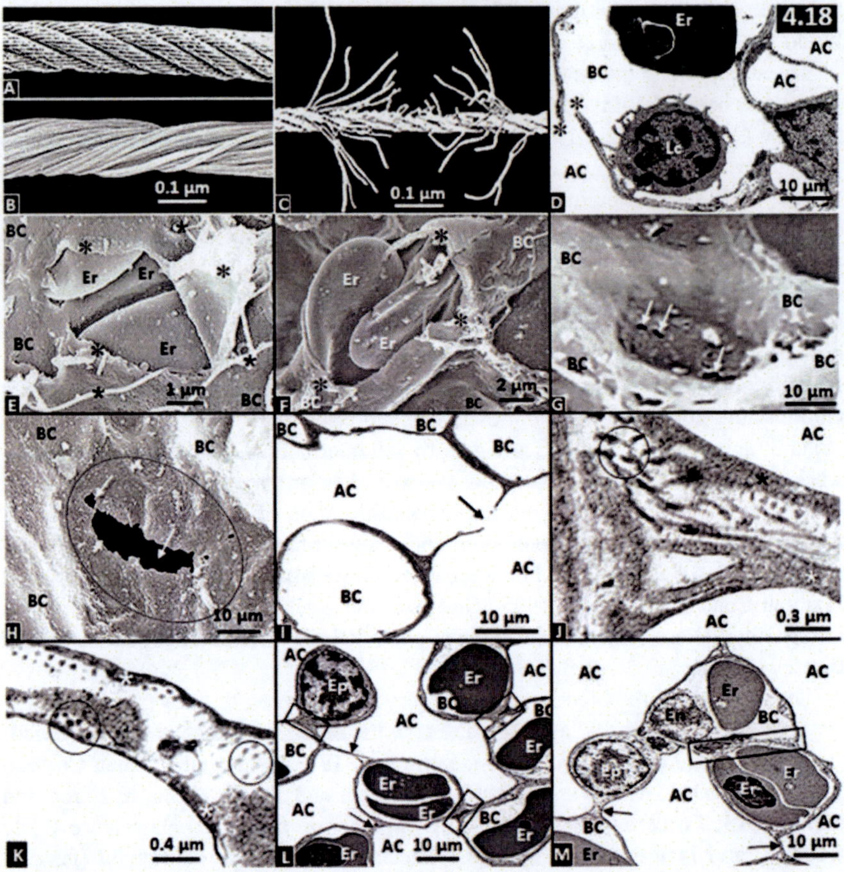

Plate 4.18 The similarity of the structural failure of a steel wire cable (**a**, **c**) with that of a compound collagen fiber (**b**). In engineering and materials science as well as for natural/biological structures fatigue failure occurs at extreme tensions, stresses, and strains when cracks, breaks, or tears result in breaking. In the lung of the Domestic Fowl (*Gallus gallus* variant *domesticus*), (**d–f**) show structural breaks of the blood-gas (tissue) barrier while (**g–i**) show the epithelial-epithelial cell connections (epithelial retinaculae) induced by exercise on a treadmill and by perfusion of lungs at various hydrostatic pressures: (**d**, **i**) are transmission electron micrographs and (**e**, **h**) are scanning electron micrographs. (**j**, **k**) To strengthen the blood capillaries of the exchange tissue of the avian lung, collagen fibers of the epithelial-epithelial cells connections are concentrated at the points where the structures connect to the blood capillaries (**j**, **k**). Circle (O) (**j**, **k**), collagen fibers; Er, erythrocytes; Lc, leukocyte; BC, blood capillary; AC, air capillary; an asterisk (*) (**h–j**), site of blood-gas barrier failure (rupture); star (▯) (**e**), collagen fiber; arrows (↓) (**g–i**), failure of the epithelial-epithelial connections

3.20 and 3.24). On account of their very small sizes (Maina and Nathaniel 2001; Maina 2005), early investigators could not precisely determine the shape of the air capillaries (e.g. Rainey 1849; Eberth 1862; Huxley 1882): they were incorrectly termed "capillaries," despite the fact they differ from classical capillaries that are much longer than they are wide. Recently, by three-dimensional serial section computer reconstruction, Woodward and Maina (2005, 2008) and Maina and Woodward (2009) showed that the air capillaries are not morphologically tubular, i.e., they are not "capillaries," but are globular, i.e., rather spherical, structures that interconnect via very narrow passageways (Plates 3.19 and 3.20).

While in the blood capillaries of the mammalian lung display notable degrees of recruitment and distension (Watson et al. 2008), those of the avian lung (Plate 3.21) behave as rigid tubes (Macklem et al. 1979; Powell et al. 1985; West et al. 2006, 2007a, b; Maina 2007a, b; Watson et al. 2008; West 2009; Maina et al. 2010; Maina and Jimoh 2013a, b). Over the equivalent range of pressure difference, in the mammalian lung, the mean width of the pulmonary capillaries increased by ~125% in dogs and 128% in cats (Glazier et al. 1969). Macklem et al. (1979) observed that internal compression of the avian lung did not cause a significant collapse of the air capillaries and the parabronchi tolerated an external pressure of 2 kPa. In ducks, doubling of pulmonary intramural blood pressure by unilateral pulmonary artery occlusion did not cause distension of the blood capillaries (Powell et al. 1985). Watson et al. (2008) noted that when the pressure outside, i.e., in the avian capillaries, was increased to 35 cm H_2O (26 mm Hg = 3.5 kPa) above that inside, i.e., in the blood capillaries, the blood vessels remained open while in the mammalian lung the blood capillaries totally collapsed. Erythrocytes tightly squeeze through the pulmonary blood capillaries as they pass through them (Plate 3.21l): as they do so, they tense the blood-gas barrier (Plate 2.1d, i, j).

Birds and mammals are the only extant endothermic-homeotherms (Koteja 2004; Grigg et al. 2004; Geiser 2008): birds, that largely lead a more highly energetic lifestyle (Sibly et al. 2012), operate at relatively higher body temperatures (40–42 ° C) (Aschoff and Pohl 1970). To meet their large metabolic demands for O_2, birds have developed large hearts (Hartman 1955) with large cardiac outputs (Grubb 1982; Bishop 1997; Seymour and Blaylock 2000) and an exceptionally efficient respiratory system (Duncker 1971; Scheid 1979; Fedde 1980; Maina 2005; Maina et al. 2017). Compared with mammals, birds generally have higher arterial blood pressures (Lenel et al. 1948; Sturkie et al. 1956; Spector 1956; Weiss and Sheahan 1958; Krista et al. 1963a, b; Speckmann and Ringer 1963; Ferguson et al. 1969; Krista et al. 1967, 1970, 1986; Seymour and Blaylock 2000; Lichtenberger 2005; Kılıçkap et al. 2018). For example, compared with that of 95 mmHg (12.7 kPa) in a mammal such as the Crab-Eating Macaque (*Macaca fascularis*) (body mass 4.60 kg), a Turkey (*Meleagris gallopavo*) of equivalent body mass (4.77 kg) has a systolic blood pressure of 150 mmHg (20 kPa) (Seymour and Blaylock 2000). The blood pressures of the carotid artery of the broad-breasted bronze turkey (*Melleagris gallopavo*) are 302 mm Hg (40.3 kPa) systolic, 204 mm Hg (27.2 kPa) diastolic and the pulse pressure is 98 mm Hg (13.1 kPa): for the species, the respective values for the tibial artery are 286 mm Hg (38 kPa), 200 mm Hg (26.7 kPa) and 85 mm Hg (11.3 kPa) (Speckmann and Ringer 1963). Comparison of the blood pressures of the broad-breasted bronze turkey (BBBT) and the Beltsville small white turkey (BSWT) (of which ages ranged from 17 to 21 months) by Ferguson et al. (1969) determined that systolic, diastolic, and pulse pressures for male (M) and female (F) specimens were as follows, respectively: BBBT (M) 270.6 mmHg (36.1 kPa), 167.1 mmHg (22.3 kPa), 103.5 mmHg (13.8 kPa) and BBBT (F) 223.4 mmHg (29.8 kPa), 170.8 mmHg (22.8 kPa), mmHg 52.5 (7.0 kPa) and BSWT (M) mmHg 235.8 (31.4 kPa), 141.6 mmHg (18.9 kPa), 94.2 mmHg (12.3 kPa) and BSWT (F) 191.2 (25.5 kPa), 146.4 (19.5 kPa), 44.8 (6.0 kPa), and the systolic and the diastolic heart rates were as follows, respectively—BBBT (M) 160 times and (F) 219.1 times and for BSWT (M) 172.8 times and (F) 193.1 times per minute.

The fast-growing broiler chickens and turkeys normally develop hypertensive complications that cause high myocardial stress that causes pathologies such as aortic aneurysms and ascites (Speckmann and Ringer 1963; Krista et al. 1967, 1970; Julian and Wilson 1986; Groves 1997; Currie 1999; Issac et al. 2010; Gupta 2011; Franciosini et al. 2012; Dosković et al. 2019). In chickens, ostriches and waterfowls, spontaneous deaths have been associated with sudden aortic rupture (Kurtz et al. 1969; Neumann and Ungar 1973; Mitchinson and Keymer 1977; Ferreras et al. 2001) which has been termed "dissecting aneurism" or "turkey heart attack": fast-growing, healthy turkeys (mostly males of between 7 and 24 weeks of age) die suddenly from bleeding after aortic rupture or aortic dissection (Collins, 1971; Neumann and Ungar 1973; Krista et al. 1986; van Veen 1999; Stenzel et al. 2008). In hypertensive turkeys, mortality rates from aortic rupture of between 20% and 43.7% have been reported (Simpson 1978; Krista et al. 1967; Stenzel et al. 2008). Commercial losses of 1–2% occur in turkey flocks (van Veen 1999). A case of spontaneous proximal aortic dissection (dissecting aortic aneurysm) was reported in a mature Ostrich (*Struthio camelus*) by Ferreras et al. (2001) and attributed to copper deficiency alongside risk factors such as high blood pressure. In the battery chicken production industry, where birds are kept in overcrowded spaces and placed on strict feeding regimens that may comprise force-feeding (Arce et al. 1992; Camacho-Fernández et al. 2002; Bölükbasi et al. 2005; Baghbanzadeh and Decuypere 2008), a management scheme meant to induce fast growth rate, stress may cause susceptibility to pathologies and diseases that involve the cardiovascular system such as ascites and hypertension (Maxwell et al. 1986a, b; Julian and Boulianne 1988; Maxwell and Robertson 1997; Martrenchar et al. 1997; Julian 1998, 2000; Xiang et al. 2002; Walton et al. 2001; Al-Mesri and Hassanzadeh 2010; Singh et al. 2011; Kalmar et al. 2013). Among factors such as age, copper deficiency, and homornal imbalance, hypertension is one of the primary causes of aortic rupture (Terpin and Roach 1987; Ferreras et al. 2001; Tsamis et al. 2013) that stems from microstructural changes of the aortic tissue, especially collagen and elastic tissue fibers. In two groups of turkeys of 14–36 weeks of age, Krista et al. (1986) determined that hydrostatic pressures of 1478 mmHg (197 kPa), 1804 mmHg (240 kPa), and 794 mmHg (106 kPa) were needed to cause rupture of respectively the most caudal abdominal aortic, the adjacent cranial abdominal, and the thoracic parts of the aorta. On autopsy, the heart of an earlier visibly healthy Cardinal (*Cardinalis cardinalis*), that that died after a short territorial spat with a congener, showed a 7 mm wide hole in the ventricle of the heart (Erickson 1994): it was reckoned to have been caused by acute blood pressure increase during the stressful episode. Given that birds generally have relatively higher systolic and diastolic arterial blood pressures than mammals and consequently higher pulmonary capillary blood pressures, it is enigmatic that birds have pulmonary blood-gas barriers that are generally ~3 times thinner than those of lungs of mammals of equivalent body mass (Gehr et al. 1981; Maina and King 1982a; Maina et al. 1989a; Maina 1989a; Maina and West 2005; Maina 2005). The strengths of particularly the lung, the parabronchi, and the air and blood capillaries may account for the existence of thin blood-gas (tissue) barriers in the lungs of birds (Maina 2007a, b; West et al. 2006, 2007a, c).

While the morphology and morphometry of the avian lung have been well-investigated (Abdalla et al. 1982; Maina and King 1982a; Vidyadaran et al. 1987, 1988, 1991; Maina 1989a; Maina et al. 1989a; Maina and West 2005; Watson et al. 2007; Maina et al. 2017), the strengths and structural failures of the blood-gas barrier were only much recently investigated—to a large extent in the lung of the Domestic Fowl (*Gallus gallus* variant *domesticus*) (West et al. 2006, 2007a, c; Watson et al. 2008; Maina et al. 2010; Maina and Jimoh 2013a, b; Jimoh and Maina 2013; Maina 2015a, b). The lung has a well-developed connective tissue continuum that mostly comprises collagen fibers, elastic tissue, and smooth muscle fibers (Plate 3.10, 3.22 and 3.23a, d–f). The connective tissue elements are particularly well-organized in the gas exchange tissue of the parabronchi (Plate 3.18a–d) and in the air and the blood capillaries (Plate 4.19). By immunogold staining (Marshall et al. 1990a, b), the presence of type-IV collagen was determined in the basement membrane of the blood-gas barrier (Plate 3.22i) and the epithelial lamina of the lung of the domestic fowl (Scheuermann et al. 1997; Jimoh and Maina 2013) (Plate 4.16d, k–m). In the exchange tissue of the avian lung, blood capillaries are literally suspended in the exchange across the epithelial lamina (Plates 3.21g, h, 3.22a–c and 4.16a, e, k–m) and only occasionally connect directly (Plate 4.15e).

The most important structural features that contribute to the strength of the air and blood capillaries of the avian lung are: (a) presence of a lipoproteinaceous substance (surfactant) that covers the respiratory surface, forming a three-dimensional web-like system that connects and reinforces the terminal respiratory units (Klika et al. 1997); (b) close entwining of air and blood capillaries, a feature termed "honeycomb arrangement" by West et al. (2006), forms compact and "strong" exchange tissue (Maina 1982a, 1988, 2015b; Woodward and Maina 2005, 2008; West et al. 2006, 2010; Watson et al. 2007, 2008; Maina and Woodward 2009) (Plates 1.7a, 3.19, 3.21a, b and 3.24), (c) presence and profuse arrangement of collagen and elastic-tissue fibres (connective tissue elements that possess various mechanical properties) that may contribute to formation of a tensegrity system in the lung (Maina 2007a, b; Maina et al. 2010) and; (d) existence of the particularly strong type-IV collagen in the basement membrane of the blood-gas barrier of the lung (Fisher and Wakely 1976; West and Mathieu-Costello 1999; Pöschl et al. 2004; West 2011; Mereness and Mariani 2021): type-IV collagen is specifically found in the lamina densa of the basement membrane of the blood-gas barrier of the lung (Kühn 1995; Maina and West 2005; Jimoh and Maina 2013). By maintaining the structural integrity of the extremely thin blood-gas barrier and therefore supporting life, West (2009) termed particularly the extremely thin lamina densa of the blood-gas barrier a "lifeline." Together with type-IV collagen (Jimoh and Maina 2013; Maina and Jimoh

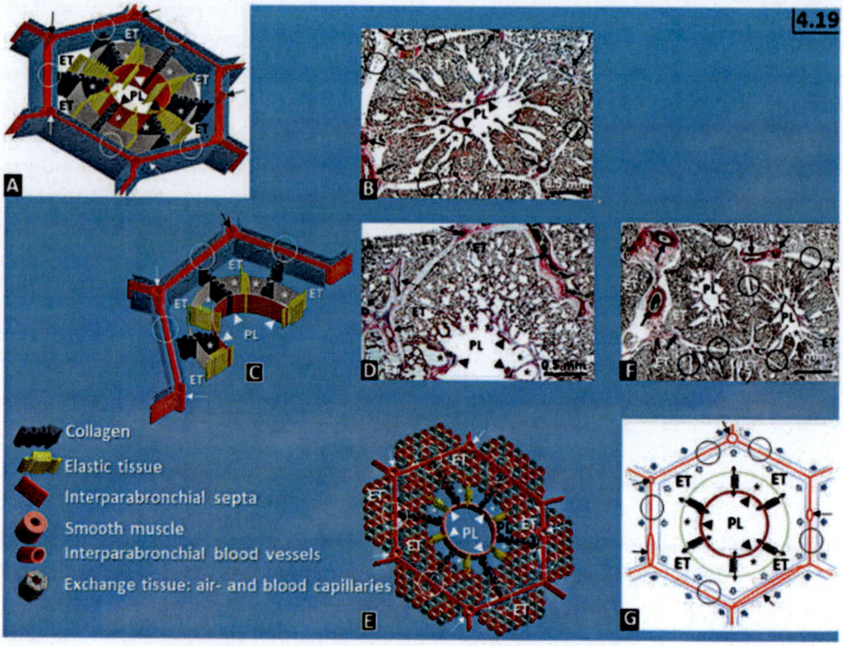

Plate 4.19 Schematic illustrations (**a**, **c**, **e**, **g**) and histological preparations (**b**, **d**, **f**) of the lung of the Domestic Fowl (*Gallus gallus* variant *domesticus*) showing the shape of the parabronchi. Connective tissue elements such as collagen, elastic tissue, and smooth muscle fibers support and/or generate forces/tensions that may create a tensegrity system that strengthens the parabronchial units. Arrows (↓), interparabronchial arteries; circles (O), interparabronchial septa; ET, exchange tissue; stars (⊕), atria; arrowheads (▶), interatrial muscles; PL, parabronchial lumen. In (**e**), the blood capillaries are shown as red-coloured circular structures and the air capillaries similar structures in cyan

2013a, b), other forms of collagen should exist in the basement membrane of the blood-gas barrier of the avian lung (Plate 4.16f–j).

In species of birds such as the Andean Goose (*Chloephaga malanoptera*) and the Domestic Fowl (*Gallus gallus* variant *domesticus*), respectively, the pleural coverings of the lung, i.e., the parietal and the viscera, are interconnected by collagen fibers (Plate 3.10a, b) and fused (Plate 3.10c–g), totally obliterating the pleural cavity. For the Domestic Fowl, the air and the blood capillaries that are located at the periphery of the lung connect directly to the pleural covering (Plate 3.10c–f). For the lung of the domestic fowl, the pleural covering is well-endowed with collagen, elastic tissue, and smooth muscle fibers (Plate 3.10d–h). For most mammalian species, interestingly, except for the African (*Loxodonta africana*) and Asian (*Elephas maximus*) elephants, where the pleural cavity is traversed or occupied by connective tissue fibers (Short 1962; Brown et al. 1997; West 2001), an open space (pleural cavity) exists between the visceral and the parietal pleurae, allowing free movement of the lung in the thoracic cavity. For the avian lung, this cannot occur because first the lung is firmly attached to the ribs and the vertebrae and secondly because an open (free) pleural cavity is lacking. Debates and controversies on the evolutionary and functional significance of the anatomy of the elephant's respiratory system, where the pleural cavity is "filled" by extrapulmonary connective tissue fibers at the later stages of life were critically discussed by West (2002). In the lung of the domestic fowl, the pleural linings fuse very early during the development of the lung (Maina 2003). The rare folds of the epithelial lamina that connect the blood capillaries in the exchange tissue of the avian lung (Plate 3.22a, b) suggest that at the gas exchange level, the avian lung may not be as suggested by Jones et al. (1985) and Makanya et al. (2021) inflexible. A continuous basement membrane is lacking in the epithelial lamina (Maina and King 1982a; Abdalla et al. 1982) (Plate 4.16d, e, k–m), a feature that suggests a strong developmental relationship between a basement membrane and the endothelial cell (Plate 4.15e): the basement membrane contains loose extracellular matrix substance that comprises collagen fibers, mostly in parts next to the blood capillaries (West et al. 2006, 2010; Maina and Jimoh 2013a, b) (Plates 4.16d, e and 4.18j, k). In the avian lung, collagen fibers are plentiful in the pleural covering of the lung (Plate 3.10), in the gas exchange tissue i.e., the periparabronchial areas (Plates 3.22d–o, 3.23a, d–f and 4.16h), around larger blood vessels (Plate 3.23d–f) and in the blood-gas barrier and the epithelial lamina (Maina et al. 2010) (Plate 4.16d–l). In the avian lung, the parabronchi and the terminal respiratory units, namely, the air and the blood capillaries, are strengthened by the codependence between tensional and compressional forces which are transmitted by connective tissue elements such as elastic, smooth muscle, and collagen fibers (Plates 3.22d–o, 3.23a, d–f and 4.16f–l): the structures may create a tensegrity (tensional integrity or floating compression) state, a self-stabilized system (Maina 2007a, b; Maina et al. 2010). A tensegrity system is a type of construction wherein the constituent structural constituents are stabilized by a balance of tensile and compression forces (Fuller 1961; Ingber 2003). For the mammalian lung (Mead 1961; Hance and Crystal 1976; Weibel 1984), collagen fibers are part of a fibrous continuum that continues from the pleural covering to the interalveolar septa (Plate 3.25b). The blood-gas barrier and the epithelial lamina of the avian lung (Plates 3.22a–c, 3.25c and 4.13) are largely subjected to three forces, namely, a) intramural pressure in the blood capillaries that derives from the beating heart and prevailing barometric pressure; b) surface tensional force on the surface of the air capillaries; c) dependent on body posture, the gravitational force that stems from the weight of the structural components of the lung and that of blood contained in it; and d) the forces generated by contractile elements such as elastic tissue and smooth muscle (Plate 4.13a–d).

Structural failure of the blood-gas barrier and the epithelial lamina of the lung of free-range chickens (*Gallus gallus* variant *domesticus*) was experimentally investigated by Jimoh (2012), Jimoh and Maina (2013) and Maina and Jimoh (2013a, b) and the spatial organizations of the connective tissue elements such as collagen, elastic, and smooth muscle fibers determined by Maina et al. (2010): The control group of birds was rested while the experimental groups were exercised by running them on a treadmill at various speeds and angles of inclinations (Plate 4.20a, b) and

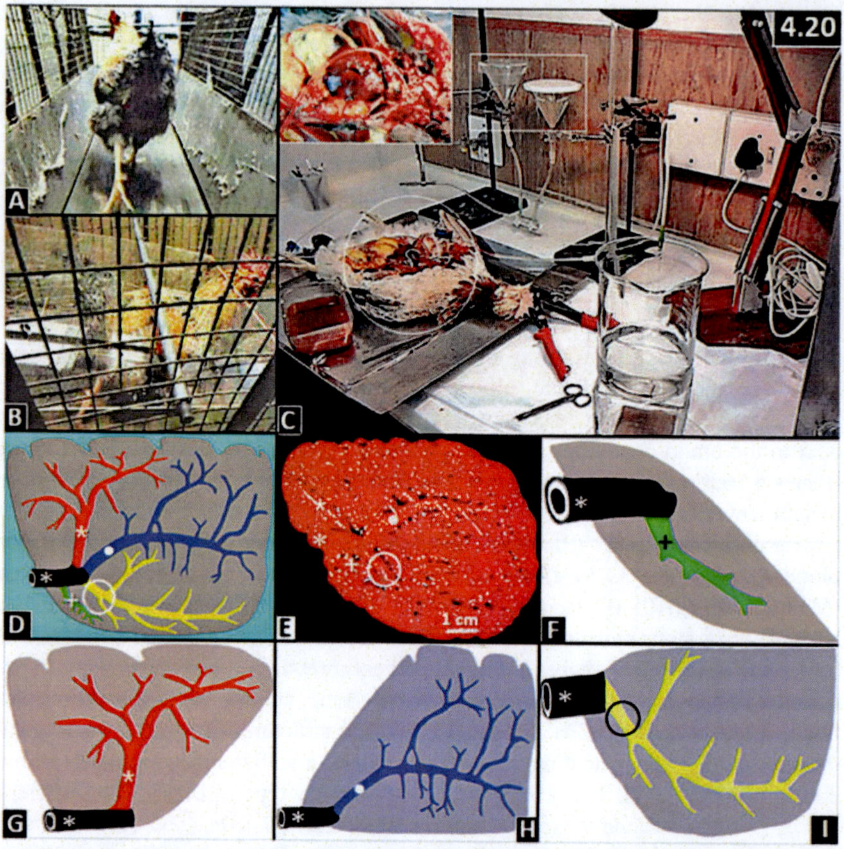

Plate 4.20 Experimental investigation of the strength of the blood-gas barrier of the lung of free range Domestic Fowl (*Gallus gallus* variant *domesticus*). In one of the experiments, birds were exercised on a treadmill at different speeds and angles of inclination (**a**, **b**) while in others (**c**), lungs (encircled area) were perfused with physiological saline at various head pressures (area shown with a square) (**c**) after correctly cannulating the heart (**c**—inset). (**d**–**i**) Four branches of the pulmonary artery supply blood to different regions of the lung. The frequency of structural failure of the blood-gas barrier and the epithelial retinaculae correlated with the point of branching and the orientation of the branches relative to the pulmonary artery. Asterisks (*), pulmonary artery; star (⨀), cranial branch; dot (•), caudomedial branch; circle (O), caudolateral branch; cross (+), accessory branch. (**e**) is a latex rubber cast preparation

perfusion of the lung with phosphate buffered physiological saline for 20 min at various inflow and outflow pressures (Plate 4.20c—inset). At the end of the experiment, the lungs were fixed for microscopy by perfusion intratracheal instillation with 2.5% glutaraldehyde buffered in phosphate solution (pH 7.4) for 10 min. Because of the technical difficulties of accurately determining pulmonary blood capillary pressure directly, the mean value of the inflow pressure (which was taken by the pressure in the pulmonary artery) and the outflow one (that in the pulmonary

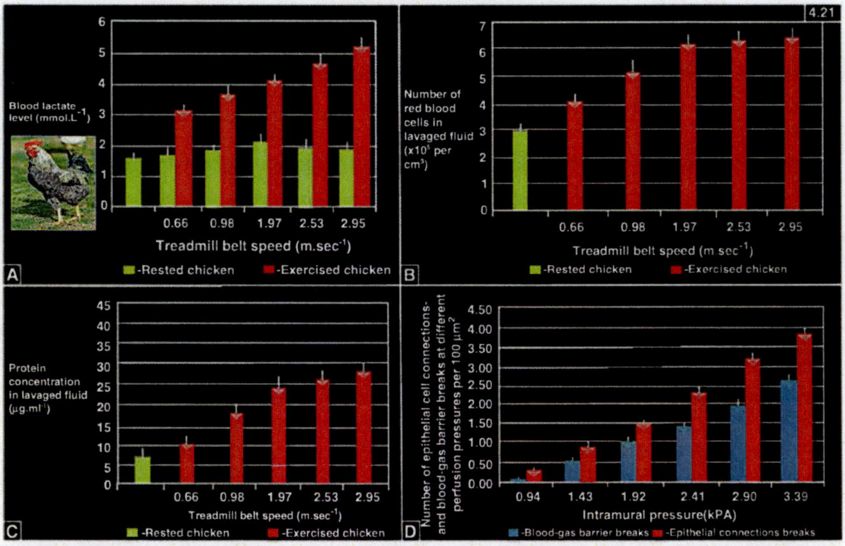

Plate 4.21 Structural failure of the blood-gas barrier of lungs of free range chickens (*Gallus gallus* variant *domesticus*). (**a**) In exercised birds, the level of lactic acid in the blood increased with exercise intensity. (**b**) In the lavaged fluid, as a result of structural failure of the blood-gas (tissue) barrier, the number of extravasated erythrocytes increased with exercise intensity. (**c**) From the exudation of blood plasma across the blood-gas barrier, protein concentration increased in the lavaged fluid with increased exercise intensity. (**d**) The number of epithelial-epithelial cell connections (epithelial retinacula) and the blood-gas barrier breaks increased with the hydrostatic (perfusion) pressures (**d**)

vein) was regarded as the intramural blood pressure of the blood capillaries. The same assumption was made for the mammalian lung (Bhattacharya et al. 1982; West 2008). Using a hemocytometer, the number of extravasated red blood cells was counted after lavaging the lungs, and the protein concentration was determined in the recovered fluid by the Lowry method (Lowry et al. 1951) while blood taken from the brachial vein was used to determine the blood lactate levels (Jimoh 2012; Jimoh and Maina 2013). From rest to exercise, at the fastest treadmill speed of 2.95 m s^{-1}, blood lactate level (mmol L^{-1}) increased fourfold (Plate 4.21a) and respectively the number of red blood cells in the lavage fluid ($\times 10^5$ per cm^3) and the protein concentration in it (μg ml^{-1}) increased with exercise intensity (Plate 4.21b, c). The blood-gas barrier and the epithelial lamina breaks increased with perfusion increase of perfusion pressure (Plate 4.21d). Showing their relative strengths and hence their vulnerability to structural failure, the number of epithelial lamina breaks, in general, surpassed those of the blood-gas barrier by a factor of 2 (Maina and Jimoh 2013a, b). The increase in the number of tissue breaks with increasing exercise and perfusion pressures is attributable to the increase in the pulmonary capillary blood pressure that stems from faster heart rates, high cardiac outputs, and systolic blood pressures (Lumeij and Ritchie 1994; Seymour and Blaylock 2000).

It is paradoxical that bird lungs possess extremely thin blood-gas barriers (harmonic mean thickness), with values of 0.090 μm in the African Rock Martin (*Hirundo fuligula*), 0.096 μm in the House Sparrow (*Passer domesticus*) (Maina 1984, 1989a, 2005; Maina et al. 1989a) and 0.099 in the Violet-Eared Hummingbird (*Colibri coruscans*) (Dubach 1981) whereas compared with mammals the group (birds) have relatively higher systolic and systemic (arterial) blood pressures (Ferguson et al. 1969; Krista et al. 1986; Lumeij and Ritchie 1994; Seymour and Blaylock 2000). For the Domestic Fowl (*Gallus gallus* variant *domesticus*), the differences between the number of breaks of the blood-gas barrier (Plate 4.18d–f) and the epithelial lamina (Plate 4.18g–i) correlate with features such as the diameters of the blood vessels that supply blood to the various parts of the lung, their points of origin from the pulmonary artery and the angles of bifurcation from the blood vessel (Maina and Jimoh 2013a, b) (Plate 4.20d–i). Maina and Jimoh (2013a, b) noted that most of the blood-gas barrier and the epithelial lamina occurred in regions supplied with blood by the accessory (Plate Plate 4.20f), the cranial (Plate 4.20g), and the caudomedial (Plate 4.20h) branches of the pulmonary artery. The accessory branch is the smallest, and the first branch to branch off from the pulmonary artery while the caudomedial branch is the largest and the most direct continuation (branch) of the pulmonary artery. The cranial branch is the second blood vessel to originate from the pulmonary artery (Plate 4.20d, e) and the caudolateral branch of the pulmonary artery (Plate 4.20i) supplies blood to most of the ventral part of the lung. The presence of erythrocytes in the lavaged fluid of the lungs of rested chickens (Maina and Jimoh 2013a, b) (Plate 4.20b) and microscopic observation of erythrocytes on the surface of atria and infundibula (Plate 3.17a–c, e) indicate that in birds, failure of the blood-gas barrier may be a common event and that when it is not serious, it may not always life-threatening. The mechanisms by which birds contend with modest injuries of the blood-gas barrier are currently unknown. Compared with the mammalian lung (West et al. 1991, 1993; Mathieu-Costello et al. 1995), the blood-gas barrier of the lung of the free-range chickens is stronger (Maina and Jimoh 2013a, b). Interestingly, in pilot studies for the experiments performed by Jimoh (2012), Maina and Jimoh (2013a, b), and Jimoh and Maina (2013) attempts to train Domestic Turkeys (*Meleagris gallopavo*), Domestic Ducks (*Anas platyrhynchos domesticus*) and Purebreed White Leghorn Chickens (*Gallus gallus* variant *domesticus*) to run on treadmill totally failed: the birds appear to have been too heavy to exercise.

4.8 Morphometric Properties of the Avian Lung: Comparative Perspectives

The functional efficiency of the avian respiratory system is largely attributable to its complex and highly specialized morphology and physiology (Duncker 1971; King 1966; Fedde 1980; Maina 2005, 2009b, 2022; Powell 2014). Determined by aspects/

factors such as phylogenetic status, lifestyle, and environment occupied, various morphometric specializations meaningfully contribute to gas exchange efficiency (Figueroa et al. 2007). Currently, available morphometric data are summarized in Tables 4.1, 4.2 and 4.3 and some of the parameters are succinctly outlined below. Comparisons of data with the other air-breathing vertebrates are made to draw attention to the adaptations and specializations that exist in the avian lungs.

4.8.1 Lung Volume

The total volume of the avian respiratory system, i.e., the volume of the lung, the air sacs and their extensions into certain bones comprise ~20% of the volume of the body (Duncker 1971): for the Mute Swan (*Cygnus olor*), a bird that has low breathing frequency and high tidal volume that surpasses the values predicted from allometric relations (Bech and Johansen 1980), the value is as high as 34% of the body volume. According to Tenney and Remmers (1963) and Tenney and Tenney (1970), the total volume of the avian lung and that of the air sacs is 3 to 5 times greater than that of the volume of the mammalian lung and 2 times larger than that of reptiles. In mammals, a strong positive correlation exists between the volume of the lung and body mass (Tenney and Remmers 1963; Gehr et al. 1981). The five species of birds that were investigated by Lasiewski and Calder (1971) showed that a strong positive correlation exists between the total volume of the respiratory system (i.e., the lung-air sac system) and that of body mass. Based on measurements of lung weight, Jürgens et al. (1981) noted that bats had relatively larger lungs compared with those of terrestrial (nonvolant = nonflying) mammals: the result was corroborated by lung volume measurements in species of bats (Maina et al. 1982a, b; Maina and King 1984; Maina et al. 1989a, 1991). The mass-specific lung volume, i.e., the volume of the lung per unit body mass, of a bat is 2 to 5 times larger than that of shrews, the smallest extant mammalian taxon (Gehr et al. 1981) and about twice that of a hummingbird (Dubach 1981). In reptiles and amphibians, Tenney and Tenney (1970) determined that lung volume correlates more strongly with metabolic rate than with body mass. For birds, lung volumes correlate very strongly with body mass ($r = 0.9970$) (Plate 4.22a): among the species that have been investigated (Table 4.3), the Andean Goose (*Chloephaga melanoptera*) (Maina et al. 2017) and the Violet Eared Hummingbird (Dubach 1981) have the highest mass-specific lung volume, values that are 42.8 and 42.9 $cm^2\ kg^{-1}$, respectively. For animals of equivalent body mass, the volume of the avian lung is 27% smaller than that of a mammal (Maina 1989a, 1993, 2005).

Table 4.1 Volume densities of the structural components of the avian lung

Order, scientific, and common English name	Number of specimens studied	Body mass (g)	Total lung volume (cm^3)	Exchange tissue (%)	Lumina of parabronchi and secondary bronchi (%)	Blood vessels larger than blood capillaries (%)	Primary bronchus (%)
STRUTHIOFORMES							
Struthio camelus							
Ostrich[a]	1	45000	1563	80.17	13.82	4.38	1.63
SPHENISCIFORMES							
Spheniscus humboldti							
Humboldt Penguin[b]	1	4500	1368	51.29	37.71	7.44	3.56
ANSERIFORMES							
Anas platyrhynchos							
Mallard Duck[c,d,e]	5	1038	30.60	40.55	50.37	6.70	2.38
Anser anser							
Graylag Goose[c,d,e]	5	38.38	95.30	40.37	50.99	6.91	1.73
Branta canadensis							
Canada Goose[f,g]	–	–	–	36	–	7.50	30.20
Cygnus olor							
Mute Swan[f,g]	–	–	–	31	–	4.90	32.50
Cairina moschata							
Muscovy Duck (dom.)[h]	5	1627	48.10	49.24	37.47	8.24	5.05
Chloephaga melanoptera							
Andean Goose[i]	3	2636	112.0	61.7	25.80	9.50	3.00
FALCONIFORMES							
Falco tinnunculus							
Common Kestrel[c,d,e]	2	66	3.10	51.85	39.03	7.33	1.79

GALLIFORMES							
Gallus gallus var. *domesticus*							
Domestic Fowl[j]	3	2141	27.0	46.36	30.56	13.65	9.34
Domestic Fowl[f,g]	–	–	–	46.7	31.80	6.80	14.7
Domestic Fowl[h]	5	1873	26.6	49.66	38.01	6.64	5.69
Domestic Fowl[k,l] (RJF)	5	478	8.7	52.71	36.05	6.61	4.63
Numida meleagris							
Guinea Fowl (dom.)[m]	–	1839	25.8	47.35	29.37	13.09	10.19
GRUIFORMES							
Amaurornis phoenicurus							
White-Breasted Water Fowl[n]	5	146	5.1	55.02	36.73	5.20	3.05
Fulica atra							
Coot[f,g]	–	–	–	50.0	31.90	7.3	10.8
CHARADRIIFORMES							
Alca torda							
Razorbill[c,d,e]	2	487	18.1	32.79	53.43	11.40	2.38
Cephus carbo							
Spectacled Guillemot[c,d,e]	9	737	24.1	33.40	47.92	13.86	4.82
Larus argentatus							
Herring Gull[c,d,e]	2	654	18.2	32.64	60.41	5.56	1.39
Larus canus							
Common Gull[c,d,e]	1	302	7.1	30.85	61.70	5.39	2.06
Larus ridibundus							
Black-Headed Gull[c,d,e]	6	253	7.5	35.83	54.88	7.17	2.12

(continued)

Table 4.1 (continued)

Order, scientific, and common English name	Number of specimens studied	Body mass (g)	Total lung volume (cm^3)	Exchange tissue (%)	Lumina of parabronchi and secondary bronchi (%)	Blood vessels larger than blood capillaries (%)	Primary bronchus (%)
COLUMBIFORMES							
Columba livia							
Rock Dove[c,d,e,o]	1	216	7.4	43.57	32.26	14.85	9.32
Columba livia							
Domestic Pigeon[c,d,e,o]	–	–	–	49.0	24.7	7.5	18.8
Streptopelia decaocto							
Collared Turtle Dove[c,d,e,o]	16	189	6.5	52.17	29.46	15.53	2.84
Streptopelia senegalensis							
Laughing Dove[c,d,e,o]	1	58	1.8	52.26	40.22	6.18	1.34
PSITTACIFORMES							
Mellopsitacus undulutus							
Budgerigar[c,d,e,o]	6	36.4	1.03	46.56	47.46	4.02	1.96
Budgerigar[c,d,e,o]	5	36.8	1.11	47.8	21.10	8.20	22.9
CUCULIFORMES							
Chrysococcyx klaas							
Klass's Cuckoo[c,d,e,o]	3	27.0	0.66	53.0	37.03	8.4	1.57
APODIFORMES							
Colibri coruscans							
Violet-Eared Hummingbird[p]	3	7.3	0.27	46.4	28.6	13.6	11.4
COLIIFORMES							
Colius striatus							

Spectacled Mousebird[c,d,e,o]	1	50.05	0.71	50.81	37.43	9.78	1.98
PICIFORMES							
Jynx ruficollis							
Red-Breasted Wryneck[c,d,e,o]	2	54.3	1.09	45.69	47.65	5.44	1.22
Pogoniulus bilineatus							
Golden-Rumped Tinkerbird[c,d,e,o]	1	14.9	0.23	54.30	35.22	8.64	1.94
CASUARIFORMES							
Dromaius novaehollandiae							
Emu[q]	1	30,000	1100	17.76	49.11	28.57	4.56
PASSERIFORMES							
Ambryospiza albifrons							
Grosbeak Weaver[c,d,e,o]	5	37.4	0.97	48.83	44.1	6.17	0.90
Cercotrichus leukophrys							
White-Winged Robin[c,d,e,o]	1	16.0	0.38	42.39	47.13	9.48	1.00
Chloropeta natalensis							
Yellow Flycatcher[c,d,e,o]	1	10.9	0.31		41.57	7.74	2.08
Cisticola cantans							
Singing Cisticola[c,d,e,o]	4	15.0	0.31	48.61	41.57	7.74	2.08
Cossypha cafra							
Robin Chat[c,d,e,o]	1	25.1	0.58	46.50	44.95	6.85	1.70
Estrilda astrid							
Waxbill[c,d,e,o]	3	7.0	0.15	59.85	30.06	8.40	1.69

(continued)

Table 4.1 (continued)

Order, scientific, and common English name	Number of specimens studied	Body mass (g)	Total lung volume (cm^3)	Exchange tissue (%)	Lumina of parabronchi and secondary bronchi (%)	Blood vessels larger than blood capillaries (%)	Primary bronchus (%)
Hirundo fuligula							
African Rock Martin[c,d,e,o]	1	13.7	0.33	51.09	41.40	6.69	0.82
Laganosticta senegala							
Red-Billed Firefinch[c,d,e,o]	1	8.0	0.223	55.92	32.95	10.02	1.11
Lanius aethiopicus							
Tropical Boubou[c,d,e,o]	1	42.8	0.89	54.57	37.90	6.62	0.91
Lanius collaris							
Fiscal Shrike[c,d,e,o]	6	32.5	0.72	52.99	39.66	6.43	0.92
Lonchura cucullata							
Bronze Mankin[c,d,e,o]	7	9.4	0.23	55.92	33.93	8.64	1.51
Nectarina kilimensis							
Bronze Sunbird[c,d,e,o]	1	16.6	0.50	56.32	32.64	8.70	2.34
Nectarina reichenowi							
Golden-Winged Sunbird[c,d,e,o]	3	13.7	0.36	58.56	31.91	8.67	0.86
Passer domesticus							
House Sparrow[c,d,e,o]	12	25.5	0.76	55.68	36.91	6.45	0.96
House Sparrow[c,d,e,o]							
Ploceus baglafecht	5	26.5	0.80	52.60	25.3	10.3	11.80
Baglafecht Weaver [c,d,e,o]							
Ploceus cucullatus	6	32.5	0.88	49.43	42.38	6.72	1.47

Black-Headed Weaver[c,d,e,o]							
Ploceus ocularis	3	34.9	0.87	57.61	35.25	5.67	1.47
Spectacled Weaver [c,d,e,o]							
Ploceus xanthops	2	26.9	0.72	47.05	45.04	6.30	1.61
Holub's Golden Weaver[c,d,e,o]							
Prinia subflava	3	39.3	1.03	49.13	41.09	7.87	1.91
Twany Prinia[c,d,e,o]							
Yellow-Fronted Canary[c,d,e,o]	1	9.1	0.21	51.88	37.37	9.08	1.49
Sturnus vulgaris							
Common Starling[c,d,e,o]	10	72.60	2.02	51.76	41.16	5.19	1.89
Turdus iliacus							
Redwing[c,d,e,o]	1	51.0	1.19	45.66	45.76	7.11	1.47
Turdus olivaceous							
Olive Thrush[c,d,e,o]	2	65.1	1.40	55.84	36.38	5.92	1.86

[a]Maina and Nathaniel (2001)
[b]Maina and King (1987)
[c,d,e]Maina and King (1982b), Maina (1987, 1989a, 2005), Maina et al. (1989a)
[f,g]Duncker (1972, 1973)
[h]Vidyadaran et al. (1990)
[i]Maina et al. (2017)
[j]Abdalla et al. (1982)
[k, l] Vidyadaran et al. (1987, 1988)
[m]Abdalla and Maina (1981), Maina et al. 1982a, b
[n]King et al. (1992)
[o]Maina (1984, 1989a), Maina et al. (1989a)
[p]Dubach (1981)
[q]Maina and King (1989)
RJF, Red Jungle Fowl

Table 4.2 Volume densities of components of the exchange tissue of the avian lungs

Order, scientific, and common English name	Air capillaries (%)	Blood capillaries (%)	Tissue of the blood-gas barrier (%)	Tissue of the exchange tissue not involved in gas exchange (%)
Ostrich	61.19	20.41	12.69	5.71
Humboldt Penguin	34.36	51.04	12.30	2.30
Mallard Duck	59.21	32.98	5.74	2.08
Graylag Goose	62.03	32.43	4.24	1.30
Canada Goose	48.5	41.5	10.0	
Mute Swan	46.0	43.1	10.9	
Muscovy Duck (dom.)	57.24	29.63	5.88	7.25
Andean Goose	63.4	28.2	8.4	
Common Kestrel	53.02	25.56	17.28	4.14
Domestic Fowl	60.90	27.92	6.30	4.88
Domestic Fowl	55.8	28.3	15.9	
Domestic Fowl	55.59	27.89	7.41	9.09
Domestic Fowl (RJF)	64.82	21.29	6.01	7.88
Guinea Fowl (dom.)	53.51	33.80	10.90	1.89
White-Breasted Water Fowl	63.20	23.29	5.89	7.61
Coot	33.3	56.3	10.4	
Razorbill	49.54	36.89	10.19	3.38
Spectacled Guillemot	47.49	39.75	9.46	3.30
Herring Gull	53.68	35.92	7.04	3.36
Common Gull	58.23	30.45	8.74	2.58
Black-Headed Gull	62.52	28.92	5.96	2.60
Rock Dove	58.70	33.63	5.90	1.77
Domestic Pigeon	43.0	44.0	13.0	–
Collared Turtle Dove	62.96	22.81	10.42	3.81
Laughing Dove	45.13	34.63	16.77	3.46
Budgerigar	52.47	33.34	11.44	2.75
Budgerigar	67.3	26.1	6.6	–
Klass's cuckoo	55.09	34.50	8.26	2.15
Violet-Eared Hummingbird	53.9	36.7	9.4	–
Spectacled Mousebird	53.75	36.43	7.49	2.33
Golden-Rumped Tinkerbird	38.21	39.02	19.51	3.25
Emu	79.15	14.2	3.4	3.25
Grosbeak Weaver	44.84	42.15	9.88	3.13

(continued)

Table 4.2 (continued)

Order, scientific, and common English name	Air capillaries (%)	Blood capillaries (%)	Tissue of the blood-gas barrier (%)	Tissue of the exchange tissue not involved in gas exchange (%)
Singing Cisticola	48.20	38.32	9.92	3.56
African Rock Martin	45.20	44.35	7.74	2.08
Fiscal Shrike	49.49	37.13	10.40	2.98
House Sparrow	45.68	38.04	12.87	3.41
House Sparrow	54.10	36.80	9.20	–
Baglafecht Weaver	50.96	36.22	8.99	3.83
Twany Prinia	50.47	37.38	9.35	2.80
Common Starling	51.68	32.55	12.55	3.22
Redwing	47.11	39.33	11.19	2.37
Olive Thrush	50.45	39.04	7.81	2.70

The orders, scientific names, number of specimens investigated, body masses, lung volumes, and sources of data for each species are given in Table 4.1. The names of the birds are listed in the same order they appear in Table 4.1

For the various species of birds, references to the works in which the investigations were published are shown in Table 4.1. RJF—Red Jungle Fowl

4.8.2 Respiratory Surface Area

Extensive respiratory surface areas of the vertebrate gas exchangers have developed by two process: (a) for the gills, such as those of the gills of crabs (Plate 4.23a, b) and the teleost ones (Plate 4.23c–h) that are evaginated (Plate 4.3b, c), large surface area has been generated by a hierarchical assemblage: stratification of the teleost gills gives rise to gill arches which in turn give rise to hundreds of gill filaments that in turn originate thousands of secondary lamellae (Plate 4.23c–f) that are intensely vascularized (Plate 4.23g, h) (Hughes and Morgan 1973; Maina 1990b) and; (b) lungs that are invaginated structures (Plate 4.3c, d) in which respiratory surface area is increased by internal subdivision (compartmentalization) into gas exchange tissue (Maina 1994, 1998, 2002a–d) (Plate 1.7). For the avian lung, because of extreme surface tension forces that would make them highly vulnerable to collapse, for the avian lung, particularly small terminal air spaces (the air capillaries) could only form and exist in a rigid, i.e., noncompliant lung: surface tension forces exist on the air-tissue interfaces of the respiratory units of lungs of air-breathing vertebrates, including those of the air capillaries of the avian lung (Maina 2014) and the alveoli of the mammalian lung (Levitzky 2013; Mason and Dobbs 2016; Knudsen and Ochs 2018) (Plate 1.7b). While in mammals the compliance of the respiratory system (except for the thoracic walls) exists in the terminal parts of the respiratory tree (mainly at the alveolar level) (Dubois et al. 1956), for the avian respiratory system, pulmonary compliance has mainly been totally relegated to the the air sacs (Scheid and Piiper 1989). Firmly fixed to the ribs and the vertebrae (Plates 1.1a, 2.6, 3.6a–d,

Table 4.3 Body mass normalized pulmonary morphometric parameters of the avian lung

Order, Scientific and Common Name	Total volume of the lung per unit body (VL) mass ($VL.kg^{-1}$) $cm^3\ kg^{-1}$	Surface area of the blood-gas (tissue) barrier (BGB) (St) per unit body mass ($St\ g^{-1}$) $cm^2\ g^{-1}$	Surface density of the blood-gas (tissue) barrier per unit volume of the exchange tissue ($St.Vx^{-1}$) $mm^2\ mm^3$	Volume of the pulmonary capillary blood (Vc) per unit surface area of the BGB (St) $Vc.St^{-1}$ $cm^3\ m^{-2}$	Arithmetic mean thickness of the BGB (τ_t) μm	Harmonic mean thickness of the BGB (τ_{ht}) μm	Pulmonary morphometric diffusing capacity of the BGB for O_2 (DtO_2) per unit body mass ($DtO_2.kg^{-1}$) ml $O_2.s^{-1}.mbar^{-1}.kg^{-1}$	Total pulmonary morphometric diffusing capacity of the lung for O_2 (DLO_2) per unit body mass ($DLO_2\ kg^{-1}$) ml $O_2\ s^{-1}\ mbar^{-1}\ kg^{-1}$
Ostrich	39.08	30.10	98	2.1	0.690	0.560	0.196	0.065
Humboldt Penguin	30.4	18.1	116	4.4	–	0.530	0.141	0.051
Mallard Duck	30	28.7	240	1.4	0.903	0.113	0.896	0.05
Greylag Goose	25	23.1	253	1.4	0.887	0.118	0.875	0.045
Muscovy Duck (dom.)	30	30.0	200	1.5	0.303	0.199	0.620	0.060
Andean Goose	42.8	96.5	330	0.8	0.474	0.222	1.777	0.119
Common Kestrel	18.7	43.4	324	0.8	1.662	0.210	1.560	0.076
Domestic Fowl	12.6	8.7	172	1.6	1.24	0.318	0.130	0.017
Domestic Fowl (RJF)	14.7	12.5	135	1.7	0.459	0.322	0.211	–
Guinea Fowl	18.1	13.0	195	1.6	0.452	0.252	–	0.017
Turkey (dom.)	14.0	7.0	188	1.7	1.12	0.320	–	–

White-Breasted Water Hen	35.1	34.8	180	1.3	0.395	0.204	0.705	0.055
Coot	–	45.8	248	–	–	–	–	–
Razorbill	37.2	29.9	237	1.6	0.803	0.230	0.526	0.060
Spectacled Guillemot	32.7	27.0	240	1.7	0.850	0.193	0.569	0.058
Herring Gull	27.8	22.1	236	1.5	1.27	0.153	0.622	0.038
Common Gull	23.5	20.8	292	1.1	0.684	0.116	0.742	0.033
Black-Headed Gull	29.6	27.6	238	1.3	0.925	0.146	0.712	0.043
Rock Dove	34.3	39.8	254	1.3	0.804	0.161	1.023	0.093
Domestic Pigeon	–	40.3	302	–	0.315	0.172	0.889	–
Collared Turtle Dove	34.4	43.1	253	1.0	0.801	0.218	1.185	0.063
Laughing Dove	31.0	47.2	296	1.2	0.986	0.227	–	0.086
Budgerigar	28.3	42.6	317	1.1	0.976	0.117	1.484	0.055
Budgerigar	29.6	48.1	301	0.9	0.210	0.118	1.576	–
Klaa's Cuckoo	24.4	31.9	274	1.4	0.679	0.157	0.963	0.074
Violet-Eared Hummingbird	42.9	87.1	389	1.0	0.183	0.099	3.562	–
Spectacled Mousebird	14.1	19.9	260	1.4	0.761	0.148	0.554	0.027
Golden-Rumped Tinkerbird	15.4	22.5	272	1.5	1.01	0.165	0.147	0.067

(continued)

Table 4.3 (continued)

Order, Scientific and Common Name	Total volume of the lung per unit body (VL) mass (VL.kg^{-1}) cm^3 kg^{-1}	Surface area of the blood-gas (tissue) barrier (BGB) (St) per unit body mass (St g^{-1}) cm^2 g^{-1}	Surface density of the blood-gas (tissue) barrier per unit volume of the exchange tissue (St.Vx^{-1}) mm^2 mm^3	Volume of the pulmonary capillary blood (Vc) per unit surface area of the BGB (St) Vc.St^{-1} cm^3 m^{-2}	Arithmetic mean thickness of the BGB (τ_t) μm	Harmonic mean thickness of the BGB (τ_{ht}) μm	Pulmonary morphometric diffusing capacity of the BGB for O_2 (DtO_2) per unit body mass (DtO_2.kg^{-1}) ml O_2.s^{-1}.mbar^{-1}. kg^{-1}	Total pulmonary morphometric diffusing capacity of the lung for O_2 (DLO_2) per unit body mass (DLO_2 kg^{-1}) ml O_2 s^{-1} mbar^{-1} kg^{-1}
Emu	36.7	5.4	82	1.7	–	0.232	0.096	0.013
Grosbeak	25.9	40.7	314	1.5	0.584	0.121	1.283	0.080
Weaver	1.40	29.3	271	1.4	0.608	0.122	1.000	0.067
Singing	24.1	86.5	353	0.7	0.613	0.090	4.015	0.219
Cisticola	22.2	37.4	300	1.2	0.638	0.170	1.015	0.063
African Rock	29.8	63.0	389	0.9	1.030	0.096	2.902	0.118
Martin	30.2	59.4	326	1.1	0.218	0.118	2.792	–
Fiscal Shrike	–	58.6	298	–	0.762	0.151	0.954	0.062
House Spar-	23.1	29.2	317	1.5	0.675	0.124	0.989	0.110
row	27.8	49.3	247	1.0	1.124	0.141	1.446	0.069
House Spar-	23.3	33.4	342	1.3	1.012	0.120	1.569	0.059
row	21.5	37.1	313	1.3	0.599	0.127	1.137	0.077
Baglafecht								
Weaver								
Twany Prinia								
Common								
Starling								
Redwing								
Olive Thrush								

The orders, scientific names, number of specimens investigated, body masses, lung volumes and sources of data for each species are given on Table 4.1. The names of birds are listed in the same order they appear on Table 4.1

For the various species of birds, references to the works in which the investigations were published are shown in Table 4.1. RJF—red jungle fowl

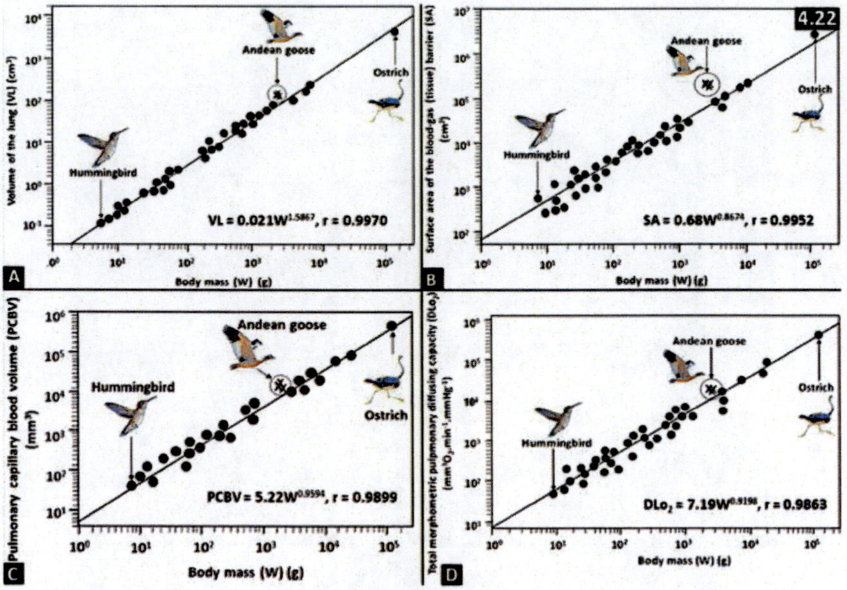

Plate 4.22 Allometric plots showing relationships between lung volume (**a**), respiratory surface area (**b**), pulmonary capillary blood volume (**c**), and pulmonary diffusing capacities of the blood-gas barrier (**d**) of lungs of species of birds that have been morphometrically investigated. The scaling factors and correlation coefficients between the structural parameters and body mass are positive and strong

f–h, 3.7a–c, e, f, h, 3.8 and 3.9), the avian lung is practically rigid (Jones et al. 1985; Makanya et al. 2021). The firm attachment of the avian lung to the boney structures of the body wall (vertebrae and ribs), the connection to the horizontal septum on the septal aspect (Plate 3.6g) and relegation of compliance to the air sacs permitted extreme subdivision of the parabronchial exchange tissue into tiny respiratory units (air capillaries) (Plate 1.7a) to occur: in a rigid avian lung, the process of compartmentalization was not extremely constrained by the need for ventilatory movements of the lung, as occurs for the mammalian lung, where the compliant structural elements are inbuilt in the organ itself. The diameters of the air capillaries range from ~3 μm in the lungs of the smallest species of birds (Duncker 1974) to ~20 μm in those of the Ostrich (*Struthio camelus*) lungs (Maina and Nathaniel 2001). In the mammalian lung, the smallest, i.e., the narrowest, alveoli, which were ~50 μm in diameter, were reported in the lung of the Etruscan Shrew (*Suncus etruscuns*) the smallest mammalian species (Fons and Sicart 1976; Fons et al. 1997; Jürgens 2002) and in an unnamed species of bat (Tenney and Remmers 1963). In the adult human lung, a total of 480 million alveoli exist and in one cubic millimeter of lung parenchyma, there are ~170 alveoli that have individual volumes of 4.2×10^6 μm^3 (Ochs et al. 2004). In an intricate manner, alveolar shape, size, and stability are maintained by surface and tissue forces (Boyle et al. 1977; Lum and Mitzner 1985; Smith and Stamenovic 1986).

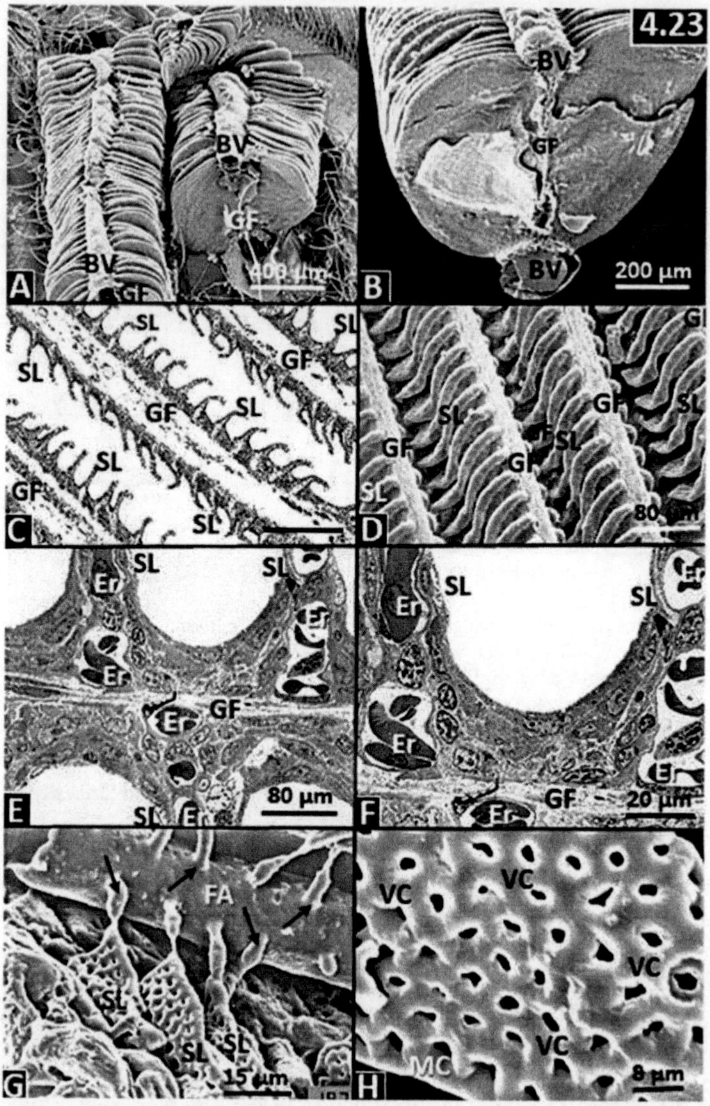

Plate 4.23 **(a–g)** Evaginated water-breathing organs such as gills achieve large respiratory surface area by a hierarchical, i.e., stratified, morphological design: **(a, b)** Gills of the crab, *Potamon niloticus* and **(c–h)** are those of a teleost fish, *Oreochromis alcalicus*. For the teleost gills, gill arches give rise to hundreds of gill filaments (GF) that in turn give rise to thousands of secondary lamella (SL) which are the terminal respiratory units. The secondary lamellae are intensely vascularized (**e–h**). BV (**a, b**), blood vessel; Er, erythrocyte; FA, gill filament artery; arrows (↓), afferent lamellar artery; MC (**h**), marginal channel; VC, vascular channels

From the minuscular sizes (diameters) of the air capillaries of the avian lung (Duncker 1971, 1972, 1974; Maina 1982a; Maina and Nathaniel 2001; Woodward and Maina 2005, 2008; Maina and Woodward 2009) (Plate 3.19), compared with the relatively larger alveoli that are more than ten times wider in diameter (Tenney and Remmers 1963), the surface tension should be considerably greater. Theoretically, if the avian lung was compliant, with the small air capillaries, much greater forces would be required to overcome surface tension forces during the ventilation of the lung. Also, the respiratory units would be greatly susceptible to collapse. While birds have lungs that are ~27% smaller and have much lower volume density of the parenchyma (~45% for the avian lungs against that of as much as ~80% in the mammalian lung) (Gehr et al. 1981; Maina 1989a, 2005; Maina et al. 1989a) (Table 4.1) compared with lungs of mammals (Maina et al. 1982a, b; Maina and King 1984; Maina 1989a; Maina et al. 1989a, 1991), the respiratory surface area, i.e., the surface area of the blood-gas barrier, is ~15% greater for animals of equivalent body mass (Maina 1989a, 1993, 2005; Maina et al. 1989a, 1991).

Among the air-breathing vertebrates, the surfactant, that lines the respiratory surface of lungs, is an important factor in pulmonary operations (Daniels et al. 1995; Sullivan et al. 1998; Daniels and Orgeig 2001, 2003). Regulation of lipid composition and maintenance of its proper fluidity is critical to the optimal function of the surfactant (Lau and Keough 1981) which is a complex mixture of lipids and proteins that lower surface tension at the air-liquid interface (Goerke 1998; Notter 2000; Orgeig and Daniels 2001; Alonso et al. 2004; Bernhard et al. 2004b): lowering surface tension reduces the work of breathing and ensures the stability of the terminal air spaces. Interestingly, the lipids that comprise the surfactant have been reported to allow oxygenation of the capillary extracellular fluid layer, promoting O_2 transfer (Olmeda et al. 2010). Although some early investigators suggested that the respiratory surface of the air capillaries of the avian lung lacked surfactant (Miller and Bondurant 1961), it was later well-authenticated (Petřík and Riedel 1968; Pattle 1978; Del Corral 1995; Bernhard et al. 2001, 2004a; Hughes and Wright 1970; Bernhard 2016) (Plates 3.25a and 4.15a–e, g). The presence of surfactant on the respiratory surface of the rigid (noncompliant) avian lung (Jones et al. 1985; Makanya et al. 2021) (Plate 2.6, 3.6 and 4.9a) is rather intriguing. After observing that the maximal surface tension of avian surfactant was lower than that of porcine surfactant, Bernhard et al. (2001) posited that avian pulmonary surfactant largely works in maintaining airflow in tubular structures rather than preventing "alveolar" (air capillary!) collapse and that surfactant mostly made of dipalmitoylphosphatidylcholine and surfactant proteins B is adequate for maintaining the patency of the air capillaries. Albeit gross morphological differences, the composition and concentration of the surfactant per unit respiratory surface of the lungs of mammals and birds are equivalent (Fujiwara et al. 1970). Two forms of surfactant, namely, the lamellated osmiophilic bodies and the trilaminar substance exist on the respiratory surface of the avian lung (McLelland 1989): the former form resembles the general air-breathing vertebrates and is produced by merocrine secretion from the type-II epithelial cells which in the avian lung are located on the surfaces of the atria and the infundibulae (Plate

2.11j, k) and not directly on the respiratory surface, i.e., on the air capillaries (Plates 2.11l and 3.17f) while the later (form of surfactant) is unique to birds (Akester 1970). Bernhard et al. (2001) observed that in the lungs of ducks and chickens, while surfactant protein B exists, tubular myelin is lacking.

In the vertebrate lung, the other important functions of the surfactant in lungs are (a) prevention of adhesion of respiratory surfaces during breathing (Daniels et al. 1998a, b; Daniels and Orgeig 2001; Foot et al. 2006); (b) protection of the respiratory surface from the injurious effects of pathogenic microorganisms (Ohmer-Schröck et al. 1995; Creuwels et al. 1997; Ferguson et al. 1999; Crouch and Wright 2001; Wright 2005; Nayak et al. 2012); (c) prevention of exudation of blood plasma onto the respiratory surface (Nicholas 1996; Daniels et al. 1998a, b; Alonso et al. 2005); and (d) relaxation of the airway smooth muscle by a prostanoid-mediated, epithelium-dependent effect that does not involve nitric oxide (NO) (Koetzler et al. 2006). For the noncompliant avian lung, it is possible that these processes are more important than lowering surface tension and stabilization of the air capillaries. Fedde (1980) suggested that the surfactant may be a leftover from the evolution of the reptilian lung (Perry and Sander 2004): it may be particularly involved in the prevention of leaking of blood plasma onto the respiratory surface rather than lowering surface tension. Pattle (1978) proposed that the trilaminar substance may be associated with roles such as the coagulation of blood that leaks across the blood-gas barrier, hydration of the surfactant, and absorption of fluid that may accumulate on the respiratory surface. After hatching, the trilaminar substance is largely located in the type-II cells of the atria and the infundibula and lasts for about 2 weeks on the surface of the air capillary after hatching (Jones and Radnor 1972a, b; Pattle 1978). While in the developing lung, the lamellated osmiophilic body form of surfactant develops before the trilaminar substance, interestingly, in some occasions, the two materials/substances coexist in the same epithelial cells (Pattle 1978). Albeit the fact that very little of it exists on the respiratory surface, Scheuermann et al. (1997, 2000) attributed the remarkable strength of the air capillaries to the presence of the trilaminar substance, especially in a mature lung. Klika et al. (1997) suggested that the trilaminar substance in the cytoplasm of the type-I cells of the avian lung forms an intercapillary "skeletal" system that comprises a supporting system.

The respiratory surface area of the avian lung correlates strongly ($r = 0.9952$) with body mass (Plate 4.22b). For the lungs of a 45 kg body mass Ostrich (*Struthio camelus*), the surface area of the air capillaries which is 183 m^2 is contained in a total lung volume of 1.6 L (Maina and Nathaniel 2001): the values are very close to that of a singles tennis court (Plate 2.3). Among the lungs of species that have been investigated, with a value of 96.5 $cm^2\ g^{-1}$, the Andean Goose (*Chloephaga melanoptera)* has the largest mass-specific respiratory surface area (Maina et al. 2017) and with values of 87.1 and 86.5 $cm^2\ g^{-1}$, respectively, the Violet-Eared Hummingbird (*Coribri coruscans*) (Dubach 1981) and the African Rock Martin (*Hirundo fuligula*) (Maina 1984, 1989a, Maina et al. 1989a) are the other bird species with considerably high mass-specific respiratory surface areas (Table 4.3). The metabolic capacities of birds, attributes that are determined by factors such as lifestyles pursued, habitat occupied and body sizes vary greatly (McNab and

Morrison 1963; Kendeigh and Blem 1974; Hails 1983; O'Conner 1996; Merola-Zwartjes and Ligon 2000; McKechnie and Wolf 2004; McKechnie et al. 2006; White et al. 2007; Figueroa et al. 2007). After comparing various morphometric parameters of the lungs of some flying and nonflying species of mammals and birds, Figueroa et al. (2007) determined that the respiratory surface density (i.e., the surface area of the blood-gas barrier per unit volume of the exchange tissue) was the most informative parameter of the energetic demands for powered flight. The Andean Goose (*Chloephaga melanoptera*) is a large bird that inhabits the high altitude of the Andes and flies to an elevation of 6000 m (Fjeldså and Krabbe 1990). Hummingbirds (Family: Trochilidae) are the smallest extant group of birds. Especially when feeding on flower nectar, they hover for substantial times. Hovering is the most highly energetically costly mode of locomotion (Winter 1998; Voigt and Winter 1999; Ward et al. 2001; Morris et al. 2010; Sholtis et al. 2015; Skandalis et al. 2017). The African Rock Martin (*Hirundo fuligula*) is a small passerine bird that flies swiftly over long distances (Turner and Rose 1989). The lowest mass-specific respiratory surface areas have been found in flightless or weakly flying species of birds (Table 4.3): some of these are the Emu (*Dromaius novaehollandiae*) (5.4 $cm^2\ g^{-1}$) (Maina and King 1989), the Domestic Fowl (*Gallus gallus* variant *domesticus*) (8.7 $cm^2\ g^{-1}$) (Abdalla et al. 1982; Vidyadaran et al. 1987, 1988), the Humbolt Penguin (*Spheniscus humboldti*) (18.1 $cm^2\ g^{-1}$) (Maina and King 1987) and the Guinea Fowl (*Numida melleagris*) (7.0 $cm^2\ g^{-1}$) (Abdalla and Maina 1981). The Emu is a large Australian bird that evolved in a habitat (a continent) with a low level of predation, especially from placental mammals that did not exist before human external settlement occurred. In nonflying mammals, the highest mass-specific respiratory surface area of 43 $cm^2\ g^{-1}$ has been reported in the shrews, *Crocidura flavescens*, and *Sorex* sp. (Gehr et al. 1980): the shrews are the smallest and most highly metabolically active mammals (Fons and Sicart 1976; Buckner 2001; Hawkins et al. 2009).

The surface density of the blood-gas barrier expresses the intensity of the subdivision (compartmentalization) of the exchange tissue and therefore the size of the terminal respiratory units (Maina 1989a, 1993, 2005; Maina et al. 1989a). Among the lungs of the avian species that have been investigated, the surface density of the blood-gas barrier correlates negatively with body mass (Maina 1989a; Maina et al. 1989a; Maina 2002d, 2005, 2006). The highest surface density of the blood-gas barrier of 330 $mm^2\ mm^3$ has been reported in the lung of the Andean goose (*Chloephaga melanoptera)* (Maina et al. 2017): the lowest values are those of 82 $mm^2\ mm^3$ in the lungs of the Emu (*Dromaius novaehollandiae*) (Maina and King 1989) (Table 4.3) and 231 $mm^2\ mm^3$ in those of the weakly flying Chilean Tinamou (*Nothoprocta perdicaria*) (Figueroa et al. 2007). For the Andean goose, the exchange tissue of the lung is exceptionally intensely subdivided and vascularized (Plates 3.15e, f, k and 3.17i–l) and unlike those of the other species of birds that have been investigated such as the Domestic Fowl (*Gallus gallus* variant *domesticus*) (Plate 3.17a–f), the infundibula of the lung of the Andean Goose are unusually well-vascularized (Plate 3.17j–l) and the air capillaries that are about the size of the erythrocytes (Plate 3.15k, l), are very small for birds of their size (2.6 kg) (Maina et al. 2017). Compared with the larger species of birds, operating at the relatively higher body temperatures of 39–44 °C (Wojciechowski and Pinshow 2009), the generally small passerine species with relatively greater metabolic rates (Petrov and Razuvaeva 2018) have larger mass-specific respiratory surface areas (Maina 1989a, 1993, 2005; Maina et al. 1989a) (Table 4.3).

4.8.3 Blood-Gas Barrier

Compared with that of the mammalian lung in which thin and thick sides of the interalaveolar septum exist (Plate 3.25b), the blood-gas barrier of the avian lung is more uniform in thickness (Plates 3.25c and 4.16c). Compared with the mammalian lungs, the blood-gas barrier of the avian lung is generally ~2.5 times thinner (West 2009) and for animals of equivalent body mass, birds have a blood gas barrier that is 56–67% thinner than that of a mammal (Maina et al. (1989a). For the lung of the Domestic Fowl (*Gallus gallus* variant *domesticus*), the epithelium, the basement membrane (lamina), and the endothelium comprise ~66.7%, 21%, and 12.3%, respectively, of the volume of the blood-gas barrier (Maina and King 1982a). The notably thin blood-gas barrier of the avian lung (Dubach 1981; Maina and King 1982a, b; Maina 1989a; Maina et al. 1989a) and generally thin the mammalian one (Weibel 1973) derives from the nonexistence of an interstitial space between individual basal laminae of the epithelial and the endothelial cells, a feature which characterizes the lungs of the lower vertebrates, i.e., the amphibian and the reptilian lungs, where connective tissue elements such as collagen and elastic tissue fibers and cellular elements such as pericytes are located (Goniakowska-Witalinska 1978, 1986, 1995; Meban 1980): in the lungs of birds and mammals, the basal lamina (membranes) of the epithelial, and the endothelial cells have fused into a common basement membrane (Weibel 1973; Maina and King 1982a) (Plates 2.1b–j, 3.25a, b, 4.15, 4.16f and 4.17a–c, e).

During the development of the blood-gas barrier of the avian lung (Plate 2.16), cellular reorganization, remodeling, and apoptotic processes lead to the configuration of a tripartite blood-gas barrier. The strategic location of most of the perikarya of the type-I and practically all the type-II cells away from the respiratory surface (i.e., the air capillaries) should contribute to the development of a thin blood-gas barrier. With their perikarya remotely located from the respiratory surface (i.e., on the surfaces of the infundibula and atria), the type-1 cells send extremely thin and long cytoplasmic extensions to the surface of the air capillaries (Plates 2.1b, j, 3.25a and 4.15). In rare cases where the perikarya of the type-I cells exist in the gas exchange tissue, the cells are located away from the blood-gas (tissue) barrier (Plate 4.18l, m).

In avian lungs, the blood-gas barrier is markedly sporadically attenuated (Meban 1980; Maina and King 1982a; Watson et al. 2007) (Plates 2.1e, j and 4.15a–f). Measurements of the thinnest parts of the blood-gas barrier have been determined in some amphibian, reptilian, mammalian, and avian lungs (Meban 1980; Maina and

King 1982a). The ratio of the arithmetic mean thickness (τ_t) and the harmonic mean thickness (τ_{ht}) of the blood-gas barrier registers the degree of corrugation (Weibel and Knight 1964; Weibel 1973; Maina and King 1982a; Maina et al. 1989a). For the lung of the Graylag Goose (*Anser anser*), while the τ_{ht} is 0.112 μm, the mean thickness of the thinnest parts of the blood-gas barrier is 0.050 μm (Maina and King 1982b; Maina 1987). Among the lungs of the species of birds that have been studied, the greater τ_t:τht_t ratio is 10.8 in the House Sparrow (*Passer domesticus*) (12 g) (Maina 1984, 1989a, 2005; Maina et al. 1989a) and the lowest is 1.2 in the Ostrich (*Struthio camelus*) (45 kg) (Maina 2005; Maina and Nathaniel 2001). The sporadic attenuation of the blood-gas barrier indicates a compromise design where the thin parts promote gas exchange while the thicker ones guarantee the mechanical integrity of the blood-gas (tissue) barrier (Weibel and Knight 1964; Weibel 1970/71; Weibel 1973; Meban 1980; Abdalla et al. 1982; Maina and King 1982a; Maina and West 2005; Leiby et al. 2020).

Among the lungs of species of birds that have been investigated (Table 4.3), the Violet-Eared Hummingbird (*Colibri coruscans*) (7.3 g) (Dubach 1981) and the African Rock Martin (*Hirundo fuligula*) (Maina 1984, 1989a, 2005; Maina et al. 1989a), with harmonic mean thicknesses (τ_{ht}) of 0.099 and 0.090 μm, respectively, the two species have the thinnest blood-gas barriers: the thickest ones occur in the lung of the Ostrich (*Struthio camelus*) (0.56 μm) (Maina and Nathaniel 2001) and the Humboldt Penguin (*Spheniscus humboldti*) (0.53 μm) (Maina and King 1987). The harmonic mean thickness of the blood-gas barriers of the virtually flightless Chilean Tinamou (*Nothoprocta perdicaria*) is 0.469 μm (Figueroa et al. 2007). Welsch and Aschauer 1986) attributed the particularly thick blood-gas barrier of the lung of the Emperor Penguin (*Aptenodytes forsteri*) to the presence of collagen fibers in the basement membrane and ascribed the thickness to the requirement of tolerance of hydrostatic pressure during underwater dives. Compared with nonflying mammals of equivalent body mass, the τ_{ht} of blood-gas barriers of the avian lung are 56 to 67% thinner (Maina 1989a, 2005; Maina et al. 1989a). The low, positive, slopes (scaling factors) of plots between body mass and thickness of the blood-gas barrier in birds ($r = 0.4512$), bats ($r = 0.3342$), and nonflying mammals ($r = 0.8115$) (Plate 4.24) show that in at least these animal taxa, the thickness of the blood-gas (tissue) barriers have been optimized for gas exchange by diffusion (Maina and King 1984; Maina 2006). For the nonvolant mammals, while the thickness of the blood-gas barrier in a 2.2 g shrew (*Suncus etruscus*) is 0.230 μm (Gehr et al. 1980), the corresponding value in the lung of a 54 tonne Bow-Head Whale (*Balaena mysticetus*) (Jefferson et al. 1993; Reeves et al. 2002) is 0.35 μm (Henk and Haldman 1990): the factorial differences between the thicknesses of the blood-gas barriers and body masses are 1.5 and 2.5.10, respectively.[7] The thickness of the blood-gas barrier in a 5 g Pipistrelle Bat (*Pipistrellus pipistrellus*) is 0.206 μm (Maina and King 1984) while that of a 900 g Flying Fox (*Pteropus poliocephalus*) (one of the largest bat species) is 0.303 μm (Maina et al. 1991): the factorial differences between the thicknesses of the blood-gas barrier and the body masses are 1.5 and 180 and in birds, the thicknesses of the blood-gas barrier range from 0.090 μm in a 7.3 g Violet-Eared Hummingbird (*Colibri coruscans*) (Dubach 1981) to 0.56 μm in a 45 kg Ostrich (*Struthio camelus*)

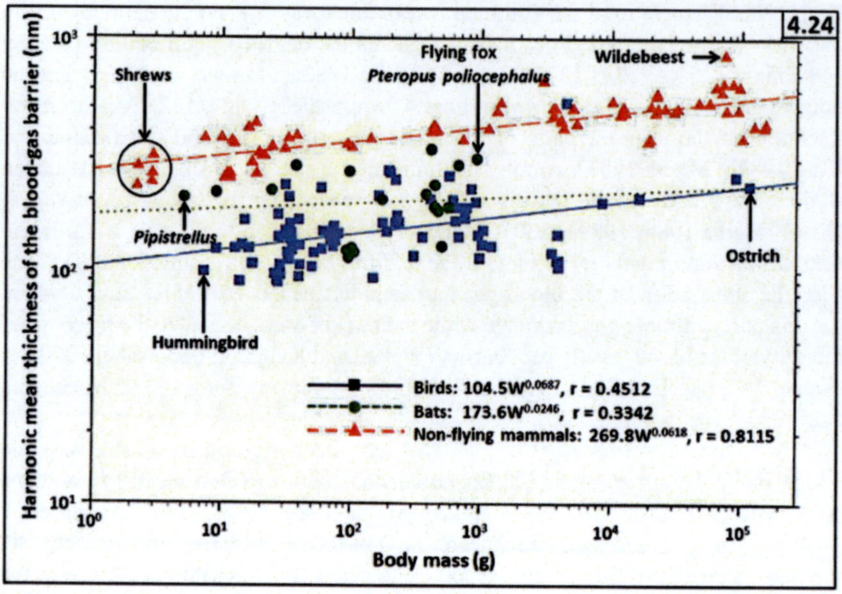

Plate 4.24 The allometric plot of the relationship between the thickness of blood-gas barrier (harmonic mean thickness) (τht) and body mass of lungs of birds, bats, and nonflying mammals. The parameter τht scales and correlates weakly with body mass

(Maina and Nathaniel 2001): the factorial differences between thicknesses of the blood-gas barrier and the body masses are are 6.2 and 6.2×10^3. On average, for the three species of animals, the average factorial difference of body mass is 2.5×10^4 and that of the thickness of the blood-gas barrier is only 3.8: it shows the allometric optimization of the thickness of the blood-gas (tissue) barrier.

4.8.4 Pulmonary Capillary Blood Volume

In gas exchangers, pulmonary efficiency is determined by the pulmonary capillary blood volume that is largely reflected in the degree of vascularization of the respiratory surface (Plate 4.25) and the manner the blood is exposed to the external respiratory fluid medium (water or air) (Plates 4.23e–h and 4.25). Protrusion of the blood capillaries from the supporting framework (Plate 4.26) increases respiratory surface area (Maina 1998, 2000a, 2004) (Plate 4.25). For the avian lung, the volume of blood comprises as much as 36% of the volume of the lung with 58–80% of it located in the blood capillaries (Duncker and Güntert 1985a, b; Maina 1989a, 2005; Maina et al. 1989a). The pulmonary capillary blood volume of the avian lung is 2.5 to 3 times larger than that in the exchange tissue (parenchyma) of a mammalian lung for animals of equivalent body mass: in the mammalian lung, 20% of blood is found

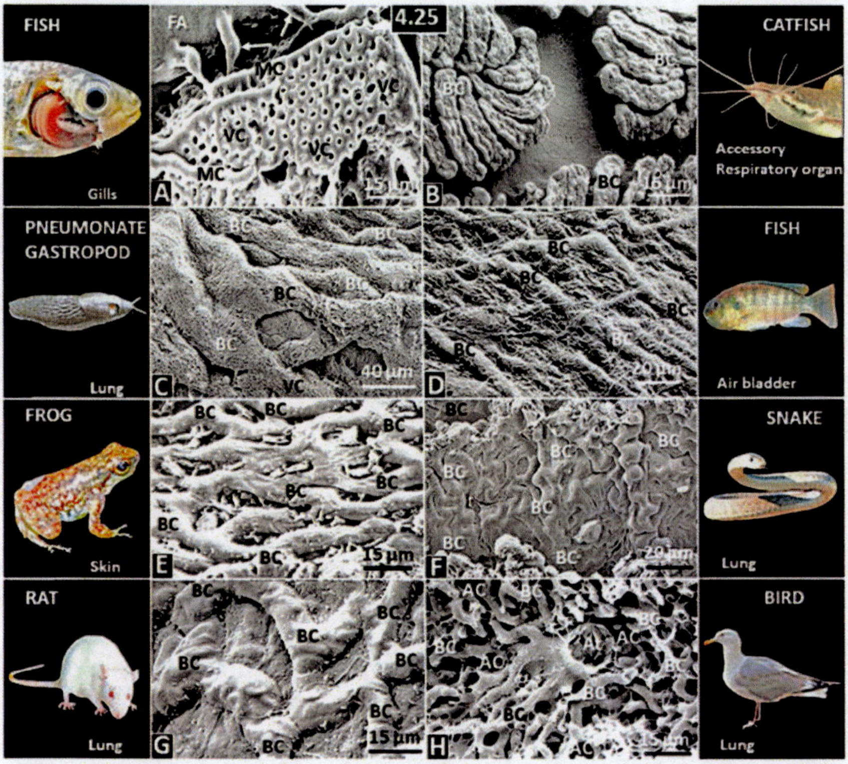

Plate 4.25 In gas exchangers, pulmonary capillary blood volume is increased by intensification of vascularization of the respiratory surface(s). This is seen in the vasculalization of the secondary lamellae of the fish, *Oreochromis alcalicus*, gills (**a**); the accessory respiratory organs of the catfish, *Clarias mossambicus* (**b**), the lung of the pneumonate gastropod, *Trichotoxon copleyi* (**c**), the air bladder of a fish, *Oreochromis alcalicus* (**d**), the amphibian skin (**e**), the lung of a snake—Black Mamba (*Dendroaspis polylepis*) (**f**), the mammalian lung—the Lesser Bushbaby (*Galago senegalensis*) and the gas exchange tissue of the lung of the avian lung—the Domestic Fowl (*Gallus gallus* variant *domesticus*) (**g**). FA, gill filament artery; arrows (↓), afferent secondary lamellae arterioles; MC (**a**), marginal channel; BC, blood capillaries; At (**h**), arteriole; AC, air capillary

in the alveolar blood capillaries (Weibel 1963). The avian lung has a pulmonary capillary blood volume that is 22% greater than that of a nonflying mammal of equivalent body mass (Maina et al. 1989a). Morphometrically, the pulmonary capillary blood volume of the avian lung correlates strongly ($r = 0.9899$) with body mass (Plate 4.22c). For the African Rock Martin (*Hirundo fuligula*), a small exceptionally energetic passerine bird (Turner and Rose 1989; Hockey et al. 2005), blood comprises 29% of the volume of the lung, with 79% of it found in the blood capillaries (Maina 1984, 2005). The intense interconnectedness (anastomoses) of blood capillaries in organs such as gas exchangers has been termed "sheet-flow model" by investigators such as Fung and Sobin (1969), Sobin et al. (1970) and

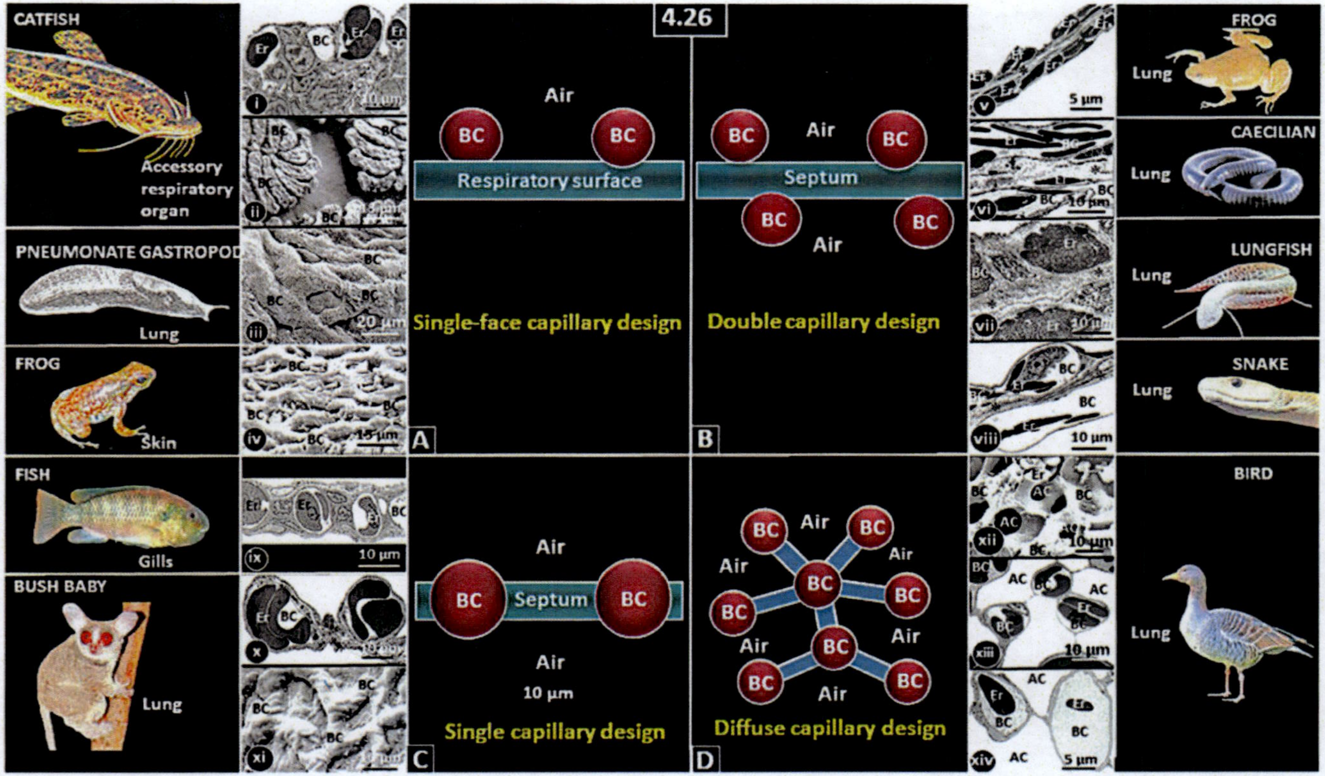

Plate 4.26 Exposure of blood in the pulmonary blood capillaries of gas exchangers to the external respiratory fluid medium (water/air). Generally, animal taxa with higher metabolic capacities display more optimal exposures. (**a**) (i–iv): "Single-face capillary system" that occurs in accessory respiratory organs of air-breathing fish (i, ii), lungs of pneumonate gastropods (iii) and amphibian skin (iv): blood is exposed to water/air (external respiratory fluid medium) only partly,

only the part of the vascular structure that protrudes above the respiratory surface. **(b)** "Double capillary system" (v–viii) occurs in lungs of animal taxa such as amphibian, reptilian, and lungfish lungs: capillary blood is exposed to air on alternative sides of the interfaveolar septa, i.e., the supporting columns. **(c)** (ix–xi): "Single capillary system" occurs in gas exchangers such as mammalian lung: capillary blood is exposed to air on both sides of the interalveolar septum. **(d)** (xii–xiv): "Diffuse capillary system" exists in the exchange tissue of the avian lung: capillary blood is exposed to air across the blood-gas barrier about all around the wall. Asterisk (*), interfaveolar septum; BC, blood capillary, AC, air capillary; Er, erythrocyte

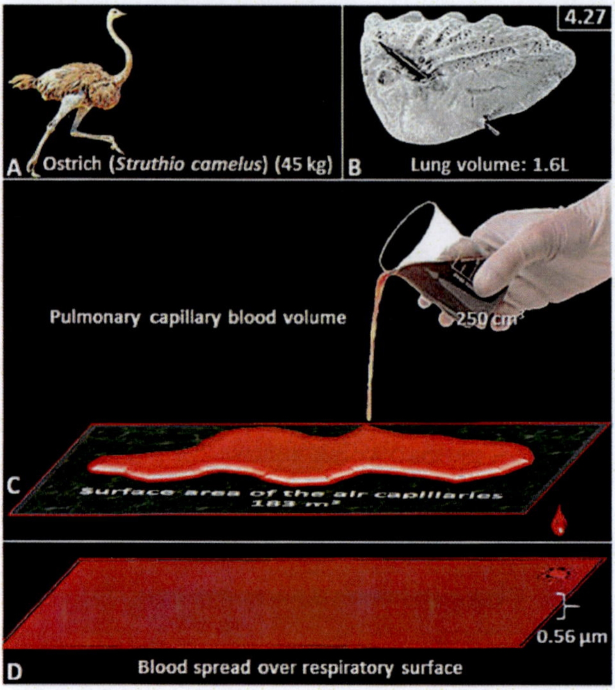

Plate 4.27 (**a**) A 45 kg (immature) Ostrich (*Struthio camelus*) (**a**) has a total lung volume of 1.6 L (**b**) that has a pulmonary capillary blood volume of 250 cm^3 (**c**). According to the "sheet flow concept," the blood is spread over an area (183 m^2) (**c, d**) that is equivalent to that of a singles tennis court: the film of blood is separated from the air in the air capillaries across a blood-gas barrier which is 0.56 µm thin (harmonic mean thickness) (**d**)

Maina (2000b) (Plates 4.23g, h and 4.25g, h). For the 1.6 L total lung volume of a 45 kg weight Ostrich (*Struthio camelus*), 250 cm^3 of blood is literally spread out across a surface area of 183 m^2 and separated from the pulmonary blood capillaries across the blood-gas barrier that is 0.56 µm thin (Maina and Nathaniel 2001) (Plate 4.27a–c).

The ratio of the volume of the pulmonary capillary blood to the respiratory surface area or the surface area of the blood-gas barrier has been designated "capillary loading" (Perry 1976, 1983, 1990a, b; Perry et al. 1994): it specifies the volume of blood that is exposed to the respiratory fluid medium per unit respiratory surface area. In the lungs of species of birds that have been investigated (Table 4.3), the lowest values are those of 0.7 and 0.8 $cm^3\ m^{-2}$, respectively, in the lungs of the African Rock Martin (*Hirundo fuligula*) (Maina 1984, 1989a; Maina et al. 1989a) and the Andean Goose (*Chloephaga melanoptera)* (Maina et al. 2017) to 1.7, 2.1, and 4.4 $cm^3\ m^{-2}$ in the lungs, respectively of the Emu (*Dromaius hollandiae*) (Maina and King 1989), the Ostrich (*Struthio camelus*) (Maina and Nathaniel 2001) and the Humboldt Penguin (*Spheniscus humboldti*) (Maina and King 1987).

Diving animals have relatively larger circulatory blood volumes in their bodies that serve an O_2-storing role (Kooyman 1973; Butler and Jones 1997; Costa et al. 1998; Davis and Kanatous 1999; Butler 2004; Panneton 2013). Small, more highly energetic birds (Maina 1984) and distinctive ones such as the high-altitude living and flying Andean Goose (*Chloephaga melanoptera*) (Maina et al. 2017) have low pulmonary capillary loading (Table 4.3). In ectotherms, capillary loadings as high as 13 $cm^3\ m^{-2}$ have been reported in the lung of the Turtle, *Pseudemys scripta elegans* (Perry 1978), and 12–19 $cm^3\ m^{-2}$ in that of the lungfish, *Lepidosiren paradoxa* (Hughes and Weibel 1976).

In gas exchangers, exposure to pulmonary capillary blood is an important attribute of respiratory efficiency. Single-face blood capillary/hemolymph exposure occurs in gas exchangers such as the accessory respiratory organs of catfishes (Maina and Maloiy 1986) (Plates 4.14b, c, 4.25b and 4.26a, i, ii), lungs of the pneumonate gastropods (Maina 1989b) (Plates 4.14d, e and 4.25c), the amphibian skin (Maina 1989c) (Plate 4.25e) and the physostomatous (respiratory) swim (air) bladders of air-breathing fish (Maina et al. 1996; Maina 2000c, 2009a) (Plates 4.14f, g and 4.25d): capillary blood is exposed to air only for a small part of the walls of the blood-capillaries, i.e., the blood-gas barrier. Double capillary exposure occurs in blood capillaries of the amphibian, the reptilian, and the lungfish lungs (Hughes and Weibel 1976; Goniakowska-Witalinska 1978, 1986, 1995; Meban 1980; Perry 1983; Maina and Maloiy 1985, 1988; Maina 1989c, d): with blood capillaries located on opposite sides of supporting structure, e.g., the intrapulmonary septa, blood is exposed to air only on one side of the wall of the blood capillary (Plates 4.25f and 4.26b, v–viii); single capillary blood exposure occurs in gas exchangers such as the secondary lamellae of fish gills (Maina 1990b) (Plates 2.1a, 4.14a, 4.25a and 4.26c, ix) and the mammalian lung (Weibel 1973, 1984; Maina and Igbokwe 2020) (Plates 3.25b, 4.17d, e and 4.26c, x, xi): blood is exposed to air on both sides of the interalveolar septum where in certain instances thin and thicker sides exist (Plates 3.25b and 4.26c, x) and; diffuse pulmonary capillary blood exposure exists in the exchange tissue of the avian lung where blood is exposed to air in the air capillaries practically all round (Plates 3.21, 3.25c, 4.16c and 4.26d, xii–xiv). An exceptionally efficient "direct delivery system" exists in insects (Maina 1989e) where trachea transport air from the atmosphere to the cells of the tissues of the body (Plate 3.1a) and in some highly metabolically active tissues, the tracheoles penetrate the cell wall and deliver O_2 directly to the mitochondria (Plates 3.2f–h and 4.3d—inset). In lungs of three small passerine species of birds of similar body mass but displaying various diurnal activities, namely, the Welcome Swallow (*Hirundo neoxena*) (an aerial forager), the Brown Honeyeater (*Lichmera indistincta*) (a nectarivorous and insectivorous species that does not very much engage in aerial foraging) and the Splendid Wren (*Malurus splendens*) (that forages mostly at ground level), morphometric analysis conducted by Vitali and Richardson (1998) showed correlation between diurnal activity and pulmonary volumetry: it was noted that the only significant difference between the species of birds was that of exchange tissue plasma volume and pulmonary hematocrit, with the ground-foraging, low activity *Malurus splendens* having significantly lower values. Between the species of birds investigated by Vitali and Richardson (1998), cardiovascular parameters were more important in determining metabolic activities in small birds compared with the properties relating to pulmonary structure.

4.8.5 *Mathematical Modeling of the Diffusing Capacity of the Avian Lung for Oxygen*

In complex life forms, O_2 transfer from the external respiratory fluid medium (water/air) to the mitochondria, where it is utilized for energy production in the form of adenosine triphosphate (ATP), happens across an elaborate pathway that consists of various tissues and spaces, the so-called "O_2cascade" (Weibel 1984; Ben-Tal 2006; Martin et al. 2008; Baffico et al. 2010). In effect, depending on the respiratory fluid medium utilized (water or blood), respectively, for gills and lungs, the pathway starts at the level of the blood-water/blood-gas barriers and terminates with O_2-utilization at the mitochondrial level (Plate 4.2): O_2consumption at the mitochondrial level drives the flow of O_2 from outside into the body. According to Fick's Law (Fick 1855), the diffusing capacity or the conductance of O_2 (DO_2) by a gas exchanger correlates directly with the respiratory surface area (S_a) and inversely with the thickness of the blood-gas barrier which is suitably expressed in the harmonic mean thickness (τht) (Weibel and Knight 1964; Weibel 1970/71; Weibel et al. 1993) (Plates 4.17a–c and 4.28a, b). Thus:

$$DO_2 = KtO_2.\Delta PO_2.Sa.\tau ht^{-1}$$

where K_tO_2 is Krogh's permeation coefficient of O_2 across the blood-gas (tissue) barrier and ΔPO_2 the partial pressure gradient of O_2 across the tissue barrier.

Fick's Law is the physiologist equivalent of Ohm's Law of electricity,

$$I = U \cdot R^{-1}$$

where:

I is the electric current; U is the potential difference, i.e., the voltage, between two points of a conductor, and R is the resistance.

Structurally, gas exchangers comprise various tissue compartments and spaces which are involved in transporting O_2 and CO_2, respectively, between the external respiratory fluid medium (milieu extérieur) (water/air) and the internal compartments of the body and between the cells of the body to the outside. The flux of O_2 occurs by passive diffusion under an existing partial pressure gradient (ΔPO_2) (Roughton and Forster 1957; Forster 1996).

Pulmonary modeling is a robust mathematical method of communicating or expressing the functional design of a gas exchanger: O_2 molecules traverse various spaces between the respiratory surface to the cytoplasm of the erythrocytes (the

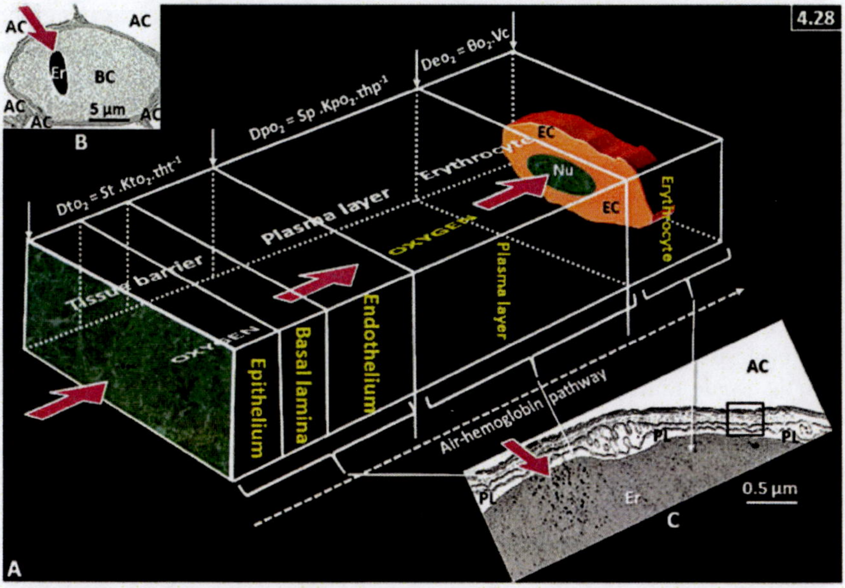

Plate 4.28 (**a**) The air-hemoglobin pathway of the avian lung (large pink arrows) comprises a thin blood-gas barrier (that consists of a surfactant lining, an epithelial cell, a basement lamina, an endothelial cell), a plasma layer (PL), and erythrocyte (Er). EC, erythrocyte cytoplasm; Nu, nucleus. From relevant morphometric parameters of the structural components that comprise the air-hemoglobin pathway and their physical constants/coefficients that determine the flow of oxygen and binding of molecular oxygen to hemoglobin, the different diffusing capacities and the total morphometric diffusing capacity of the lung for oxygen can be determined. The diffusing capacity of the blood-gas (tissue) barrier is calculated from the surface area of the blood-gas barrier (St), the oxygen diffusing coefficient of O_2 through the blood-gas barrier (Kto_2), and the harmonic mean thickness of the blood-gas barrier (τht); the diffusing capacity of the plasma layer for oxygen (DpO_2) is calculated from oxygen diffusing coefficient of the plasma layer (KpO_2), the surface area of the plasma layer (Sp) and the harmonic mean thickness of the plasma layer (τhp); DeO_2 or the oxygen diffusing capacity for the erythrocyte (Er) is calculated from the oxygen uptake coefficient of the whole blood (θO_2) and the volume of the pulmonary capillary blood (Vc). The total morphometric pulmonary diffusing capacity of the lung (DLO_2) is calculated from the reciprocals of the diffusing capacities of the blood-gas barrier, the plasma layer (DpO_2), and the erythrocytes (DeO_2). Insets—(**b**) and (**c**) Cross-section of a pulmonary blood capillary (**b**) and electron micrograph (**c**) showing the air-hemoglobin pathway (large pink arrow). Boxed (□) area (**c**), blood-gas (tissue) barrier; plasma layer (PL); erythrocyte (Er); AC, air capillary; BC, blood capillary; EC (**a**), erythrocyte cytoplasm; Nu (**a**), nucleus

so-called "air-hemoglobin pathway"), where it (O_2) is biochemically bound to hemoglobin (Plates 4.17a–c and 4.28a, b). Essentially, the air-hemoglobin pathway comprises the following parts and spaces: a very thin surfactant lining, the blood-gas (tissue) barrier (that comprises an epithelial cell, a basement membrane, and an endothelial cell), a plasma layer, and the erythrocyte cytoplasm that contains hemoglobin (Plates 2.1b–j, 4.15, 4.16c, e, 4.17a, b, d, e and 4.28b). With the structures and spaces of the air-hemoglobin pathway arranged in series, O_2 molecules pass in succession across them (Plates 4.17 and 4.28a, b). The structural components of the air-hemoglobin pathway comprise materials of different constructions and consistencies and hence present various resistances or conductances to O_2diffusion. The avian erythrocytes, that are nucleated (Plates 2.13h, i, 2.14c, d–f, h, i, 2.16a–f, 3.22a, h and 4.18d, m) and contain mitochondria in their cytoplasm (Staub et al. 1962; Schweiger 1962). Compared with the anucleate ones of mammals (at maturity and under normal conditions), the erythrocytes of birds are metabolically more active (Lutz et al. 1973; Boutilier and Ferguson 1989).

Biological models define the mechanisms and processes that direct the working or functioning of biological systems. For the gas exchangers, an instructive mathematical model "separates" and identifies the most functionally germane structural parts that comprise the air-hemoglobin pathway and then "integrates/assembles" them into the whole structure. Gutmann and Bonik (1981) termed a mathematical model as "an abstraction of a real situation that describes only the essential aspects of the situation"; Scheid (1987) designated a model as "an image of part of the physical or mental world apt to explain or predict observations"; Scheid (1987) stated that "models are not only helpful but often indispensable in quantitative biology" while; Powell and Scheid (1989) pointed out that "in an attempt at deriving a functional model for gas exchange from morphologic evidence, the physiologist has to identify the simplest functional subunit in the gas exchange organ" and added that "many anatomical details have to be left out [of model] to allow quantitative treatment of gas exchange."

Gas exchangers are elaborate constructions in which the constituent parts are structurally and functionally highly integrated (Weibel 1984; Maina 1998, 2011). Mathematical models have been developed and used to understand the functional design of especially the mammalian (Weibel 1970/71; Weibel et al. 1983; Gehr et al. 1981; Maina et al. 1991; Maina 1993) and the avian lungs (Maina 1989a; Maina et al. 1989a, 2017). Morphometric parameters such as respiratory surface area, pulmonary capillary blood volume, and thickness of the blood-gas barrier (Gehr et al. 1981; Weibel et al. 1983; Vidyadaran et al. 1987, 1988, 1991; Maina 1989a) are applied to model gas exchangers: modifications and refinements can occur anywhere along the air-hemoglobin pathway. This was determined for the lung of the Humboldti Penguin (*Spheniscus humboldti*) by Maina and King (1987) who found that the blood-gas (tissue) barrier was particularly thick (0.53 μm), a feature that gave very low morphometric diffusing capacity of the tissue barrier (DtO_2) but the high pulmonary capillary blood volume afforded particularly large diffusing capacity of the erythrocytes (DeO_2). On the whole, compared with other species of birds of equivalent body mass, because of specialization at the level of the erythrocyte, the penguin lung had relatively higher total morphometric pulmonary diffusing capacity (DLO_{2m}) (Maina and King 1987) (Table 4.3). The observation on the lung of the penguin (Maina and King 1987) shows that determining and modeling only one part of the elaborate O_2diffusion pathway provides inadequate and in some cases misleading data: for the penguin, if only the anatomical diffusing factor (ADF) of the blood-gas (tissue) barrier was determined, it would wrongly have shown that compared with those of other birds, the penguin's lung was functionally inferior.

The ADF for O_2 ($ADFO_2$) is the ratio of the surface area of the blood-gas (tissue) barrier to its thickness. It has been used by some investigators to specify and compare the functional capacities of gas exchangers (Strazny and Perry 1984; Perry 1989b, c, 1990a, b; Perry et al. 1994; Luis da Cruz et al. 2009; Starck et al. 2015). Because it concerns only part of the air-hemoglobin pathway, i.e., the blood-gas (tissue) barrier, ADF falls short of fully defining the efficiency of a gas exchanger. For the Humboldti Penguin (Maina and King 1987), like for the lungs of diving air-breathing vertebrates in general and birds in particular (Irving et al. 1935; Irving 1939; Lenfant et al. 1970; Snyder 1983; Butler and Jones 1997; Hassrick et al. 2010; Blix 2018), adaptively, high pulmonary capillary blood volume per unit respiratory surface area (capillary loading) promotes the morphometric diffusing capacity of the lung (DLO_2m) (Table 4.3).

The diffusing capacities of the various parts of the air-hemoglobin pathway are determined from the relevant morphometric parameters, i.e., respiratory surface area, pulmonary capillary blood volume and thicknesses of the blood-gas barrier, the plasma layer and the physicochemical coefficients of O_2 transfer across the tissue barriers and spaces and the O_2 uptake by hemoglobin (θo_2). In a slightly revised model, Weibel et al. (1993) advanced the calculation of the diffusing capacity of the blood-gas (tissue) barrier (DtO_2) and the plasma layer (DpO_2) as the membrane diffusing capacity (DMO_2). Krogh's permeation coefficient of the blood-gas barrier (KtO_2) is the product of solubility (α) and diffusion coefficients (D) of O_2 (Roughton and Forster 1957; Forster 1996). For the reason that α and D are affected by temperature in opposite directions, i.e., solubility of O_2 decreases at higher temperatures while diffusion increases at higher ones: the permeation constant (K) (KtO_2) is not affected much by changes in temperature.

From the unique structure of the avian lung (Duncker 1971; King 1966; McLelland 1989; Maina 2005), certain modifications and assumptions are necessary for the mathematical modeling of the avian lung (Maina 1982b, 1989a, 2005; Maina et al. 1989a) and application of mammalian permeation constants (K) is unavoidable because equivalent values on avian tissues are largely lacking: practically, all the existing values have been determined on mammalian tissues (Krogh 1918/19; Grote 1967; Grote et al. 1977; Gertz and Loeschke 1954). An O_2-uptake coefficient (θO_2) of the chicken erythrocytes is, however, available (Nguyen-Phu et al. (1986). In the earliest modeling study of the avian lung performed by Abdalla et al. (1982) (before a value of θO_2 was reported on the chicken blood), adjustment of the venous hematocrit was made by mathematically "subtracting" the volume of the nucleus (Plates 3.10f, 3.22a, h and 4.18d, m): the amendment gave a value that was equivalent to the volume of the cytoplasm in the anucleate mammalian erythrocytes and the value was in turn used to correct the mammalian based θO_2 that is based on a mean venous hematocrit of 45%. After the publication of avian θO_2, the total morphometric pulmonary diffusing capacity for O_2 (DLO_{2m}) that was determined using the avian θO_2 reported by Nguyen-Phu et al. (1986): the values were lower than those determined by mathematically correcting the mammalian value (θO_2) (Maina et al. 1989a).

For the mammalian lung, the physiological diffusing capacity of the lung (DLO_{2p}) is generally lower than the DLO_{2m} (Siegwart et al. 1971; Gehr et al. 1981; Crapo and Crapo 1983; Weibel 1984; Weibel et al. 1993; Maina 2011): similar relationship appears to exist for birds (Meyer et al. 1977; Scheid et al. 1977; Burger et al. 1979). Because the mammalian lung intratracheal instillation with fixative (fluid) at pressure is commonly used for the preparation of tissues for morphometric analysis, the alveolar surface area is increased to about the functional residual capacity (Siegwart et al. 1971; Weibel 1973). The process of fixation is preferred because it preserves the intravascular components of the air-hemoglobin pathway, especially the erythrocytes in the blood capillaries, that can be analyzed. On the downside, the surfactant lining is removed, i.e., dissolved away. However, in the morphometric modeling studies of the lung, because the surfactant lining is very thin, it should not significantly affect the diffusing capacity of the blood-gas barrier for oxygen (DtO_2). The DLO_{2m} specifies that the entire blood-gas barrier is involved in gas exchange under ideal conditions where inequalities of ventilation and perfusion are lacking, i.e., when the lung is operating under an optimal partial pressure gradient. In the normal mammalian lung, at maximum O_2consumption (VO_{2max}), DLO_{2p} approximates DLO_{2m} (Weibel 1990). Compared with larger species, small birds have greater DLO_{2m} (Maina 1989a, 2005; Maina et al. 1989a) (Table 4.3), and for birds in general, the DLO_{2m} scales ($b = 0.9198$) and correlates ($r = 0.9863$) strongly with body mass (Plate 4.22d).

References

Aaronson PI, Robertson TP, Ward JP (2002) Endothelium-derived mediators and hypoxic pulmonary vasoconstriction. Respir Physiol Neurobiol 132:107–120

Aaronson PI, Robertson TP, Knock GA, Becker S, Lewis TH, Snetkov V, Ward JP (2006) Hypoxic pulmonary vasoconstriction: mechanisms and controversies. J Physiol 570:53–58

Abdalla MA (1989) The blood supply to the lung. In: King AS, McLelland J (eds) Form and function in birds, vol 4. Academic Press, London, pp 281–306

Abdalla MA, King AS (1975) The functional anatomy of the pulmonary circulation of the domestic fowl. Respir Physiol 23:267–290

Abdalla MA, King AS (1976a) Pulmonary arteriovenous anastomoses in the avian lung: do they exist? Respir Physiol 27:187–191

Abdalla MA, King AS (1976b) The functional anatomy of the bronchial circulation of the domestic fowl. J Anat 121:537–550

Abdalla MA, Maina JN (1981) Quantitative analysis of the exchange tissue of the avian lung (Galliformes). J Anat 134:677–680

Abdalla MA, Maina JN, King AS, King DZ, Henry J (1982) Morphometrics of the avian lung. 1. The domestic fowl, *Gallus domesticus*. Respir Physiol 47:267–278

Agrawal N, Kumar A, Aggarwal P, Jamshed N (2016) Sympathetic crashing acute pulmonary edema. Indian J Crit Care Med 20:719–723

Akester A (1960) The comparative anatomy of the respiratory pathways in the domestic fowl (*Gallus domesticus*), pigeon (*Columba livia*) and domestic duck (*Anas platyrhyncha*). J Anat 4: 487–505

Akester AR (1970) Osmiophilic inclusion bodies as the sources of laminated membrane in the epithelial lining of avian tertiary bronchi. J Anat 107:189–190

Alcorn JL (2017) Pulmonary surfactant trafficking and homeostasis. In: Sidhaye VK, Koval M (eds) Lung epithelial biology in the pathogenesis of pulmonary disease. Academic Press, Boston (MA), pp 59–75

Alexander RM (1967) functional design in fishes. Hutchison, London

Al-Mesri F, Hassanzadeh M (2010) Anatomical parameters of cardiopulmonary system, venous blood PCO_2 and PO_2 tensions and the development of ascites syndrome in two genetic line chickens that are differing in their growth rate. Iranian J Vet Res 11:214–221

Alonso C, Alig T, Yoon J, Bringezu F, Warriner H, Zasadzinski JA (2004) More than a monolayer: relating lung surfactant structure and mechanics to composition. Biophys J 87:4188–4202

Alonso C, Waring A, Zasadzinski JA (2005) Keeping lung surfactant where it belongs: protein regulation of two-dimensional viscosity. Biophys J 89:266–273

Anderson JD (2016) Some reflections on the history of fluid dynamics. In: Johnson RW (ed) Handbook of fluid dynamics, 2nd edn. CRC Press, Boca Raton (FL), pp 13–28

Andrade JS, Alencar AM, Almeida MP, Mendes F, Buldyrev JSV, Zapperi S, Stanley HE (1998) Asymmetric flow in symmetric branched structure. Phys Rev Lett 81:926–929

Aota A, Nonaka M, Hibara A, Kitamori T (2007a) Countercurrent laminar microflow for highly efficient solvent extraction. Angewandte Chemie 119:896–898

Aota A, Akihide H, Kyosuke S, Yasuhiko S, Koji O, Kitamori T (2007b) Flow velocity profile of micro counter-current flows. Analytical Sci 23:131–133

Araújo PM, Viegas I, Rocha AD, Villegas A, Mendonça L, Ramos JA, Masero JA, Alves JA (2019) Understanding how birds rebuild fat stores during migration: insights from an experimental study. Sci Rep 9:10065. https://doi.org/10.1038/s41598-019-46487-z

Arce J, Berger M, Lopez Coello C (1992) Control of ascites syndrome by feed restriction techniques. J Appl Poult Res 1:1–5

Archer SL, Weir EK, Reeve HL, Michelakis E (2000) Molecular identification of O_2 sensors and O_2-sensitive potassium channels in the pulmonary circulation. Adv Exp Med Biol 475:219–240

Aschoff J, Pohl H (1970) Rhythmic variations in energy metabolism. Fed Proc 29:1541–1552

Baffico L, Grandmont C, Maury B (2010) Multiscale modelling of the respiratory tract. Mathematical models and methods in applied sciences. World Scientific Publishing, pp 59–93. https://doi.org/10.1142/S0218202510004155

Baghbanzadeh A, Decuypere E (2008) Ascites syndrome in broilers: physiological and nutritional perspectives. Avian Pathol 37:117–126

Bai C, She J, Goolaerts A, Song Y, Shen C, Shen J, Hong Q (2010) Stress failure plays a major role in the development of high-altitude pulmonary oedema in rats. Eur Respir J 35:584–591

Banzett RB, Nations CS, Barnas JL, Lehr JL, Jones JH (1987) Inspiratory aerodynamic valving in goose lungs depends on gas density and velocity. Respir Physiol 70:287–300

Banzett RB, Nations CS, Wang N, Fredberg JJ, Butler PJ (1991) Pressure profiles show features essential to aerodynamic valving in geese. Respir Physiol 84:295–309

Barja G (1998) Mitochondrial free radical production and aging in mammals and birds. Ann NY Acad Sci 854:224–238

Barnas G, Mather FB, Fedde MR (1978) Response of avian intrapulmonary smooth muscle to changes in carbon dioxide concentration. Poult Sci 57:1400–1407

Bartlett RG, Specht H (1957) Enery cost of breathing determined with a simplified technique. J Appl Physiol 11:84–86

Bastacky J, Lee C, Goerke J, Koushafar H, Yager D, Kenaga L, Speed TP, Chen Y, Clements JA (1995) Alveolar lining layer is thin and continuous: low-temperature scanning electron microscopy of rat lung. J Appl Physiol 79:1615–1628

Batchelor GK (2000) An introduction to fluid dynamics. Cambridge University Press, Cambridge

Bech C, Johansen K (1980) Ventilation and gas exchange in the mute swan, *Cygnus olor*. Respir Physiol 39:285–295

Beddard FE (1898) the structure and classification of birds. Longmans, Green and Co, New York

Bemis WE, Burggren WW, Kemp NE (eds) (1987) The Biology and evolution of lungfishes. Allan Liss, New York

Ben-Tal A (2006) Simplified models for gas exchange in the human lungs. J Theor Biol 238:474–495

Berger AJ (1960) Some anatomical characters of the Cuculidae and the Musophagidae. Wilson Bull 72:60–104

Berner RA, Canfield DE (1989) A new model for atmospheric oxygen over Phanerozoic time. Am J Sci 289:333–361

Bernhard W (2016) Lung surfactant: function and composition in the context of development and respiratory physiology. Ann Anat 208:146–150

Bernhard W, Gebert A, Vieten G, Rau GA, Hohlfeld JM, Postle AD, Freihorst J (2001) Pulmonary surfactant in birds: coping with surface tension in a tubular lung. Am J Physiol Regul Integr Comp Physiol 281:R327–R337

Bernhard W, Haslam PL, Floros J (2004a) From birds to humans: new concepts on airways relative to alveolar surfactant. Am J Respir Cell Mol Biol 30:6–11

Bernhard W, Pynn CJ, Jaworski A, Rau GA, Hohlfeld JM, Freihorst J, Poets CF, Stoll D, Postle AD (2004b) Mass spectrometric analysis of surfactant metabolism in human volunteers using deuteriated choline. Am J Respir Crit Care Med 170:54–58

Bhattacharya J, Nanjo S, Staub NC (1982) Micropuncture measurement of lung microvascular pressure during 5-HT infusion. J Appl Physiol Respir Environ Exerc Physiol 52:634–637

Biggs PM, King AS (1957) A new experimental approach to the problem of the air pathway within the avian lung. J Physiol (Lond) 138:282–289

Birks EK, Mathieu-Costello O, Fu Z, Tyler WS, West JB (1994) Comparative aspects of the strength of pulmonary capillaries in rabbit, dog, and horse. Respir Physiol 97:235–246

Birks EK, Mathieu-Costello O, Fu Z, Tyler WS, West JB (1997) Very high pressures are required to cause stress failure of pulmonary capillaries in thoroughbred racehorses. J Appl Physiol 82: 1584–1592

Biro GP (2013) From the atmosphere to the mitochondrion: the oxygen cascade. In: Kim H, Greenburg A (eds) Hemoglobin-based oxygen carriers as red cell substitutes and oxygen therapeutics. Springer, Berlin, pp 27–53

Bishop CM (1997) Heart mass and the maximum cardiac output of birds and mammals: implications for estimating the maximum aerobic power input of flying animals. Philos Trans R Soc (Lond) B 352:447–456

Bishop CM (1999) The maximum oxygen consumption and aerobic scope of birds and mammals: getting to the heart of the matter. Proc R Soc (Lond) B 266:2275–2281

Bissonnette EY, Lauzon-Joset JF, Debley JS, Ziegler SF (2020) Cross-talk between alveolar macrophages and lung epithelial cells is essential to maintain lung homeostasis. Front Immunol 11:583042. https://doi.org/10.3389/fimmu.2020.583042

Black CP, Tenney SM (1980) Oxygen transport during progressive hypoxia in high altitude and sea level water-fowl. Respir Physiol 39:217–239

Blake RW (ed) (1991) Efficiency and economy in animal physiology. Cambridge University Press, Cambridge

Blix AS (2018) Adaptations to deep and prolonged diving in phocid seals. J Exp Biol 221: jeb182972. https://doi.org/10.1242/jeb.182972

Boccia CK, Swierk L, Ayala-Varela FP, Boccia J, Borges IL, Estupiñán CA, Martin AM, Martínez-Grimaldo RE, Ovalle S, Senthivasan S et al (2021) Repeated evolution of underwater rebreathing in diving *Anolis* lizards. Curr Biol 31:2947–2954

Bölükbasi SC, Aktas MS, Güzel M (2005) The effect of feed regimen on ascites induced by cold temperatures and growth performance in male broilers. Int J Poult Sci 4:326–329

Bosworth RCL, Groden CM (1964) Optimum thermal efficiency in countercurrent heat exchangers. Aust J Phys 17:37–44

Boutilier RG, Ferguson RA (1989) Nucleated red cell function: metabolism and pH regulation. Can J Zool 67:2986–2993

Boyle J, Englestein ES, Sinoway LI (1977) Mean air space diameter, lung surface area and alveolar surface tension. Respiration 34:241–249

Brackenbury JH (1972) Lung-air-sac anatomy and respiratory pressure in the bird. J Exp Biol 57: 543–550

Brackenbury JH (1973) Respiratory mechanics in the bird. Comp Biochem Physiol 44:599–611

Brainerd EL (1994) The evolution of lung-gill bimodal breathing and the homology of vertebrate respiratory pumps. Am Zool 34:289–299

Brainerd EL (1999) New perspectives on the evolution of lung ventilation mechanisms in vertebrates. EBO 4:1–28

Bremer JL (1940) The pneumatization of the humerus in the common fowl and the associated activity of theelin. Anat Rec 77:197–211

Bretz WL, Schmidt-Nielsen K (1971) Bird respiration: flow patterns in the duck lung. J Exp Biol 54:103–118

Brimioulle S, Julien V, Gust R, Kozlowski JK, Naeije R, Schuster DP (2002) Importance of hypoxic vasoconstriction in maintaining oxygenation during acute lung injury. Crit Care Med 30:874–880

Brown RE, Kovacs CE, Butler JP, Wang N, Lehr J, Banzett RB (1995) The avian lung: is there an aerodynamic expiratory valve? J Exp Biol 198:2349–2357

Brown RE, Butler JP, Godleski JJ, Loring SH (1997) The elephant's respiratory system: adaptations to gravitational stress. Respir Physiol 109:177–194

Brunelle JK, Chandel NS (2002) Oxygen deprivation induced cell death. An update. Apoptosis 7: 475–482

Bubel A (1983) A transmission and scanning electron microscopy study of the cuticle and epidermis of *Pomatoceros lamarckii* (Polychaeta: Serpulidae). J Zool (Lond) 36:217–268

Buckner CH (2001) Metabolism, food capacity, and feeding behavior in four species of shrews. Can J Zool 42:259–279

Burger RE, Meyer M, Werner G, Scheid P (1979) Gas exchange in the parabronchail lung of birds: experiments in unidirectionally ventilated ducks. Respir Physiol 36:19–37

Burggren WW, West NH (1982) Changing respiratory importance of gills, lungs and skin during metamorphosis in the bullfrog *Rana catesbeiana*. Respir Physiol 47:151–164

Burri PH (1985) Morphology and respiratory function of the alveolar unit. Int Arch Allergy Appl Immunol 76:2–12

Butler PJ (1970) The effect of progressive hypoxia on the respiratory and cardiovascular systems of the pigeon and duck. J Physiol 211:527–538

Butler PJ (1991) Exercise in birds. J Exp Biol 160:233–262

Butler PJ (2004) Metabolic regulation in diving birds and mammals. Respir Physiol Neurobiol 141: 297–315

Butler PJ (2016) The physiological basis of bird flight. Philos Transac R Soc (Lond) B 371: 20150384. https://doi.org/10.1098/rstb.2015.0384

Butler PJ, Jones DR (1997) Physiology of diving of birds and mammals. Physiol Rev 77:837–899

Butler JP, Banzett RB, Fredberg JJ (1988) Inspiratory valving in avian bronchi: aerodynamic considerations. Respir Physiol 73:241–256

Buttemer WA, Abele D, Costantini D (2010) From bivalves to birds: oxidative stress and longevity. Funct Ecol 24:971–983

Butterworth RF (2006) Metabolic encephalopathies. In: Siegel G, Albers RW, Brady S, Price D (eds) Basic neurochemistry: molecular, cellular, and medical aspects. Elsevier, Amsterdam, pp 291–302

Camacho-Fernández D, López C, Ávila E, Arce J (2002) Evaluation of different dietary treatments to reduce ascites syndrome and their effects on corporal characteristics in broiler chickens. J Appl Poult Res 11:164–174

Campana JC (1875) Physiologie de la respiration chez les oiseaux. Anatomie de l'appareil pneumatique-pulmonaire, de faux diaphragmes, des sereuses et de Fintestin chez le poulet. Masson et Cie, Paris

Canals M, Atala C, Olivares R, Guajardo F, Figueroa DP, Sabat P, Rosenmann M (2005) Functional and structural optimization of the respiratory system of the bat *Tadarida brasiliensis* (Chiroptera, Molossidae): does airway geometry matter? J Exp Biol 208:3987–3995

Canals M, Sabat P, Veloso C (2008) The proximal airway of the bat *Tadarida brasiliensis*: a minimum entropy production design. J Comp Physiol B 178:377–384

Canals M, Figueroa D, Sabat P (2010) Symmorphosis in the proximal pathway for oxygen in the leaf-eared mouse, Phyllotis darwini. Biol Res 43:75–81

Canney M, O'Hara PV, McEvoy CM, Medani S, Connaughton DM, Abdalla AA, Doyle R, Stack AG, O'Seaghdha CM, Clarkson MR et al (2016) Spatial and temporal clustering of anti-glomerular basement membrane disease. Clin J Am Soc Nephrol 11:1392–1399

Casas AI, Geuss E, Kleikers PWM, Mencl S, Herrmann AM, Buendia I, Egea J, Meuth SG, Lopez MG, Kleinschnitz C et al (2017) NOX4-dependent neuronal autotoxicity and BBB breakdown explain the superior sensitivity of the brain to ischemic damage. Proc Natl Acad Sci (USA) 114: 12315–12320

Castiglione GM, Xu Z, Zhou L, Duh EJ (2020) Adapatation of the master antioxidant response connects metabolism, lifespan and feather development pathways in birds. Nat Commun 11:1–15

Castresana J, Saraste M (1995) Evolution of energetic metabolism: the respiration-early hypothesis. Trends Biochem Sci 20:443–448

Champini BG, da Cruz AL, Klein W (2022) Heart position and pulmonary vasculture in snakes with different lung morphologies. Anim Biol 72. https://doi.org/10.1163/15707563-bja10075

Chan ET, Quon GT, Chua G, Babak T, Trochesset M, Zirngibl RA, Aubin J, Ratcliffe MJH, Wilde A, Brudno M, Morris QD, Hughes TR (2009) Conservation of core gene expression in vertebrate tissues. J Biol 8(33). https://doi.org/10.1186/jbiol130

Charolidi N, Carroll VA (2017) Hypoxia and pulmonary hypertension. In: Zheng J, Zhou C (eds) Hypoxia and human diseases. IntechOpen, London, pp 211–225

Cherniack NS, Longobardo GS (1970) Oxygen and carbon dioxide stores in the body. Physiol Rev 50:196–243

Clarke A, Rothery P (2008) Scaling of body temperature in mammals and birds. Funct Ecol 22:58–67

Clement AM, Long JA (2010) Air breathing adaptation in a marine Devonian lungfish. Biol Lett 6: 509–512

Codd JR (2010) Uncinate processes in birds: morphology, physiology and function. Comp Biochem Physiol A Mol Integr Physiol 156:303–308

Codd JR, Boggs DF, Perry SF, Carrier DR (2005) Activity of three muscles associated with the uncinate processes of the giant Canada goose *Branta canadensis maximus*. J Exp Biol 208:849–857

Codd JR, Manning PL, Norell MA, Perry SF (2008) Avian-like breathing mechanics in maniraptoran dinosaurs. Proc R Soc (Lond) B 275:157–161

Cohn JE, Shannon R (1968) Respiration in unanaethetised geese. Respir Physiol 5:259–268

Coitier V (1573) (cited by Campana 1875) Anatomia avium. In: Externum et internarum praecipalium humani corporis partium tabulae arque anatomicae exercitationes. Nuremberg, pp 130–133

Collins JJ (1971) Dissecting aneurysms in turkeys and man. Arch Surg 102:159–160

Comroe JH (1974) Physiology of respiration: an introductory text, 2nd edn. Year Book Medical Publishers, Chicago

Connett RJ, Honig CR, Gayeski TE, Brooks GA (1990) Defining hypoxia: a systems view of VO_2, glycolysis, energetics, and intracellular PO_2. J Appl Physiol 68:833–842

Coronado-Zamora M, Salvador-Martínez I, Castellano D, Barbadilla A, Salazar-Ciudad I (2019) Adaptation and conservation throughout the *Drosophila melanogaster* life-cycle. Genome Biol Evol 11:1463–1482

Costa LE, Méndez G, Boveris A (1997) Oxygen dependence of mitochondrial function measured by high-resolution respirometry in long-term hypoxic rats. Am J Physiol 273:C852–C858

Costa DP, Crocker DE, Gales NJ (1998) Blood volume and diving ability of the New Zealand sea lion, *Phocarctas hookeri*. Physiol Zool 71:208–213

Cottrell AH (1981) The mechanical properties of matter. Krieger, Malabar (FL)

Crapo JD, Crapo RO (1983) Comparison of total lung diffusion capacity and the membrane component of diffusion capacity as determined by physiologic and by morphometric techniques. Respir Physiol 51:183–194

Craven J (2021) Geodesic domes and space-frame structures. ThoughtCo, October 18, 2021. Retrieved from https://www.thoughtco.com/what-is-a-geodesic-dome-177713. Accessed 20 Feb 2022

Creuwels LA, van Golde LM, Haagsman HP (1997) The pulmonary surfactant system: biochemical and clinical aspects. Lung 175:1–39

Crossley D, Altimiras J, Wang T (1998) Hypoxia elicits an increase in pulmonary vasculature resistance in anaesthetised turtles (*Trachemys scripta*). J Exp Biol 201:3367–3375

Crouch E, Wright JR (2001) Surfactant proteins A and D and pulmonary host defense. Annu Rev Physiol 63:521–554

Cupello C, Meunier FJ, Herbin M, Clément G, Brito PM (2017) Lung anatomy and histology of the extant coelacanth shed light on the loss of air-breathing during deep-water adaptation in actinistians. R Soc Open Sci 4:161030. https://doi.org/10.1098/rsos.161030

Cupello C, Hirasawa T, Tatsumi N, Yabumoto Y, Gueriau P, Isogai S, Matsumoto R, Saruwatari T, King A, Hoshino M et al (2022) Lung evolution in vertebrates and the water-to-land transition. bioRxiv preprint. Accessed 28-06-22. https://doi.org/10.1101/2022.02.14.480340

Currie RJW (1999) Ascites in poultry: recent investigations. Avian Pathol 28:313–326

Daniels CB, Orgeig S (2001) The comparative biology of pulmonary surfactant: past, present and future. Comp Biochem Physiol A Mol Integr Physiol 129:9–36

Daniels CB, Orgeig S (2003) Pulmonary surfactant: the key to the evolution of air breathing. News Physiol Sci 18:151–157

Daniels CB, Orgeig S, Smits AW (1995) Invited perspective: the evolution of the vertebrate pulmonary surfactant system. Physiol Zool 68:539–566

Daniels CB, Orgeig S, Wood PG, Sullivan LC, Lopatko OV, Smits AW (1998a) The changing state of surfactant lipids: new insights from ancient animals. Am Zool 38:305–320

Daniels CB, Lopatko OV, Orgeig S (1998b) Evolution of surface activity related functions of vertebrate pulmonary surfactant. Clin Exp Pharmacol Physiol 25:716–721

Daniels CB, Orgeig S, Sullivan LC, Ling N, Bennett MB, Schürch S, Luis Val A, Brauner CJ (2004) The origin and evolution of the surfactant system in fish: insights into the evolution of lungs and swim bladders. Physiol Biochem Zool 77:732–749

Davies KJ (1995) Oxidative stress: the paradox of aerobic life. Biochem Soc Symp 61:1–31

Davis RW, Kanatous SB (1999) Convective oxygen transport and tissue oxygen consumption in Weddell seals during aerobic dives. J Exp Biol 202:1091–1113

Dejours P (1981) Principles of comparative respiratory physiology, 2nd edn. Elsevier, New York

Dejours P (1988) Respiration in water and air: adaptations, regulation, evolution. Elsevier, Amsterdam

Dejours P (1989) Current concepts in comparative physiology of respiration. In: Wood SC (ed) Comparative pulmonary physiology: current concepts. Marcel Dekker, New York, pp 1–10

Del Corral JPD (1995) Anatomy and histology of the lung and air sacs of birds. In: Pastor LM (ed) Histology, ultrastructure and immunohistochemistry of the respiratory organs in non-mammalian vertebrtaes. Publicaciones de la Universitatd de University of Murcia, Murcia, pp 179–233

Dock DS, Kraus WL, McGuire LB, Hyland JW, Haynes FW, Dexter L (1961) The pulmonary blood volume in man. J Clin Invest 40:317–328

Dodd GAA, Scott GR, Milsom WK (2007) Ventilatory roll off during sustained hypercapnia is gender specific in pekin ducks. Respir Physiol Neurobiol 156:47–60

Dosković V, Bogosavljević-Bošković S̆S, Lukić M, Milošević B, Rakonjac S, Petričević V (2019) Nutrition-related aspects of ascites in poultry. Acta Agricult Serbica 48:119–1331

Dotterweich H (1930) Versuch überden Weg der Atemluft in der Vogellunge. Zeitsch Vergl Physiol 11:271–284

Dotterweich H (1933) Ein weiterer Beitrag zur Atmungsphysiologie der Vögel. Zeitsch vergleich. Physiol 18:803–809

Dotterweich H (1936) Die Atmung der Vögel. Zeitscht vergleich Physiol 23:744–770

Dreyfuss D, Saumon G (1997) Experimental changes in the alveolo-capillary barrier induced by artificial ventilation. Schweiz Med Wochenschr 127:1023–1029

Dubach M (1981) Quantitative analysis of the respiratory system of the house sparrow, budgerigar, and violet-eared hummingbird. Respir Physiol 46:43–60

Dubois A, Brody AW, Lewis DH, Burgess F (1956) Oscillation mechanics of lungs and chest in man. J Appl Physiol 8:587–594

Dudley R, Gans C (1991) A critique of symmorphosis and optimality models in physiology. Physiol Zool 64:627–637

Duncker HR (1971) The lung-air sac system of birds. A contribution to the functional anatomy of the respiratory apparatus. Ergeb Anat Entwicklung 45:1–171

Duncker HR (1972) Structure of the avian lung. Respir Physiol 14:4–63

Duncker HR (1973) Der quantitative Aufbau des Lungenluftsacksystems der Vögel. Verhandlungen Anat Ges 67:197–204

Duncker H-R (1974) Structure of the avian respiratory tract. Respir Physiol 22:1–34

Duncker HR, Güntert M (1985a) The quantitative design of the avian respiratory system: from hummingbird to the mute swan. In: Nachtigall W (ed) BIONA Report No. 3. Gustav-Fischer, Stuttgart, pp 361–378

Duncker HR, Güntert M (1985b) Morphometric analysis of the avian respiratory system. In: Duncker HR, Fleischer G (eds) Vertebrate morphology. Gustav-Fischer, Stuttgart, pp 383–387

Dunham-Snary KJ, Wu D, Sykes EA, Thakrar A, Parlow L, Mewburn JD, Parlow JL, Archer SL (2017) Hypoxic pulmonary vasoconstriction: from molecular mechanisms to medicine. Chest 151:181–192

Dunnill MS (1962) Quantitative methods in the study of pulmonary morphology. Thorax 17:320–328

Dusenbery DB (1973) Countercurrent separation: a new method for studying behavior of small aquatic organisms. Proc Natl Acad Sci (USA) 70:1349–1352

Dzal YA, Jenkin SEM, Lague SL, Reichert MN, York JM, Pamenter ME (2015) Oxygen in demand: how oxygen has shaped vertebrate physiology. Comp Biochem Physiol A 186:4–26

Eberth CJ (1862) Uber den feineren Bau der Lunge. Z Wiss Zool 12:427–454

Egbert MG (2013) Measuring mitochondrial oxygen tension: from basic principles to application in humans. Anaethesia Analgesia 117:834–846

Ege R (1915) On the respiratory function of the air stores carried by some aquatic insects. Z Allg Physiol 17:81–124

Eikenaar C, Hegemann A, Packmor F, Kleudgen I, Isaksson C (2020) Not just fuel: energy stores are correlated with immune function and oxidative damage in a long-distance migrant. Curr Zool 66:21–28

Epp TS, McDonough P, Padilla DJ, Gentile JM, Edwards KL, Erickson HH, Poole DC (2006) Exercise-induced pulmonary harmmorhage during submaximal exercise. Equine Vet J Suppl 36:502–507

Erickson LL (1994) For the birds: an uncommon guide. Pfeifer-Hamilton, Duluth (MN)

Erickson HH, Lowe BS (1994) Mechanisms of exercise-induced pulmonary hemorrhage in the equine athlete. Biomed Sci Instrum 30:33–38

Euler US, Liljestrand G (1946) Observations on the pulmonary arterial blood pressure in the cat. Acta Physiol Scand 12:301–320

Faraci FM (1986) Circulation during hypoxia in birds. Comp Biochem Physiol A 85:613–620

Faraci FM (1991) Adaptations to hypoxia in birds: how to fly high. Ann Rev Physiol 53:59–70

Faraci FM, Fedde MR (1986) Regional circulatory responses to hypocapnia and hypercapnia in bar-headed geese. Am J Physiol 250:R499–R504

Faraci FM, Kilgore DL, Fedde MR (1984) Oxygen delivery to the heart and brain during hypoxia: pekin duck vs. bar-headed goose. Am J Physiol 247:R69–R75

Faraci FM, Kilgore DL Jr, Fedde MR (1985) Blood flow distribution during hypocapnic hypoxia in Pekin ducks and bar-headed geese. Respir Physiol 61:21–30

Farhi L (1964) Gas stores of the body. In: Fenn WO, Rahn H (eds) Handbook of physiology, section 3: respiration, vol I. American Physiological Society, Washington, DC, pp 873–924

Farhi L, Rahn H (1955) Gas stores of the body and the unsteady state. J Appl Physiol 7:472–484

Farmer CG, Jackson DC (1998) Air-breathing during activity in the fishes *Amia calva* and *Lepisosteus oculatus*. J Exp Biol 201:943–948

Farner DS (1970) Some glimpses of comparative avian physiology. Fed Proc 29:1649–1663

Fedde MR (1980) The structure and gas flow pattern in the avian lung. Poult Sci 59:2642–2653

Fedde MR, Burger RE, Kitchell RL (1963) Localization of vagal afferents involved in the maintenance of normal avian respiration. Poult Sci 42:1224–1236

Fedde MR, Burger RE, Kitchell RL (1964a) Anatomic and electromyographic studies of the costopulmonary muscles in the cock. Poult Sci 43:1177–1184

Fedde MR, Burger RE, Kitchell RL (1964b) Electromyographic studies of the effects of bodily position and anaethesia on the activity of the respiratory muscles of the domestic fowl. Poult Sci 43:839–846

Fedde MR, Burger RE, Kitchell RL (1964c) Electromyographic studies of the effects of bilateral vervical vagotomy on the action of the respiratory muscles of the domestic cock. Poult Sci 43: 1119–1125

Ferguson TM, Miller DH, Bradley JW, Atkinson RL (1969) Blood pressure and heart rate of turkeys, 17-21 months of age. Poult Sci 48:1478–1481

Ferguson J, Voelker D, McCormack F, Schlesinger L (1999) Surfactant protein D binds to *Mycobacterium tuberculosis* bacilli and lipoarabinomannan via carbohydrate-lectin interactions resulting in reduced phagocytosis of the bacteria by macrophages. J Immunol 163:312–321

Ferreras MC, González J, Pérez V, Reyes LE, Gómez N, Pérez C, Corpa JM, García-Marín JF (2001) Proximal aortic dissection (dissecting aortic aneurysm) in a mature ostrich. Avian Dis 45: 251–256

Fick A (1855) Ueber diffusion. Ann Physik 94:59–86

Figueroa D, Olivares R, Salaberry M, Sabat P, Canals M (2007) Interplay between the morphometry of the lungs and the mode of locomotion in birds and mammals. Biol Res 40:193–201

Fisher RF, Wakely J (1976) The elastic constants and ultrastructural organization of a basement membrane (lens capsule). Proc R Soc Lond B Biol Sci 193:335–358

Fisher S, Grice EA, Vinton RM, Bessling SL, McCallion AS (2006) Conservation of RET regulatory function from human to zebrafish without sequence similarity. Science 312:276–279

Fjeldså J, Krabbe N (1990) Birds of the high Andes: a manual to the birds of the temperate zone of the Andes and Patagonia, South America. Zoological Museum of the University of Copenhagen, Copenhagen

Fletcher OJ (1980) Pathology of the avian respiratory system. Poult Sci 59:2666–2679

Flynn MR, Bush JWM (2008) Underwater breathing: the mechanics of plastron respiration. J Fluid Mech 608:275–296

Fons R, Sicart R (1976) Contribution à la connaissance du métabolisme énergétique chez deux Crocidurinae: *Suncus etruscus* (Savi 1822) et *Crocidura russula* (Herman 1780), Insectivora, Soricidae, Mammalia. Mammalia 40:229–311

Fons R, Sender S, Peters T, Jürgens KD (1997) Rates of rewarming, heart and respiratory rates and their significance for oxygen transport during arousal from torpor in the smallest mammal, the Etruscan shrew, *Suncus etruscus*. J Exp Biol 200:1451–1458

Foot NJ, Orgeig S, Daniels CB (2006) The evolution of a physiological system: the pulmonary surfactant system in diving mammals. Respir Physiol Neurobiol 154:118–138

Forster RE (1996) Transfer of gas by diffusion and chemical reaction in pulmonary capillaries. In: West JB (ed) Respiration physiology: peoples and ideas. Oxford University Press, New York, pp 49–74

Fowler ME, Cubas ZS (eds) (2001) Biology, medicine, and surgery of South American wild animals. Wiley-Blackwell, Hoboken (NJ)

Franciosini MP, Tacconi G, Leonardi L (2012) Ascites syndrome in broiler chickens. Vet Sci Res 3: 60–66

Franklin SH, Van Erck-Westergren E, Bayly WM (2012) Respiratory responses to exercise in the horse. Equine Vet J 44:726–732

Fu Z, Costello ML, Tsukimoto K, Prediletto R, Elliott AR, Mathieu-Costello O, West JB (1985) High lung volume increases stress failure in pulmonary capillaries. J Appl Physiol 73:123–133

Fujiwara T, Adams FH, Nozaki M, Dermer GB (1970) Pulmonary surfactant phospholipids from turkey lung: comparison with rabbit lung. Am J Physiol 218:218–225

Fuller B (1961) Tensegrity. Portfolio Artnews Ann 4:112–127

Fuller A, Mitchell D, Maloney SK, Hetem RS (2016) Towards a mechanistic understanding of the responses of large terrestrial mammals to heat and aridity associated with climate change. Clim Chang Respon 3:10. https://doi.org/10.1186/s40665-016-0024-1

Fung YC, Sobin SS (1969) Theory of sheet flow in lung alveoli. J Appl Physiol 26:472–488

Gans C (1991) Efficiency, effectiveness, perfection, optimization: their use in understanding vertebrate evolution. In: Blake R (ed) Efficiency and economy in animal physiology. Cambridge University Press, Cambridge, pp 1–12

Gardiner B (1980) Tetrapod ancestry: a reappraisal. In: Panchen AL (ed) The terrestrial environment. Academic Press, London, pp 177–185

Garg SG, Martin WF (2016) Mitochondria, the cell cycle, and the origin of sex via a syncytial eukaryote common ancestor. Genome Biol Evol 8:1950–1970

Gayeski TE, Connett RJ, Honig CR (1987) Minimum intracellular PO_2 for maximum cytochrome turnover in red muscle *in situ*. Am J Physiol Heart Circ Physiol 252:H906–H915

Gehr P, Sehovic S, Burri PH, Classen H, Weibel ER (1980) The lung of shrews: morphometric estimation of diffusion capacity. Respir Physiol 44:61–86

Gehr P, Mwangi DK, Amman A, Maloiy GMO, Taylor CR, Weibel ER (1981) Design of the mammalian respiratory system. V. Scaling morphometric diffusing capacity to body mass: wild and domestic animals. Respir Physiol 44:41–86

Geiser F (2008) Ontogeny and phylogeny of endothermy and torpor in mammals and birds. Comp Biochem Physiol A Mol Integr Physiol 150:176–180

Gertz KH, Loeschke HH (1954) Bestimmung der Diffusionskoeffizienten von H_2, O_2, N_2, und He in Wasser und Blutserum bei konstant gehaltener Konvektion. Z Naturforsch 9b:1–9

Gifford JR, Garten RS, Nelson AD, Trinity JD, Layec G, Witman MA, Weavil JC, Mangum T, Hart C, Etheredge C et al (2016) Symmorphosis and skeletal muscle VO_2 max: *in vivo* and *in vitro* measures reveal differing constraints in the exercise-trained and untrained human. J Physiol 594:1741–1751

Gilbert PW (1939) The avian lung and air-sac system. Auk 56:57–63

Ginhoux F, Jung S (2014) Monocytes and macrophages: developmental pathways and tissue homeostasis. Nat Rev Immunol 14:392–404

Glazier JB, Hughes JMB, Maloney JE, West JB (1969) Measurement of capillary dimensions and blood volume in rapidly frozen lungs. J Appl Physiol 38:23–35

Goerke J (1998) Pulmonary surfactant: functions and molecular composition. Biochim Biophys Acta 1408:79–89

Goniakowska-Witalinska L (1978) Ultrastructure and morphometric study of the lung of the European salamander, *Salamandra salamandra* L. Cell Tissue Res 191:43–356

Goniakowska-Witalinska L (1986) Lung of the tree frog, *Hyla arborea*: a scanning and transmission electron microscope study. Anat Embryol 174:379–389

Goniakowska-Witalinska L (1995) The histology and ultrastructure of the amphibian lung. In: Pastor LM (ed) Histology, ultrastructure and immunohistochemistry of the respiratory organs in non-mammalian vertebrates. Publicaciones Universidad de Murcia 1995, Murcia, pp 77–112

Goodpasture EW (2009) Landmark publication from the American Journal of the Medical Sciences: the significance of certain pulmonary lesions in relation to the etiology of influenza. Am J Med Sci 338:148–151

Gordon MS, Olson EC (1994) The invasions of the land. Columbia University Press, New York (NY)

Gottschalk CW, Myle M (1958) Evidence that the mammalian nephron functions as a countercurrent multiplier system. Science 128:594

Gottschalk CW, Myle M (1959) Micropuncture study of the mammalian urinary concentrating mechanism: evidence for the countercurrent hypothesis. Am J Physiol 196:927–936

Goujet D (2011) 'Lungs' in placoderms, a persistent palaeobiological myth related to environmental preconceived interpretations. C R Palevol 10:323–329

Goymann W, Spina F, Ferri A, Fusani L (2010) Body fat influences departure from stopover sites in migratory birds: evidence from whole-island telemetry. Biol Lett 23:478–481

Graham JB (1997) Air breathing fishes: evolution, diversity and adaptation. Academic Press, San Diego

Graham JB, Wegner N (2010) Breathing air in water and in air: the air-breathing fishes. In: Nilsson G (ed) Respiratory physiology of vertebrates: life with and without oxygen. Cambridge University Press, Cambridge, pp 174–221

Graham JB, Aguilar N, Dudley R, Gans CG (1995) Implications of the late Palaeozoic oxygen pulse for physiology and evolution. Nature 375:117–120

Gray MW (2012) Mitochondrial evolution. Cold Spring Harb Perspect Biol 4:a011403. https://doi.org/10.1101/cshperspect.a011403

Greco A, Rizzo MI, De Virgilio A, Gallo A, Fusconi M, Pagliuca G, Martellucci S, Turchetta R, Longo L, De Vincentiis M (2015) Goodpasture's syndrome: a clinical update. Autoimmun Rev 14:246–253

Greenlee KJ, Henry JR, Kirkton SD, Westneat MW, Fezzaa K, Lee WK, Harrison JF (2009) Synchrotron imaging of the grasshopper tracheal system: morphological and physiological components of tracheal hypermetry. Am J Physiol Regul Integr Comp Physiol 297:R1343–R1350

Grigg GC, Beard LA, Augee ML (2004) The evolution of endothermy and its diversity in mammals and birds. Physiol Biochem Zool 77:982–997

Grote J (1967) Die Sauerstoffdiffusionskonstanten im Lungengewebe und Wasser und ihre Temperaturabhängigkeit. Pflügers Arch 295:245–254

Grote J, Susskind R, Vaupel P (1977) Oxygen diffusivity in tumor tissue (DS-carcinosarcoma) under temperature conditions within the range of 20-40 degrees C. Pflugers Arch 1:37–42

Groves P (1997) Broiler ascites syndrome: a review. Proc Austr Poult Sci Symp:85–91

Grubb BR (1982) Cardiac output and stroke volume in exercising ducks and pigeons. J Appl Physiol 53:203–211

Grubb B, Mills CD, Colacino JM, Schmidt-Nielsen K (1977) Effect of arterial carbon dioxide on cerebral blood flow in ducks. Am J Physiol 232:H596–H601

Grubb B, Colacino JM, Schmidt-Nielsen K (1978) Cerebral blood flow in birds: effect of hypoxia. Am J Physiol 234:H230–H234

Guglielmo CG (2018) Obese super athletes: fat-fueled migration in birds and bats. J Exp Biol 221: jeb165753. https://doi.org/10.1242/jeb.165753

Guglielmo CG, Piersma T, Williams TD (2001) A sport-physiological perspective on bird migration: evidence for flight-induced muscle damage. J Exp Biol 2690:2683–2690

Gupta AR (2011) Ascites syndrome in poultry: a review. World's Poult Sci J 67:457–467

Gutmann WF, Bonik K (1981) Kritische Evolutionstheorie: ein Beitrag zur Überwindung altdarwinistischer Dogmen. Hildesheim, Gerstenberg

Gythiel W, Mommeyer C, Raymaekers T, Schevenels M (2020) A comparative study of the structural performance of different types of reticulated dome subjected to distributed loads. Front Built Environ 6:56. https://doi.org/10.3389/fbuil.2020.00056

Hachenberg T, Rettig R (1998) Stress failure of the blood-gas barrier. Curr Opin Anaesthesiol 11: 37–44

Hails CJ (1983) The metabolic rate of tropical birds. Condor 85:61–65

Hammarlund EU (2020) Harnessing hypoxia as an evolutionary driver of complex multicellularity. Interface Focus 10:20190101. https://doi.org/10.1098/rsfs.2019.0101

Hammarlund EU, Flashman E, Mohlin S, Licausi F (2020) Oxygensensing mechanisms across eukaryotic kingdoms and their roles in complex multicellularity. Science 370:eaba3512. https://doi.org/10.1126/science.aba3512

Hance AJ, Crystal RG (1976) Collagen. In: Crystal RG (ed) The biochemical basis of pulmonary function. Dekker, New York, pp 215–271

Hanel B, Law I, Mortensen JJ (2003) Maximal rowing has an acute effect on the blood-gas barrier in elite athletes. J Appl Physiol 95:1076–1082

Hansen NB, Brubakk AM, Bratlid D, Oh W, Stonestreet BS (1984) Theeffects of variations in $PaCO_2$ on brain blood flow and cardiacoutput in the newborn piglet. Pediatr Res 18:1132–1136

Harkema JR, Nikula KJ, Haschek WM (2018) Respiratory system. In: Wallig MA, Haschek WM, Rousseaux CG, Bolon B (eds) Fundamentals of toxicologic pathology, 3rd edn. Academic Press, New York, pp 351–393

Hartman FA (1955) Heart weight in birds. Condor 57:221–238

Hartung DK, Kirkton SD, Harrison JF (2004) Ontogeny of tracheal system structure: a light and electron-microscopy study of the metathoracic femur of the American locust, Schistocerca americana. J Morphol 262:800–812

Harvey EP, Ben-Tal A (2016) Robust unidirectional airflow through avian lungs: new insights from a piecewise linear mathematical model. PLoS Comput Biol 12:e1004637. https://doi.org/10.1371/journal.pcbi.1004637

Hassrick JL, Crocker DE, Teutschel NM, McDonald BI, Robinson PW, Simmons SE, Costa DP (2010) Condition and mass impact oxygen stores and dive duration in adult female northern elephant seals. J Exp Biol 213:585–592

Hawkins AE, Jewell PA, Tomlinson G (2009) The metabolism of some British shrews. J Zool (Lond) 135:99–103

Hazelhoff EH (1943) Bouw en functie van de vogellong. Verslag van de gewone vergaderingen der Afdeeling Natuurkunde. Kon Ned Akad Wet 52:391–400

Henk WG, Haldman JT (1990) Microanatomy of the lung of the bowhead whale *Balaena mysticetus*. Anat Rec 226:187–197

Hills BA, Hughes GM (1970) A dimensional analysis of oxygen transfer in the fish gill. Respir Physiol 9:126–140

Hlastala MP, Berger AJ (1996) Physiology of respiration. Oxford University Press, New York

Hockey PAR, Dean WRJ, Ryan PG (2005) Roberts - birds of southern Africa, 7th edn. The Trustees of the John Voelcker Bird Book Fund, Cape Town

Hogg DA (1984) The development of pneumatisation in the postcranial skeleton of the domestic fowl. J Anat 139:105–113

Holle JP, Heisler N, Scheid P (1978) Blood flow distribution in the bird lung and its controls by respiratory gases. Am J Physiol 234:R146–R154

Hopkins SR, Schoene RB, Henderson WR, Spragg RG, Martin TR, West JB (1997) Intense exercise impairs the integrity of the pulmonary blood–gas barrier in elite athletes. Am J Respir Crit Care Med 155:1090–1094

Hopkins SR, Schoene RB, Henderson WR, Spragg RG, West JB (1998) Sustained submaximal exercise does not alter the integrity of the lung blood-gas barrier in elite athletes. J Appl Physiol 84:1185–1189

Hou JB, Hughes NC, Hopkins MJ (2021) The trilobite upper limb branch is a well-developed gill. Sci Adv 7:eabe7377. https://doi.org/10.1126/sciadv.abe7377

Hsia CC, Schmitz A, Lambertz M, Perry SF, Maina JN (2013) Evolution of air breathing: oxygen homeostasis and the transitions from water to land and sky. Compr Physiol 3:849–915

Hughes GM, Morgan M (1973) The structure of fish gills in relation to their respiratory function. Biol Rev 48:419–475

Hughes GM, Weibel ER (1976) Morphometry of fish lungs. In: Hughes GM (ed) Respiration of amphibious vertebrates. Academic Press, London, pp 213–232

Hughes GM, Weibel ER (1978) Visualization of layers lining the lung of the South American lungfish, *Lepidosiren paradoxa* and a comparison with the frog and the rat. Tissue Cell 10:343–353

Hughes GM, Wright DE (1970) A comparative study of the ultrastructure of the water/blood pathway in the secondary lamellae of teleost and elasmobranch fishes - benthic forms. Z Zellforsch Mikrosk Anat 104:478–493

Hull D, Bacon DJ (2001) Introduction to dislocations. Butterworth Heinmann, Oxford

Husak JF, Lailvaux SP (2022) Conserved and convergent mechanisms underlying performance - life-history trade-offs. J Exp Biol 225:jeb243351. https://doi.org/10.1242/jeb.243351

Huxley TH (1882) On the respiratory organs of *Apteryx*. Proc Zool Soc (Lond) 1882:560–569

Icardo JM, Colvee E, Lauriano ER, Capillo G, Guerrera MC, Zaccone G (2015) The structure of the gas bladder of the spotted gar, *Lepisosteus oculatus*. J Morphol 276:90–101

Ingber DE (2003) Tensegrity I. Cell structure and hierarchical systems biology. J Cell Sci 116: 1157–1173

Irving L (1939) Respiration in diving mammals. Physiol Rev 19:112–134

Irving L, Solandt OM, Solandt DY, Fisher KC (1935) Respiratory characteristics of the blood of the seal. J Cell Comp Physiol 6:393–403

Issac YM, Abraham J, Sreeparvathy GJ, Balusami C (2010) Managemental practices to control ascites in a flock. Vet World 3:250–252

James AE, Hutchins G, Bush M, Natarajan TK, Burns B (1976) How birds breathe: correlation of radiographic and anatomical and pathological studies. J Vet Radiol Soc 17:77–86

Jammes Y (1975) Etude comparée des échangeurs respiratoires de l'oiseau et du mammifère [Comparative study of respiratory exchanging surfaces in birds and mammals]. Poumon Coeur 31:165–168

Javed F, He Q, Davidson LE, Thornton JC, Albu J, Boxt L, Krasnow N, Elia M, Kang P, Heshka S et al (2010) Brain and high metabolic rate organ mass: contributions to resting energy expenditure beyond fat-free mass. Am J Clin Nutrit 91:907–912

Jefferson TA, Leatherwood S, Webber MA (1993) FAO species identification guide, marine mammals of the world. FAO, Rome

Jenni-Eiermann S (2017) Energy metabolism during endurance flight and the post-flight recovery phase. J Comp Physiol A 203:431–438

Jenni-Eiermann S, Jenni L, Smith S, Costantini D (2014) Oxidative stress in endurance flight: an unconsidered factor in bird migration. PLoS One 9:e97650. https://doi.org/10.1371/journal.pone.0097650

Jimenez AG (2018) The same thing that makes you live can kill you in the end: exploring the effects of growth rates and longevity on cellular metabolic rates and oxidative stress in mammals and birds. Integr Comp Biol 58:544–548

Jimenez AG, O'Connor ES, Tobin KJ, Anderson KN, Winward JD, Fleming A, Winner C, Chinchilli E, Maya A, Carlson K et al (2019) Does cellular metabolism from primary fibroblasts and oxidative stress in blood differ between mammals and birds? The (lack-thereof) scaling of oxidative stress. Integr Comp Biol 59:953–969

Jimoh SA (2012) Study of the basis of the strength of the pulmonary blood-gas barrier of the domestic fowl, *gallus gallus* variant *domesticus*. PhD Thesis, University of the Witwatersrand (Johannesburg - South Africa)

Jimoh SA, Maina JN (2013) Immuno-localization of type-IV collagen in the blood-gas barrier and the epithelial - epithelial cell connections of the avian lung. Biol Lett 9:20120951. https://doi.org/10.1098/rsbl.2012.0951

Jones JH (1998) Symmorphosis and the mammalian respiratory system: what is optimal design and does it exist? In: Weibel ER, Taylor CR, Bolis L (eds) Principles of animal design: the optimization and symmorphosis debate. Cambridge University Press, Cambridge, pp 241–248

Jones AW, Radnor CJP (1972a) The development of the chick tertiary bronchus. I. General development and the mode of production of the osmiophilic inclusion body. J Anat 113:303–324

Jones AW, Radnor CJP (1972b) The development of the chick tertiary bronchus. II. The origin of the surface lining system. J Anat 113:325–340

Jones JH, Effmann EL, Schmidt-Nielsen K (1981) Control of air-flow in bird lungs: radiographic studies. Respir Physiol 45:121–131

Jones JH, Effmann EL, Schmidt-Nielsen K (1985) Lung volume changes during respiration in ducks. Respir Physiol 59:15–25

Juillet A (1911) Recherches anatomiques, embryologiques, histologiques et comparatives sur le poumon des oiseaux. Arch Zool Exp Gen IX:207–371

Julian RJ (1998) Rapid growth problems: ascites and skeletal deformities in broilers. Poult Sci 77: 1773–1780

Julian RJ (2000) Physiological, management and environmental triggers of the ascites syndrome: a review. Avian Pathol 29:519–527

Julian RJ, Boulianne M (1988) Natural and experimental lung pathology causing pulmonary hypertension, right ventricular hypertrophy, right ventricular failure and ascites in broiler chickens. Am J Vet Res 192:1781(abstr. 209)

Julian RJ, Wilson JB (1986) Right ventricular failure as a cause of ascites in broiler and rooster chickens. In: Proceedings of 4th international symposium of veterinary laboratory diagnosticians, pp 608–611

Jürgens KD (2002) Etruscan shrew muscle: the consequences of being small. J Exp Biol 205:2161–2166

Jürgens JD, Gros G (2002) Phylogeny of gas exchange systems. Anasthesiol Intensivmed Notfallmed Schmerzther 37:185–198

Jürgens JD, Bartels H, Bartels R (1981) Blood oxygen transport and organ weight of small bats and small nonflying mammals. Respir Physiol 45:243–260

Kadono H, Okada T (1962) Electromyographic studies on the respiratory muscles of the domestic fowl. Jap J Vet Sci 24:215–222

Kadono H, Okada T, Ono K (1963) Electromyographic studies on the respiratory muscles of the chicken. Poult Sci 42:121–128

Kalmar ID, Vanrompay D, Janssens GPJ (2013) Broiler ascites syndrome: collateral damage from efficient feed to meat conversion. Vet J 197:169–174

Kandyel RM, El Basyouny HA, El Nahas EE, Madkour F, Haddad S, Ma D (2021) A histological and immunohistochemical study on the parabronchial epithelium of the domestic fowl's (*Gallus gallus domesticus*) lung with special reference to its scanning and transmission electron microscopic characteristics. Microsc Res Tech 2021:1–12. https://doi.org/10.1002/jemt.23980

Kardong KV (1988) Vertebrates: comparative anatomy, function, evolution. McGraw-Hill, New York, pp 278–279

Kasting JF (1993) Earth's early atmosphere. Science 259:920–926

Kendeigh CS, Blem CR (1974) Metabolic adaptation to local climate in birds. Comp Biochem Physiol A: Physiol 48:175–187

Kern A, Eble J, Golbik R, Kühn K (1993) Interaction of type IV collagen with the isolated integrins a1b1 and a2b1. Eur J Biochem 215:151–159

Kılıçkap M, Barçın C, Göksülük H, Karaaslan D, Özer N, Kayıkçıoğlu M, Ural D, Yılmaz MB, Abacı A, Arıcı M et al (2018) Systematic review, meta-analysis and meta-regression of epidemiological studies on cardiovascular risk factors. Turk Kardiyol Dern Ars 46:525–545

King AS (1957) The aerated bones of *Gallus domesticus*. Acta Anat 31:220–230

King AS (1966) Structural and functional aspects of the avian lung and its air sacs. Intern Rev Gen Exp Zool 2:171–267

King AS, Cowie AF (1969) The functional anatomy of the bronchial muscle of the bird. J. Anat 105:323–336

King AS, McLelland J (1984) Birds: their structure and function. Baillière Tindall, Eat Sussex

King AS, Molony V (1971) The anatomy of respiration. In: Bell DF, Freeman BM (eds) Physiology and biochemistry of the domestic fowl, vol 1. Academic Press, London, pp 347–384

King AS, Vidyadaran MK, Kassim H (1992) Quantitative pulmonary anatomy of a ground-dwelling bird, the white-breasted water-hen (*Amaurornis Phoenicurus*). J Zool (Lond) 227: 185–191

Kirkton SD (2007) Effects of insect body size on tracheal structure and function. Adv Exp Med Biol 618:221–228

Klabunde RE (2005) Cardiovascular physiology concepts. Lippincott Williams and Wilkins, Philadelphia

Klika E, Scheuermann DW, De Groodt-Lasseel MH, Bazantova I, Switka A (1997) Anchoring and support system of pulmonary gas exchange tissue in four bird species. Acta Anat 159:30–41

Klingele TG, Staub NC (1965) Alveolar shape changes with volume in isolated, air filled lobes of cat lung. J Appl Physiol 28:411–423

Knudsen L, Ochs M (2018) The micromechanics of lung alveoli: structure and function of surfactant and tissue components. Histochem Cell Biol 150:661–676

Koetzler R, Saifeddine M, Yu Z, Schüurch S, Hollenberg MD, Green FHY (2006) Surfactant as an airway smooth muscle relaxant. Am J Respir Cell Mol Biol 34:609–615

Kooyman GL (1973) Respiratory adaptations in marine mammals. Amer Zool 13:457–468

Koteja P (2004) The evolution of concepts on the evolution of endothermy in birds and mammals. Physiol Biochem Zool 77:1043–1050

Krista LM, Burger RE, Waibel PE (1963a) Blood pressure and heart rate in the turkey as measured by the indirect method and their modifications by pharmacological agents. Poult Sci 42:646–652

Krista LM, Burger RE, Waibel PE (1963b) Comparison of indirect method to direct arterial blood pressure measurements in the turkey. Poult Sci 42:1282

Krista LM, Waibel PE, Shoffner RN, Sautter JH (1967) Natural dissecting aneurysm (aortic rupture) and blood pressure in the turkey. Nature 214:1162–1163

Krista LM, Waibel PE, Shoffner RN, Sautter JH (1970) A study ofaortic rupture and performance as influenced by selection for hypertension andhypotension in the turkey. Poult Sci 49:405–411

Krista LM, Beckett SD, McDaniel GR, Patterson RM, Mora EC (1986) Rupture pressure of elastic and muscular aortic segments from hypertensive and hypotensive turkeys. Br Poult Sci 27:207–213

Krogh A (1908) Some new methods for the tonometric determination of gas tensions in fluids. Skand Arch Physiol 20:259–278

Krogh A (1910) On the mechanism of gas exchange in the lungs. Skand Arch Physiol 23:192–199

Krogh A (1918/1919) The rate of diffusion of gases through animal tissues, with some remarks on the coefficient of invasion. J Physiol (Lond) 52:391–408

Krogh A, Krogh M (1909a) On the tensions of gases in the arterial blood. Skand Arch Physiol 23: 179–192

Krogh A, Krogh M (1909b) Rate of diffusion into lungs of man. Skand Arch Physiol 23:236–247

Kuethe DO (1988) Fluid mechanical valving of airflow in bird lungs. J Exp Biol 136:1–12

Kühn K (1995) Basement membrane (type IV) collagen. Matrix Biol 14:439–445

Kunchala SR, van Dijk A, Veldhuizen EJA, Donnellan SC, Haagsman HP, Orgeig S (2023) Avian surfactant protein (SP)-A2 first arose in an early tetrapod before the divergence of amphibians and gradually lost the collagen domain. Dev Comp Immunol 139:104582. https://doi.org/10.1016/j.dci.2022.104582

Kuno K, Staub NC (1968) Acute mechanical effects of lung volume changes on artificial microholes in alveolar walls. J Appl Physiol 24:83–98

Kurtz HJ, Krista LM, Katter MA, Sautter JH (1969) Aortic rupture in turkeys: quantitative evaluations of histopathologic changes in aorta as influenced by dietary and genetic factors. Am J Vet Res 30:91–100

Laguë SL (1985) High-altitude champions: birds that live and migrate at altitude. J Appl Physiol 123:942–950

Lasiewski RC, Calder WA (1971) A preliminary allometric analysis of respiratory variables in resting birds. Respir Physiol 11:152–166

Lau MJ, Keough KWW (1981) Lipid com- position of lung and lung lavage fluid from map turtles (*Malaclemys geographica*) maintained at different environmental temperatures. Can J Biochem 59:208–219

Lawson SL, Jones MB, Moate RM (2019) Structural variability and distribution of cells in a posterior gill of *Carcinus maenas* (Decapoda: Brachyura). J Mar Biol Ass (UK) 74:771–785

Lee DY, Wexler AS, Fanucchi MV, Plopper CG (2008) Expiration rate drives human airway design. J Theor Biol 253:381–387

Leiby KL, Raredon MSB, Niklason LE (2020) Bioengineering the blood-gas barrier. Compr Physiol 10:415–452

Leitner MM, Roumy M (1974) Vagal afferent activities related to the respiratory cycle in the duck: sensitivity to mechanical, chemical and electrical stimuli. Respir Physiol 22:41–56

Lenel R, Katz LN, Rodbard S (1948) Arterial hypertension in the chicken. Am J Physiol 152:557–562

Lenfant C, Johansen K, Torrance JD (1970) Gas transport and oxygen storage capacity in some pinnipeds and the sea otter. Respir Physiol 9:277–286

Levitzky MG (2013) Pulmonary physiology. McGraw-Hill Companies, New York (NY)

Lichtenberger M (2005) Determination of indirect blood pressure in the companion bird. Semn Avian Exotic Pet Med 14:149–152

Liem KF (1988) Form and function of lungs: the evolution of air breathing mechanisms. Am Zool 28:739–759

Liem KF, Walker WF (eds) (1994) Functional anatomy of the vertebrates: an evolutionary perspective, 3rd edn. Saunders College Publishing, New York

Lin T, Smith T, Pinnock C, Mowatt C (eds) (2009) Fundamentals of anaethesia, 4th edn. Cambridge University Press, Cambridge (MA)

Little C (1990) The terrestrial invasion. Cambridge University Press, Cambridge

Liu J, Guo Y, Zhang C, Zeng Y, Luo Y, Wang G (2022) Clearance systems in the brain, from structure to function. Front Cell Neurosci 15:729706. https://doi.org/10.3389/fncel.2021.729706

López-García P, Eme L, Moreira D (2017) Symbiosis in eukaryotic evolution. J Theor Biol 434:20–33

Loscertales M, Nicolaou F, Jeanne M, Longoni M, Gould DB, Sun Y, Maalouf FI, Nagy N, Donahoe PK (2016) Type-IV collagen drives alveolar epithelial-endothelial association and the morphogenetic movements of septation. BMC Biol 14:59. https://doi.org/10.1186/s12915-016-0281-2

Lowry OH, Rosbrough NJ, Farr AL, Randall RJ (1951) Protein measurement with the Folin phenol reagent. J Biol Chem *193*:265–275

Luis da Cruz A, Pedretti ACE, Fernandes MN (2009) Stereological estimation of the surface area and oxygen diffusing capacity of the respiratory stomach of the air-breathing armored catfish, *Pterygoplichthys anisitsi* (Teleostei: Loricariidae). J Morphol 270:601–614

Lum H, Mitzner W (1985) A species comparison of alveolar size and surface forces. J Appl Physiol 62:1865–1871

Lumeij JT, Ritchie BW (1994) Cardiology. In: Ritchie BW, Harrison GJ, Harrison LR (eds) Avian medicine: principles and application. Wingers Publishing, Lake Worth (FL), pp 695–722

Lutz PL, Longmuir IS, Tuttle JV, Schmidt-Nielsen K (1973) Dissociation curve of bird blood and effect of red cell oxygen consumption. Respir Physiol 17:269–275

Lyons TW, Reinhard CT, Planavsky NJ (2014) The rise of oxygen in Earth's early ocean and atmosphere. Nature 506:307–315

Maarek JMI, Chang HK (1991) Pulsatile pulmonary microvascular pressure measured with vascular occlusion techniques. J Appl Physiol 70:998–1005

Mackelprang R, Goller F (2013) Ventilation patterns of the songbird lung/air sac system during different behaviors. J Exp Biol 216:3611–3619
Macklem P, Bouverot P, Scheid P (1979) Measurement of the distensibility of the parabronchi in duck lungs. Respir Physiol 33:23–35
Magnussen H, Willmer H, Scheid P (1976) Gas exchange in the air sacs: contribution to respiratory gas exchange in ducks. Respir Physiol 26:129–146
Maina JN (1982a) A scanning electron microscopic study of the air- and blood capillaries of the lung of the domestic fowl (*Gallus domesticus*). Experientia 35:614–616
Maina JN (1982b) Qualitative and quantitative observations on the lungs of aves with comments on the lung of a species of chiroptera. A morphological study. PhDThesis, University of Liverpool
Maina JN (1984) Morphometrics of the avian lung. 3. The structural design of the passerine lung. Respir Physiol 55:291–309
Maina JN (1987) Morphometrics of the avian lung. 4. The structural design of the charadriiform lung. Respir Physiol 68:99–119
Maina JN (1988) Scanning electron microscopic study of the spatial organization of the air- and blood conducting components of the avian lung *(Gallus gallus domesticus)*. Anat Rec 222:145–153
Maina JN (1989a) The morphometry of the avian lung. In: King AS, McLelland J (eds) Form and function in birds, vol 4. Academic Press, London, pp 307–368
Maina JN (1989b) The morphology of the lung of a tropical terrestrial slug, *Trichotoxon copleyi* (Mollusca: Gastropoda: Pulmonata): a scanning and transmission electron microscopic study. J Zool (Lond) 217:355–366
Maina JN (1989c) The morphology of the lung of the East African tree frog *Chiromantis petersi* with observations on the skin and the buccal cavity as secondary gas exchange organs: a TEM and SEM study. J Anat 165:29–43
Maina JN (1989d) The morphology of the lung of the black mamba *Dendroaspis polylepis* (Reptilia: Ophidia: Elapidae): a scanning and transmission electron microscopic study. J Anat 167:31–46
Maina JN (1989e) A scanning and transmission electron microscopic study of the tracheal air-sac system in a grasshopper (*Chrotogonus senegalensis*, Kraus)- (Orthoptera: Acrididae: Pygomorphinae). Anat Rec 223:393–405
Maina JN (1990a) The morphology of the gills of the freshwater African crab *Potamon niloticus* (Crustacea: Brachyura: Potamonidae): a scanning and transmission electron microscopic study. J Zool (Lond) 221:499–515
Maina JN (1990b) A study of the morphology of the gills of an extreme alkalinity and hyperosmotic adapted teleost *Oreochromis alcalicus grahami* (Boulenger) with particular emphasis on the ultrastructure of the chloride cells and their modifications with water dilution. A SEM and TEM study. Anat Embryol 181:83–98
Maina JN (1993) Morphometries of the avian lung: the structural-functional correlations in the design of the lungs of birds. Comp Biochem Physiol A: Physiol 105:397–410
Maina JN (1994) Comparative respiratory morphology and morphometry: the functional design of the respiratory systems. In: Gilles R (ed) Advances in comparative and environmental physiology. Springer, Berlin, pp 111–232
Maina JN (1998) The gas exchangers: structure, function, and evolution of the respiratory processes. Springer, Heidelberg
Maina JN (2000a) Comparative respiratory morphology: themes and principles in the design and construction of the gas exchangers. Anat Rec 261:25–44
Maina JN (2000b) Is the sheet-flow design a ‘frozen core’ (a Bauplan) of the gas exchangers? Comparative functional morphology of the respiratory microvascular systems: illustration of the geometry and rationalization of the fractal properties. Comp Biochem Physiol A Mol Integr Physiol 126:491–515

Maina JN (2000c) Functional morphology of the gas-gland cells of the air-bladder of *Oreochromis alcalicus grahami* (teleostei: cichlidae): an ultrastructural study on a fish adapted to a severe, highly alkaline environment. Tissue Cell 32:117–132

Maina JN (2002a) Fundamental structural aspects and features in the bioengineering of the gas exchangers: comparative perspectives. Adv Anat Embryol Cell Biol 163:1–108

Maina JN (2002b) Structure, function and evolution of the gas exchangers: comparative perspectives. J Anat 201:281–304

Maina JN (2002c) Functional morphology of the vertebrate respiratory systems. Science, Enfield (NH)

Maina JN (2002d) Some recent advances of the study and understanding of the functional design of the avian lung: morphological and morphometric perspectives. Biol Rev 77:97–152

Maina JN (2003) Developmental dynamics of the bronchial (airway)- and air sac systems of the avian respiratory system from days 3 to 26 of life: a scanning electron microscopic study of the domestic fowl, *Gallus gallus* variant *domesticus*. Anat Embryol 207:119–134

Maina JN (2004) Structure and function of nonmammalian vertebrate lung. In: Massaro DJ, Massaro GC, Chambon P (eds) Lung development and regeneration. Marcel Dekker, New York, pp 319–354

Maina JN (2005) The lung-air sac system of birds: development, structure, and function. Springer, Berlin

Maina JN (2006) Development, structure, and function of a novel respiratory organ, the lung-air sac system of birds: to go where no other vertebrate has gone. Biol Rev 81:545–579

Maina JN (2007a) Spectacularly robust! Tensegrity principle explains the mechanical strength of the avian lung. Respir Physiol Neurobiol 155:1–10

Maina JN (2007b) Minutialization at its extreme best! The underpinnings of the remarkable strengths of the air- and the blood capillaries of the avian lung: a conundrum. Respir Physiol Neurobiol 159:141–145

Maina JN (2009a) The morphology of the lung of a tropical terrestrial slug *Trichotoxon copleyi* (Mollusca: Gastropoda: Pulmonata): a scanning and transmission electron microscopic study. J Zool (Lond) 217:355–366

Maina JN (2009b) Functional morphology of the avian respiratory system, the lung-air sac system: efficiency built on complexity. Ostrich 79:117–132

Maina JN (2011) Bioengineering aspects in the design of gas exchangers: comparative evolutionary, morphological, functional, and molecular perspectives. Springer, Berlin

Maina JN (2014) Comparative respiratory physiology: the fundamental mechanisms and the functional designs of the gas exchangers. Open Access Anim Physiol 6:53–66

Maina JN (2015a) Structural and biomechanical properties of the exchange tissue of the avian lung. Anat Rec 298:1673–1688

Maina JN (2015b) The design of the avian respiratory system: development, morphology and function. J Ornithol 156:41–63

Maina JN (2017a) Pivotal debates and controversies on the structure and function of the avian respiratory system: setting the record straight. Biol Rev 92:1475–1504

Maina JN (2017b) Functional design of the mature avian respiratory system. In: Maina JN (ed) The biology of the avian respiratory system: evolution, development, structure and function. Springer, Cham, pp 191–218

Maina JN (2017c) Critical appraisal of some factors pertinent to the functional designs of the gas exchangers. Cell Tissue Res 367:747–767

Maina JN (2022) Perspectives on the structure and function of the avian respiratory system: functional efficiency built on structural complexity. Front Anim Sci 3:851574. https://doi.org/10.3389/fanim.2022.851574

Maina JN, Africa M (2000) Inspiratory aerodynamic valving in the avian lung: functional morphological study of the extrapulmonary primary bronchus. J Exp Biol 203:2865–2876

Maina JN, Igbokwe CO (2020) Comparative morphometric analysis of lungs of the semifossorial giant pouched rat (*Cricetomys gambianus*) and the subterranean Nigerian mole rat (*Cryptomys foxi*). Sci Rep 10:5244. https://doi.org/10.1038/s41598-020-61873-8

Maina JN, Jimoh SA (2013a) Structural failures of the blood-gas barrier and the epithelial-epithelial connections in the different vascular regions of the lung of the domestic fowl, *Gallus gallus* variant *domesticus*, at rest and during exercise. Biol Open 2:267–276

Maina JN, Jimoh SA (2013b) Study of stress induced failure of the blood-gas barrier and the epithelial-epithelial cells connections of the lung of the domestic fowl, *Gallus gallus* variant *domesticus* after vascular perfusion. Biomed Engineer Comput Biol 5:77–88

Maina JN, King AS (1982a) The thickness of the avian blood-gas barrier: qualitative and quantitative observations. J Anat 134:553–562

Maina JN, King AS (1982b) Morphometrics of the avian lung. 2. The wild mallard (*Anas platyrhynchos)* and graylag goose *(Anser anser)*. Respir Physiol 50:299–313

Maina JN, King AS (1984) The structural functional correlation in the design of the bat lung. A morphometric study. J Exp Biol 111:43–63

Maina JN, King AS (1987) A morphometric study of the lung of a humboldt penguin *(Spheniscus humboldti)*. Zentralb Vet Med C Anat Histo Embryol 16:293–297

Maina JN, King AS (1989) The lung of the emu, *Dromaius novaehollandiae*: a microscopic and morphometric study. J Anat 163:67–74

Maina JN, Maloiy GMO (1985) The morphometry of the lung of the lungfish (*Protopterus aethiopicus*): its structural-functional correlations. Proc R Soc (Lond) B 244:399–420

Maina JN, Maloiy GMO (1986) The morphology of the respiratory organs of the African air-breathing catfish (*Clarias mossambicus*) (Peters): a light and electron microscopic study, with morphometric observations. J Zool (Lond) 209A:421–445

Maina JN, Maloiy GMO (1988) A scanning and transmission electron microscopic study of the lung of a caecilian *Boulengerula taitanus*. J Zool (Lond) 215:739–751

Maina JN, Nathaniel C (2001) A qualitative and quantitative study of the lung of an ostrich, *Struthio camelus*. J Exp Biol 204:2313–2330

Maina JN, van Gils P (2001) Morphometric characterization of the airway and vascular systems of the lung of the domestic pig, *Sus scrofa*: comparison of the airway, arterial, and venous systems. Comp Biochem Physiol 130A:781–798

Maina JN, West JB (2005) Thin and strong! The bioengineering dilemma in the structural and functional design of the blood-gas barrier: comparative and evolutionary perspectives. Physiol Rev 85:811–844

Maina JN, Woodward JD (2009) Three-dimensional serial section computer reconstruction of the arrangement of the structural componenets of the parabronchus of the ostrich, *Struthio camelus*. Anat Rec 292:1685–1698

Maina JN, Abdalla MA, King AS (1982a) Light microscopic morphometry of the lungs of 19 avian species. Acta Anat 112:264–270

Maina JN, King AS, King DZ (1982b) A morphometric analysis of the lung of a species of bat. Respir Physiol 50:1–11

Maina JN, King AS, Settle G (1989a) An allometric study of the pulmonary morphometric parameters in birds, with mammalian comparison. Philos Trans R Soc (Lond) 326B:1–57

Maina JN, Maloiy GMO, Warui CN, Njogu EK, Kokwaro ED (1989b) A scanning electron microscopic study of the morphology of the reptilian lung: the savanna monitor lizard *(Varanus exanthematicus)* and the pancake tortoise *(Malacochersus tornieri)*. Anat Rec 224:514–522

Maina JN, Thomas SP, Dallas DM (1991) A morphometric study of bats of different size: correlations between structure and function of the chiropteran lung. Philos Trans R Soc (Lond) B 333:31–50

Maina JN, Jimoh SA, Hosie M (2010) Implicit mechanistic role of the collagen, smooth muscle, and elastic tissue components in strengthening the air- and blood capillaries of the avian lung. J Anat 217:597–608

Maina JN, McCracken KG, Chua B, York JM, Milsom WK (2017) Morphological and morphometric specializations of the lung of the Andean goose, *Chloephaga melanoptera*: a lifelong high-altitude resident. PLoS One 12(3):e0174395. https://doi.org/10.1371/journal.pone.0174395

Maina JN, Singh P, Moss EA (2009) Inspiratory aerodynamic valving occurs in the ostrich, *Struthio camelus* lung: a computational fluid dynamics study under resting unsteady state inhalation. Respir Physiol Neurobiol 169:262–270

Maina JN, Veltcamp CJ, Henry J (1999) A study of the spatial organization of the gas-exchange components of a snake lung - the sandboa *Eryx colubrinus* (Reptilia: Ophidia; Corubridae) by latex casting. J Zool (Lond) 247:81–90

Maina JN, Wood CM, Narahara A, Bergman HL, Laurent P, Walsh P (1996) Morphology of the swim (air) bladder of a cichlid teleost: *Oreochromis alcalicus grahami* (Trewavas, 1983), a fish adapted to a hyperosmotic, alkaline and hypoxic environment: a brief outline of the structure and function of the swimbladder. In: Dutta HH, Munshi JSD (eds) Recent advances in fish biology, 1st edn. Oxford University Press and BHD Press, New Delhi, pp 27–45

Makanya A (ed) (2015) The vertebrate blood-gas barrier in health and disease: structure, development and remodeling. Springer, Cham

Makanya A, Anagnostopoulou A, Djonov V (2013) Development and remodeling of the vertebrate blood-gas barrier. BioMed Res Internat 2013:101597. https://doi.org/10.1155/2013/101597

Makanya AN, Kavoi BM, Kihurani DO (2021) Slight volume changes in the duck lung do not imply a fundamental change in the structure of the parenchyma. Anat Histol Embryol 50:169–174

Malte H, Lomholt JP (1998) Ventilation and gas exchange. In: Jergensen JM, Lomholt JP, Weber RE, Malte H (eds) The biology of hagfishes. Chapman and Hall, London, pp 223–234

Maltepe E, Saugstad OD (2009) Oxygen in health and disease: regulation of oxygen homeostasis - clinical implications. Pediatr Res 65:261–268

Manohar M, Hutchens E, Coney E (1993) Pulmonary haemodynamics in the exercising horse and their relationship to exercise-induced pulmonary haemorrhage. Br Vet J 149:419–428

Margulis L (1981) Symbiosis in cell evolution. Freeman, San Francisco

Markley JS, Carrier DR (2010) The cost of ventilation in birds measured via unidirectional artificial ventilation. Comp Biochem Physiol A, Mol Integr Physiol 155:146–153

Marshall GE, Konstas AG, Lee WR (1990a) Immunogold localization of type-IV collagen and laminin in the aging human outflow system. Exp Eye Res 51:691–699

Marshall GE, Konstas AG, Lee WR (1990b) Collagens in ocular tissues. Br J Ophthalm 77:515–524

Martin S, Maury B, Similowski T, Straus C (2008) Modelling of respiratory system mechanics involving gas exchange in the human lungs. In: Boudin L, Grandmont C, Maday Y, Maury B, Sapoval B, Gerbeau JF (eds) ESAIM: Proceedings, pp 23, 30–47. https://doi.org/10.1051/proc:082303

Martin WF, Tielens AGM, Mentel M, Garg SG, Gould SB (2017) The physiology of phagocytosis in the context of mitochondrial origin. Microbiol Mol Biol Rev 81:8–17

Martin WF, Tielens AG, Mentel M (2021a) Mitochondria and anaerobic energy metabolism in eukaryotes. De Gruyter, Berlin

Martin FP, Jacqueline C, Poschmann J, Roquilly A (2021b) Alveolar macrophages: adaptation to their anatomic niche during and after inflammation. Cells 10:2720. https://doi.org/10.3390/cells10102720

Martrenchar A, Morisse JP, Huonnic D, Cotte JP (1997) Influence of stocking density on some behavioural, physiological and productivity traits of broilers. Vet Res 28:473–480

Marza E, Barthe C, Andre M, Villeneuve L, Helou C, Babin PJ (2005) Developmental expression and nutritional regulation of a zebrafish gene homologous to mammalian microsomal triglyceride transfer protein large subunit. Dev Dyn 232:506–518

Mason RJ, Dobbs LG (2016) Alveolar epithelium and pulmonary surfactant. In: Murray and Nadel's textbook of respiratory medicine, pp 134–149.e5. https://doi.org/10.1016/B978-1-4557-3383-5.00008-7

Mathieu-Costello O, Willford DC, Fu Z, Garden RM, West JB (1995) Pulmonary capillaries are more resistant to stress failure in dogs than in rabbits. J Appl Physiol 79:908–917

Mauroy B, Filoche M, Weibel ER, Sapoval B (2004) An optimal bronchial tree may be dangerous. Nature 427:633–636

Maxwell MH, Robertson GW (1997) World broiler ascites survey. Poult Internat 36:16–30

Maxwell MH, Robertson GW, Spence S (1986a) Studies on an ascetic syndrome in young broilers. I. Hematology and pathology. Avian Pathol 15:511–524

Maxwell MH, Robertson GW, Spence S (1986b) Studies on an ascitic syndrome in young broilers. II. Ultrastructure. Avian Pathol 15:525–538

Mbanga B, van Dyk C, Maina JN (2018) Morphometric and morphological study of the respiratory organs of the bimodally-breathing African sharptooth catfish (*Clarias gariepinus*): Burchell (1822). Zoology 130:6–18

McKechnie AE, Freckleton RP, Jetz W (2006) Phenotypic plasticity in the scaling of avian basal metabolic rate. Proc R Soc (Lond) B 273:931–937

McKechnie AE, Wolf BO (2004) The allometry of avian basal metabolic rate: good predictions need good data. Physiol Biochem Zool 77:502–521

McNab BK (1966) An analysis of the body temperature of birds. Condor 68:47–55

McNab BK (2002) The physiological ecology of vertebrates: a view from energetics. Cornell University Press, Ithaca (NY)

McNab BK, Morrison P (1963) Body temperature and metabolism in subspecies of *Peromyscus* from arid and mesic environments. Ecol Monogr 33:63–82

McLelland J (1989) Anatomy of the lungs and air sacs. In: King AS, McLelland J (eds) Form and function in birds, vol IV. Academic Press, London, pp 221–279

McWilliams SR, Guglielmo C, Pierce B, Klaassen M (2004) Flying, fasting, and feeding in birds during migration: a nutritional and physiological ecology perspective. J Avian Biol 35:377–393

Mead J (1961) Mechnical properties of lungs. Physiol Rev 45:48–79

Meban C (1980) Thicknesses of the air-blood barriers in vertebrate lungs. J Anat 131:299–307

Mereness JA, Mariani TJ (2021) The critical role of collagen VI in lung development and chronic lung disease. Matrix Biol Plus 10:100058. https://doi.org/10.1016/j.mbplus.2021.100058

Merola-Zwartjes M, Ligon JD (2000) Ecological energetics of the Puerto Rican tody: heterothermy, torpor, and intra-island variation. Ecology 81:990–1003

Meyer MR, Burger RE, Scheid P, Piiper J (1977) Analyses des échanges gazeux dans les poumons d'oiseaux. J Physiol, Paris 73:9A

Miller DN, Bondurant S (1961) Surface characteristics of vertebrate lung extracts. J Appl Physiol 16:1075–1077

Miller MA, Zachary JF (2017) Mechanisms and morphology of cellular injury, adaptation, and death. Pathol Basis Veterinary Dis 2-43:e19. https://doi.org/10.1016/B978-0-323-35775-3.00001-1

Milnor WR (1982) Hemodynamics. Williams & Wilkins, Baltimore (MD)

Milsom WK (1991) Respiration in air breathing vertebrates: optimization and efficiency in design and function. In: Blake R (ed) Efficiency and economy in animal physiology. Cambridge University Press, Cambridge, pp 133–158

Milsom WK (2002) Phylogeny of CO_2/H^+ chemoreception in vertebrates. Respir Physiol Neurobiol 131:29–41

Mirastschijski U, Dembinski R, Maedler K (2020) Lung surfactant for pulmonary barrier restoration in patients with COVID-19 pneumonia. Front Med 7:254. https://doi.org/10.3389/fmed.2020.00254

Mitchinson MJ, Keymer IF (1977) Aortic rupture in ostriches (*Struthio camelus*) - a comparative study. J Comp Pathol 87:27–33

Miura K (1990) Tension truss concept and formation of geodesic shapes in space. Intern J Space Struct 5:331–340

Mojzsis SJ, Arrhenius G, McKeegan KD, Harrison TM, Nutman AP, Friend CRL (1996) Evidence for life on Earth before 3800 million years ago. Nature 270:43–45

Molony V, Graf W, Scheid P (1976) Effects of CO_2 on pulmonary flow resistance in the duck. Respir Physiol 26:333–349

Monahan-Earley R, Dvorak AM, Aird WC (2013) Evolutionary origins of the blood vascular system and endothelium. J Thromb Haemost 11:46–66

Mondrinos MJ, Lelkes PI, Samadikuchaksaraei I, Mantalaris A, Polak JM (2014) lungs. In: Robert L, Robert L, Joseph V (eds) Principles of tissue engineering, 4th edn. Elsevier/Academic Press, Boston, pp 1561–1577

Morrell MJ (2015) One hundred years of pulmonary function testing: a perspective on 'The diffusion of gases through the lungs of man' by Marie Krogh. J Physiol 593:351–352

Morris CR, Nelson FE, Askew GN (2010) The metabolic power requirements of flight and estimations of flight muscle efficiency in the cockatiel (*Nymphicus hollandicus*). J Exp Biol 213:2788–2796

Moudgil R, Michelakis ED, Archer SL (2006) The role of K+ channels in determining pulmonary vascular tone, oxygen sensing, cell proliferation, and apoptosis: implications in hypoxic pulmonary vasoconstriction and pulmonary arterial hypertension. Microcirculation 8:615–632

Müller B (1908) The air sacs of the pigeon. Smithson misc Colls 50:365–414

Müller-Redetzky HC, Lienan J, Witzenrath M (2015) The lung endothelial barrier in acute inflammation. In: Makanya AN (ed) The vertebrate blood-gas barrier in health and disease: structure, development and remodeling. Springer, Cham, pp 159–187

Munshi JSD, Hughes GM (1992) Air breathing fishes of India: their structure, function and lifehistory. AA Balkema Uitgevers BV, Rotterdam

Munshi-South J, Wilkinson GS (2010) Bats and birds: exceptional longevity despite high metabolic rates. Ageing Res Rev 9:12–19

Murray CD (1926) The physiological principle of minimum work. I. The vacular system and the cost of blood volume. Proc Natl Acad Sci (USA) 12:207–214

Nayak A, Dodagatta-Marri E, Tsolaki AG, Kishore U (2012) An insight into the diverse roles of surfactant proteins, SP-A and SP-D in innate and adaptive immunity. Front Immunol 3:131. https://doi.org/10.3389/fimmu.2012.00131

Newton A (1896) A dictionary of birds. Adam and Charles Black, London

Neumann F, Ungar H (1973) Spontaneous aortic rupture in turkeys and the vascularization of the aortic wall. Can Vet J 14:136–138

Nganpiep L, Maina JN (2002) Composite cellular defense stratagem in the avian respiratory system: functional morphology of the free (surface) macrophages and specialized pulmonary epithelia. J Anat 200:499–516

Nguyen QM, Oza AU, Abouezzi J, Sun G, Childress S, Frederick C, Ristroph L (2021) Flow rectification in loopy network models of bird lungs. Phys Rev Lett 126:114501. https://doi.org/10.1103/PhysRevLett.126.114501

Nguyen-Phu D, Yamaguchi K, Scheid P, Piiper J (1986) Kinetics of oxygen uptake and release by erythrocytes of the chicken. J Exp Biol 125:15–27

Nicholas TE (1996) Pulmonary surfactant: no mere paint on the alveolar wall. Respirology 4:247–257

Notter RH (ed) (2000) Lung surfactant: basic science and clinical applications. Marcel Dekker, New York

Nutman AP, Bennett VC, Friend CRL (2015) The emergence of the Eoarchaean proto298 arc: evolution of a c. 3700 Ma convergent plate boundary at Isua, southern West 299 Greenland. In: Roberts NMW, Van Kranendonk M, Parman S, Shirey S, Clift PD (eds) Continent formation through time, vol 389. The Geological Society, London. (Special Publications), pp 113–133

Nutman AP, Bennett VC, Friend CRL, Van Kranendonk MJ, Chivas AR (2016) Rapid emergence of life shown by discovery of 3,700-million-year-old microbial structures. Nature 537:535–538

Ochs M, Nyengaard JR, Jung A, Knudsen L, Voigt M, Wahlers T, Richter J, Gundersen JG (2004) The number of alveoli in the human lung. Am J Respir Crit Care Med 169:120–124

O'Conner TP (1996) Geographic variation in metabolic seasonal acclimatization in house finches. Condor 98:371–381

Ohmer-Schröck D, Schlatterer C, Plattner H, Schlepper-Schäfer J (1995) Lung surfactant protein A (SP-A) activates a phosphoinositide/calcium signaling pathway in alveolar macrophages. J Cell Sci 108:3695–3702

Ohtani O, Nakatani T (1994) Spatial organization of the collagen and elastin fibers of the lung in the Japanese monkeys, Macaca fascata. Anthropol Sci 102:181–187

Olmeda B, Villén L, Cruz A, Orellana G, Perez-Gil J (2010) Pulmonary surfactant layers accelerate O(2) diffusion through the air-water interface. Biochim Biophys Acta 1798:1281–1284

Orgeig S, Daniels CB (2001) The roles of cholesterol in pulmonary surfactant: insights from comparative and evolutionary studies. Comp Biochem Physiol A Mol Integr Physiol 129:75–89

Oro J (1983) Chemical evolution and the origin of life. Adv Space Res 3:77–94

Othilia G (2013) Geodesic structures. Internat J Edu Res 1(5). https://ijern.com/images/May-2013/23.pdf

Otis AB (1954) The work of breathing. Physiol Rev 34:449–458

Panchen AL (1980) The terrestrial environment and the origin of land vertebrates. Academic Press, London

Pandey AK, Praveen PK, Ganguly S, Para PA, Wakchaure R, Saroj P, Mahajan T (2015) Avian respiratory anatomy and physiology with its interspecies variations: a review. World J Pharmaceut Life Sci 1:137–148

Panneton WM (2013) The mammalian diving response: an enigmatic reflex to preserve life? Physiology 28:284–297

Parr N, Wilkes M, Hawkes LA (2019) Natural climbers: insights from avian physiology at high altitude. High Alt Med Biol 20(4). https://doi.org/10.1089/ham.2019.0032

Parry K, Yates MS (1979) Observations on the avian pulmonary and bronchial circulation using labeled microspheres. Respir Physiol 38:131–140

Pascoe JR, Ferraro GL, Cannon JH, Arthur RM, Wheat JD (1981) Exercise-induced pulmonary hemorrhage in racing thoroughbreds: a preliminary study. Am J Vet Res 42:703–707

Pattle RE (1978) Lung surfactant and lung lining in birds. In: Piiper J (ed) Respiratory function in birds, adult and embryonic. Springer, Berlin, pp 23–32

Paul GS (1988) Predatory dinosaurs of the world: complete illustrated guide. Simon and Schuster, New York

Pedersen O, Colmer TD (2012) Physical gills prevent drowning of many wetland insects, spiders and plants. J Exp Biol 215:705–709

Peretó J (2005) Controversies on the origin of life. Int Microbiol 8:23–31

Perry SF (1976) Model of exchange barrier and respiratory surface area in the lung of the tortoise (*Testudo graeca*) and its practical application. Mikroskopie 32:282–293

Perry SF (1978) Quantitative anatomy of the lungs of the red-eared turtle, *Pseudemys scripta elegans*. Respir Physiol 35:245–262

Perry SF (1983) Reptilian lungs: functional anatomy and evolution. Adv Anat Embryol Cell Biol 79:1–81

Perry SF (1988) Functional morphology of the lungs of the Nile crocodile *Crocodylus niloticus*: non-respiratory parameters. J Exp Biol 143:99–117

Perry SF (1989a) Morphometry of crocodilian lungs. Fortschr Zool 35:546–549

Perry SF (1989b) Mainstreams in the evolution of vertebrate respiratory structures. In: King AS, McLelland J (eds) Form and function in birds, vol V. Academic Press, London, pp 1–67

Perry SF (1989c) Structure and function of the reptilian respiratory system. In: Wood SC (ed) Comparative pulmonary physiology: current concepts. Marcel Dekker, New York, pp 193–236

Perry SF (1990a) Recent advances and trends in the comparative morphometry of vertebrate gas exchange organs. In: Boutilier RG (ed) Vertebrate gas exchange: advances in comparative and environmental physiology, vol 6. Springer, Berlin, pp 45–71

Perry SF (1990b) Gas exchange strategy in the Nile crocodile: a morphometric study. J Comp Physiol B 159:761–769

Perry SF, Hein J, van Dieken E (1994) Gas exchange morphometry of the lungs of the tokay, *Gekko gecko* L. (Reptilia: Squamata: Gekkonidae). J Comp Physiol B 164:206–214

Perry SF, Sander M (2004) Reconstructing the evolution of the respiratory apparatus in tetrapods. Resp Physiol Neurobiol 144:125–139

Petřík P, Riedel B (1968) An osmiophilic bilaminar lining film at the respiratory surfaces of avian lungs. Z Zellforsch 88:204–219

Petrov AI, Razuvaeva MV (2018) Effect of temperature on metabolism and lifespan in several homeothermic animals. Tech Phys 63:1410–1414

Phillips JR, Hewes AE, Womack MC, Schwenk K (2022) The mechanics of air breathing in African clawed frog tadpoles, *Xenopus* laevis (Anura: Pipidae). J Exp Biol 225:jeb243102. https://doi.org/10.1242/jeb.243102

Piiper J, Scheid P (1972) Maximum gas transfer efficacy of models for fish gills, avian lungs and mammalian lungs. Respir Physiol 14:115–124

Piiper J, Scheid P (1973) Gas exchange in the avian lung: model and experimental evidence. In: Bolis L, Schmidt-Nielsen K, Maddrell SHP (eds) Comparative physiology. Elsevier, Amasterdam, pp 161–185

Piiper J, Scheid P (1975) Gas transport efficiency of gills, lungs and skin: theory and experimental data. Respir Physiol 13:292–304

Pittman RN (2011) Regulation of tissue oxygenation. Morgan and Claypool Life Sciences, San Rafael (CA). https://www.ncbi.nlm.nih.gov/books/NBK54110/

Planavsky NJ, Reinhard CT, Wang X, Thomson D, McGoldrick P, Rainbird RH, Johnson T, Fischer WW, Lyons TW (2014) Low Mid-Proterozoic atmospheric oxygen levels and the delayed rise of animals. Science 346:635–638

Plankensteiner K, Reiner H, Rode BM (2004) From earth's primitive atmosphere to chiral peptides - the origin of precursors for life. Chem Biodivers 1:1308–1315

Poole DC, Erickson DH (2011) Highly athletic terrestrial mammals: horses and dogs. Compr Physiol 1:1–37

Portier P (1928) Sur le role physiologigue des sacs aeriens des oiseaux. C r Séanc Soc Biol 99: 1327–1329

Pöschl E, Schlötzer-Schrehardt U, Brachvogel B, Saito K, Ninomiya Y, Mayer U (2004) Collagen IV is essential for basement membrane stability but dispensable for initiation of its assembly during early development. Development 131:1619–1628

Powell FL (2014) Respiration. In: Scanes CG (ed) Sturkie's avian physiology, 6th edn. Academic Press, San Diego, pp 301–336

Powell FL, Geiser J, Gratz RK, Scheid P (1981) Air flow in the avian respiratory tract: variation of O_2 and CO_2 concentrations in the bronchi of the duck. Respir Physiol 44:195–213

Powell FL, Hastings RH, Mazzone RW (1985) Pulmonary vascular resistance during unilateral pulmonary artery occlusion in ducks. Am J Physiol 249:R34–R43

Powell FL, Hopkins SR (2004) Comparative physiology of lung complexity: implications for gas exchange. News Physiol Sci 19:55–60

Powell FL, Milsom WK, Mitchell GS (1998) Time domains of the hypoxic ventilatory response. Respir Physiol 112:123–134

Powell FL, Scheid P (1989) Physiology of gas exchange in the avian respiratory system. In: King AS, McLelland J (eds) Form and function of the avian lung, vol 4. Academic Press, London, pp 393–437

Powers KA, Dhamoon AS (2022) Physiology, pulmonary ventilation and perfusion. In: StatPearls [Internet] Treasure Island (FL): StatPearls Publishing

Prinzinger R, Preßmar A, Schleucher E (1991) Body temperature in birds. Comp Biochem Physiol A 99:499–506

Pushpavanam S, Malengier B (2012) comparison of co-current and counter-current flow fields on extraction performance in micro-channels. Adv Chem Engineer Sci 2:309–320

Rahn H (1964) Oxygen stores of man. In: Dickens F, Neil E (eds) Proceedings of a symposium held under the joint auspices of the International Union of Biochemistry and the International Union of Physiological Sciences, London, 1963. Pergamon, London, pp 609–619

Rainey G (1849) On the minute anatomy of the lung of the bird, considered chiefly in relation to the structure with which the air is in contact whilst traversing the ultimate subdivisions of the air passages. Med Chir Trans 82:47–58

Ramirez JM, Folkow LP, Blix AS (2007) Hypoxia tolerance in mammals and birds: from the wilderness to the clinic. Annu Rev Physiol 69:113–143

Randall DJ, Burggren WW, Farrell AP, Haswell MS (1981) The evolution of air breathing in vertebrates. Cambridge University Press, Cambridge

Rawal UM (1976) Nerves in the avian air sacs. Pavo 14:57–60

Ray PJ, Fedde MR (1969) Responses to alterations in respiratory PO_2, and PCO_2, in the chicken. Respir Physiol 6:135–143

Reeves RR, Brent SS, Phillip JC, James AP (2002) National Audubon Society guide to marine mammals of the world. Chanticleer, New York

Rink C, Khanna S (2011) Significance of brain tissue oxygenation and the arachidonic acid cascade in stroke. Antioxid Redox Signal 14:1889–1903

Rodriguez C, Prieto GI, Vega IA, Castro-Vazquez A (2019) Functional and evolutionary perspectives on gill structures of an obligate air-breathing, aquatic snail. PeerJ 7:e7342. https://doi.org/10.7717/peerj.7342

Rosenberg AA (1988) Response of the cerebral circulation to profoundhypocarbia in neonatal lambs. Stroke 19:1365–1370

Ross MJ (1899) Special structural features in the air sacs of birds. Trans Amer Micr Soc 20:29–40

Roughton FJW, Forster RE (1957) Relative importance of diffusion and chemical reaction rates in determining rate of exchange of gases in the human lung, with special reference to true diffusing capacity of pulmonary membrane and volume of blood in the lung capillaries. J Appl Physiol 11:290–302

Roux E (2002) Origine et évolution de l'appareil respiratoire aérien des Vertébrés. Rev Mal Respir 19:601–615

Rutten MG (1970) The history of atmospheric oxygen. Origins Life Evol Biosph 2:5–17

Scanes CG, Witt J, Ebeling M, Schaller S, Baier V, Bone AJ, Preuss TG, Heckmann D (2022) Quantitative comparison of avian and mammalian physiologies for parameterization of physiologically based kinetic models. Front Physiol 13:858386. https://doi.org/10.3389/fphys.2022.85838

Scheid P (1979) Mechanisms of gas exchange in bird lungs. Rev Physiol Biochem Pharmacol 86: 137–186

Scheid P (1987) The use of models in physiological studies. In: Feder ME, Bennett AF, Burggrenn WW, Huey RB (eds) New direction in ecological physiology. Cambridge University Press, Cambridge, pp 275–288

Scheid P (1990) Avian respiratory system and gas exchange. In: Sutton JR, Coates G, Remmers JE (eds) Hypoxia: the adaptations. BC Decker, Burlington, pp 4–7

Scheid P, Hook C, Piiper J (1986) Model for analysis of counter-current gas transfer in fish gills. Respir Physiol 64:365–374

Scheid P, Piiper J (1969) Volume, ventilation and compliance of the respiratory system in the domestic fowl. Respir Physiol 6:298–308

Scheid P, Piiper J (1971) Direct measurement of the pathway of of respired gas in duck lung lungs. Respir Physiol 11:308–314

Scheid P, Piiper J (1972) Cross-currrent gas exchange in the avian lungs: effects of reversed parabronchial air flow in ducks. Respir Physiol 16:304–312

Scheid P, Piiper J (1989) Respiratory mechnics and air flow in birds. In: King AS, McLelland J (eds) Form and function in birds, vol 4. Academic Press, London, pp 364–391
Scheid P, Slama H, Piiper J (1972) Mechanisms of unidirectional flow in parabronchi of avian lungs: measurements in duck lung preparations. Respir Physiol 14:83–95
Scheid P, Worth H, Holle JP, Meyer M (1977) Effects of oscillating and intermittent ventilatory flow on efficacy of pulmonary oxygen transfer in the duck. Respir Physiol 31:251–258
Scheuermann DW, Klika E, De Groodt-Lasseel MH, Bazantova I, Switka A (2000) Lamellar inclusions and trilaminar substance in the parabronchial epithelium of the quail (*Coturnix coturnix*). Ann Anat 182:221–233
Scheuermann DW, Klika E, Lasseel DG, Bazantova I, Switka A (1997) An electron microscopic study of the parabronchial epithelium in the mature lung of four bird species. Anat Rec 249: 213–225
Schmidt-Nielsen B (1979) Urinary concentrating processes in vertebrates. Yale J Biol Med 52:545–561
Schmidt-Nielsen K (1981) Countercurrent systems in animals. Sci Am 244:118–128
Schmidt-Nielsen K (1983) Animal physiology: adaptation and environment, 3rd edn. Cambridge University Press, Cambridge
Schmidt-Nielsen K, Kanwisher J, Lasiewski RC, Cohn JE, Bretz WL (1969) Temperature regulation and respiration in the ostrich. Condor 71:341–352
Schopf JW (1978) The evolution of the earliest cells. Sci Am 239:85–103
Schumann D, Piiper J (1966) Oxygen requirement in fish respiration according to measurements on the anesthetized tench (*Tinca tinca*). Pflugers Arch Gesamte Physiol Menschen Tiere 288:15–26
Scott GR (2011) Elevated performance: the unique physiology of birds that fly at high altitudes. J Exp Biol 214:2455–2462
Scott GR, Dawson NJ (2017) Flying high: the unique physiology of birds that fly at high altitudes. In: Maina JN (ed) The biology of the avian respiratotu system: evolution, development, structure and function. Springer, Cham, pp 113–128
Scott GR, Milsom WK (2007) Control of breathing in birds: implications for high-altitude flight. In: Glass M, Wood S (eds) Cardio-respiratory control in vertebrates. Springer, Berlin, pp 429–448
Schweiger HG (1962) Pathways of metabolism in nucleate and anucleate erythrocytes. Int Rev Cytol 13:135–201
Selenka E (1866) Beltrag zur Entwickelungsgeschichte der Luffsache des Huhn. Zeitsehr fur wiss Zool 16:178–182
Seymour RS, Blaylock AJ (2000) The principle of Laplace and scaling of ventricular wall stress and blood pressure in mammals and birds. Physiol Biochem Zool 73:389–405
Seymour RS, Matthews PGD (2013) Physical gills in diving insects and spiders: theory and experiment. J Exp Biol 216:164–170
Sharp F, Bernaudin M (2004) HIF1 and oxygen sensing in the brain. Nat Rev Neurosci 5:437–448
Sholtis KM, Shelton RM, Hedrick TL (2015) Field flight dynamics of hummingbirds during territory encroachment and defense. PLoS One 10(6):e0125659. https://doi.org/10.1371/journal.pone.0125659
Short RV (1962) The peculiar lungs of the elephant. New Scientist 316:570–572
Sibly RM, Witt CC, Wright NA, Venditti V, Jetze W, Brown JH (2012) Energetics, lifestyle, and reproduction in birds. Proc Natl Acad Sci (USA) 109:10937–10941
Siefert E (1896) Ueber die Athmung der Reptilien und Vögel. Pflügers Arch ges Physiol 64:321–506.i
Siegwart B, Gehr P, Gil J, Weibel ER (1971) Morphometric estimation of pulmonary diffusion capacity. IV. The normal dog lung. Respir Physiol 13:141–159
Simpson CF (1978) Relation of hemodynamics to the incidence of diethylostilboestrol-inducted aortic ruptures in hypertensive and hypotensive lines of turkeys. Atherosclerosis 30:249–254
Singh PK, Shekhar P, Kumar K (2011) Nutritional and managemental control of ascites syndrome in poultry. Internat J Livestock Prod 2:117–123

Skandalis DA, Segre PS, Bahlman JW, Groom DJE, Welch KC Jr, Witt CC, McGuire JA, Dudley R, Lentink D, Altshuler DL (2017) The biomechanical origin of extreme wing allometry in hummingbirds. Nat Commun 8(1):1047. https://doi.org/10.1038/s41467-017-01223-x

Skovgaard N, Abe AS, Andrade DV, Wang T (2005) Hypoxic pulmonary vasoconstriction in reptiles: a comparative study of four species with different lung structures and pulmonary blood pressures. Am J Physiol Regul Integr Comp Physiol 289:R1280–R1288

Skovgaard N, Wang T (2006) Local control of pulmonary blood flow and lung structure in reptiles: implications for ventilation perfusion matching. Respir Physiol Neurobiol 154:107–117

Smith JC, Stamenovic D (1986) Surface forces in lungs. I. Alveolar surface tension-lung volume relationships. J Appl Physiol 60:1341–1350

Snelling EP, Seymour RS, Runciman S, Matthews PGD, White CR (2011) Symmorphosis and the insect respiratory system: allometric variation. J Exp Biol 214:3225–3237

Snelling EP, Seymour RS, Runciman S, Matthews PG, White CR (2012) Symmorphosis and the insect respiratory system: a comparison between flight and hopping muscle. J Exp Biol 215: 3324–3333

Snyder GK (1983) Respiratory adaptations in diving mammals. Respir Physiol 54:269–294

Sobin SS, Tremer HM, Fung YC (1970) Morphometric basis of the sheet-flow concept of the pulmonary alveolar microcirculation in the cat. Circ Res 26:397–414

Sommer N, Dietrich A, Schermuly RT, Ghofrani HA, Gudermann T, Schulz R, Seeger W, Grimminger F, Weissmann N (2008) Regulation of hypoxic pulmonary vasoconstriction: basic mechanisms. Eur Respir J 32:1639–1651

Sommer N, Strielkov I, Pak O, Weissmann N (2016) Oxygen sensing and signal transduction in hypoxic pulmonary vasoconstriction. Eur Respir J 47:288–303

Sosa Torres ME, Saucedo-Vázquez JP, Kroneck PMH (2015) The magic of dioxygen. In: Kroneck P, Sosa Torres M (eds) Sustaining life on planet earth: metalloenzymes mastering dioxygen and other chewy gases, Metal ions in life sciences, vol 15. Springer, Cham, pp 1–12

Soum JM (1896) Recherches physioloqiques sur l'appareil respiratoire des oiseaux. Ann Univ Lyon 28:1–126

Speckmann EW, Ringer RK (1963) The cardiac output and carotid and tibial blood pressure of the turkey. Can J Biochem Physiol 41:2337–2354

Spector WS (1956) Handbook of biological data. Saunders, Philadelphia (PA)

Sperry JS, Hacke UG, Pittermann J (2006) Size and function in conifer tracheids and angiosperm vessels. Am J Bot 93:1490–1500

Starck JM, Weimer I, Aupperle H, Müller K, Marschang RE, Kiefer I, Pees M (2015) Morphological pulmonary diffusion capacity for oxygen of Burmese pythons (*Python molurus*): a comparison of animals in healthy condition and with different pulmonary infections. J Comp Pathol 153:333–351

Stark JM, Mueller GA (2012) Lung defenses: intrinsic, innate, and adaptive. In: Wilmott RW, Boat TF, Bush A, Chernick V, Deterding RR, Ratjen F (eds) Kendig and Chernick's disorders of the respiratory tract in children, 8th edn. WB Saunders, Philadelphia, pp 89–109

Staub NC, Bishop JM, Forster RE (1962) Importance of diffusion and chemical reaction rates in O_2 uptake in the lung. J Appl Physiol 17:21–27

Stenzel T, Tykałowski B, Koncicki A (2008) Cardiovascular system diseases in turkeys. Polish J Vet Sci 11:245–250

Stinner JN (1982) Functional anatomy of the lung of the snake *Pituophis melanoleucus*. Am J Physiol Regul Integr Comp Physiol 243:R251–R257

Strazny F, Perry SF (1984) Morphometric diffusing capacity and functional anatomy of the book lungs in the spider *Tegenaria* spp. (Agelenidae). J Morphol 182:339–354

Stübel H (1910) Beiträge zur Kenntnis der Physiologie des Blutkreislaufes bei verschiedenen Vögelarten. Pflügers Arch ges Physiol 135:249–365

Sturkie PD, Ringer RK, Weiss HS (1956) Relationship of blood pressure to mortality in chickens. Proc Soc Exp Biol Med 92:301–303

Suarez RK (1992) Hummingbird flight: sustaining the highest mass-specific metabolic rates among vertebrates. Experientia 48:565–570
Suarez RK, Brown GS, Hochachka PW (1986) Mitochondrial respiration in hummingbird muscles. Am J Physiol 251:R537–R542
Suarez RK, Lighton JRB, Brown GS, Mathieu-Costello O (1991) Mitochondrial respiration in hummingbird flight muscles. Proc Natl Acad Sci (USA) 87:9207–9210
Suarez RK, Welch KC Jr, Hanna SK, Herrera MLG (2009) Flight muscle enzymes and metabolic flux rates during hovering flight of the nectar bat, *Glossophaga soricina*: further evidence of convergence with hummingbirds. Comp Biochem Physiol A Mol Integr Physiol 153:136–140
Sullivan LC, Daniels CB, Phillips ID, Orgeig S, Whitsett JA (1998) Conservation of surfactant protein A: evidence for a single origin for vertebrate pulmonary surfactant. J Mol Evol 46:131–138
Suresh K, Shimoda LA (2016) Lung circulation. Comp Physiol 6:897–943
Sweeney CR, Soma LR, Bucan CA, Ray SG (1984) Exercise-induced pulmonary hemorrhage in exercising thoroughbreds: preliminary results with pre-exercise medication. Cornell Vet 74: 263–268
Swenson ER, Bärtsch P (2012) High-altitude pulmonary edema. Compr Physiol 2:2753–2773
Sylvester JT, Shimoda LA, Aaronson PI, Ward JPT (2012) Hypoxic pulmonary vasoconstriction. Physiol Rev 92:367–520
Tarnai T, Iijima S, Hare J, Fowler P (1993) Geodesic domes and fullerenes [and Discussion]. Philos Trans Phys Sci Engineer 343:145–154
Tarry D, Powell M (2017) Hypoxic pulmonary vasoconstriction. BJA Educ 17:208–213
Tatsumi N, Kobayashi R, Yano T, Noda M, Fujimura K, Okada N, Okabe M (2016) Molecular developmental mechanism in polypterid fish provides insight into the origin of vertebrate lungs. Sci Rep 6:30580. https://doi.org/10.1038/srep30580
Tavolzhanova TI, Rozova KV, Dubova MH, Petrenko OV (2004) Correlation between the thickness of air-blood barrier of the lung and main indices of respiration during changes in oxygen concentration. Fiziol Zh 50:24–29
Taylor CR, Weibel ER (1981) Design of the mammalian respiratory system. I. Problem and strategy. Respir Physiol 44:1–10
Taylor HH, Greenaway P (1979) The structure of the gills and lungs of the arid-zone crab, *Holthuisana (Austrothelphusa) transversa* (Brachyura: Sundathelphusidae) including observations on arterial vessels within the gills. J Zool (Lond) 189:359–384
Tenney SM, Remmers JE (1963) Comparative quantitative morphology of the mammalian lung: diffusing area. Nature (Lond) 197:54–56
Tenney SM, Tenney JB (1970) Quantitative morphology of cold-blooded lungs: amphibia and Reptilia. Respir Physiol 9:197–215
Terpin T, Roach MR (1987) A biophysical and histological analysis of factors that lead to aortic rupture in normal and lathyritic turkeys. Can J Physiol Pharmacol 65:395–400
Theissen IL, Meissner A (1996) Hypoxic pulmonary vasoconstriction. Anaesthesist 45:643–652
Thibaud V, Rioux-Leclercq N, Vigneau C, Morice S (2019) Recurrence of Goodpasture syndrome without circulating anto-glomerular basement membrane antibodies after kidney transplant, a case report. BMC Nephrol 20:6. https://doi.org/10.1186/s12882-018-1197-6
Thorpe WH (1950) Plastron respiration in aquatic insects. Biol Rev 25:344–390
Thorpe WH, Crisp DJ (1941) Studies on plastron respiration. II. The respiratory efficiency of the plastron in *Amphelocheirus*. J Exp Biol 24:270–303
Tickle PG (2010) Breathing and locomotion in birds. PhD Thesis, University of Manchester (UK)
Tickle PG, Codd JR (2009) Ontogenetic development of the uncinate processes in the domestic turkey (*Meleagris gallopavo*). Poultry Sci 88:179–184
Tickle PG, Ennos AR, Lennox LE, Perry SF, Codd JR (2007) Functional significance of the uncinate processes in birds. J Exp Biol 210:3955–3961
Tickle P, Nudds R, Codd J (2009) Uncinate process length in birds scales with resting metabolic rate. PLoS One 4:e5667. https://doi.org/10.1371/journal.pone.0005667

Timoshenko SP (1953) History of strength of materials. McGraw Hill, New York

Toshima M, Ohtani Y, Ohtani O (2004) Three-dimensional architecture of elastin and collagen fiber networks in the human and rat lung. Arch Histolol Jpn 67:31–40

Townsley MI (2012) Structure and composition of pulmonary arteries, capillaries, and veins. Compr Physiol 2:675–709

Tsamis A, Krawiec JT, Vorp DA (2013) Elastin and collagen fibre microstructure of the human aorta in ageing and disease: a review. J R Soc Interface 10(83):20121004. https://doi.org/10.1098/rsif.2012.1004

Tsukimoto K, Mathieu-Costello O, Prediletto R, Elliott AR, West JB (1991) Ultrastructural appearances of pulmonary capillaries at high transmural pressures. J Appl Physiol 71:573–582

Turner AK, Rose C (1989) A handbook to the swallows and martins of the world. Christopher Helm, London

Untersee P, Gil J, Weibel ER (1972) Diffusion capacity in rat lungs fixed at different degrees of inflation. Experientia 28:732–742

Van Erck-Westergren E, Franklin SH, Bayly WH (2013) Respiratory diseases and their effects on respiratory function and exercise capacity. Equine Vet J 45:376–387

van Veen L (1999) Aortaruptuur bij pluimvee: een overzicht [Aortic rupture in poultry: a review]. Tijdschr Diergeneeskd 124:244–247

Vella CA, Marks D, Robergs RA (2006) Oxygen cost of ventilation during incremental exercise to VO_{2max}. Respirology 11:175–181

Venturas MD, Sperry JS, Hacke UG (2017) Plant xylem hydraulics: what we understand, current research, and future challenges. J Integr Plant Biol 59:356–389

Vidyadaran MK, King AS, Kassim H (1987) Deficient anatomical capacity for oxygen uptake of the developing lung of the female domestic fowl when compared with red-jungle fowl. Schweizer Archiv Tierheilkund 129:225–237

Vidyadaran MK, King AS, Kassim H (1988) Quantitative studies of the lung of the domestic fowl (*Gallus gallus* var. *domesticus*). Perlanika 11:229–238

Vidyadaran MK, King AS, Kassim H (1990) Quantitative comparisons of lung structure of adult domestic fowl and red jungle fowl, with reference to broiler ascites. Avian Pathol 19:51–58

Vidyadaran MK, King AS, Kassim H (1991) Quantitative observations on the pulmonary anatomy of the domestic Muscovy duck (*Cairina moschata*). J Zool (Lond) 224:113–119

Vitali SD, Richardson KC (1998) Evaluation of pulmonary volumetric morphometry at the light and electron microscopy level in several species of passerine birds. J Anat 193:573–580

Voigt CC, Winter Y (1999) Energetic cost of hovering flight in nectar-feeding bats (Phyllostomidae: Glossophaginae) and its scaling in moths, birds and bats. J Comp Physiol B 169:38–48

Vos HJ (1934) Über die Vege der Atemluft in der Entenlunge. Z vergl Physiol 21:552–578

Vos HJ (1935) Über die wege der Atemluft in der Entenlunge. Zeisch Vergl Physiol 21:552–578

Wagner GP (1989) The origin of morphological characters and the biological basis of homology. Evolution 43:1157–1171

Wagner GP (1998) Complexity matters. Science 279:1158–1159

Wallach V (1998) The lungs of snakes. In: Gans C, Gaunt AS (eds) Biology of the reptilia, Morphology G. Visceral organs, vol 19. Society for the Study Amphibians and Reptiles, Ithaca, pp 93–295

Walton JP, Julian RJ, Squires EJ (2001) The effect of dietary flax oil and antioxidant on ascites and pulmonary hypertension in broilers using a low temperature model. Br Poult Sci 42:123–129

Wang N, Banzett RB, Butler JP, Fredberg JJ (1988) Bird lung models show that convective inertia effects inspiratory aerodynamic valving. Respir Physiol 73:111–124

Wang N, Banzett RB, Nations CS, Jenkins EA (1992) An aerodynamic valve in the avian primary bronchus. J Exp Biol 262:441–445

Wang ZM (1997) Human body composition models and methodology: theory and experiment PhD Thesis, Landbouw Universiteit Wangeningen (The Netherlands)

Ward S, Möller U, Rayner JM, Jackson DM, Bilo D, Nachtigall W, Speakman JR (2001) Metabolic power, mechanical power and efficiency during wind tunnel flight by the European starling *Sturnus vulgaris*. J Exp Biol 204:3311–3322

Warrell DA, Evans JW, Clarke RO, Kingaby GP, West JB (1972) Pattern of filling in the pulmonary capillary bed. Pattern of filling in the pulmonary capillary bed. J Appl Physiol 32:346–356

Watson RR, Fu Z, West JB (2007) Morphometry of the extremely thin pulmonary blood-gas barrier in the chicken lung. Am J Physiol Lung Cell Mol Physiol 292:L769–L777

Watson RR, Fu Z, West JB (2008) Minimal distensibility of pulmonary capillaries in avian lungs compared with mammalian lungs. Respir Physiol Neurobiol 160:208–214

Weibel ER (1963) Morphometry of the human lung. Springer, Berlin

Weibel ER (1970/71) Morphometric estimation of pulmonary diffusion capacity. I. Model and method. Respir Physiol 11:54–75.

Weibel ER (1973) Morphological basis of the alveolar-capillary gas exchange. Physiol Rev 53: 419–495

Weibel ER (1984) The pathways for oxygen. Harvard University Press, Harvard (MA)

Weibel ER (1989) Lung morphometry and models in respiratory physiology. In: Chang HK, Paiva M (eds) Respiratory physiology: an analytical approach. Dekker, New York, pp 1–55

Weibel ER (1990) Morphometry: stereological theory and practical methods. In: Gill J (ed) Models of lung disease: microscopy and structural methods. Marcel Dekker, New York, pp 199–251

Weibel ER (2000) Symmorphosis: on form and function in shaping life. Harvard University Press, Cambridge (MA)

Weibel ER, Federspiel WJ, Fryder-Doffey F, Hsia CW, König M, Stalder-Navarro V, Vock R (1993) Morphometric model for pulmonary diffusing capacity I. Membrane diffusing capacity. Respir Physiol 93:125–149

Weibel ER, Knight BW (1964) A morphometric study on the thickness of the pulmonary air-blood barrier. J Cell Biol 21:367–384

Weibel ER, Taylor CR, Hoppeler H (1991) The concept of symmorphosis: a testable hypothesis of structure-function relationship. Proc Natl Acad Aci (USA) 88:10357–10361

Weibel ER, Taylor CR, Bolis L (eds) (1998) Principles of animal design: the optimization and symmorphosis debate. Cambridge University Press, Cambridge

Weibel ER, Taylor CR, O'Neil JJ, Leith DE, Gehr P, Hoppeler H, Langman V, Baudinette RV (1983) Maximal oxygen consumption and pulmonary diffusing capacity: a direct comparison of physiologic and morphometric measurements in canids. Respir Physiol 54:173–188

Weiss HD, Sheahan M (1958) The influence of maturity and sex on the blood pressure of the turkey. Am J Vet Res 19:209–211

Welch KC, Myrka AM, Ali RS, Dick MF (2018) The metabolic flexibility of hovering vertebrate nectarivores. Physiology 33:127–137

Welsch U, Aschauer B (1986) Ultrastructural observations on the lung of the emperor penguin (*Aptenodytes forsteri*). Cell Tissue Res 243:137–144

Welty JC (1979) The life of birds, 2nd edn. Saunders, Philadelphia

Wenger RH (2006) Mitochondria: oxygen sinks rather than sensors? Medical Hypotheses 66:380–383

West JB (1985) Invited review: pulmonary capillary stress failure. J Appl Physiol 89:2483–2489

West JB (2001) Snorkel breathing in the elephant explains the unique anatomy of its pleura. Respir Physiol 126:1–8

West JB (2002) Why doesn't the elephant have a pleural space? News Physiol Sci 17:47–50

West JB (2003) Thoughts on the pulmonary blood-gas barrier. Am J Physiol Lung Cell Mol Physiol 285:L501–L513

West JB (2004) A century of pulmonary gas exchange. Am J Respir Crit Care Med 169:897–902

West JB (2008) Respiratory physiology: the essentials. Lippincott Williams and Wilkins, Philadelphia

West JB (2009) Comparative physiology of the pulmonary blood-gas barrier: the unique avian solution. Am J Physiol Regul Integr Comp Physiol 297:R1625–R1634

West JB (2011) Comparative physiology of the pulmonary circulation. Compr Physiol 1:1525–1539

West JB (2013) Role of the fragility of the pulmonary blood-gas barrier in the evolution of the pulmonary circulation. Am J Physiol Regul Integr Comp Physiol 304:R171–R176

West JB, Fu Z, Deerinck TJ, Mackey MR, Obayashi JT, Ellisman MH (2010) Structure-function studies of blood and air capillaries in chicken lung using 3D electron microscopy. Respir Physiol Neurobiol 170:202–209

West JB, Colice GL, Lee YJ, Namba Y, Kurdak SS, Fu Z, Ou LC, Mathieu-Costello O (1997) Pathogenesis of high-altitude pulmonary oedema: direct evidence of stress failure of pulmonary capillaries. Eur Respir J 8:523–529

West JB, Mathews FL (1978) Stresses, strains, and surface pressures in the lung caused by its weight. J Appl Physiol 32:332–345

West JB, Mathieu-Costello O (1992a) Stress failure of pulmonary capillaries in the intensive care setting. Schweiz Med Wochenschr 122:751–757

West JB, Mathieu-Costello O (1992b) High altitude pulmonary edema is caused by stress failure of pulmonary capillaries. Int J Sports Med 1:S54–S58

West JB, Mathieu-Costello O (1992c) Stress failure of pulmonary capillaries: role in lung and heart disease. Lancet 340:762–767

West JB, Mathieu-Costello O (1992d) Strength of the pulmonary blood-gas barrier. Respir Physiol 88:141–148

West JB, Mathieu-Costello O (1994) Stress failure of pulmonary capillaries as a mechanism for exercise induced pulmonary haemorrhage in the horse. Equine Vet J 26:441–447

West JB, Mathieu-Costello O (1995) Stress failure of pulmonary capillaries as a limiting factor for maximal exercise. Eur J Appl Physiol Occup Physiol 70:99–108

West JB, Mathieu-Costello O (1998) Stress-induced injury of pulmonary capillaries. Proc Assoc Am Physicians 110:506–512

West JB, Mathieu-Costello O (1999) Structure, strength, failure, and remodeling of the pulmonary blood-gas barrier. Annu Rev Physiol 61:543–572

West JB, Mathieu-Costello O, Jones JH, Birks EK, Logerman RB, Pascoe JR, Tyler WS (1993) Stress failure of pulmonary capillaries in racehorses with exercise-induced pulmonary haemorrhage. J Appl Physiol 75:1097–1109

West JB, Tsukimoto K, Mathieu-Costello O, Prediletto R (1991) Stress failure in pulmonary capillaries. J Appl Physiol 70:1731–1742

West JB, Watson RR, Fu Z (2006) The honeycomb-like structure of the bird lung allows a uniquely thin blood-gas barrier. Respir Physiol Neurobiol 152:115–118

West JB, Watson RR, Fu Z (2007a) Major differences in the pulmonary circulation between birds and mammals. Respir Physiol Neurobiol 157:382–390

West JB, Watson RR, Fu Z (2007b) The human lung: did evolution get it wrong? Eur Respir J 29: 11–17

West JB, Watson RR, Fu Z (2007c) Support of pulmonary capillaries in avian lung. Respir Physiol Neurobiol 159:146. https://doi.org/10.1016/j.resp.2007.08.006

West NH, Bamford OS, Jones DR (1977) A scanning electron microscope study of the microvasculature of the avian lung. Cell Tissue Res 176:553–564

White CR, Blackburn TM, Martin GR, Butler PJ (2007) Basal metabolic rate of birds is associated with habitat temperature and precipitation, not primary productivity. Proc R Soc (Lond) B 274: 287–293

Wieland C (1978) *Archeopteryx*. Creation 1, 12–16.

Wienner F, Morkin E, Skalak R, Fishman AP (1966) Wave propagation in the pulmonary circulation. Circ Res 19:834–850

Wilson CB, Dixon FJ (1973) Anti-glomerular basement membrane antibody-induced glomerulonephritis. Kidney Int 3:74–89

Winter Y (1998) Energetic cost of hovering flight in a nectar-feeding bat measured with fast-response respirometry. J Comp Physiol B 168:434–444

Wojciechowski MS, Pinshow B (2009) Heterothermy in small, migrating passerine birds during stopover: use of hypothermia at rest accelerates fuel accumulation. J Exp Biol 212:3068–3075
Woo YD, Jeong D, Chung DH (2021) Development and functions of alveolar macrophages. Mol Cells 44:292–300
Wood SC, Lenfant C (1979) Oxygen transport and oxygen delivery. In: Wood SC, Lenfant C (eds) Evolution of respiratory processes: a comparative approach: lung biology in health and disease. Marcel Dekker, New York, pp 193–223
Woodward JD, Maina JN (2005) A 3D digital reconstruction of the components of the gas exchange tissue of the lung of the muscovy duck, *Cairina moschata*. J Anat 206:477–492
Woodward JD, Maina JN (2008) Study of the structure of the air and blood capillaries of the gas exchange tissue of the avian lung by serial section three-dimensional reconstruction. J Microsc 230:84–93
Wright JR (2005) Immunoregulatory functions of surfactant proteins. Nat Rev Immunol 1:58–68
Xiang RP, Sun WD, Wang JY, Wang XL (2002) Effect of vitamin C on pulmonary hypertension and muscularization of pulmonary arterioles in broiler. Brit Poult Sci 43:705–712
Yong B, Wei-Liang J, Yong B, Wei-Liang J (2016) Marine structural design, 2nd edn. Butterworth-Heinemann, Oxford, pp 689–705
York JM, Chua BA, Ivy CM, Alza L, Cheek R, Scott GR, McCracken KG, Frappell PB, Dawson NJ, Laguë SL et al (2017) Respiratory mechanics of eleven avian species resident at high and low altitude. J Exp Biol 220:1079–1089
Young RA (1976) Fat, energy and mammalian survival. Amer Zool 16:699–710
Zachar I, Boza G (2020) Endosymbiosis before eukaryotes: mitochondrial establishment in protoeukaryotes. Cell Mol Life Sci 77:3503–3523
Zakynthinos S, Roussos C (1991) Oxygen cost of breathing. In: Gutierrez G, Vincent JL (eds) Tissue oxygen utilization. Update in intensive care and emergency medicine, vol 12. Springer, Berlin, pp 171–184
Zehnder AM, Hawkins MG, Pascoe PJ (2014) Avian anatomy and physiology. In: Heard D, Caulkett N (eds) Zoo animal and wildlife immobilization and anaethesia, 2nd edn. Wiley, New York, pp 389–398
Zheng J, Zhou C (eds) (2017) Hypoxia and human diseases. IntechOpen, London. https://doi.org/10.5772/62960
Zeuthen E (1942) The ventilation of the respiratory tract in birds. K danske Vidensk Selsk Skr 17:1–50
Zhang F, Zhoghe Z, Lianhai H, Gang G (2001) Early diversification of birds: evidence from a new opposite bird. Chinese Sci Bull 46:945–949
Zimmer K (1935) Beiträge zur Mechanik der Atmung bei den Vögeln in Stand und Flug. Zoologica Stuttg 35:1–69
Zhu H, Traore K, Santo A, Trush MA, Li YR (2016) Oxygen and oxygen toxicity: the birth of concepts. React Oxyg Species 1:1–8

Index

A
Abdominal air sac, 15, 43, 133, 135, 214, 227, 241–246, 279, 283
Abdominal pumping, 197
Accessory respiratory organ, 276, 278, 299, 300, 335, 336, 339
Acid-base balance, 107
Active flight, 44
Adaptation, 16, 20, 22, 26, 28, 38, 44, 118, 202, 277, 289, 299, 315
Adaptive radiation, 20, 38
Adenosine triphosphate (ATP), 269–271, 340
Aerodynamic, 13, 206, 283, 285, 286
Aerodynamic forces, 20, 284
Aerodynamic valving, 193, 194, 287
African Chameleon, 193
Air breathers, 195, 276
Air breathing, 39, 110, 123, 134, 273, 276, 277, 296, 299
Air breathing fish, 107, 299, 336, 339
Air breathing vertebrate, 3, 34, 38, 42, 43, 102, 107, 111, 191, 192, 198, 206, 235, 274, 276, 278, 280, 281, 288, 289, 297, 300, 315, 323, 329, 343
Air capillaries, 25, 36, 43, 44, 113, 116, 118, 123, 125, 126, 130, 132–134, 194, 202, 203, 212, 214, 221–225, 229–235, 237, 238, 273, 290–295, 297, 298, 301, 302, 304–307, 310, 311, 323, 327, 329–332, 335, 337–339, 341
Airflow, 14, 192, 195, 201, 207, 214, 215, 227, 280–285, 287, 288, 292–294, 329
Air sac, 14, 15, 43, 44, 114, 115, 118, 137, 142, 145, 192–194, 196, 197, 207, 214, 216, 233, 242, 271, 272, 279–281, 283, 287, 315, 323
Airway, 14, 15, 100, 104, 108–116, 119–123, 127, 128, 136–138, 142–145, 147, 148, 150–152, 193, 194, 201–203, 206–210, 213–218, 222, 227, 228, 236, 238–240, 273, 277, 281–288, 292, 330
Albatross, 17, 22, 24
Alligator, 193
Allometric relation, 315
Altitude, 3, 20, 23–26, 289
Altricial, 116, 126, 127
Altricial species, 126, 219
Alveoli, 43, 108, 109, 113–116, 132, 142, 214, 231, 233, 235, 238, 275, 323, 327, 329
American Alligator, 192
American Crows, 12
American Golden Eagle, 28
American Golden Plover, 29
Amphibian, 39, 107, 276, 279, 304, 315, 332, 335–337, 339
Amur Falcon, 29
Anatomical valves, 281, 284
Andean Condor, 6, 28
Andean Goose, 3, 24–26, 212, 221, 223–226, 229, 230, 290, 310, 315, 322, 324, 330, 331, 338, 339
Andes, 25, 331
Angioblast, 129, 131, 149–151
Angioblastic cells, 127–129, 144, 149, 150
Angiogenesis, 111, 130, 131, 142, 145, 151, 152
Angiogenetic cells, 128
Anoxia, 105

© The Author(s), under exclusive license to Springer Nature Switzerland AG 2023
J. N. Maina, *Current Perspectives on the Functional Design of the Avian Respiratory System*, Zoological Monographs 8, https://doi.org/10.1007/978-3-031-35180-8

Anthropogenic activities, 5, 31–34
Aortic arches, 111
Aortic rupture, 308
Apoptosis, 40, 100, 128, 130–134, 149
Archaeopteryx, 10, 37
Archosauria, 44, 193
Arctic Tern, 30
Arterial blood pressures, 289, 307, 308
Arterial partial pressure, 275, 289
Arteriole, 113, 232, 239, 292, 297, 335
Arteriovenous anastomoses, 241, 296
Ascites, 308
Atria, 122, 123, 125, 126, 219, 221–226, 229–230, 233–236, 238, 292–295, 297, 310, 314, 329, 330, 332
Augury, 7
Australian pelican, 116, 219
Avian lung, 41–44, 108, 111, 115, 116, 118, 125, 126, 131, 132, 145, 148, 152, 192, 194, 207, 210, 213–215, 217, 218, 221–223, 225–231, 233–239, 245, 272, 273, 277, 281–284, 287, 288, 290–298, 300, 302–344
Avian respiratory system, 38–45, 107, 108, 114, 136, 137, 191–194, 198–246, 272, 279, 281–296, 314, 315, 323

B
Bald Eagle, 6
Bar-Headed Goose, 3, 24, 25, 289
Barnacle Geese, 26
Barometric pressure, 24, 311
Bar-Tailed Godwit, 29
Basement membrane, 102, 110, 123, 124, 133, 134, 137, 139, 144, 235, 238, 245, 300–305, 309–311, 332, 333, 341
Bat, 200, 231, 315, 327, 333
Bauplan, 17–19
Behaviour, 4, 6, 7, 16, 17, 103, 202
Bernoulli's Principle, 287
Bichirs, 40
Bifurcation, 43, 112, 141, 148, 204, 205, 207, 213, 314
Biodiversity, 31, 32
Biological wheels, 105
Biomechanics, 18
Bipedalism, 37
Bird, 1, 107, 191, 271
Bird song, 16
Bird-of-paradise, 6
Black-Chinned Hummingbird, 38
Black Jacobin Hummingbirds, 16
Black Mamba, 41, 279, 335
Blood, 25, 100, 195, 273
Blood capillaries, 25, 36, 115, 119, 126, 131–134, 203, 211, 212, 221, 223, 225, 228–235, 237–241, 273, 284, 292–298, 300–304, 306, 307, 309–313, 320, 323, 334–339, 341, 344
Blood chemistry, 38
Blood flow, 27, 228, 241, 288, 289, 292, 294
Blood gas barrier, 102, 111, 115, 116, 123, 126, 128, 130, 133, 134, 151, 203, 228, 232–235, 237, 238, 273, 284, 292, 297–314, 323, 327, 329–334, 337–344
Blood lactate levels, 313
Blood plasma, 111, 235, 237, 275, 313, 330
Blood pressure, 297, 307, 308, 314
Blood vessels, 27, 100, 104, 106, 110–113, 116, 120–125, 127–131, 138–140, 144, 148–152, 195, 200, 202, 203, 207–209, 213, 221, 223, 226, 232, 234–236, 238, 240, 241, 283, 285, 286, 292, 293, 298, 307, 311, 314, 320, 328
Blood volume, 107, 150, 223, 230, 237, 291, 297, 327, 334–340, 342, 343
Blue Crane, 6
Blue Manakin, 6
Blue-Throated Hummingbird, 16
Body mass, 3, 19, 20, 22, 23, 28, 31, 43–45, 113, 126, 127, 201, 241, 275, 307, 309, 315, 320, 323, 326, 327, 329–335, 339, 342, 344
Bone morphogenic proteins (BMPs), 136, 137
Bones, 17–20, 27, 35, 37, 44, 133, 193, 198, 243, 244, 281, 315
Bony fish, 38
Brain, 11–14, 27, 33, 197, 276, 278, 289
Branching, 35, 104, 110, 113, 115, 128, 137, 142, 144, 150, 203, 312
Branching morphogenesis, 100, 108, 110–113, 141, 142, 148, 152
Breathing frequency, 270, 315
Bronchial system, 43, 128, 213, 239
Bronchioalveolar, 41, 43, 45, 114, 116, 197, 233
Bronchioalveolar tree, 214
Budgerigar, 108, 322, 325

C
Caching, 12
Caecilian, 41, 42
Canary, 12
Canonical WNT signaling, 144

Capillary blood pressures, 300, 304, 309, 313
Carbon monoxide (CO), 241, 242, 283
Cardiac output, 296, 297, 307, 313
Cassowary, 3
Catfishes, 276, 300, 335, 339
Caudal air sacs, 226–228, 279, 280, 282, 284, 287
Caudothoracic air sacs, 217, 242, 244–246
Cell polarity, 107, 148
Cell-to-cell signaling, 110
Cerebral blood flow, 27, 289
Cervical air sacs (CS), 15, 207, 242–244
Chemoreceptors, 278, 288, 289
Chick, 10–12, 14, 16, 111, 114, 116, 123, 125, 126, 129, 131, 133, 137, 142, 144–146, 148, 151, 202, 211, 242
Chicken, 6, 8, 9, 14, 15, 40, 112, 116, 118, 123, 133, 137, 144, 150, 281, 296, 308, 311, 313, 314, 330, 343
Chilean Tinamou, 331, 333
Chiropteran lung, 197
Cichlid, 2
Circulatory system, 127, 195, 197, 237, 239, 271
Clavicle, 18
Clavicular air sacs (CIS), 15, 133, 244, 245
Coelacanth, 35, 39, 41
CO_2 elimination, 27
Coelomic cavity, 118, 133, 135, 193, 209, 241, 244, 245, 271, 273, 297
Cognitive, 11–13, 197
Collateral ventilation, 233
Common Coot, 244
Common Crow, 207
Common Gull, 210, 322, 325
Common Moorhen, 244
Common Quail, 108
Common Raven, 34
Communication, 16, 203, 204
Complexity, 16, 104, 105, 109, 111, 202, 208, 209, 214, 270, 277, 279, 281, 287, 290–296
Compliance, 110, 273, 284, 287, 296, 323, 327
Computer reconstruction, 209–211, 220, 230, 232, 233, 237, 240, 241, 295, 307
Conductance of O_2, 340
Coot, 218, 229, 231, 317, 322
Coracoid, 18
Cormorant, 4, 214, 218, 228
Coronary blood flow, 289
Coronet Hummingbird, 16
Costal impressions/sulci, 118, 147, 208, 209, 213
Costal sulci, 118, 194, 209–211, 220, 242
Cost of breathing, 276
Countercurrent, 238, 292–295
CO_2 washout, 272
Crab Eating Macaque, 307
Cranes, 202, 221, 228
Cranial air sacs, 242, 279, 282, 287
Cranial kinetic movement, 37
Craniothoracic air sacs, 216, 244, 246, 283
Crested Crane, 6
Crested Duck, 25
Cretaceous, 14, 27, 35, 36, 44, 205
Crosscurrent, 45, 115, 238, 292–295
Cutaneous air sacs, 233, 243
Cutaneous respiration, 197
Cysteine proteases, 131
Cytokine, 39, 100, 104

D

Dalmatian Pelican, 28
Darwin's finches, 3
Dead space, 44, 201, 229
Development, 11, 14, 37, 39–41, 44, 116–152, 197, 205, 211, 216, 219, 224, 227, 229, 242, 270, 274, 276, 277, 281, 287, 289, 297, 300, 332
Developmental stages, 114
Devonian, 36, 38–40, 110
Diaphragm, 213, 273
Dichotomous branching, 114, 115, 148, 213
Diffusing capacity, 26, 102, 277, 292, 303, 340–344
Diffusion, 25, 26, 102, 111, 115, 140–142, 195, 197, 277, 278, 292, 294, 298, 303, 304, 333, 340, 342, 343
Dinosaur, 35, 37, 38, 44, 45
Diving, 17, 245, 274, 276, 338, 343
DNA, 9, 38, 100, 137
Domestication, 8–9
Domestic Fowl, 36, 108, 111, 115–121, 123–125, 127–136, 138–140, 142, 143, 146, 147, 149, 150, 201, 203, 204, 206–213, 216–226, 229, 231, 232, 234, 238–240, 242–245, 272, 273, 279, 282, 283, 285–287, 290, 291, 293, 296, 301, 302, 306, 309–312, 314, 322, 324, 331, 332, 335
Dynamic soaring, 22, 23

E

Earth, 2, 4–6, 19, 28, 30, 32, 33, 101, 269
Ecosystem, 1, 2, 5, 8, 31, 32
Edentulism, 2, 37

Egg, 6, 9–11, 13, 33, 40, 99, 111, 126, 127
Egg production, 32, 127
Egg-size, 127
Egyptian Vulture, 13
Electromyographic studies, 272
Elephant bird, 3
Embryo, 105, 106, 116, 123, 127–129, 142, 145, 146, 148, 150, 151, 202, 242
Embryonic, 39, 40, 99, 100, 103, 105, 106, 109–111, 114, 116, 123, 126–131, 133, 137, 140–142, 144, 147, 148, 150, 211, 279
Embryonic chick lung, 137, 142, 148, 151
Endothelial cells, 102, 129–131, 133, 134, 142, 145, 149–152, 235, 238, 300–304, 311, 332, 341
Endotherm, 44
Energetic cost, 191, 275, 276
Environment, 5, 22, 30, 31, 34, 103, 105, 106, 152, 192, 269, 270, 277, 302, 315
Environmental pollution, 31
Eocene, 11, 20
Epidermal growth factor (EGF), 137
Epigenetics, 100, 103
Epithelial cells, 102, 106, 109, 110, 117, 118, 122–124, 132–134, 137, 139, 140, 142, 144, 145, 147, 148, 151, 152, 202, 214, 215, 234, 235, 238, 245, 300–303, 329, 330, 341
Epithelium, 106, 109, 111, 120–122, 124, 125, 136, 139, 141, 142, 145, 202, 203, 214, 215, 218, 229, 245, 302, 304, 332
Erythroblasts, 127–129, 150
Erythrocytes, 102, 125, 129, 130, 132, 134, 149, 203, 212, 225, 232, 233, 235, 238, 298, 300, 302, 303, 306, 307, 313, 314, 328, 331, 337, 340–344
Etruscan Shrew, 327
Eukaryotic cells, 270
Eupneic breathing, 272, 286, 288
European Robins, 30
Evagination, 107, 122, 123, 125, 277, 278
Evolution, 2, 101, 192, 275
Exchange tissue mantle, 217–219, 221, 223, 229
Expiration, 207, 213, 227, 271, 272
Expiratory aerodynamic valving, 279, 284, 287
Extinction, 5, 32–34, 277
Extracellular matrix (ECM), 103, 104, 144, 150, 151, 300, 311
Extraembryonic yolk sac, 128
Extrapulmonary primary bronchi (EP), 14, 15, 204–208, 243

F
Feathers, 2, 32, 34, 35, 37, 44, 126, 200
Fibroblast growth factor (FGF), 136, 137, 141–144, 148, 151
Fick's Law, 102, 340
Fish, 2, 7, 11, 28, 39–41, 235, 276, 292, 299, 300, 304, 328, 335, 339
Flapping flight, 22, 23
Flight, 2, 8, 10, 11, 19–24, 27–31, 33, 37, 38, 126, 192, 200, 201, 224, 229, 243, 273, 275, 287
Flight efficiency, 28
Flight muscles, 17–20, 22, 23, 25–28, 31, 44, 196
Fluid dynamics, 282, 286
Fluid valve theory, 284
Foraging, 10, 12, 16, 29, 339
Foregut, 11, 40, 106, 145
Fossil, 3, 10, 14, 34, 35, 37, 40, 44, 205, 274
Fossilization, 35
Fractal, 101
Frequency, 5, 15, 16, 19, 23, 201, 205, 271, 312
Functional design, 101, 193, 194, 269–277, 302, 340, 342
Functional residual capacity, 272, 275, 344

G
Galápagos Islands, 3
Galliform, 221, 222, 228, 229
Garden Warblers, 275
Gas exchanger, 102, 110, 114, 191–198, 214, 235, 269–281, 283, 292, 296, 297, 299, 300, 302, 304, 323, 334–337, 339, 340, 342, 343
Gasotransmitter, 26
Gastrointestinal system, 44, 104, 109, 118, 120, 143, 147, 196
Gene, 6, 8, 11, 28, 38, 103, 104, 107, 109, 116, 136, 137, 144, 145, 150, 152
Genome, 8, 28, 38, 41, 193
Giant Teratorn, 27
Gills, 41, 102, 195, 235, 276–278, 280, 292, 296, 297, 299, 300, 323, 328, 335, 339, 340
Giraffe, 200
Glandular structure, 112
Gliders, 11
Gliding flight, 23
Gliding Frog, 11
Global warming, 32, 33
Glucocorticoids, 112, 136
Goblet cells, 122, 202, 215, 245

Granular pneumocyte, 109, 198, 229, 297
Granulocytes, 128
Grasshopper, 196, 197
Great Bustard, 28
Greater Flamingo, 244
Greater Rhea, 244
Great Oxidation Event, 277
Great Snipe, 29
Great Tit, 5, 30
Greylag Goose, 25, 218, 324
Growth factor (GF), 104, 109, 110, 116, 141, 144, 151, 152
Guillemot, 17, 18, 322, 325
Guinea Fowl, 108, 239, 276, 317, 322, 324, 331

H

Habitat, 3, 5, 6, 12, 17, 28, 31–33, 38, 102, 270, 276, 330, 331
Harmonic mean thickness, 116, 314, 326, 333, 334, 338, 340, 341
Harpy Eagle, 32
Hatching, 10, 33, 111, 112, 114–116, 118, 123, 126, 128, 132, 133, 135, 219, 273, 330
Heart, 22, 23, 31, 111, 127, 193, 213, 236, 241, 243, 244, 270, 283, 289, 296–298, 307, 308, 311, 312
Heart rate, 25, 289, 308, 313
Hemangioblast, 129–131, 139, 149, 151
Hemangiogenetic cells, 127
Hematocrit, 25, 26, 339, 343
Hematogenetic cells, 129, 150
Hemoglobin, 25, 26, 127, 129, 130, 150, 275, 341
Hemorrhage, 304, 305
Heparin sulphate proteoglycan (HSPG), 140
Heterokairy, 112
Hexagon, 290–292
High altitude, 24–27, 224, 287, 289, 305, 331, 339
High elevation, 20, 24–27
Hilum, 207, 210, 211, 214, 216, 239, 241
Himalayas, 24
Hoatzin, 6, 10–11, 213
Holy Bible, 4
Holy Qur'an, 4
Homeostasis, 100, 103, 141, 144, 192
Homeotherm, 23, 192, 280, 307
Honeyguide, 4, 8
Hoopoe, 6
Horizontal septum (HS), 118, 133, 135, 207, 208, 210, 243, 244, 273, 327
House sparrow, 108, 207, 244, 314, 323, 326, 333
Hovering, 22, 23, 331
Hovering flight, 31
HOXb genes, 137
Hummingbird, 3, 6, 11, 13, 16, 19, 22, 23, 27, 31, 38, 126, 135, 218, 244, 245, 315, 331
Hypercapnia, 288, 289, 304
Hyperventilation, 289
Hypocapnia, 27, 288–290
Hypoxia, 20, 24–26, 106, 112, 123, 150, 288–290
Hypoxia induced factor (HIF), 290
Hypoxic pulmonary vasoconstriction, 288–290

I

Icelandic Whimbrels, 29
Immunogold staining, 309
Incubation, 10, 33, 106, 111, 116–120, 122–125, 127, 128, 131, 135, 137, 144, 146, 147, 149, 150, 202, 243
Incubation period, 119, 121, 126, 127, 137, 144
Infundibula, 26, 123, 125, 126, 219, 221, 222, 224, 225, 229–231, 233, 235, 236, 292, 294, 295, 297, 314, 330–332
Insects, 13, 19, 28, 112, 114, 141, 194–197, 276, 299, 339
Inspiration, 227, 271, 272, 274, 284
Inspiratory aerodynamic valving, 279, 284–288
Inspiratory muscles, 272
Interalveolar pores, 231, 233
Interalveolar septa, 133, 311
Interatrial septa, 123, 224, 225, 229, 234, 240, 273
Intermittent flight, 24
Internal pipping, 111, 114, 131, 134
Interparabronchial septa, 123, 219, 221–224, 226, 229, 233, 290, 291, 310
Intramural blood pressure, 304, 307, 313
Intramural pressure, 311
Intraparabronchial arterioles, 231
Intrapulmonary primary bronchus (IPPB), 108, 119, 120, 122, 133, 138, 143, 202, 203, 208–211, 213–218, 227, 228, 239, 245, 246, 280–287
Intussusceptive microvascular growth, 131, 151
Invagination, 214, 277

J

Japanese Quail, 123
Joint locking, 24
Jungle Fowl, 9, 321, 323, 326
Jurassic, 19, 34, 37, 194

K

Kiwi, 3, 6, 20, 127, 244
Kleptoparasitism, 12
Kori Bustard, 28
Krebs cycle, 116
Krogh's permeation coefficient, 340, 343

L

Larynx, 14, 15, 198–199, 201, 204–206, 287
Laterodorsal secondary bronchi (LDSB), 108, 213, 215–217, 220, 221, 227, 228, 246, 290
Lateroventral secondary bronchi (LVSB), 209–211, 213, 216, 217, 220, 221, 228, 246
Latex rubber cast, 15, 108, 208, 219, 220, 224, 232, 238, 242, 279, 282, 296, 312
Laws of thermodynamics, 105
Lifestyle, 3, 20, 22, 192, 197, 270, 276, 289, 302, 307, 315, 330
Line of anastomoses, 220, 221
Locomotion, 19–28, 192, 274, 331
Locusts, 197
Lovebirds, 7
Lung, 25, 100, 192, 271
Lung-air sac system, 43, 45, 114, 197–246, 273, 277, 279, 280, 315
Lung bud, 110, 117, 118, 137, 148
Lung development, 41, 107, 109–111, 116, 136, 141, 142, 145
Lung explants, 118
Lungfish, 41–43, 107, 279, 337, 339
Lungless, 39
Lung morphogenesis, 137, 148
Lung volume, 26, 113, 133, 219, 315, 320, 323, 326, 327, 330, 338

M

Mallard Duck, 126, 207, 210, 217–219, 229, 239, 283, 322, 324
Mammalian lung, 27, 42, 44, 107–114, 118, 127, 128, 132, 136, 137, 148, 192, 197, 202, 213, 214, 218, 231, 235, 236, 238–240, 272, 273, 282, 283, 291, 292, 296, 303, 305, 307, 311, 313–315, 323, 327, 329, 332, 334, 335, 337, 339, 344
Mammals, 5, 11, 12, 14, 17, 18, 28, 37, 41, 43, 45, 105, 107, 114, 192, 193, 197, 200–202, 204, 241, 271–273, 275, 276, 280, 289, 296, 299, 300, 307, 309, 314, 315, 323, 329, 331–335, 342
Manatees, 200
Mathematical model, 342
Mechanoreceptors, 288
Mediodorsal secondary bronchi (MDSB), 108, 208–211, 213, 215–221, 227, 228, 246, 279, 280, 282, 284, 288
Medioventral secondary bronchi (MVSB), 108, 208–211, 213, 215–217, 220, 221, 227, 239, 245, 246, 280–288
Megapode, 9–10, 20
Membrane diffusing capacity, 343
Mesenchymal cell, 104, 110, 112, 116–125, 127–132, 134, 136–140, 142–144, 147, 150
Mesenchymal tissue, 122–124, 132, 139
Mesenchyme, 104, 109–112, 141, 142, 214
Mesobronchus, 214
Mesoderm, 110, 129
Mesozoic, 34, 44
Metabolic biomarkers, 116
Metabolic capacities, 26, 38, 194, 197, 330, 336
Metabolic cost, 22, 276
Metabolic demands, 7, 26, 197, 307
Metabolic energy, 28
Metabolic rate, 23, 31, 126, 195, 275, 276, 280, 300, 315, 332
Metabolism, 31, 45, 104, 105, 116, 126, 141, 269, 275, 277
Migration, 24, 25, 28–31, 100, 106, 131, 148, 151, 152, 274, 275
Miocene, 27, 200
Mitochondria, 27, 195–197, 270, 271, 299, 339, 340, 342
Modeling, 45, 281, 340–344
Molecular biology, 100, 101
Molecular cues, 100, 113
Molecular factor, 41, 100, 103, 105, 107, 113, 131, 136, 137, 145, 151
Molecular motors, 103
Molecular signaling factor, 113, 136
Molecular wheels, 104
Monitor Lizard, 41, 193, 279
Monopodial budding, 114
Monopodial sprouting, 213
Morphogenesis, 100, 102, 107, 148
Morphogenetic processes, 103, 106, 107
Morphogens, 100, 136, 140, 148
Morphometric specialization, 315
Mound-builders, 10, 126
Multicameral, 41, 107
Multicapillary serial arterialization systems, 293–296
Muscle power, 23

Muscovy Duck, 108, 233, 239, 316, 322, 324
Mute Swan, 28, 218, 221, 315, 322
Myoepithelial cells, 112
Myoglobin, 275

N
Natural knockouts, 38
Natural selection, 3, 101, 104, 126
Necrotic cell death, 131
Nectar, 11
Neopulmo, 217, 218, 221, 225–228, 246
Neornithes, 9, 37
Nest, 11–13, 200
Neural plasticity, 4
Neurogenesis, 4
New Caledonian Crows, 12, 13
Nile crocodile, 279
Noncanonical WNT signaling, 145
Northern Cardinal, 207

O
Object dropping, 12
Oblique septum, 193, 244
Ohm's law, 340
Ordovician, 194
Ornithischian dinosaurs, 45
Ornithology, 7, 8
Osmiophilic lamellated bodies, 125
Ostia, 15, 108, 133, 245–246, 279, 280
Ostrich, 3, 8, 13, 17, 20, 33, 38, 113, 121–123, 133, 193, 208, 210, 213, 216, 218, 221, 223, 224, 228, 229, 231, 242–244, 282, 285, 287, 308, 322, 324, 327, 330, 333, 338
Oxygen (O_2) affinity, 25, 26
Oxygen (O_2) availability, 26
Oxygen (O_2) cascade, 195, 340
Oxygen (O_2) consumption, 23, 25, 271, 276, 340, 344
Oxygen (O_2) sensing, 289
Oxygen (O_2) sink, 195, 270
Oxygen (O_2) uptake coefficient, 341, 343

P
Paleogene, 34
Paleopulmo, 218, 225, 227, 228
Paleopulmonic parabronchi (PP), 117, 192, 209, 220, 222, 227, 228, 280, 283, 288, 296
Pancake Tortoise, 41, 279
Parabronchi, 44, 108, 118–126, 133, 138–140, 144, 147, 148, 193, 194, 208, 209, 211–215, 217–224, 226–228, 238, 241, 245, 246, 273, 280, 282–284, 290, 291, 293, 295, 296, 307, 309–311, 320
Parabronchial lumina, 36, 115, 120, 122–125, 139, 140, 203, 219, 221–223, 226, 229, 236, 238, 292–295, 310
Parabronchial lung, 41, 43, 193, 280
Parenchyma, 41, 118, 123, 193, 273, 327, 329, 334
Parrot, 7, 200, 205, 228
Partial pressure, 20, 111, 195, 196, 270, 271, 277, 280, 288, 292, 295, 296, 298, 299, 340, 344
Passerine, 2, 11, 19, 30, 228, 229, 331, 335
Passerine species, 16, 218, 221, 223, 224, 228, 231, 244, 332, 339
Pasteur Point, 269, 290
Pectoral girdle, 18, 243
Pectoralis, 17, 19, 27
Pekin Ducks, 25, 289
Penguins, 3, 16, 20, 126, 200, 202, 213, 214, 216, 217, 224, 227, 228, 231, 244–246, 322, 324, 331, 333, 338, 342, 343
Peregrine Falcons, 28
Perfusion pressure, 298, 313
Permian, 192
Peruvian Andes, 26
Pessulus, 204, 206, 207
Pharynx, 40, 106
pH compensation, 289
Phonation, 14
Phosphatidylcholine (PC), 111
Photosynthesis, 112
Phylogenetic diversity, 105
Physiology, 7, 8, 32, 38, 105, 191, 292, 314
Pigeon, 3, 9, 322, 325
Placenta, 127
Plastron, 276
Pleistocene, 27
Plesiomorphic, 44, 45
Pleurae, 211, 272, 310
Pleural cavity, 211, 212, 310, 311
Pleural covering, 110, 211, 212, 310, 311
Pneumatization, 44, 193
Pneumonate gastropods, 198, 299, 300, 335, 336, 339
Pneumothorax, 291, 292
Pollution, 5, 6, 32, 33
Postcranial pneumaticity, 44
Poultry, 9, 32, 33

Powered flight, 2, 7, 11, 17, 19, 20, 23, 26, 37, 38, 44, 191, 192, 197, 200, 205, 331
Prebiotic broth, 269
Precocial, 10, 114, 115, 126, 127, 219
Precursor cells, 128–130, 151
Primary bronchus, 43, 120, 144, 273, 279, 282, 285–288, 320
Primates, 11, 13, 14
Primeval soup, 269
Primitive foregut, 106, 109, 110, 116
Primitive pharynx, 40, 117, 137, 214
Primitive streak, 116
Primordia, 116, 133, 243
Pseudostratified columnar epithelium, 120, 121, 199, 202, 214
Pulmonary artery, 42, 127, 128, 207, 210, 211, 236, 238, 239, 241, 296, 297, 307, 312, 314
Pulmonary blood, 128, 151, 208, 213, 237, 288, 297, 298, 307, 312, 336, 338, 341
Pulmonary capillary blood, 26, 107, 111, 150, 223, 230, 235, 237, 291, 300, 304, 309, 313, 326, 327, 334–343
Pulmonary circulation, 236–241, 288, 296
Pulmonary compliance, 323
Pulmonary diffusing capacity, 223, 230, 327, 341–343
Pulmonary edema, 27, 305
Pulmonary vascular tone, 288
Pulmonary vein, 128, 210, 211, 236, 239–241, 296, 312
Pump-handle movement, 271

Q

Quaternary, 3
Quetzalcoatlus northropii, 27

R

Radical pairs, 30
Rare Island rail, 28
Razorbills, 18, 322, 325
Reactive species, 275
Receptors, 30, 103, 111, 136, 141, 142, 144, 145, 148, 150, 151, 272
Red blood cells, 128, 223, 313
Red Crowned Crane, 6
Red knot, 30
Reptiles, 10, 41, 107, 193, 204, 315
Reptilian lung, 27, 192–194, 330, 332
Respiratory alkalosis, 202, 228, 289
Respiratory bronchiole, 109
Respiratory cycles, 280, 283
Respiratory distress, 24, 305
Respiratory efficiency, 270, 276, 339
Respiratory fluid, 102, 195, 269, 270, 276, 277, 280, 292, 297, 334, 336, 338, 340
Respiratory muscles, 19, 272, 273, 296
Respiratory organ, 272, 283, 299
Respiratory rate, 201, 271, 286
Respiratory surface, 25, 41, 109, 111, 195, 196, 202, 235, 237, 297, 309, 329–332, 334, 335, 337, 340
Respiratory surface area, 26, 102, 107, 113, 123, 127, 128, 150, 151, 223, 230, 277, 291, 297, 303, 323–332, 334, 338, 340, 342, 343
Respiratory systems, 15, 22, 26, 27, 33, 35, 43–45, 114, 126, 136, 137, 152, 193, 194, 196, 197, 207, 214, 233, 241, 271, 275, 276, 279, 281, 292, 304, 307, 311, 315, 323
Respiratory tree, 110, 112, 214, 323
Rhea, 3, 20
Rib cage, 17, 18, 273
Ribs, 17, 18, 44, 117, 118, 122, 194, 208–211, 213, 271–273, 297, 305, 311, 323, 327
Rock Dove, 108, 126, 229, 243–245, 322, 325
Rotary motor, 105
Ruddy Turnstones, 29
Rufous Griffon Vulture, 3
Ruminant, 10–11
Ruppell's Griffon Vulture, 24

S

Salamander, 35, 39, 40
Sandboa, 279
Sauropod dinosaurs, 45
Scapula, 18
Screamers, 17, 274
Secondary bronchi (SB), 43, 119–122, 133, 135, 138, 142–144, 147, 148, 193, 202, 203, 209, 211–215, 218, 219, 221, 273, 284, 320
Secondary lamellae, 323, 328, 335, 339
Segmentum accelerans, 286, 287
Senegal Bichir, 39
Signal transduction, 148, 289
Signaling, 26, 30, 100, 103, 111, 112, 114, 136, 137, 141, 142, 144, 145, 148, 150, 151
Signaling network, 109
Silurian, 40, 299
Sixth mass extinction, 33
Skeletal muscle, 274, 276
Skeletal system, 18, 20
Skull, 34

Smooth muscle, 110, 112, 130, 212, 233, 234, 236, 241, 243, 245, 273, 284, 288, 289, 298, 309–311, 330
Soaring flight, 23, 24
Somite, 116
Song, 2, 12, 14–16
Songbird, 11, 13, 15, 205, 229, 231
Sonic hedgehog (SHH), 136, 137, 142, 148
Sound, 13–16, 29, 40, 202–207, 244
Speckled Hummingbird, 16
Spectacled Flowerpecker, 3
Speech, 14
Spiracles, 196, 197
Splanchnic mesoderm/mesenchyme, 106, 109, 116
Starling, 2, 30, 323, 326
Sternum, 17, 27, 243, 271, 273
Stomach, 106, 133
Stress, 6, 8, 9, 17, 19, 33, 228, 300, 304, 306, 308
Stroke volume, 25, 150, 237
Structural failure, 304–306, 309, 311–313
Super birds, 24
Super Condor, 27
Supracoracoideus, 17–19, 27
Surface area, 5, 108, 110, 113, 116, 202, 303, 323, 326, 329–331, 338, 341, 343, 344
Surface density, 326, 331
Surface tension, 126, 235, 297, 298, 323, 329, 330
Surfactant, 40, 109, 111, 112, 123, 125, 126, 198, 235, 297, 300, 301, 304, 309, 329, 330, 341, 344
Swallow, 17, 126
Swan, 200, 201, 221, 231
Swift, 16, 17, 30
Swim bladder, 39, 40, 198, 304
Symbiotic, 4
Synsacrum, 244
Syringeal muscles, 15, 205–207
Syrinx, 14, 15, 199, 201, 203–208, 243, 287
Systolic blood pressure, 307

T
Temperature, 10, 20, 24, 25, 32, 33, 105, 126, 191, 192, 300, 307, 332, 343
Tennis court, 113, 330, 338
Tensegrity, 309–311
Terminal respiratory units, 43, 111, 113, 116, 131, 223, 231, 237, 291, 309, 311, 328, 331
Terrestrial, 9, 27, 32, 34, 35, 38, 43, 110, 197, 315
Terrestrial locomotion, 17, 19, 201
Tertiary bronchi, 108, 120, 121, 139, 147, 213, 217–224, 273, 290
Tetrapod, 20, 37–39, 41, 192, 200, 204, 276, 299, 304
Thermoregulation, 32, 274
Theropod dinosaur, 34, 37, 43, 44, 194
Thoracic cage, 272
Thyroid gland, 106, 109
Tibetan chicken, 24
Tidal volume, 25, 196, 201, 270–272, 315
Tool, 11–13, 281
Trachea, 14, 15, 42, 108, 110, 114, 120, 122, 128, 137, 147, 148, 196–202, 204–206, 208, 242, 243, 272, 279, 280, 282, 284, 285, 287, 299, 339
Tracheal cartilages, 110, 201
Tracheal diverticulum, 202
Tracheal system, 141, 195–197, 276–278
Tracheoles, 196, 197, 339
Transcription factor, 106, 110, 114, 136, 137, 148
Transforming growth factor, 39
Trilaminar substance, 126, 329, 330
Trumpeter Swan, 28, 201
Trumpet Manucode, 201
Turkey, 108, 126, 210, 239, 243, 244, 273, 307, 308, 314, 324
Tympanum, 206, 207
Type-II cells, 109, 123, 125, 126, 198, 229, 235, 297, 330, 332
Type-IV collagen, 305, 309
Typhaniform membrane, 206

U
Uncinate processes, 17, 18, 273, 274
Undulating flight, 24
Unicameral, 41
Unidirectional airflow, 43, 192–194, 283, 287
Urbanization, 5, 6

V
Vascular endothelial growth factor (VEGF), 136, 142, 148–152
Vascular smooth muscle, 288
Vasculature, 27, 110, 111, 127, 150, 151
Vasculoendothelial cell, 129
Vasculogenesis, 111, 130, 131, 144, 150, 151
Vasculogenetic process, 127

Vasoconstriction, 26, 288
Vasoformative cells, 128
Ventilation, 25, 192–195, 201, 241, 270, 272, 276, 283, 288, 305, 329, 344
Ventilation-perfusion inequalities, 288
Ventilatory airflow, 272
Ventilatory phases, 272
Venules, 113, 240
Vertebral arteries, 201
Vertebrate, 7, 9, 11, 31, 33, 34, 36–39, 107, 136, 195, 203, 204, 206, 233, 236, 274, 275, 280, 292, 296, 297, 299, 303, 304, 323, 332
Vertebrate lungs, 39–41, 104, 106, 107, 109, 114, 141, 150, 198, 298, 299, 330
Violet-Tailed Sylph Hummingbird, 16
Vocalization, 12–16, 199, 203–205, 207
Vocal learning, 13, 14, 205
Volancy, 2, 20, 21, 28, 37, 44
Volume density, 123, 223, 316, 322, 329

W

Wandering Albatrosses, 23, 24
Water, 4, 12, 18, 25, 29, 32, 33, 38–40, 102, 127, 195, 200, 243, 269, 274–277, 280, 292, 296, 297, 300, 317, 322, 325, 334, 336, 340
Water-blood (tissue) barrier, 271, 296, 297, 299, 300
Water-breathing organ, 276, 277, 328
White Bellbirds, 16
White-Cheeked Bulbul, 6
Whooper Swan, 28
Whooping Crane, 201
Wing, 2, 10, 18–20, 22–24, 26, 27, 31, 37, 197, 199, 200, 273
Wingbeat, 20, 23
Wing flapping, 17, 23–25
Wingless and Int (WNTs), 136, 137, 144–145
Wing loading, 27
Wing morphology, 20, 31
Wingspan, 8, 27
WNT signaling, 144, 145
Woodpecker Finch, 13

Y

Yolk sac, 127

Z

Zebra finch, 3, 5
Zebrafish, 39